HARCOURT Science

Harcourt School Publishers
Orlando • Boston • Dallas • Chicago • San Diego
www.harcourtschool.com

School Permissions and Copyrights, Harcourt, Inc.
6277 Sea Harbor Drive
Orlando, FL 32887-6777
Fax: 407-345-2418

Printed in the United States of America

ISBN 0-15-323701-5

3 4 5 6 7 8 9 10 030 2004 2003 2002

HARCOURT Science

TEACHER'S EDITION CONTENTS

ABOUT THE PROGRAM

THE UNITS

PUPIL EDITION REFERENCES

TEACHER REFERENCES

Authors

Robert M. Jones, Ed.D. Professor of Education and Environmental Education Affiliate of the Environmental Institute of Houston, he has served at the University of Houston–Clear Lake for 25 years. Dr. Jones has also served on the Board of Directors of the Association for the Education of Teachers in Science and as Director of the Southwest Region of that organization. His areas of interest include aerospace and motion engineering, cooperative learning, and environmental education.

Joyce C. McLeod Visiting Professor, Rollins College, Winter Park, Florida, she has focused on training teachers in how to use the results of brain research to match instructional strategies to the curriculum and to the capabilities and learning styles of their students. As former Harcourt Editor-in-Chief of Science and Mathematics, she guided the development of programs that implemented national standards and reflected the latest research in teaching and learning. Her areas of interest include research on teaching and learning, primary education, teacher professional development, assessment, and connections between science, mathematics, and reading.

Gerald H. Krockover, Ph.D. Professor of Earth and Atmospheric Science Education, he has served as chair of the Elementary Education program area at Purdue University for 10 years. He was a member of the Executive Committee and the Board of Directors of the National Science Teachers Association. He is the coauthor of the two-volume *Creative Sciencing: Ideas and Activities for Primary and Intermediate Teachers and Children.* His areas of interest include earth-space science education, equity in science education, the integration of mathematics and science, and the use of technology and telecommunications in the classroom.

Marjorie Slavick Frank A former member of the adjunct faculty at Hunter, Brooklyn, and Manhattan Colleges, New York, she is a specialist in literacy and language development as well as a composer and lyricist. Her areas of interest include elementary science and reading instruction; children's literature; and strategies for content-area reading.

Mozell P. Lang Science Education Consultant in the Curriculum Development Program of the Michigan Department of Education, she provides technical assistance and support to schools and organizations for the implementation of professional development in teaching, learning, assessment, and multicultural efforts. She also coordinates statewide recognition programs, such as the Presidential Awards. Her areas of interest include the physical sciences.

Carol J. Valenta Vice President—Exhibits, Education, and Programs at the St. Louis Science Center, she coordinates the Science Center and Outreach Programs with school districts in Illinois and Missouri. She was a classroom teacher, an elementary school principal, and the Coordinator of Science Center Instructional Programs for the Los Angeles Unified School District. Her areas of interest include physical science, integrated curriculum, teacher empowerment and training, video-based instruction, community-based education, family education, and urban education.

Barry A. Van Deman Section Head and Program Director of Informal Science Education at the National Science Foundation, he was an award-winning science teacher, school administrator, university instructor, and leader in science museums. He served as president of the Council for Elementary Science International and board member of the National Science Teachers Association. He is the author of *Nuts and Bolts: A Matter of Fact Guide to Science Fair Projects*. His areas of interest include life sciences, environmental education, museum exhibits, Web-based activities, and teacher professional development.

Dear Educator,

The National Science Education Standards state, "In a world filled with the products of scientific inquiry, scientific literacy has become a necessity for everyone."

Harcourt Science is based on the following principles of scientific literacy:

- All children can investigate and learn science concepts and can experience success in science.
- Children must develop knowledge of and the ability to use the tools and processes of scientific inquiry.
- Children experience success in science when they develop age-appropriate knowledge and understanding of the life, earth, and physical sciences and when they learn about the history and nature of science.
- While engaged in the study of science, children should have the opportunity to build success in other curricular areas.
- Science content should be presented to children in an interesting, comprehensible, and clearly organized format.
- Children's competence in the concepts and processes of science should be assessed through a variety of tools that are consistent, authentic, and fair.

Harcourt Science provides children with the opportunity to

- **Investigate** the natural world,
- **Learn About** interesting, relevant, and exciting science ideas, and
- **Link** science to mathematics, writing, technology, and all other aspects of the elementary curriculum.

Harcourt Science gives you the resources to promote scientific literacy by nurturing in your students a lifelong fascination with the natural world.

Sincerely,

The Authors

Senior Editorial Advisor

Napoleon Adebola Bryant, Jr.
Professor Emeritus of Education
Xavier University
Cincinnati, Ohio

Program Advisor and Activities Writer

Barbara ten Brink
Science Director
Round Rock Independent School District
Round Rock, Texas

Program Advisors

Michael J. Bell
Assistant Professor of Early Childhood Education
School of Education
West Chester University
West Chester, Pennsylvania

George W. Bright
Professor of Mathematics Education
The University of North Carolina at Greensboro
Greensboro, North Carolina

Pansy Cowder
Science Specialist
Tampa, Florida

Robert H. Fronk
Head, Science/Mathematics Education Department
Florida Institute of Technology
Melbourne, Florida

Bernard A. Harris, Jr.
Physician and Former Astronaut
(STS 55-Space Shuttle *Columbia*,
STS 63 Space Shuttle *Discovery*)
President, The Harris Foundation
Houston, Texas

Lois Harrison-Jones
Education and Management Consultant
Dallas, Texas

Kenneth R. Mechling
Professor of Biology and Science Education
Clarion University of Pennsylvania
Clarion, Pennsylvania

Nancy Roser
Professor of Language and Literacy Studies
University of Texas, Austin
Austin, Texas

Dr. Harris invites students to participate in an exciting program that involves them in space exploration. See the article titled “Soaring into Space” on pages T78–79.

Classroom Reviewers and Contributors

Sidney Jay Abramowitz
District Administrator for Mathematics, Science & Technology
Stamford Public Schools
Stamford, Connecticut

Rose G. Baublitz
Director Curriculum and Staff Development
Granville Exempted Village School District
Granville, Ohio

Melcene Beasley
Eastside Elementary School
Douglas, Georgia

Cathy Boles
Sawnee Primary School
Cumming, Georgia

Sharon W. Bowers
Kemps Landing Magnet School
Virginia Beach, Virginia

Susan Busenlehner
Snow Rogers Elementary School
Gardendale, Alabama

Patricia L. Bush
Jesse Stuart Elementary School
Madisonville, Kentucky

Gary Callahan
Goodland Elementary School
Racine, Wisconsin

Edward J. Carroll
Thomas E. Bowe School
Glassboro, New Jersey

Lisa Crooks
Black Bob Elementary School
Olathe, Kansas

Kimberly Gay
Mango Elementary School
Seffner, Florida

Joe E. Hart
Science Lead Teacher
Clayton County Public Schools
Morrow, Georgia

Kathryn Henry
Teacher Trainer
Public School CS 200
New York, New York

Jacqueline Howard
Millard Hensley School
Salyersville, Kentucky

Scott Hudson
Covedale Elementary
Cincinnati, Ohio

Denise Hunt
Cumming Elementary School
Cumming, Georgia

Donna Kavanaugh
Pleasant Hill Elementary School
Cincinnati, Ohio

Faye Kelley
Gardendale Elementary School
Gardendale, Alabama

Kimberly Knause
Brantner Elementary School
Cincinnati, Ohio

Angela Langenkamp
St. Gabriel School
Louisville, Kentucky

Faye McCollum
Instructional Specialist
Muscogee County School District
Columbus, Georgia

Margie H. McCoy
Salyer Elementary School
Salyersville, Kentucky

Marie A. McDermott
Kingsbury School
Waterbury, Connecticut

Leah McGriff
Garrison Elementary School
Savannah, Georgia

Beverly W. McNair
Hollis Hand Elementary School
LaGrange, Georgia

Chester H. Melcher
Supervisor of Curriculum and Instruction—Science
Racine Unified School District
Racine, Wisconsin

Philip J. Natoli
Geneseo Central School
Geneseo, New York

Clyde Partner
Science/Health Curriculum Coordinator
Evanston S.D. #65
Evanston, Illinois

Michael F. Ryan
Educational Technology Specialist
Lake County Schools
Tavares, Florida

Arnold E. Serotsky
Teacher/Science Department Coordinator
Greece Athena Middle School
Rochester, New York

Michelle Bonk Severtson
Olson Elementary
Bloomington, Minnesota

Ann Starek
Memorial School
Natick, Massachusetts

Sandra Seim Tauer
Math/Science Instructional Coordinator
Educational Support Center
Derby, Kansas

Sharon Till
Chattahoochee Elementary School
Cumming, Georgia

Ross A. Vandercook
Principal
Hiawatha Elementary School
Okemos, Michigan

Stanley J. Wegrzynowski
Director of Science
Buffalo Public Schools
Buffalo, New York

Lee Ann White
Morgantown Elementary School
Morgantown, Kentucky

Lynda Wood
Science Coordinator
Southfield Education Center
Southfield, Michigan

Implementing the National Science

At the request of teacher associations, professional societies, and government officials, the National Research Council established a National Committee on Science Education Standards and Assessment to be responsible for the development of national science education standards. The standards were published by the National Academy Press in 1996. The goals for school science that underlie the National Science Education Standards are to educate students so that they are able to

- experience the richness and excitement of knowing about and understanding the natural world;
- use appropriate scientific processes and principles in making personal decisions;
- engage intelligently in public discourse and debate about matters of scientific and technological concern; and
- increase their economic productivity through the use of the knowledge, understanding, and skills of the scientifically literate person in their careers.

"Learning science is something students do, not something that is done to them."

National Science Education Standards

Education Standards

The Science Content and Assessment Standards, which are within the National Science Education Standards, outline what students should know, understand, and be able to do in the natural sciences. They include these categories:

- Unifying concepts and processes in science
- Science as inquiry
- Life science
- Earth and space science
- Physical science
- Science and technology
- Science in personal and social perspective
- History and nature of science
- Assessment that is consistent, fair, and provided in a variety of contexts

Statue of Albert Einstein, Washington, D.C.

Harcourt Science was developed to meet the goals and content objectives of the National Science Education Standards, and it does so in the following ways:

- Standards-based life, earth, and physical science content is provided at each grade.
- Students use the tools and processes of scientific inquiry in every lesson, which always begins with a hands-on investigation.
- Exciting and informative aspects of science and technology are included within the life, earth, and physical science content.
- The history and nature of science and the work of scientists are featured.
- Students are assessed in a variety of ways. (See Assessment Guide.)

A correlation of *Harcourt Science* to the National Science Education Standards is provided in the back of this Teacher's Edition.

Skills for Lifelong Learning

Science Process Skills

Harcourt Science provides many opportunities for students to develop and maintain the essential skills that form the basis for lifelong learning. The science process skills are important inquiry tools and are essential for investigating the natural world. Opportunities for developing the process skills are embedded throughout *Harcourt Science*.

Science Process Skill	Description
Observe	using one or more of the senses to perceive properties of objects and events; can be done directly with the senses or indirectly through the use of simple or complex instruments
Compare	identifying common and distinguishing characteristics among objects or events
Classify/Order	grouping or organizing objects or events into categories based on specific criteria
Gather, Record, Display, or Interpret Data	making observations of objects or events from which to make inferences or predictions; writing down the observations on paper as notes or displaying the data in charts, in tables, or in graphs; making predictions, inferences, and hypotheses from a set of data
Use Numbers	estimating or quantifying data
Communicate	transmitting observable data or ideas visually, orally, or electronically
Plan and Conduct Simple Investigations/ Experiments	using one or more of the process skills to find the answer to a question or the solution to a problem
Measure	making quantitative observations using both nonstandard and standard measures
Predict	anticipating outcomes of future events, based on patterns or experience
Infer	using logical reasoning to make conclusions based on observations
Draw Conclusions	interpreting data to make conclusions; the final step of an investigation
Use Time/Space Relationships	estimating the relationships of moving and of nonmoving objects to one another; includes sequencing
Hypothesize	posing a testable explanation for observations or events and stating it as the expected outcome of an experiment
Formulate or Use Models	making a mental or physical representation of a process or structure or using a model that someone else has provided
Identify and Control Variables	stating or controlling factors that affect the outcome of an experiment
Experiment	designing procedures for gathering data to test hypotheses under conditions in which variables are controlled or manipulated

Reading, Writing, and Math Skills

Effective science instruction integrates science content and experiences with all areas of the elementary curriculum. *Harcourt Science* provides students with many opportunities to develop reading, writing, and math skills through meaningful activities and strategies.

Reading Skills

Reading underlies everything students do in school, and many of the reading skills you are teaching are reinforced in *Harcourt Science*. The Pupil Edition helps students become strategic readers by providing the following:

- a clear text structure, including patterns of clear and consistent headings and subheadings
- highlighted vocabulary
- graphics to support the learning
- internal features and questions that guide concept attainment
- lesson summaries and reviews that consolidate learning
- a useful set of science reading strategies in the Reference section (p. R38–R49)

Strategies for developing the following reading skills are provided at point of use throughout the Teacher's Edition. Follow-up practice is also provided in the Workbook.

- ☑ use prefixes and suffixes to determine word meanings
- ☑ use context to determine/confirm word meaning
- ☑ identify supporting facts and details
- ☑ arrange events in sequence
- ☑ identify the main idea
- ☑ summarize and paraphrase a selection
- ☑ identify cause and effect
- ☑ predict probable future actions and outcomes
- ☑ use graphic sources for information
- ☑ draw logical conclusions
- ☑ compare and contrast
- ☑ distinguish fact and opinion
- ☑ make generalizations
- ☑ use reference sources

Writing Skills

Opportunities for students to express their ideas in writing are provided throughout *Harcourt Science*. In the Pupil Editions, the Writing Link at the end of each lesson provides a prompt for writing about lesson concepts in an interesting and meaningful way. The prompts include these types of purposes for writing:

- **Informative**
- **Persuasive**
- **Narrative**
- **Expressive**

A variety of types of organizations, such as How-to, Narration, Report, Poem, and Friendly Letter, are also included. Writing Models are provided in the **Teaching Resources** book.

Math Skills

Opportunities to practice mathematics skills, solve problems, and connect mathematics and science are provided throughout the Pupil and Teacher's Editions. Look for the Math Link in the Pupil Edition at the end of every lesson. Additional Math Links are also provided at point of use in the Teacher's Edition.

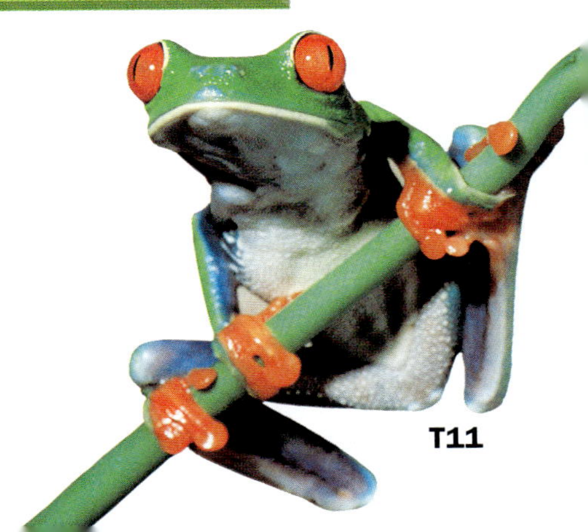

Assessment Program

Overview

In *Harcourt Science,* the Assessment Program, like the instruction, is student-centered. By allowing all learners to show what they know and can do, the Assessment Program provides you with ongoing information about each student's understanding of science. The Assessment Program also involves the student in self-assessment, offering you strategies for helping students evaluate their own growth.

The *Harcourt Science* assessment program is based on the following Assessment Model. The model's framework shows the multidimensional aspect of the program, with five assessment components supported by teacher-based and student-based assessment tools. More information about the Assessment Model is provided in the **Assessment Guide**.

Harcourt Science Assessment Model

(**Key**: PE=Pupil Edition; TE=Teacher's Edition; AG=Assessment Guide; WB=Workbook)

Formal Assessment

- **Chapter Review and Test Preparation,** PE
- **Chapter Test,** AG
- **Unit Test,** AG

Ongoing Assessment

- **✓Questions**, PE
- **Lesson Review**, PE and WB
- **Informal Assessment Strategies**
  - **Observation,** TE
  - **Performance,** TE
  - **Portfolio,** TE
- **Observation Checklist,** AG

Student Self-Assessment

- **Self-Assessment—Investigate,** AG
- **Self-Assessment — Learn About,** AG
- **Experiment/Project Summary Sheet,** AG

Performance Assessment

- **Chapter Review,** PE
- **Chapter Performance Task**, AG
- **Experiment/Project Evaluation Checklist**, AG

Portfolio Assessment

- **Science Experiences Record**, AG
- **Guide to My Science Portfolio**, AG
- **Portfolio Evaluation Checklist,** AG

Formal Assessment

Chapter Review and Test Preparation • Chapter Test • Unit Test

Research into the learning process has shown the positive effects of periodic review. To help you reinforce and assess mastery of chapter objectives, *Harcourt Science* includes both reviews and tests. You will find the Chapter Review and Test Preparation in the Pupil Edition and the Chapter Test and Unit Test in the **Assessment Guide.** Answers to these assessments, including sample responses to open-ended items, are given in the Answer Key.

Performance Assessment

Chapter Review • Chapter Performance Task • Experiment/Project Evaluation Checklist

Science literacy involves more than just what students know. It is also concerned with how they think and how they do things. Performance assessment provides evidence of students' ability to use science process skills and critical thinking skills to complete an authentic problem-solving task. At the chapter level, you will find a performance task included in Chapter Review and Test Preparation. Another is located in the **Assessment Guide,** following each Chapter Test. Each includes teacher's directions and a scoring rubric. Also in the **Assessment Guide** is the Experiment/Project Evaluation Checklist (p. AGxviii), a measure for evaluating students' science projects and unit experiments.

Ongoing Assessment

✓Questions • Lesson Review • Informal Assessment Strategies • Observation Checklist

There are many opportunities to observe and evaluate student growth during regular classroom instruction in science. *Harcourt Science* supports ongoing informal assessment in three ways. In the Pupil Edition, there are boldface ✓Questions at the end of sections to help you assess students' immediate recall of information. Then, at the end of each lesson, you will find a Lesson Review to help you find out how well students grasped the concepts taught. The Lesson Review also includes a "test prep" question. In the Teacher's Edition, you will find Informal Assessment Strategies to support classroom observation, performance assessment, and portfolio assessment. In the **Assessment Guide,** you will find yet another assessment tool, the Observation Checklist (p. AGxiv), on which you can record classroom observations. Ultimately, it is your experienced eye that will provide the most comprehensive assessment of student progress.

Student Self-Assessment

Self-Assessment — Investigate • Self-Assessment — Learn About • Project Summary Sheet

Student self-assessment encourages students to reflect on and monitor changes in their science knowledge, skills, and attitudes. The **Assessment Guide** includes two self-assessment checklists for this purpose, one for use after students complete Investigate (p. AGxvi) and one for use after instruction in Learn About (p. AGxvii). Following the checklists, you will find the Experiment/Project Summary Sheet (p. AGxix), a form on which students can describe and evaluate their efforts on a science project or unit experiment.

Portfolio Assessment

Science Experiences Record • Guide to My Science Portfolio • Portfolio Evaluation Checklist

In *Harcourt Science,* students may create their own portfolios. The portfolio holds self-selected work samples that the student feels are representative of gains in his or her understanding of science concepts and use of science processes. The portfolio may also contain a few required or teacher-selected papers. Support materials are included in the **Assessment Guide** (pp. AGxx–AGxxiv) to assist you and your students in developing portfolios and in using them to evaluate growth in science.

Using Technology

Harcourt Science **includes an array of technology products that**

- bring interactive experiences to the classroom.
- add depth to chapter concepts.
- meet individual needs.

Technology Links—in the Student Edition at the end of each lesson for Grades 3–6 and throughout the Teacher's Edition for Grades 1–2—help you integrate technology at the point of its most effective use.

Computer Software

Harcourt Science Instant Readers CD-ROM, Grades K–2

These CD-ROMs are designed to accompany the Harcourt Science Instant Readers. They provide a variety of interactive on-screen activities and reinforce science vocabulary and concepts. Point-of-use references to the Harcourt Science Instant Readers and the CD-ROMs appear in the Teacher's Editions.

Harcourt Science Explorations CD-ROM, Grades 3–6

These engaging CD-ROMs provide interactive, child-centered science explorations. Activities enhance and expand key chapter concepts through simulations and investigations. The activities are designed so that students can complete them with minimal teacher intervention. Multiple outcomes will give students reasons to explore the activities again and again.

Data ToolKit, Grade 6

Data ToolKit is a tool-based program that enables students to organize, communicate, and interpret data they have collected during their investigations. This software program enables students to make graphs more easily and accurately than they could by hand. Students can compare the different ways data can be represented—by symbols, bars, points on a line, or sections of a circle. They can also customize each graph by changing the font, style, and size of the text or by varying colors or grid lines. Suggestions for using Data ToolKit appear in the Math Links in the Student Edition.

Videos

Harcourt Science Newsroom Videos, Grades 3–6

Harcourt has joined forces with the leaders in news broadcasting—CNN and Turner Learning—to bring timely science news events to your classroom. For each chapter in Grades 3–6, there is a **Harcourt Science Newsroom Video** segment that helps students see the relevance of science to their lives. A CNN news reporter introduces each segment and sets the stage for viewing. The videos also add interesting depth to chapter concepts.

Harcourt Science Activity Videos, Grades 1–6

The Activity Videos show students doing the Investigates for each lesson and can be used to help you prepare for upcoming chapters. The videos can also be used to preview and model the investigation process for students, to review and discuss hands-on experiences, and to help students who may have been absent when hands-on activities were conducted. Students with special needs may also benefit from viewing the videos.

Internet

Smithsonian Institution

With *Harcourt Science*, it's easy to make the Smithsonian Institution a part of the science classroom. Through a special agreement, you can visit this Internet site to see correlated links to the Web sites of Smithsonian museums, galleries, and the National Zoological Park. Especially selected by Harcourt to complement textbook lessons, the Web sites include intriguing virtual tours, such as Marine Ecosystems from the National Museum of Natural History, extraordinary online exhibits like How Things Fly, from the National Air and Space Museum, and a world of exceptional images and information.

Providing Web access to such resources is a valuable outreach service from the Smithsonian. And helping teachers find and use these resources at precisely the right time—when they are directly relevant to school learning—is a service to classroom teachers from Harcourt. That's why the Student Editions at Grades 3–6 include Smithsonian references in the Technology Links that appear at the end of each lesson. Point-of-use suggestions for teacher background information and activity ideas appear in the Teacher's Editions for Grades 1 and 2.

www.si.edu/harcourt/science

SciLinks

These special Internet links are provided by the National Science Teachers Association (NSTA) in partnership with Harcourt. The *Sci*Links Web page connects students to Web sites that were specially selected to enhance and enrich *Harcourt Science* chapter content. With SciLinks, your students will have the best Internet science resources, all selected by NSTA's network of teacher-webwatchers.

www.scilinks.org/harcourt

The Learning Site

The Learning Site, a product of Harcourt, is a user-friendly Web site that provides interactive learning for students, professional development and Webliography resources for teachers, and school-home resources. Visit The Learning Site for a world of science resources, including science Expeditions, interactive learning games and activities tied to chapter content, and a multimedia Science Glossary. Point-of-use references to The Learning Site activities appear in the Student Edition for Grades 3–6 and the Teacher's Editions for Grades 1–2. But feel free to visit The Learning Site at any time!

www.harcourtschool.com

Curriculum Integration

Harcourt Science is designed to allow you to integrate science into your day through the use of links to all curriculum areas. These ideas promote a broader development of science concepts and skills while showing students examples of how science connects to all parts of their lives. Look for *How to Integrate Your Day* at the beginning of each unit.

Curriculum Integration

Health and Physical Education

Investigating healthful practices encourages students to apply these ideas to their daily lives.

Writing

Writing Links appear in the Pupil Edition at the end of every lesson and chapter. These prompts invite students to apply their science learning in writing as they use the four basic purposes for writing and various types of organizers.

Math

Practice in measurement, data, and computational skills reinforces for students the many connections between math and science. Math Links appear in the Pupil Edition and Teacher's Edition for every lesson and chapter.

Language Arts

Language Arts Links provide varied opportunities for students to communicate their knowledge of science concepts.

Fine Arts

Through music, dance, and drawing, students can express their understandings of the natural world. These Links appear throughout the Pupil and Teacher's Editions.

Literature

Exposure to literature provides ways for students to apply their knowledge of science concepts and to build vocabulary. A list of suggested science trade books is provided for every chapter. Be sure to also consult the Bibliography of Science Trade Books at the back of this Teacher's Edition.

Social Studies

Experience in exploring the ways science is interwoven with history, geography, and culture helps students connect science with the ancient and modern worlds. Suggestions for making these connections are provided throughout the Pupil and Teacher's Editions.

Family Support

A strong partnership between home and school helps students succeed in science. *Harcourt Science* encourages that partnership in a variety of ways.

SCHOOL-HOME CONNECTION

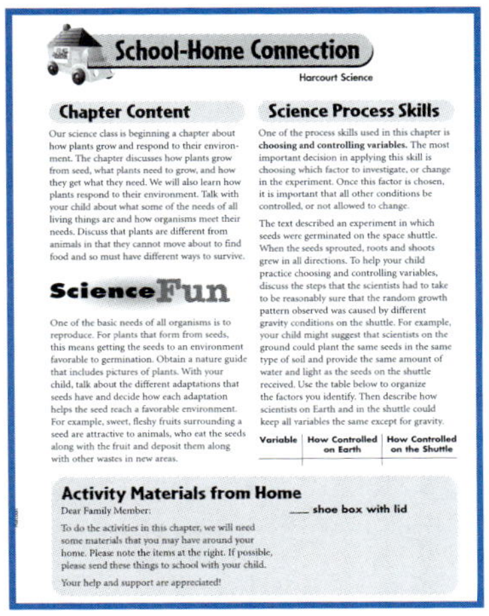

An informative family letter—School-Home Connection—is provided for each chapter. It describes the chapter content, gives tips to parents about science process skills, and includes Science Fun. It also provides a list of materials family members can send to school for the investigations in the Pupil Edition.

ACTIVITIES FOR HOME OR SCHOOL

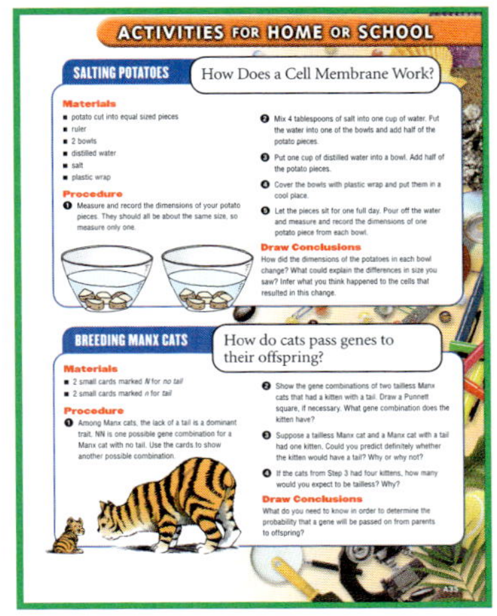

These activities appear in the Pupil Edition and can be assigned for students to complete at home. Reproducible copies of the activities are provided in the Teaching Resources book.

TAKE-HOME BOOKS

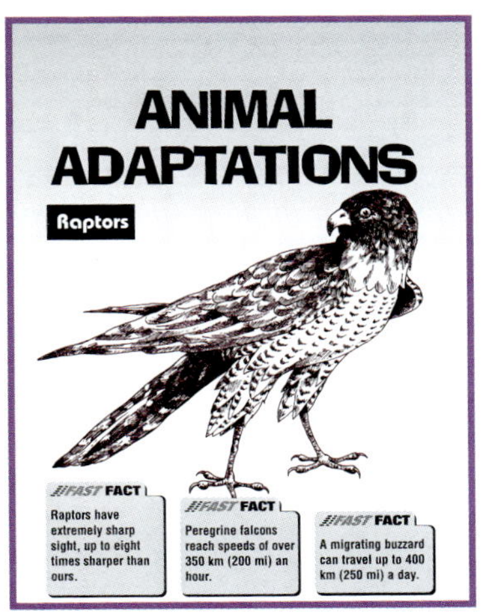

These delightful books are designed to provide additional chapter content and activities. Students can take them home to read with family members. The Take-Home Books can also involve students and their family members in hands-on activities, vocabulary reinforcement, and science fun.

THE LEARNING SITE

Harcourt's user-friendly Web site provides a world of resources for students, teachers, and parents. A clearly organized menu leads family members to interactive learning games, Science News Breaks, and online exhibits and virtual tours from the Smithsonian Institution. At the Multimedia Science Glossary, families can see illustrations of *Harcourt Science* vocabulary terms and hear them pronounced. There is also a special page with tips and ideas for parents.

www.harcourtschool.com

Program Components

Harcourt Science Components	1	2	3	4	5	6
Pupil Editions	•	•	•	•	•	•
Teacher's Editions	•	•	•	•	•	•
Unit Big Books	•	•				
Workbook	•	•	•	•	•	•
Workbook, Teacher's Editions	•	•	•	•	•	•
Assessment Guides	•	•	•	•	•	•
Science Instant Readers	•	•				
Science Take-Home Books	•	•	•	•	•	•
Teaching Transparencies	•	•	•	•	•	•
Science Handbook Transparencies	•	•				
Inside Story Transparencies			•	•	•	•
Health Activity Book	•	•	•	•	•	•
Picture Cards	•	•	•			
Teaching Resources	•	•	•	•	•	•
Text on Tape Audiotapes	•	•	•	•	•	•
Science Songs CD	•	•				
Materials Kits	•	•	•	•	•	•

Technology	1	2	3	4	5	6
Activity Videos	•	•	•	•	•	•
Harcourt Science Newsroom Videos—CNN			•	•	•	•
Science Instant Readers CD-ROMs	•	•				
Explorations CD-ROMs			•	•	•	•
Electronic Test System			•	•	•	•
Planning Resources CD-ROM	•	•	•	•	•	•
Smithsonian Links	•	•	•	•	•	•
NSTA SciLinks	•	•	•	•	•	•
The Learning Site Activities and Resources	•	•	•	•	•	•

For Kindergarten

- Big Book
- Teacher's Edition
- Activity Cards
- Picture Cards
- Activity Book
- Teaching Resources
- Science Instant Readers
- Science Instant Readers CD-ROMs
- The Learning Site Activities and Resources
- Materials Kit
- Science Songs CD
- Planning Resources CD-ROM

Materials List

Quantities are indicated for a class of 30 students working individually or in groups of 5, depending on the nature of the activity. Where shared equipment is suggested, a smaller number of items is specified. Quantities are also listed for those materials included in the Materials Kit.

Nonconsumable Materials	Class Quantity Needed	Kit Quantity	Activity Page
apron, lab	30	Safety Kit	E28, E50
balance, two-pan	1*	1	B1, D1, E44, F4
barometer	1*	1	C1
binder, loose leaf	6		E114
block, wood	6		D54, F56
block, wood (cube)	30	30	E4, E44
board, 1 m	6		E74, F22, F48, F56
books	30		C20, D1, D46, E74, F12, F48
book, large	6		F4
bottle, spray	6		E122
bucket	6	6	D1
cart with wheels	3*	3	E74, F22, F48, F56
cellophane, 4 colors	6 sheets of each color	6 sheets of each color	B1
chair	6		F56
chopsticks	3 pair*	3 pair	C68
clamp, screw	1	1	D46
clock or watch with second hand	1		D46, D92
compass	6	6	F12
compass, drawing	30		D20, D76
container, clear plastic w/lids	6	6	C20
copper, strip	6	6	E90
corn, purple-and-white	6	6	A22
cup, measuring	6	6	B1, B4, C28, E1
dowel	6	6	C28
dropper	6	6	E28, E50
eggbeater	1	1	C28
flashlight, large with batteries	3*	3	D114, E122
flashlight, small with batteries	6*	6	D92, D114
galvanometer	1*	1	E82, E90
goggles, safety	30	Safety Kit	B32, C28, C44, D128, E28, E50
graduate	3*	3	E44
graphite	6 pieces	6 pieces	E12
hand lens	6	6	B32, B42, B50, B74, D38, D62, E44
hook, eye	12		F56
hygrometer	1*		C1

*Student groups should share this equipment.

Nonconsumable Materials	Class Quantity Needed	Kit Quantity	Activity Page
jar, plastic	6	6	A4
knife, plastic	6	1 pkg of 24	B50, E50
lamp with 60-watt bulb	6		D92
magnet	6	6	D38, E12, E58, E82
marbles	6	1 bag of 20	E4, F12
meterstick	12		D1, E74, F1, F38, F56
Microslide Viewer	6	6	A12, B74, C52
Microslide, hydra	6	6	B74, C52
Microslide, plant mitosis	6	6	A12
mineral collection (6 pieces of 11 minerals)	1	1	D38
mirror, small	12	12	E122
nail, iron thin	6	1 pkg of 8	E12
nails, steel	6	8	D38
pail, plastic	6	2	C98, D46
pan, plastic	1*	1	C28
paper clips, large	18	1 box of 100	F4, F38, F56
penny or nickel	6		D84
pictures, mosses	6 sets		B26
pulley (or spool or grommet)	6		F56
pushpins	12	1 box of 100	D84
ring stand with ring	1*	1	F38
rock kit (6 pieces of 6 igneous rocks)	1	1	D62
rolling pins (dowel)	6	6	D54
rubberband, large	12	1 bag	E114, F4
ruler, metric	6		B1, C68, C92, D4, D12, D76, D84, D92, E44
ruler, metric with groove	6	6	F12, F48

Nonconsumable Materials	Class Quantity Needed	Kit Quantity	Activity Page
scissors	6		C68, D12, D100
screen, window	6	6	C28
shell, clam	3*	3	B84
shell, sea urchin	3*	3	B84
shell, snail	3*	3	B84
shell, spirula	3*	3	B84
shell, starfish	3*	3	B84
sieve, plastic	3*	3	E58
sponge, natural	6 sm. pieces	3 med. pieces	B74
spoon, measuring	6	6	C28, E1
spoon, metal	6	6	C20
spring scale	6	6	F38, F48, F56
spring toy	3*	3	E106
stirring rod	6	6	E1
stopwatch	1*	1	D1, E106, F12
stream table	1*	1	D1, D46
sulfur	6 pieces	6 pieces	E12
tablespoon	6	6	A1
tape measure, metric	6	6	F12
thermometer	12	12	D92, E1, E20
tray, large	6		E1
tubing	3 pieces	3 pieces	D46
washer, metal	12	12	F38
watch or clock	6		E1, E20
watering can	1*	1	C20, D1
weather station	3*	3	C1, C106
zinc, strip	6	6	E90

*Student groups should share this equipment.

Consumable Materials	Class Quantity Needed	Kit Quantity	Activity Page
aluminum foil	1 roll	1 roll	B1, E44
ammonia-containing cleaner	1 bottle		E28
antacid liquid	1 bottle		E28
apples	12		A1, E50
bag, plastic zip-top	6	1 box of 50	A4
baking soda	1 box	1 box	E50
ball, plastic foam	1*	1	D100
balloons	6	1 bag of 10	D128
balloons, round	12	1 pkg of 35	A4
beans, mixed	1 bag	1 bag	A40
bottle, 0.5 L with cap	6		C98
bottle, 2-L plastic	6		D128
box, cardboard	12		B66, F1
bubble wrap	6 pieces	6 pieces	F1
cardboard (large sheet) or posterboard	6 pieces		E4
chalk	6 pieces		D128
chart, bone sorting	6	1 pad of 15	C44
chart, hurricane tracking	6	See TR	C114
clay, brown modeling	3 boxes	3 boxes	D4, D54
clay, cream modeling	3 boxes	3 boxes	D12, D128
clay, green modeling	5 boxes	5 boxes	A4, D4
clay, red modeling	4 boxes	4 boxes	A4, D4, D54
clay, yellow modeling	3 boxes	3 boxes	D4, D54
cone, pine	6	6	B32
cone, spruce	6	6	B32
conifer needles	12		B32
cotton fabric	6 pieces		F1
cups, foam	72	75	B1
cups, plastic	60	2 pkgs of 50	B4, B12, E20, E50, E58
detergent, powdered	1 box		E28
diagram, fish	30	1 pkg of 50	B102
diagram, frog	30	1 pkg of 50	B102
diagram, grasshopper	30	1 pad of 50	B84
diagram, skeleton of reptile, bird, and mammal	30	See TR	B112
eggs, fresh	6		F1
epsom salts	1 box	1 box	E1
fern frond	6		B32
filter paper	6 pieces	1 pkg of 100	E58
flower	6		B42
food coloring	1 bottle	1 bottle	C20
glue	1 bottle		D122
gravel	14 bags	14 bags	C20, D1, E58
ice chips	about 3 cups		E20
index cards	36	1 pkg of 100	A4, C44
indicator paper, wide-range	1 vial	1 vial	E28
iron filings	2 containers	2 containers	E58
items, everyday small	18		F4
lemon juice	1 bottle		E28
litmus paper, blue	1 vial	1 vial	E28
litmus paper, red	1 vial	1 vial	E28
map of the western United States	30	See TR	D20
marker, black	6		A1, D100, E1
marker, yellow	6		D122
milk	1 bottle		E28
newspaper	48 sheets		C28, D4, D84
owl pellets	6	1 pkg of 6	C44

Consumable Materials	Class Quantity Needed	Kit Quantity	Activity Page
paper towels	2 rolls		B4, B12, B42, B50, C44, E28, E50, E58
paper, black	12 sheets		D92
paper, black construction	6 sheets		E122
paper, construction	6 sheets		D84
paper, graph	102 sheets	See TR	C4, C68, C92, D92
paper, notebook	24 sheets		A22, A50, C12, F4, F22
paper, scrap	6 sheets		C28
paper, shredded	enough to fill 6 boxes		F1
paper, white	45 sheets		D12, D100, D122, E122
paper, white large	12 sheets		D76, D122
paper, yellow construction	1 sheet		D100
pellets, plastic foam	3 bags	3 bags	D54, F1
pencil	6		A22, A50, B1, C12, F12, F22
pencils, colored	30 sets		B112, C4, C114, D76, D92
planaria	6	coupon	B74
plastic wrap	1 roll	1 roll	B1
plates, paper	18	50	A1
posterboard	1 pkg		A40, C68
potato	6		E90
rock salt	1 box	1 box	E1
salt	1 box	1 box	A1, E58
sand	4 bags	4 bags	A4, C20, D1, E58
sand	about 30 lb		D1, D46
sand, coarse	1 bag	1 bag	D1
seeds, bean	1 pkg	1 pkg	B4
seeds, corn	1 pkg	1 pkg	B50
seeds, lima bean	1 pkg	1 pkg	B50
shoe box	30		B1, B4, D12
shoe box with lid	6		C68
soda, carbonated	1 bottle		E28
soil, clay	2 bags	2 bags	C20
soil, potting	4 bags	4 bags	B1, C20
spoon, plastic	6	1 pkg of 24	E58
starch	1 bottle	1 bottle	C28
straw, plastic	12	1 pkg of 50	D4, D128
streak plate	6	8	D38
string	2 rolls	2 rolls	D76, D84, F38, F56
sugar	1 bag	1 bag	A1
supplies, diorama	6 sets		B66
tape, adhesive	1 roll		E82, E90
tape, masking	1 roll		B1, B4, C68, D100, F4, F38
tape, transparent	1 roll		A40
tea	1 bottle		E28
toothpick	1 box	1 box of 250	C44, D100
vinegar	2 bottles	2 bottles	E1, E28, E50
wax paper	1 roll	1 roll	D54
wire, copper enameled	1 roll	1 roll	E12, E82, E90
world map of biomes	30	See TR	C12
yarn	6 pieces	1 skein	E106

Conducting Live-Animal Activities

Several Investigates and Activities within *Harcourt Science* require the use of live animals, such as mealworms and butterflies. You may wish to keep these animals in the classroom after the conclusion of the activities for students' continued observations. Instructions for the care of these animals are included in the directions that accompany them when they arrive from the scientific supply house.

If animals are no longer needed or wanted in the classroom, they must be dealt with in an environmentally responsible manner. Any animals that have been collected locally may be released. It is, however, best to consider the season and its effects on the animal before release. Most animals ordered from supply houses are not native species and ***must not*** be released into the environment. These animals may not be able to survive in your area and would needlessly suffer before they died. Other non-native species can survive and reproduce. Their existence in a new environment may threaten native species or cause environmental damage. Oftentimes, these animals can be given to a pet store as a source of food for store animals. Otherwise, the animals must be euthanized. Please consult the instructions that accompanied these animals for advice on how to deal with these issues.

UNIT A

LIFE SCIENCE

Understanding Living Things

UNIT B

LIFE SCIENCE

Living Things Grow and Respond

UNIT C

EARTH SCIENCE

The Living Planet

UNIT D

EARTH SCIENCE

Cycles in Earth and Space

UNIT E

PHYSICAL SCIENCE

Matter and Energy

UNIT F

Physical Science

Forces and Machines

How Scientists Work

PLANNING AN INVESTIGATION

Harcourt Science provides students with many opportunities for scientific inquiry, including unit experiments, lesson investigations, activities for home or school, and science fair projects. Help your students plan their investigations by discussing the steps of the scientific method, shown on the *Pupil Edition* pages.

The following materials are needed if students choose to do the experiment on pages x–xii:

- **1 bar magnet**
- **4 large paper clips**
- **20–25 small paper clips**
- **freezer**

Scientific Method Students should use these steps or steps like these as they plan and conduct their experiments.

- **Observe, and ask questions.** Invite students to think of questions they might ask about magnetism such as: How are magnets made? Can any substance be magnetized? How cold can a magnet be and still be a magnet?
- **Form a hypothesis.** Ask students to form a hypothesis to answer their question. The hypothesis should be phrased as a statement. For example: "Magnets that are put into the freezer will stop being magnets."

HOW SCIENTISTS WORK

Planning an Investigation

How do scientists answer a question or solve a problem they have identified? They use one of several different organized ways called **scientific methods** to plan and conduct a study. They use science process skills to help them gather, organize, analyze, and present their information.

Melissa is using this scientific method for experimenting to find an answer to her question. You can use these steps too.

STEP 1 Observe and ask questions.

- Use your senses to make observations.
- Record **one** question that you would like to answer.
- Write down what you already know about the topic of your question.
- Decide what other information you need.
- Do research to find more information about your topic.

x

STEP 2 Form a hypothesis.

- Write a hypothesis for your experiment. A **hypothesis** is a possible answer to a question. You must be able to test a hypothesis scientifically.
- Write your hypothesis in a complete sentence.

My hypothesis is: A magnetized paper clip will lose its magnetic strength when it is in a freezer.

STEP 3 Plan an experiment.

- Decide how to conduct a fair test of your hypothesis by controlling variables. **Variables** are factors that can affect the outcome of the investigation.
- Write down the steps you will follow to do your test.
- List the equipment you will need.
- Decide how you will gather and record your data.

I'll magnetize four identical paper clips by stroking each 20 times with the same magnet. I'll test the strength of each clip. Then I'll put two of them in the freezer. After 20 minutes, I'll test the strength of each again. That way I control the strength of the original magnet and the things that are being magnetized. I let temperature be a variable.

STEP 4 Conduct the experiment.

- Follow the steps you wrote.
- Observe and measure carefully.
- Record everything that happens.
- Organize your data so you can study it carefully.

I'll record the number of paper clips each magnetized paper clip picks up. I'll do this both before and after the 20 minutes of time.

How Scientists Work

PLANNING AN INVESTIGATION

▶ **Plan an experiment.** Ask students to plan a way to test their hypothesis. They should write down the steps they intend to use. Lead them to an understanding of a fair test—that is, a test that takes into account one or more variables and that has a way to control the variables. In the case of the sample investigation, variables are controlled by using the same sized paper clips as magnets, stroking each the same number of times with the magnet and using the same magnet to do the magnetization.

▶ **Conduct the experiment.** Let students test their hypothesis according to their plan. They should follow the steps they wrote, record what they observe, and organize their data so that they can study it before drawing a conclusion.

How Scientists Work

PLANNING AN INVESTIGATION

- **Draw conclusions and communicate results.** Tell students that when they draw a conclusion, they compare the results of their test with their hypothesis. They decide whether their hypothesis was supported or shown to be incorrect. They write their conclusion and describe the test and the results.

Suggest to students that what they learn during their experiment might raise other questions, or it might help them think of ways to change their experiment. Encourage them to continue the experiment in any way they choose.

Experimenting Further Here are some questions your students might enjoy investigating independently.

- **How might acid rain affect plants?**
- **Will a room without air conditioning be warmer or cooler with the lights turned off?**
- **Are in-line skates faster than regular roller skates?**
- **What is the best watering schedule to conserve water and keep plants healthy?**
- **Do earthworms or sow bugs respond to different colored light differently?**
- **How can you maximize the efficiency of a machine?**
- **What are some safe ways to adjust the pH of a highly acidic environment?**
- **How could you design a toy car or a model car to move most rapidly downhill?**
- **How can you adjust your camera or film to take pictures in different lighting conditions? What are some of the tradeoffs?**

Unit Experiments are also provided at the beginning of each unit. (See pages A1–A1j, B1–B1j, C1–C1j, D1–D1j, E1–E1j, and F1–F1j.) Students will use the steps for Planning an Investigation as they conduct each Unit Experiment.

HOW SCIENTISTS WORK

STEP 5 Draw conclusions and communicate results.

- Analyze the data you gathered.
- Make charts, tables, or graphs to show your data.
- Write a conclusion. Describe the evidence you used to determine whether your test supported your hypothesis.
- Decide whether your hypothesis is supported by the experiment.

INVESTIGATE FURTHER

If your hypothesis is supported . . .

You may want to ask another question about your topic that you can test.

If your hypothesis is incorrect . . .

You may want to revise your hypothesis and do a new experiment as a test.

Do you think Melissa's new hypothesis is correct? Plan and conduct an experiment to find out!

PARTICIPATING IN A SCHOOL SCIENCE FAIR

These reproducible pages, which appear in the **Teaching Resources** book, provide helpful information for students and their families.

Using Science Process Skills

When scientists try to find an answer to a question or do an experiment, they use thinking tools called **process skills.** You use many of the process skills whenever you speak, listen, read, write, or think. Think about how these students use process skills to help them answer questions, do experiments, and investigate the world around them.

How Scientists Work

USING SCIENCE PROCESS SKILLS

Science process skills are important inquiry tools and are essential for investigating the natural world. Opportunities for developing process skills are provided throughout *Harcourt Science.* You may wish to use pages xiii–xvii at the beginning of the school year to introduce students to the processes. Students may also refer to these pages throughout the year as they conduct unit experiments and lesson investigations.

Each lesson investigation in *Harcourt Science* also develops the science processes by providing a Process Skill Tip on the *Pupil Edition* page. You can use the Process Skill Tip as a springboard for a lesson that focuses on that skill. A mini-lesson for teaching the process skill appears in the *Teacher's Edition* beneath the corresponding *Pupil Edition* page. An overhead transparency is provided for the mini-lesson, and an opportunity for practicing the process skill is provided in the *Workbook.*

How Scientists Work

USING SCIENCE PROCESS SKILLS

Observe
Compare
Classify

- **What senses does Eileen use to observe the feathers?** sight, perhaps touch
- **What characteristics could Eileen use to compare the feathers?** color, texture, weight, size
- **How might Eileen classify the feathers?** by using the characteristics; she could make a group of dark and a group of light-colored feathers, she could group them by length.

Try This Suggest that students practice **observing, comparing,** and **classifying** by identifying a group of objects, such as shells that they wish to learn more about. Remind them to use their senses to observe them carefully and to record what they observe so that they can later compare and classify the shells.

HOW SCIENTISTS WORK

What Eileen plans to investigate

Eileen is collecting feathers. She wants to collect them from as many different kinds of birds as she can. She looks for feathers of different sizes, shapes, and colors.

Process Skills

Observe—use the senses to learn about objects and events

Compare—identify characteristics of things or events to find out how they are alike and different

Classify—group or organize objects or events in categories based on specific characteristics

How Eileen uses process skills

She **observes** the feathers carefully and **compares** them to see how they are alike and different. Then she **classifies** them into groups based on their colors.

xiv

What Brett plans to investigate

Brett wants to improve the yield of his vegetable garden. He has observed that tomato plants seem to grow strong and healthy on the left side of his back yard. On the right side, the plants are small and thin. He also has observed that the amount of light is about equal on the two sides of the yard. He concludes that the difference in plant growth may have something to do with the quality of the soil.

How Brett uses process skills

He takes 500 grams of soil from the left side of the back yard, places it in a jar, and adds a liter of water. He screws the top on the jar, shakes the jar, and lets the soil settle. He does the same things with soil from the right side of the back yard. He **gathers** and **records data** by observing and by taking careful notes, as well as by drawing pictures. Later, he will improve his rough drawings and use them to **display data.**

After the soil has settled, Brett observes the jars. He notes that the soil has settled in layers. In the jar from the left side of the back yard, he sees a thick layer of black dirt over a thick layer of sand and clay and a thin layer of pebbles. In the jar from the right side, he see no black dirt at all and a thick layer of sand over a thick layer of pebbles. He **predicts** that if he plants in soil with black dirt, his vegetables will grow better.

Process Skills

Gather, Record, Display, or Interpret Data

- ► gather data—make observations that will be useful for making inferences or predictions
- ► record data—write down or draw the observations in a table, graph, or notebook
- ► display data—make tables, charts, or graphs
- ► interpret data—draw conclusions about what the data shows

Predict—form an idea of an expected outcome, based on observations, prior knowledge, and experience

How Scientists Work

USING SCIENCE PROCESS SKILLS

Gather, Record, Display, or Interpret Data

Predict

- ► **What does he use to gather and record data?** his senses, paper and pencil
- ► **How does he record data?** by making drawings and taking careful notes
- ► **How does he display data?** by improving his rough drawings
- ► **What does he use to make his predictions?** the data he has gathered and his previous knowledge of the situation

Try This Have students **gather, record,** and **display** information about the number of cars in the school parking lot at various times of day. Ask them to use their data to **predict** at what time the parking lot will have the maximum number of cars.

How Scientists Work

USING SCIENCE PROCESS SKILLS

Hypothesize

Plan and Conduct a Simple Investigation

Infer

- **What is another hypothesis Tamara could test?** There would be no difference between the dark and the light jacket.
- **How could you conduct a simple investigation to test the alternative hypothesis?** Tamara's investigation would also serve as a test for the alternative hypothesis.
- **What is the difference between inference and observation?** In observation, you sense things, but in inference you use your observations and prior knowledge to come to a conclusion about why something has occurred.
- **What is the difference between an inference and a hypothesis?** An inference uses logical reasoning to explain observations and events. A hypothesis is a statement of expected outcomes, or a prediction of cause and effect based on observations and prior knowledge. A hypothesis should be a testable statement.

Try This Suggest that students practice these skills by **planning and conducting a simple investigation** to use shadows to determine whether days are getting longer or shorter in the present season. Students can measure the shadow of a meterstick at noon on several days for two weeks. Lengthening shadows generally mean days are getting shorter. Shorter shadows generally indicate longer days.

Process Skills

Hypothesize—make a statement about an expected outcome, based on observation, knowledge, and experience

Plan and Conduct a Simple Investigation—identify and perform the steps necessary to find the answer to a question, using appropriate tools and recording and analyzing the data collected

Infer—use logical reasoning to explain events and draw conclusions based on observations

What Tamara plans to investigate

Tamara wants to buy the warmest jacket she can for the coming winter. The style she likes comes in either black or white. Tamara wants to know which color would be warmer.

How Tamara uses process skills

Tamara **hypothesizes** that the black jacket could be warmer than the white one. She **plans and conducts a simple investigation** to test her hypothesis. She places a black sheet of paper and a white sheet of paper in the sun. She puts a thermometer in the center of each sheet to measure the temperature of the paper. She sets a stopwatch for 30 minutes. At the end of 30 minutes, Tamara checks the thermometers. The thermometer on the black paper shows 28°C, and the one on the white paper shows 23°C. Tamara **infers** that black objects absorb more light than white objects, so the black jacket would be warmer.

What Jake plans to investigate

Jake wants to package a glass vase to send to his grandmother. He wants to make sure it will not be damaged in the mail. He read an ad that said bubbled plastic wrap would protect the vase best. He plans an experiment to test the bubble wrap against other protective padding.

How Jake uses process skills

First, Jake **identifies and controls variables** that will affect his experiment.

- He finds three boxes that are exactly the same.
- He finds three eggs that are as similar as possible.
- He places all materials on a table he will use to control the height of a fall.
- He gets bubbled plastic wrap, a bag of foam peanuts, and some newspapers.

The boxes, the eggs, and the height of the table are the variables he will keep the same. The type of packaging material is the variable he will change.

Then, he **experiments.**

- He wraps an egg in the bubble wrap and places it in a box, making sure that the egg and the wrap fill the box.
- He half-fills a box with newspaper, adds an egg, and then fills the box with more newspaper.
- He places some foam peanuts in a box, adds an egg, and then adds more foam peanuts to fill the box.
- One by one he places the boxes near the edge of the table. He pushes the boxes over the edge.
- He opens the boxes to find that only the egg in newspaper is broken. He infers that both foam peanuts and bubbled plastic wrap protect the egg better than newspaper.

Process Skills

Identify and Control Variables—identify and control factors that affect the outcome of an experiment so that only one variable is changed

Experiment—design ways to collect data under controlled conditions to test a hypothesis

How Scientists Work

USING SCIENCE PROCESS SKILLS

Identify and Control Variables

Experiment

- ▶ **What is a variable?** a factor of the experiment that changes or can be changed
- ▶ **What are the variables in Jake's experiment?** Possible answers: the size of the eggs, the height of the drop, the size of the containers, the packing material for each container.
- ▶ **Why is it important to control variables?** You can only be certain that you are testing what you want to test if all other conditions of the experiment are held constant.
- ▶ **What is an experiment?** An experiment is a fair test that is designed to gather data under controlled conditions to test a hypothesis.
- ▶ **It is often said that an experiment should be repeatable, that is, you should be able to conduct it again and have the same results. Why might this be an important aspect of experimenting?** If the experimental results cannot be duplicated, then the conclusions from the initial experiment may be in doubt.
- ▶ **Why is it important to document both your procedure and your results when experimenting?** to assure that anyone attempting to repeat the experiment knows what you did and what you found. In addition, it helps those looking at the experiment to determine whether some other condition might have affected the results.
- ▶ **Why is it better to say that experimental results support a hypothesis rather than the hypothesis is proven?** An experiment is usually designed to test only one question or one aspect of a problem. Other variables might affect the results.

Try This Have students plan and conduct an **experiment** to test claims for a product. Have them form a hypothesis and tell what they infer from the results. Remind them to plan how they will **control the variables** in their experiment.

How Scientists Work

READING TO LEARN

Discuss with students how scientists use reading in their work.

- **Why is having good reading skills important to a scientist?** Scientists need to read to find out about things they are investigating. They also need to read to check facts to make sure their findings agree with the findings of others.
- **What kinds of work-related materials do you think scientists read?** Possible responses: encyclopedias, scientific magazines and journals, reports on the Internet, reports from other scientists, letters and messages from other scientists

Discuss the "Before Reading" strategies students can use.

- **How does using the Find Out statement help you prepare to read?** It helps me set a purpose for reading. It helps me know what to look for as I read.
- **What are ways you can preview the vocabulary words?** I can look in the Glossary to find out how to pronounce each word and to find out what each word means.
- **Why is it important to preview the titles and headings in a section?** The titles and headings help me know what I will be reading about.

Harcourt Science provides many opportunities for students to develop science vocabulary. The following is a list of additional vocabulary features in the program.

- **Vocabulary Preview** first page of each chapter in the *Pupil Edition*
- **Building a Science Vocabulary** in the Reference Section, pp. R44–49
- **Develop Science Vocabulary** throughout the *Teacher's Edition*, such as on pp. A8–9
- **Vocabulary Strategies and Activities** in the *Teaching Resources* book
- **Vocabulary Cards** in the *Teaching Resources* book
- **Vocabulary Review** for each chapter, in the *Workbook*

HOW SCIENTISTS WORK

Reading to Learn

Scientists use reading, writing, and numbers in their work. They **read** to find out everything they can about a topic they are investigating. So it is important that scientists know the meanings of science vocabulary and that they understand what they read. Use the following strategies to help you become a good science reader!

Before Reading

- Read the Find Out statement to help you know what to look for as you read.
- Think: I need to find out how igneous rocks form and how people use them.

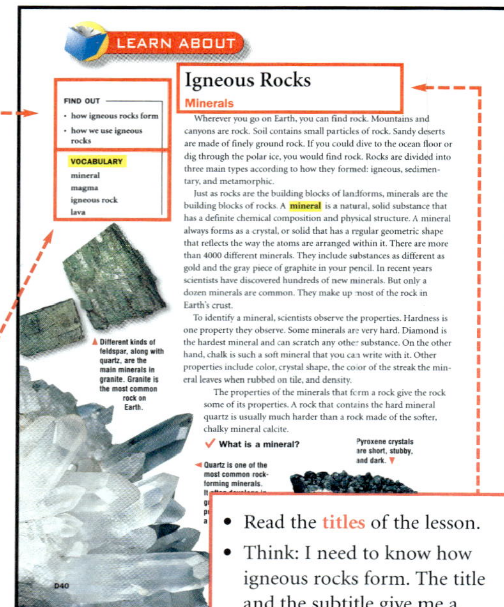

- Look at the Vocabulary words.
- Be sure you can pronounce each word.
- Look up each word in the Glossary.
- Say the definition to yourself. Use the word in a sentence to show its meaning.

- Read the titles of the lesson.
- Think: I need to know how igneous rocks form. The title and the subtitle give me a clue that igneous rocks are somehow composed of minerals.

xviii

During Reading

Find the **main idea** in a paragraph.

- A mineral is a natural, solid substance that has a definite chemical composition and physical structure.

Find **details** in the paragraph that support the main idea.

- Minerals always form as crystals, or solids that have a regular geometric shape.
- They include substances as different as gold and the gray piece of graphite in your pencil.
- A few minerals make up most of the rock in Earth's crust.

Read another paragraph to find out more about the **main idea**.

- To identify a mineral, scientists observe the properties.

Check your understanding of what you have read.

- Answer the question at the end of the section.
- If you're not sure of the answer, reread the section and look for the answer to the question.

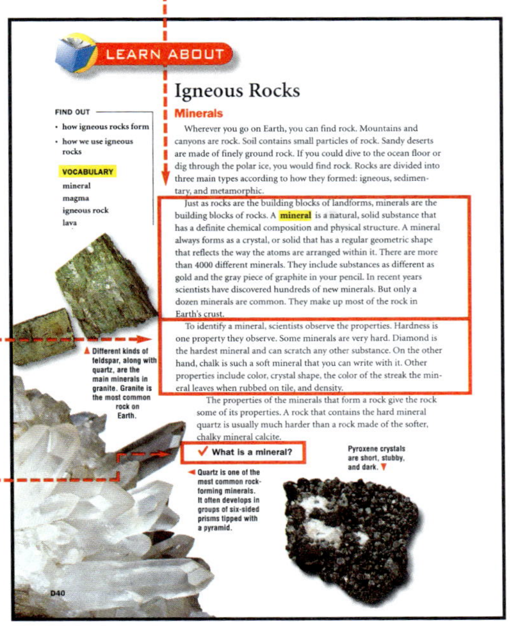
LEARN ABOUT

FIND OUT
- how igneous rocks form
- how we use igneous rocks

VOCABULARY
mineral
magma
igneous rock
lava

Igneous Rocks

Minerals

Wherever you go on Earth, you can find rock. Mountains and canyons are rock. Soil contains small particles of rock. Sandy deserts are made of finely ground rock. If you could dive to the ocean floor or dig through the polar ice, you would find rock. Rocks are divided into three main types according to how they formed: igneous, sedimentary, and metamorphic.

Just as rocks are the building blocks of landforms, minerals are the building blocks of rocks. A **mineral** is a natural, solid substance that has a definite chemical composition and physical structure. A mineral always forms as a crystal, or solid that has a regular geometric shape that reflects the way the atoms are arranged within it. There are more than 4000 different minerals. They include substances as different as gold and the gray piece of graphite in your pencil. In recent years scientists have discovered hundreds of new minerals. But only a dozen minerals are common. They make up most of the rock in Earth's crust.

To identify a mineral, scientists observe the properties. Hardness is one property they observe. Some minerals are very hard. Diamond is the hardest mineral and can scratch any other substance. On the other hand, chalk is such a soft mineral that you can write with it. Other properties include color, crystal shape, the color of the streak the mineral leaves when rubbed on tile, and density.

The properties of the minerals that form a rock give the rock some of its properties. A rock that contains the hard mineral quartz is usually much harder than a rock made of the softer, chalky mineral calcite.

✔ What is a mineral?

▲ Different kinds of feldspar, along with quartz, are the main minerals in granite. Granite is the most common rock on Earth.

◄ Quartz is one of the most common rock-forming minerals. It often develops in groups of six-sided prisms tipped with a pyramid.

Pyroxene crystals are short, stubby, and dark. ▼

D40

How Scientists Work

READING TO LEARN

Discuss the "During Reading" strategies students can use.

- ▶ **Why is it important to find the main idea in a section of text?** When I find the main idea, then I know the most important idea I need to remember.
- ▶ **Why are the supporting details important?** They help give the main idea meaning. They help me remember the main idea because they provide examples.

Harcourt Science provides many opportunities for students to become strategic readers and to practice and apply reading skills. The following features in *Harcourt Science* help students become better science readers.

- ▶ Consistent support for reading via text features, such as Find Out, highlighted vocabulary words, section review questions, Summary, and Review
- ▶ **Using Science Reading Strategies** in the Reference Section on pp. R38–43
- ▶ **Graphic Organizer for Chapter Concepts** one per chapter in the *Workbook* and on an overhead transparency
- ▶ **Reading Mini-Lesson** one in each lesson of the *Teacher's Edition*, each with an accompanying overhead transparency
- ▶ **Reading Skills Practice** in the *Workbook*

How Scientists Work

READING TO LEARN

Discuss the "After Reading" strategies students can use.

▶ **How do photographs and illustrations help you understand the main idea of a section?** They provide examples of the science ideas. They also help make the text interesting.

▶ **Why is it important to summarize what you have read?** Summarizing helps me review the main ideas and prepare for reviews and tests.

Suggest students look through their textbooks to become familiar with the text features and to see that the features appear in each lesson of the program.

HOW SCIENTISTS WORK

After Reading

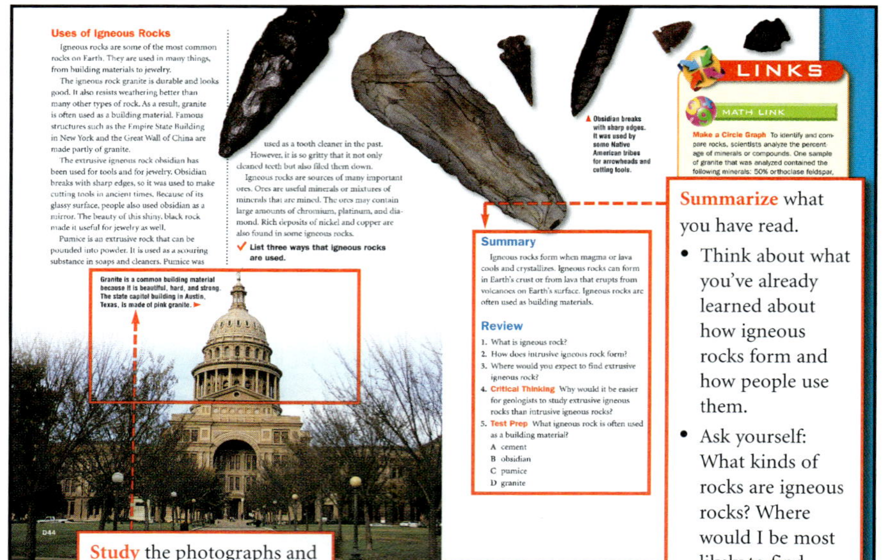
Uses of Igneous Rocks

Igneous rocks are some of the most common rocks on Earth. They are used in many things, from building materials to jewelry.

The igneous rock granite is durable and looks good. It also resists weathering better than many other types of rock. As a result, granite is often used as a building material. Famous structures such as the Empire State Building in New York and the Great Wall of China are made partly of granite.

The extrusive igneous rock obsidian has been used for tools and for jewelry. Obsidian breaks with sharp edges, so it was used to make cutting tools in ancient times. Because of its glassy surface, people also used obsidian as a mirror. The beauty of this shiny, black rock made it useful for jewelry as well.

Pumice is an extrusive rock that can be pounded into powder. It is used as a scouring substance in soaps and cleaners. Pumice was used as a tooth cleaner in the past. However, it is so gritty that it not only cleaned teeth but also filed them down.

Igneous rocks are sources of many important ores. Ores are useful minerals or mixtures of minerals that are mined. The ores may contain large amounts of chromium, platinum, and diamond. Rich deposits of nickel and copper are also found in some igneous rocks.

✔ **List three ways that igneous rocks are used.**

Granite is a common building material because it is beautiful, hard, and strong. The state capitol building in Austin, Texas, is made of pink granite. ►

▲ **Obsidian breaks with sharp edges. It was used by some Native American tribes for arrowheads and cutting tools.**

LINKS

MATH LINK

Make a Circle Graph To identify and compare rocks, scientists analyze the percentage of minerals or compounds. One sample of granite that was analyzed contained the following minerals: 50% orthoclase feldspar,

Summary

Igneous rocks form when magma or lava cools and crystallizes. Igneous rocks can form in Earth's crust or from lava that erupts from volcanoes on Earth's surface. Igneous rocks are often used as building materials.

Review

1. What is igneous rock?
2. How does intrusive igneous rock form?
3. Where would you expect to find extrusive igneous rock?
4. **Critical Thinking** Why would it be easier for geologists to study extrusive igneous rocks than intrusive igneous rocks?
5. **Test Prep** What igneous rock is often used as a building material?
   A cement
   B obsidian
   C pumice
   D granite

344

Study the photographs and diagrams.

- Read the caption and the labels on each photograph or diagram.
- Think: What kinds of things am I seeing in each picture? What does this tell me about how igneous rocks are used?

Summarize what you have read.

- Think about what you've already learned about how igneous rocks form and how people use them.
- Ask yourself: What kinds of rocks are igneous rocks? Where would I be most likely to find them? What is one main use for igneous rocks?

For more reading strategies and tips, see pages R38–R49.

Reading about science helps you understand your conclusions from your investigations.

xx

Writing to Communicate

Writing about what you are learning helps you connect the new ideas to what you already know. Scientists **write** about what they learn in their research and investigations to help others understand the work they have done. As you work like a scientist, you will use the following kinds of writing to describe what you are doing and learning.

In **informative writing**, you may

- describe your observations, inferences, and conclusions.
- tell how to do an experiment.

In **narrative writing**, you may

- describe something, give examples, or tell a story.

In **expressive writing**, you may

- write letters, poems, or songs.

In **persuasive writing**, you may

- write letters about important issues in science.
- write essays expressing your opinions about science issues.

Writing about what you have learned about science helps others understand your thinking.

xxi

How Scientists Work

WRITING TO COMMUNICATE

Discuss with students how scientists use writing in their work.

- **Why is having good writing skills important to a scientist?** Scientists need to be able to tell others about the results of their work. They need to be able to communicate what they have learned.

Harcourt Science provides many opportunities for students to develop and apply writing skills. The following is a list of features in the program that address writing skills.

- **Writing Links** in the *Pupil Edition* at the end of each lesson
- **Writing Practice** for each chapter, in the *Workbook*
- **Writing Models** in the *Teaching Resources* book

How Scientists Work

USING NUMBERS

Discuss with students how scientists use mathematics in their work.

- **What kinds of measuring tools do you think scientists use in their work?** Possible responses: thermometers, beakers, graduates, scales for measuring weight and mass
- **Why is it important to measure accurately while gathering data?** If measurements are not accurate, then you cannot draw accurate conclusions based on your data.

Remind students to look on pages R2–6 for more information about measuring tools and measurement.

HOW SCIENTISTS WORK

Using Numbers

Scientists **use numbers** when they collect, display, and interpret their data. Understanding numbers and using them correctly to show the results of investigations are important skills that a scientist must have. As you work like a scientist, you will use numbers in the following ways:

Measuring

Scientists make accurate measurements as they gather data. They use measuring instruments, such as thermometers, clocks and timers, rulers, a spring scale, and a balance, and they use beakers and other containers to measure liquids.

For more information about using measuring tools, see pages R2–R6.

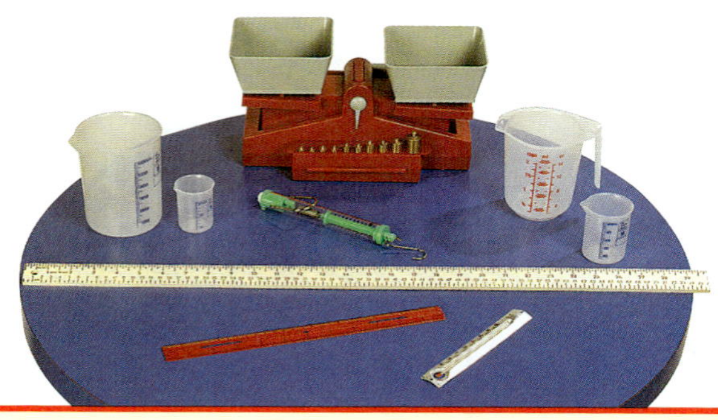

xxii

Interpreting Data

Scientists collect, organize, display, and interpret data as they do investigations. Scientists choose a way to display data that helps others understand what they have learned. Tables, charts, and graphs are good ways to display data so that it can be interpreted by others.

Using Number Sense

Scientists must understand what the numbers they use represent. They compare and order numbers, estimate and calculate with numbers, read and understand the numbers shown on graphs, use and understand percents and probability, and read the scales on thermometers, measuring cups, beakers, and other tools.

Good scientists apply their math skills to help them display and interpret the data they collect.

Weighing In Around the Solar System

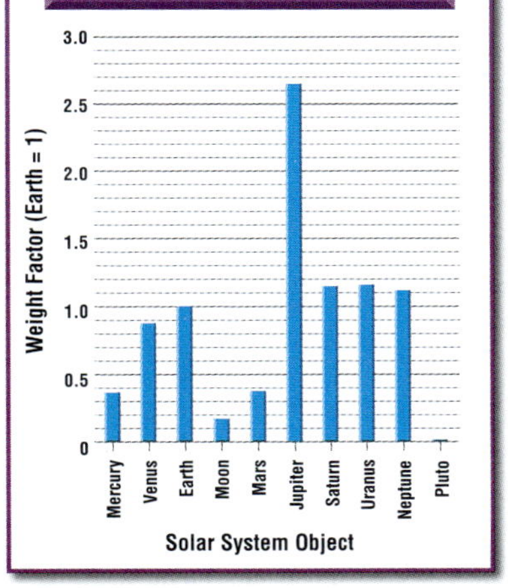

Numbers of Species in Vertebrate Groups

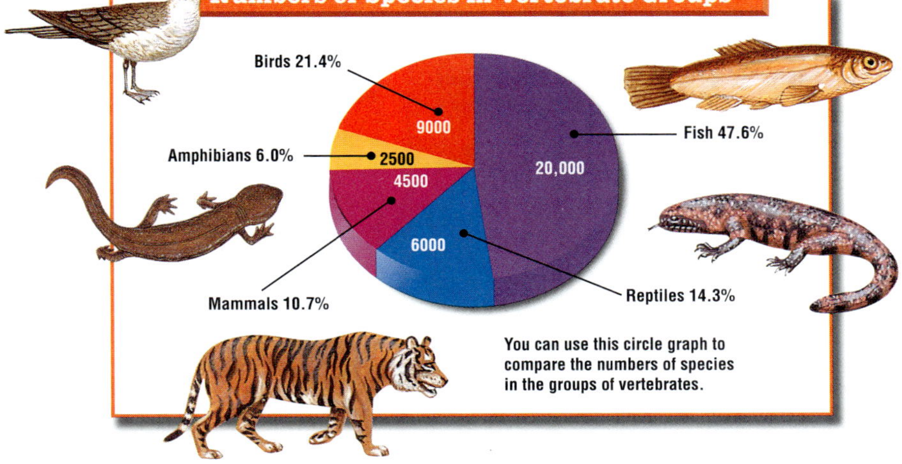

You can use this circle graph to compare the numbers of species in the groups of vertebrates.

In *Harcourt Science*, you will have many opportunities to work like a scientist. An exciting year of discovery lies ahead!

xxiii

How Scientists Work

INTERPRETING DATA

- **Why are charts and graphs important to the work of a scientist?** Charts and graphs can be used to display data in ways that make the data easier to interpret. They can show patterns and relationships.

USING NUMBER SENSE

- **How might scientists use calculation in their work?** Scientists might need to know an average speed or population. They might want to calculate work, power, or energy.
- **How might scientists use estimation in their work?** Scientists might want to estimate the number of seeds available from a field of seed corn or the population of organisms in a community.

Harcourt Science provides many opportunities for students to develop and apply mathematics skills. Many of the unit experiments, lesson investigations, and other activities involve using numbers to do things such as take measurements and record data. Numerous charts and graphs appear throughout the program for students to interpret. A Math Link is also provided at the end of each lesson.

Safety in Science

Work Like a Scientist When students first begin science explorations, it is important for them to develop a sense of what is and what is not appropriate in the laboratory. Beginning these lessons at this grade level, in which experiments are relatively safe, will prepare students for experiments in upper grades when safety rules must be observed more closely and the potential dangers are greater. Emphasize to students that as they study science, they will be working like scientists and that it is important for scientists to work safely.

The most important rule for students at this level is to behave appropriately in the laboratory. Emphasize that students must not run, push others, throw objects, or engage in any other behavior that might cause accidents.

Help students become familiar with the safety signs and caution statements that appear where they are needed throughout *Harcourt Science.* Discuss each safety rule and ask students to predict what can happen if a rule is not followed.

A copy of the Safety in Science page and a Safety Checklist are provided in the *Workbook.*

HOW SCIENTISTS WORK

Safety in Science

Doing investigations in science can be fun, but you need to do them safely. Here are some rules to follow.

1. **Think ahead.** Study the steps of the investigation so you know what to expect. If you have any questions, ask your teacher. Be sure you understand any safety symbols that are shown.
2. **Be neat.** Keep your work area clean. If you have long hair, pull it back so it doesn't get in the way. Roll or push up long sleeves to keep them away from your experiment.
3. **Oops!** If you should spill or break something or get cut, tell your teacher right away.
4. **Watch your eyes.** Wear safety goggles anytime you are directed to do so. If you get anything in your eyes, tell your teacher right away.
5. **Yuck!** Never eat or drink anything during a science activity unless you are told to do so by your teacher.
6. **Don't get shocked.** Be especially careful if an electric appliance is used. Be sure that electric cords are in a safe place where you can't trip over them. Don't ever pull a plug out of an outlet by pulling on the cord.
7. **Keep it clean.** Always clean up when you have finished. Put everything away and wipe your work area. Wash your hands.

In some activities you will see these symbols. They are signs for what you need to do to be safe.

CAUTION **Be especially careful.**

CAUTION **Wear safety goggles.**

CAUTION **Be careful with sharp objects.**

CAUTION **Don't get burned.**

CAUTION **Protect your clothes.**

CAUTION **Protect your hands with mitts.**

CAUTION **Be careful with electricity.**

xxiv

Understanding Living Things

UNIT A

Unit Overview

Unit Theme The scale and structure of living things is one key to both classification and understanding growth and development. All living things start as a single cell, and many groups of living things grow through the addition of millions upon millions of these tiny building blocks

CHAPTER 1

Cells, Genetics, and Heredity A2

All of the characteristics of an organism, its color, shape, size, and how it functions, are dictated by a molecule that is a very small part of a very tiny cell—DNA.

CHAPTER 2

Classification A38

Organisms are grouped by the things they have in common: size, shape, structure. As scientists discover more about the structures of living things, classification systems change.

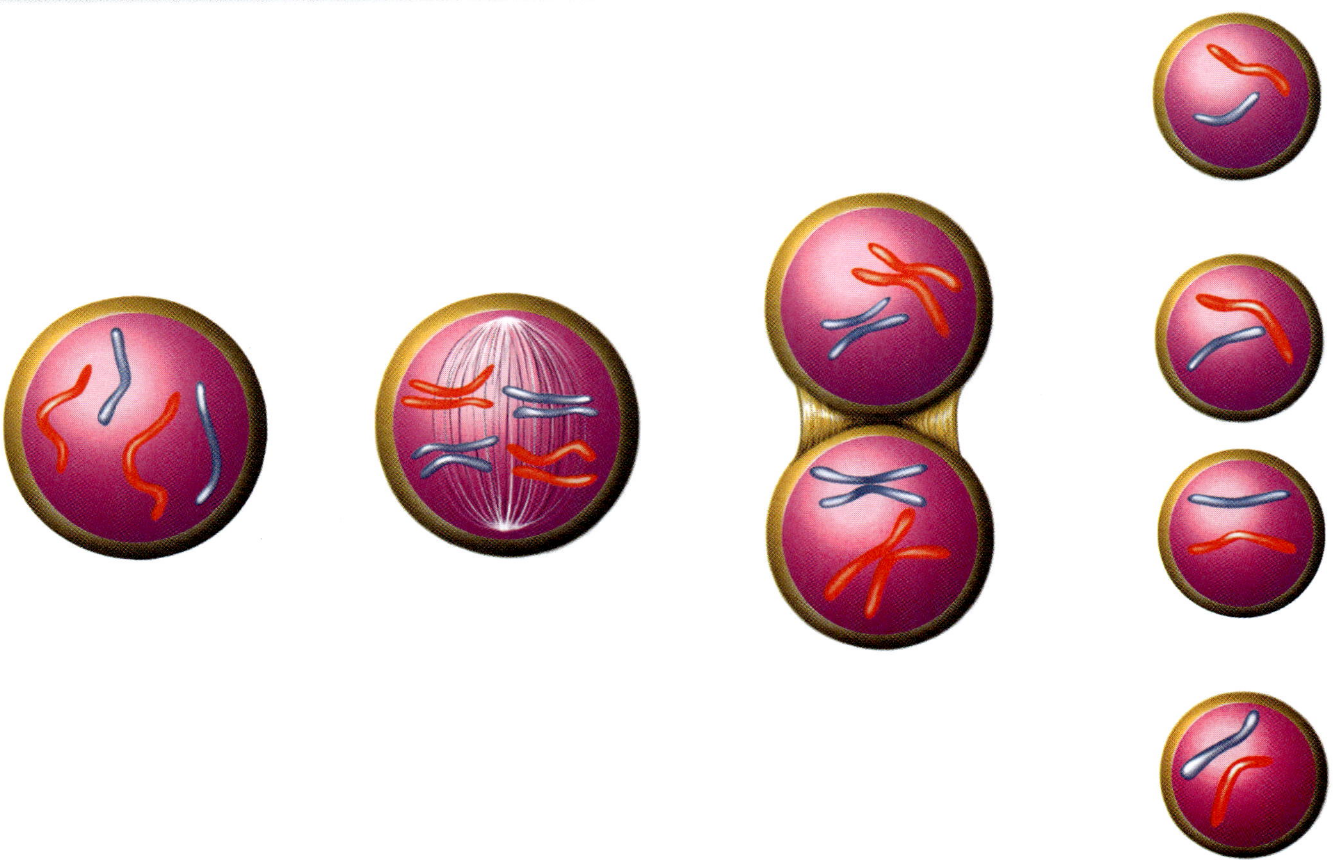

UNIT A

Skills for Lifelong Learning

Science Process Skills

The Science Process Skills are important inquiry tools and are essential for investigating the natural world. Opportunities for developing process skills are provided throughout *Harcourt Science*. The pages shown in this chart indicate where the process skills receive special emphasis.

Science Process Skill	Skill Tip and Mini-Lesson	Skills Practice (Workbook)	Reinforcement
Observe			A1j, A5, A7, A13, A16, A23, A28, A41, A45, A46
Compare	A13	WB9	A14, A20, A27, A41, A51, A53, A55
Classify/Order	A41	WB22	
Gather, Record, Display, or Interpret Data	A51	WB27	A1j, A13, A23
Use Numbers	A23	WB14	
Communicate			A1j
Plan and Conduct Simple Investigations			A1j
Measure			
Predict			
Infer	A5, A23	WB14	A7, A13, A14, A16, A18, A20, A45, A46, A53, A55
Draw Conclusions			A1j, A5, A13, A23, A35, A41, A51, A61
Use Time/Space Relationships	A13	WB9	
Formulate or Use Models	A5	WB4	A5
Hypothesize			A1j
Identify and Control Variables			
Experiment			

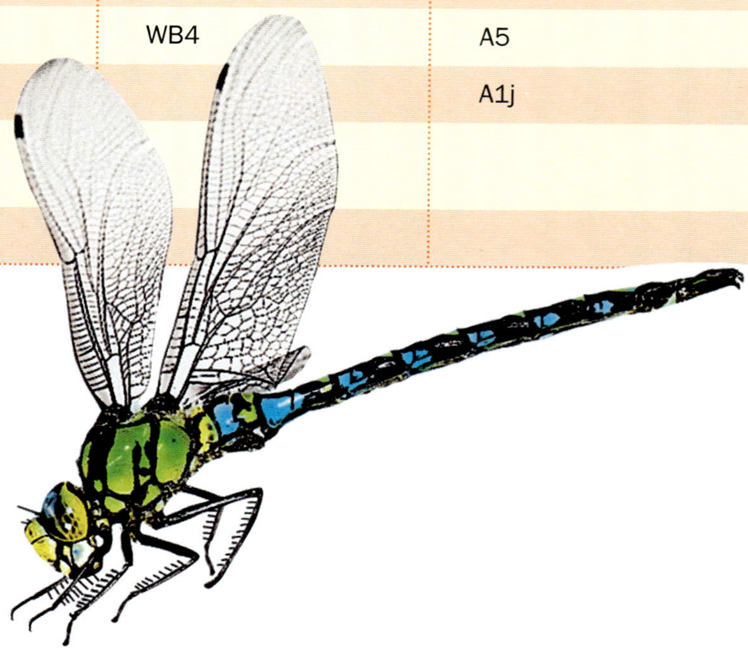

Reading, Writing and Math Skills

Effective science instruction integrates science content and experiences with all areas of the elementary curriculum. *Harcourt Science* provides students with many opportunities to develop reading, writing, and math skills via meaningful activities and strategies built into every lesson in this unit.

Reading Skills Checklist

Strategies for developing the following reading skills are provided in this unit.

- ☑ **use prefixes and suffixes to determine word meanings** A32
- ☑ **use context to determine/confirm word meaning** A58
- ☑ **identify supporting facts and details** A6, A8, A10, A14, A16, A17, A20, A24, A25, A26, A28, A30, A42, A45, A46, A47, A48, A52, A54, A55
- ☑ **arrange events in sequence** A16
- ☑ **identify the main idea** A6, A8, A10, A14, A16, A17, A20, A24, A25, A26, A28, A30, A42, A45, A46, A47, A48, A52, A54, A55
- ☑ **summarize and paraphrase a selection** A46
- ☑ **identify cause and effect** A26
- ☑ **predict probable future actions and outcomes** A29
- ☑ **use graphic sources for information** A26, A43, A48, A53
- ☑ **draw logical conclusions** A7, A15, A42, A55
- ☑ **compare and contrast** A8, A29
- ☑ **distinguish fact and opinion** A55
- ☑ **make generalizations** A7
- ☑ **use reference sources** A9, A10, A47, A49

Writing Links

Prompts that provide opportunities for students to express their ideas in writing are provided in every lesson in this unit.

Informative Writing—Compare and Contrast A11
Informative Writing—Description A21, A49
Informative Writing—Explanation A31
Expressive Writing—Poem or Song Lyrics A57
Writing About Science A37, A63

Math Links

Opportunities to practice math skills, solve problems, and connect math and science are provided in every lesson.

Find Possible Outcomes A3
Divide A3
Find the Average A3
Solve Problems A11, A17
Use Number Patterns A21
Compute Probability A29
Determine Probability A31
Estimate A39, A49
Find the Range A39
Compute Percents A57

UNIT A

How to Integrate Your Day

Use these topics to help you integrate science into your daily planning.

Curriculum Integration

Fine Arts

A Very Small City A11

Health

Genetic Disorder A9
Genes and Diseases A21
Malaria A47
Organisms and Disease A49

Writing

Informative Writing—Compare and Contrast A11
Informative Writing—Description A21, A49
Informative Writing—Explanation A31
Expressive Writing—Poem or Song Lyrics A57
Writing About Science A37, A63

Math

Find Possible Outcomes A3
Divide A3
Find the Average A3
Solve Problems A11, A17
Use Number Patterns A21
Compute Probability A29
Determine Probability A31
Estimate A39, A49
Find the Range A39
Compute Percents A57

Language Arts

Word Origins A21
Latin A43
Mnemonics A53
Naming a Species A57

Social Studies

History of Science A11
Traits in History A31

Using Technology

***Harcourt Science* includes an array of technology products that**

- bring interactive experiences to the classroom.
- add depth to chapter concepts.
- meet individual needs.

Technology Links at the end of each lesson in the student's book help you integrate technology at the point of its most effective use.

Computer Software

Harcourt Science Explorations CD-ROM

These engaging CD-ROMs provide interactive science explorations. Activities enhance and expand key chapter concepts through simulations and investigations.

Data ToolKit

Data ToolKit is a tool-based program that enables students to organize, communicate, and interpret data they have collected during their investigations.

Videos

Harcourt Science Newsroom Videos

These videos bring exciting science news events to your classroom from the leaders in news broadcasting—CNN and Turner Learning. For each chapter in this unit, there is a **Harcourt Science Newsroom Video** segment that helps students see the relevance of science to their lives. The videos also add interesting depth to chapter concepts.

Internet

GO ONLINE

Smithsonian Institution®

With *Harcourt Science*, it's easy to make the Smithsonian Institution a part of the science classroom. Through a special agreement, you can visit this Internet site to see correlated links to the Web sites of Smithsonian museums, galleries, and the National Zoological Park. Resources include virtual tours, on-line exhibits, and hands-on investigations—all especially selected to complement and enrich *Harcourt Science.*

www.si.edu/harcourt/science

The Learning Site

Visit The Learning Site for a world of science resources, including science News Breaks, interactive learning games and activities, school-home resources, and an animated Science Glossary.

www.harcourtschool.com

NSTA SciLinks

These special Internet links are provided by the National Science Teachers Association in partnership with Harcourt. The SciLinks Web page connects students to selected Web sites that enhance and enrich *Harcourt Science* chapter content.

www.scilinks.org/harcourt (for students)

www.scilinks.org (for teachers)

Other Technology Resources

Software

Steck-Vaughn Interactive Science Encyclopedia

This CD-ROM is a storehouse of articles, images, video clips, animations, and activities. Students can use this encyclopedia to gather additional information about chapter concepts and conduct investigations and simulations.

Cell City

Cambridge Educational (WIN CD/MAC CD) An innovative multimedia CD-ROM helps students understand the operation of a cell by revealing its similarities to a city. Concepts such as energy generation and supply, manufacturing, communications, waste disposal and recycling are all looked at in this unique way.

Video

Cells and Life

CLEARVUE/eav, 1991. (17 minutes) Presents a complete review of living cells.

The Complete Five - Kingdom Series

Ward's Natural Science Establishment, 1990. (95 minutes) Presents a comprehensive set of six parts which discusses the classification in the five kingdoms of life.

UNIT A

Reaching All Learners

Strategies for Multi-Age Classrooms

Develop Science Process Skills

- Help higher-level students reinforce their own understanding of the science process skill of *using a model* by placing them in groups with younger or less-advanced students. For example, in Chapter 1, Lesson 1, the higher-level students can coach the less-advanced students in making the cell.

Build Science Concepts

- When students in different grade levels are working on similar science concepts, such as classifying living things, you may want to group them to present a visual, kinesthetic, or tactile summary of the materials covered. Let the students decide which approach they want to use.
- Have advanced students adapt the investigation in Chapter 1, Lesson 3, to make a learning center for younger or less-advanced students. The advanced students should provide appropriate reference books and activity materials at the center.

Strategies for ESL Students

Build Concepts

Graphic Organizer Use this graphic organizer to summarize and reinforce the most important concepts in the unit. Where possible, repeat the use of a graphic organizer in different lessons.

Classification

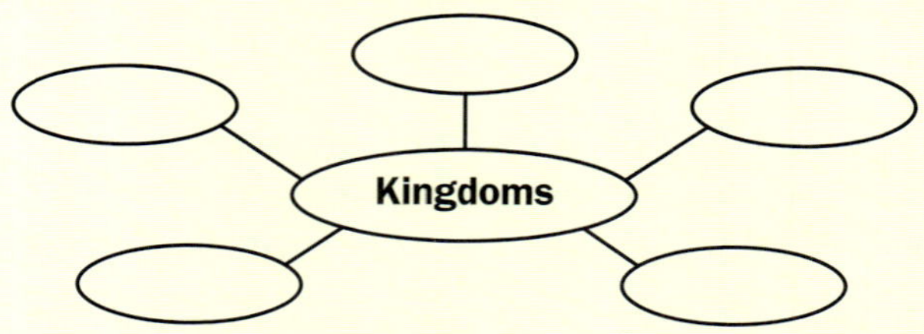

Chapter Graphic Organizer A Graphic Organizer for Chapter Concepts is provided for each chapter in this unit. ESL students can use this to summarize major concepts and reinforce chapter vocabulary. The Graphic Organizer for Chapter Concepts appears in the workbook and is also provided as an overhead transparency.

Develop Oral Language

Reread and Role-Play Pair ESL students, and ask them to reread sections of the unit and act them out for each other. Then have them discuss the content of the unit.

Voluntary Reading Have ESL students choose nonfiction books about cells, genetics, and heredity and then read and discuss them in pairs or small groups.

ESL Activities

Point-of-use activities for building concepts and developing oral language are provided in the lessons throughout the unit.

Strategies for Special Needs Students

Learning Difficulties

- Students who are working below grade level will benefit from focusing on several questions before reading or rereading. On the board, write three to five questions that you want students to find answers to.
- Let students working below grade level make a visual summary of the unit. For example, they might form collages, make maps, draw pictures, invent cartoons or comic strips, or paint a mural.
- Students with learning difficulties may also benefit from using the Graphic Organizer for Chapter Concepts to summarize key ideas. The Graphic Organizer appears in the workbook and is also provided as an overhead transparency.

Visually Impaired

- Provide hand lenses for visually impaired students to use to see details. For example, a hand lens would be helpful in examining the illustrations of plant and animal cells in Chapter 1.
- These students may benefit from listening to the Text on Tape audiocassettes.

Hearing Impaired

- Provide as many additional photographs and models as possible to reinforce the investigation of classification.

Strategies for Advanced Learners

The following strategies provide an opportunity for advanced learners to study the content of the unit in greater depth.

Chapter 1 Cells, Genetics, and Heredity

Invite students to make a model of an animal cell and its organelles. Remind students that much of the cell is made of water. With that in mind, have students make a model of the cell using water-based gelatin dessert. Have them determine appropriate food items to represent the organelles. For example, students might use grapes to show mitochondria. Have students also make a model of a plant cell, this time including a cell wall and chloroplasts.

Chapter 2 Classification

Challenge students to investigate methods taxonomists use today to classify organisms. Have students consider how the development of technology has impacted scientists' ability to identify living things. Have students research how the fossil record has played a part in relating living things to their ancestors. Students should also consider how the chemical makeup of an organism sometimes provides clues when scientists are classifying a living thing. Students should consider how the invention and advancement of the microscope have made it possible for taxonomists to look at an organism's DNA to find relationships among species of living things.

UNIT A

Unit Materials List

Quantities are indicated for a class of 30 students working individually or in groups of 5, depending on the nature of the activity. Where shared equipment is suggested, a smaller number of items is specified. Quantities are also listed for those materials included in the Materials Kit.

Nonconsumable Materials	Class Quantity Needed	Kit Quantity	Activity Page
corn, purple-and-white	6	6	A22
jar, plastic	6	6	A4
Microslide Viewer	6	6	A12
Microslide, plant mitosis	6	6	A12
tablespoon	6	6	A1

Consumable Materials	Class Quantity Needed	Kit Quantity	Activity Page
apple slices	18		A1
bag, plastic zip-top	6	1 box of 50	A4
balloon, round	12	1 pkg of 35	A4
beans, mixed	1 bag	1 bag	A40
clay, green modeling	4 boxes	4 boxes	A4
clay, red modeling	1 box	1 box	A4
index cards	30	1 pkg of 100	A4
marker	6		A1
paper, notebook	12 sheets		A22, A50
pencil	6		A22, A50
plates, paper	18	50	A1
posterboard	6 pieces		A40
salt	1 box	1 box	A1
sand	1 bag	1 bag	A4
sugar	1 bag	1 bag	A1
tape, transparent	1 roll		A40

Additional kit options are available. Contact your sales representative for details.

UNIT A LIFE SCIENCE

Understanding Living Things

UNIT EXPERIMENT

Cell Responses

Cells are the building blocks of all living things. However, each cell in a body exists in a slightly different environment. While you study this unit, you can conduct a long-term experiment related to how cells respond to their environments. Here are some questions to think about. How do cells respond to the amount of a chemical in their environment? For example, what will happen to cells if they are exposed to more salt or sugar than usual? Plan and conduct an experiment to find answers to these or other questions you have about cell responses. See pages x–xvii for help in designing your experiment.

Informal Assessment

Research shows that Informal Assessment is an effective way to monitor and evaluate students. Performance assessment and teacher observation are particularly useful for assessing projects and other, similar tasks.

Performance Together with students, develop a rubric that will be used to evaluate their completed projects. Use *Developing Your Own Rubric* on page AGxi of the Assessment Guide. Challenge and guide students to produce their best work. Then use the rubric to assess their performance.

Classroom Observation You can gain insight into students' learning by observing how they use scientific processes, collect and record information, and select and use appropriate tools to plan and implement their project. While students are working on their projects, use the *Project Evaluation Checklist* on page AGxviii of the Assessment Guide for ongoing assessment.

UNIT A EXPERIMENT

Objectives

- Promote scientific inquiry.
- Use a scientific method to plan and conduct a long-term investigation.
- Design an experiment to show how cells respond to their environments.

A lesson plan for guiding students through the Unit Experiment is provided on pages A1i–A1j.

Science Fair Project Ideas

In addition to the Unit Experiment, you may wish to allow students to plan and carry out these science fair projects related to the chapters in Unit A. Students can use the framework on the Unit Experiment pages to help them plan and conduct experiments.

Chapter 1 Model Cells After comparing animal and plant cells, students can make models of the cells and the organelles within the cells. Review the properties of a model. Challenge students to make "edible cells." It's fun for students to share these models later. Or have students make room-sized models. Hang something like a clear shower curtain from the ceiling of your classroom. This makes the cell membrane. Challenge students to make organelles using three-dimensional geometric shapes. Hang these from the cell membrane. You can even use this model when you get to cell division, making spindle fibers from which to hang your duplicating chromosomes.

Chapter 2 A Little Protozoa After studying the five kingdoms of living things, students can study and classify protists through microscope observations. Have students gather water samples from stagnant sources and compare the protists that they see. Hay or cut grass in spring water creates a hay infusion which propagates protists. What physical features could be used to classify these creatures? Suggest students make a classification key to help identify these creatures.

UNIT A EXPERIMENT

Choose from these options to help students complete the unit experiment.

Option 1 Independent Inquiry Assign the experiment for students to complete independently. Students should use the prompt as a springboard for writing their own questions and designing their own experiment. They can write a hypothesis, design a procedure, select materials, and conduct the experiment. Students can refer to pages x–xxiv in the Pupil Edition for guidance.

Option 2 Guided Inquiry Suggest students use the Experiment Log (Workbook pp. WB371–374) to help them plan their experiment. The Experiment Log pages shown below appear in the Workbook, but without answers. Use the lesson plan that follows to guide students as they design the experiment.

Option 3 Structured Inquiry Have students complete the experiment by testing the hypothesis and using the procedure that have been provided for them. Display the overhead transparencies shown below (or provide photocopies from the transparency package), and have students copy the hypothesis, variables, and procedure into the Experiment Log (Workbook pp. WB371–374). Students then conduct the experiment and gather and record their data. Use the lesson plan on page A1 to guide students as they complete the experiment.

BACKGROUND

Webliography GO ONLINE

Keyword cells

www.harcourtschool.com

Chemical Stimuli The apple slices used in the investigation are composed of plant cells that respond to chemicals in their immediate environment. Sucrose, the chemical name for table sugar, is a carbohydrate. Plants produce this organic compound through photosynthesis. The glazed covering that develops on the sugar-covered apple slice forms in response to the cells' interaction with sucrose.

Salt is an inorganic compound that triggers the movement of water across the plant cells' membranes. When salt was placed on the apple slice, a greater number of water molecules were present inside the cells than outside the cells. As a result, the water molecules diffused from an area of higher concentration to an area of lower concentration. A puddle of water formed around the apple slice as water molecules left its cells.

LESSON PLAN

Resources

Experiment Log, Unit A, Workbook pp. WB371–374

Experiment Transparencies, Unit A

Time 30 minutes to plan and set up; 20 minutes the following day

Safety Remind students that they should never taste food items used in an investigation.

Expected Results On Day 2, the slice saturated with sugar likely will be covered with a clear glaze. The slice saturated with salt will rest in a liquid puddle and will appear wilted. The control slice will display the least variation.

Suggested Materials

- three freshly cut apple slices
- three paper plates
- tablespoon
- sugar
- salt
- marker

Underlined items above are provided in the equipment kits (available separately).

Preparation Tips Use fresh apples to avoid presenting students with slices that are discolored because of natural causes. Slicing the apples just prior to the investigation will conserve time.

Transparency Exp A, p. 1
Workbook p. WB371

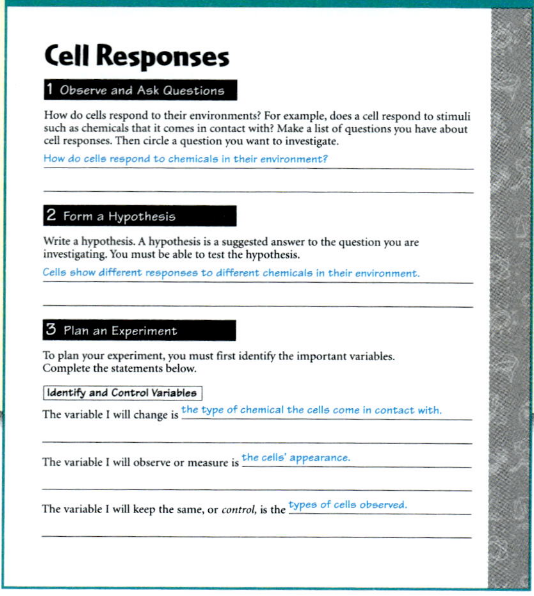

Cell Responses

1 Observe and Ask Questions

How do cells respond to their environments? For example, does a cell respond to stimuli such as chemicals that it comes in contact with? Make a list of questions you have about cell responses. Then circle a question you want to investigate.

How do cells respond to chemicals in their environment?

2 Form a Hypothesis

Write a hypothesis. A hypothesis is a suggested answer to the question you are investigating. You must be able to test the hypothesis.

Cells show different responses to different chemicals in their environment.

3 Plan an Experiment

To plan your experiment, you must first identify the important variables. Complete the statements below.

Identify and Control Variables

The variable I will change is *the type of chemical the cells come in contact with.*

The variable I will observe or measure is *the cells' appearance.*

The variable I will keep the same, or *control*, is the *types of cells observed.*

1 Observe and Ask Questions

Have students discuss the prompt and then brainstorm a list of other questions they may have about the topic. Students can record their questions in the Experiment Log and then circle the question they are going to test.

2 Form a Hypothesis

Guide students in forming a testable hypothesis.

- **Look at the question you have chosen to investigate. *How do cells respond to chemicals in their environment?* What do you predict will be an answer to this question?** Possible answer: Cells respond differently to different stimuli.
- **Form a complete statement from your answer. Your statement should include all the things you are testing. This statement will be your hypothesis.** Possible hypothesis: Cells show different responses to different chemicals in their environment.

3 Plan an Experiment

Identify and Control Variables Remind students that an experiment is a fair test of a hypothesis. Identifying and controlling the variables will ensure that the test is conducted fairly. Have students complete the statements in the Experiment Log.

Develop a Procedure and Gather Materials Have students write a detailed procedure or have them use the procedure that appears on the transparency.

4 Conduct the Experiment

Gather and Record Data Students can record their data in the Experiment Log.

Common Error Alert Students may think that the puddle of liquid surrounding the salted slice came from an outside source. Guide students in recognizing that if this were true, such a puddle probably would surround each apple slice. Since a puddle appears around only the salted slice, the liquid must have come from the apple slice.

Interpret Data Students should observe the apple slices on Days 1 and 2. After they have finished gathering their data, have students make a sketch of each slice. Then have them compare the sketches to form a Venn diagram comparing and contrasting the physical traits of the slices.

5 Draw Conclusions and Communicate Results

Have students answer the questions in the Experiment Log.

Scientists at Work Emphasize that the results of one experiment often trigger other, related investigations. Reinforce this concept by having students use their findings to form a hypothesis describing an environment that would cause an apple slice to take in a liquid rather than release a liquid.

Investigate Further Encourage students to build upon this experiment to develop a science fair project or other investigation.

Transparency Exp A, p. 2 Workbook p. WB372

Develop a Procedure and Gather Materials

Write the steps you will follow to set up an experiment and collect data.

1. Observe three freshly cut apple slices. Record observations in the log.
2. Label one paper plate "Sugar," a second plate "Salt," and a third plate "Control." Put an apple slice in the middle of each plate.
3. Sprinkle 2 tablespoons of sugar on the apple slice on the plate labeled "Sugar." Sprinkle 2 tablespoons of salt on the apple slice on the plate labeled "Salt."
4. Observe the apple slices the next day. Note any changes in color and appearance. Record observations in the log.

Use extra sheets of blank paper if you need to write down more steps.

Materials List Look carefully at all the steps of your procedure, and list all the materials you will use. Be sure that your teacher approves your plan and your materials list before you begin.

three freshly cut apple slices, three paper plates, tablespoon, sugar, salt, marker

Transparency Exp A, p. 3 Workbook p. WB373

4 Conduct the Experiment

Gather and Record Data Follow your plan and collect data. Use the chart below or a chart you design to record your data. **Observe** carefully. **Record** your observations and be sure to note anything unusual or unexpected.

Observation Log

	Day 1	Day 2
Slice with Sugar		
Slice with Salt		
Control Slice		

Other Observations

Transparency Exp A, p. 4 Workbook p. WB374

Interpret Data

Make sketches that illustrate the appearance of the apple slices on Day 2. Use the sketches to make a Venn diagram comparing and contrasting the physical traits of the apple slices.

5 Draw Conclusions and Communicate Results

Compare the **hypothesis** with the data and chart, and then answer these questions.

1. Was your hypothesis supported by the results of the experiment? Explain. Students should justify their responses with data and observations.
2. How would you revise the hypothesis? Explain. Students may revise their hypothesis to match the results of the experiment more closely.
3. What else did you **observe** during the experiment? Students should use their notes and be as specific as possible.

Prepare a presentation for your classmates to **communicate** what you have learned. Display your data table and Venn diagram.

Investigate Further Write another hypothesis that you might investigate.

Responses will vary depending on experiment outcomes and student interests.

CHAPTER 1 Planning Guide Cells, Genetics, and Heredity

LESSON	PACING	OBJECTIVES	MATERIALS
1 How Do Plant and Animal Cells Differ? A4–11 	3 days	▶ Recognize that all living things are made up of cells. ▶ Compare the functions and structures of plant and animal cells. ▶ Describe how the nucleus of a cell directs its functions.	**Investigate** plastic jar, water, 2 balloons, index cards, plastic bag, sand, red and green clay
2 How Do Cells Reproduce? A12–21 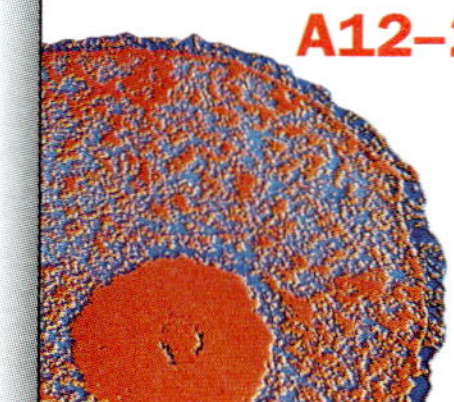	3 days	▶ Compare the processes of mitosis and meiosis. ▶ Recognize that different combinations of parental DNA produce variation in new organisms.	**Investigate** Microslide Viewer, Microslide of onion root tip
3 How Are Traits Inherited? A22–31 	3 days	▶ Recognize that a plant's or animal's traits are inherited from its parents. ▶ Identify genes as the inherited factors that determine traits. ▶ Conclude that traits of dominant genes appear in the offspring regardless of whether there is a recessive gene or not, and those of recessive genes appear only when both genes for a trait are recessive.	**Investigate** purple-and-white corn, paper and pencil
End of Chapter A32–37 **Science Through Time** **People in Science** **Activities for Home or School** **Chapter Review and Test Preparation**		▶ Connect chapter concepts with the history of science. ▶ Connect chapter concepts with the contributions of scientists. ▶ Extend chapter concepts with hands-on investigations. ▶ Review chapter concepts.	

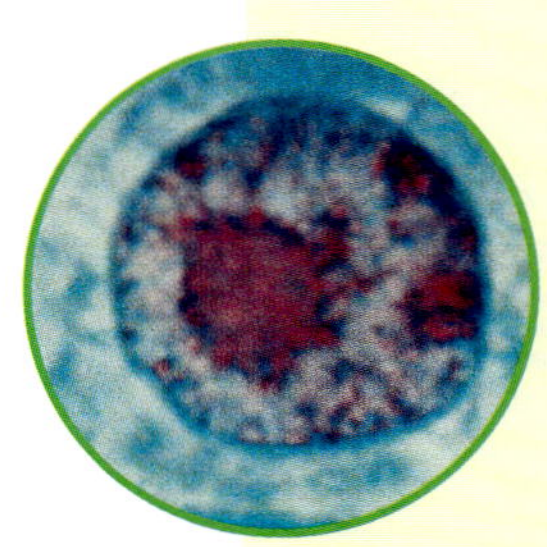

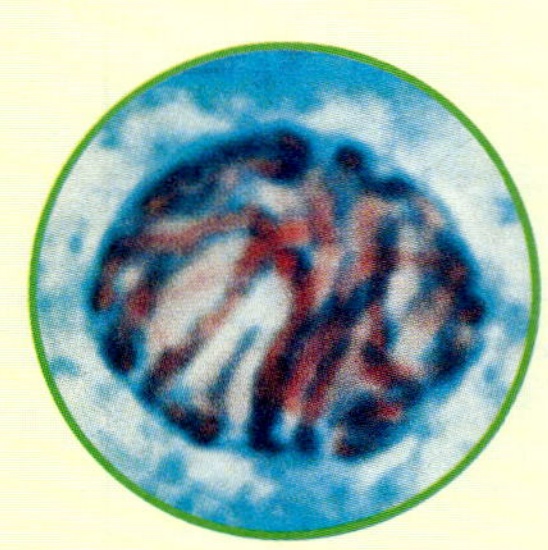

PROCESS SKILLS	VOCABULARY	RESOURCES AND TECHNOLOGY	REACHING ALL LEARNERS
Process Skill Tip use a model, infer **Other Process Skills** observe, make a model	**cell membrane** **cell wall** **cytoplasm** **mitochondria** **chloroplast** **vacuole** **nucleus** **chromosome** **nuclear membrane** **DNA**	**Workbook,** pp. WB1–6 **Vocabulary Cards,** pp. TR93–98 **Transparencies** A1, A1-1 **Harcourt Science Activity Video** **Internet Site**	**Kinesthetic Learners**, p. A10 **ESL Activity**, p. A9 **Investigation Challenge**, p. A10 **Advanced Learners**, p. A1g
Process Skill Tip compare, use time/space relationships **Other Process Skills** observe, infer, record data	**gene** **mitosis** **meiosis**	**Workbook,** pp. WB7–11 **Transparencies** A1, A1-2 **Harcourt Science Activity Video** **Harcourt Science Newsroom Video**	**Visually Impaired**, p. A20 **ESL Activity**, pp. A15, A17 **Investigation Challenge**, p. A20
Process Skill Tip use numbers, infer **Other Process Skills** observe, record data	**sexual reproduction** **dominant** **recessive** **Punnett square**	**Workbook,** pp. WB12–18 **Transparencies** A1, A1-3 **Harcourt Science Activity Video** **Harcourt Science Explorations CD-ROM**	**Visually Impaired**, p. A28 **ESL Activity**, p. A29 **Investigation Challenge**, p. A30
		Take-Home Book, pp. TH3–6 *Cracking the Code of Life* **Activities for Home or School,** p. TR65 **Chapter Test**, pp. AG1–5	**Performance Assessment**, p. AG6 **Portfolio Evaluation**, p. AGxxiv

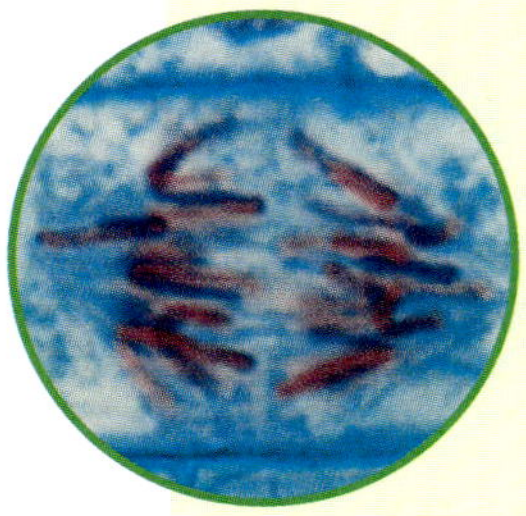

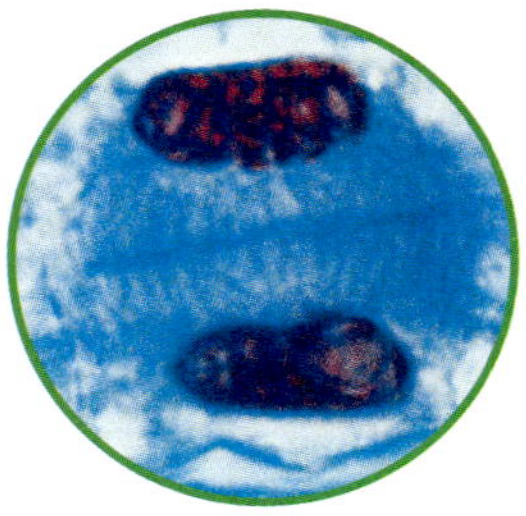

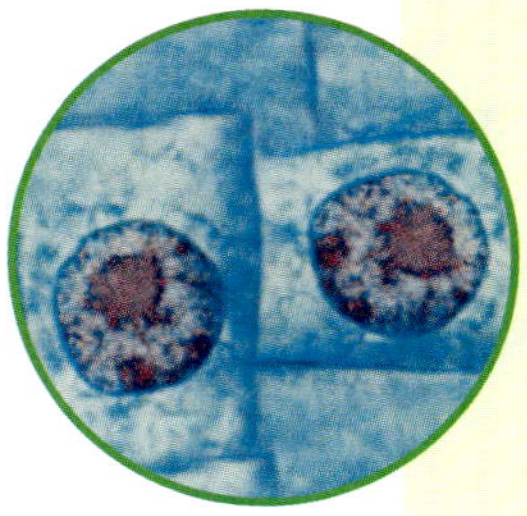

Prepare for Activities

Use this page to help you organize and prepare materials for the Investigations, which begin each lesson of this chapter.

LESSON INVESTIGATION	MATERIALS

1 The Structure of Cells
A4–5

PURPOSE Observe photographs of plant cells as seen through a microscope, and use the observations to make a model.

PREPARATION TIPS To find red and green clay, check your local hobby and craft stores. If they don't have red and green clay, any two colors will suffice. If you prefer a more open-ended lab or have other materials available, encourage students to choose alternative materials.

TIME 30–40 minutes

EXPECTED RESULTS Students' models should be three-dimensional and contain a cell wall, a cell membrane, a nucleus, and various organelles.

MATERIALS
- plastic jar
- water
- 2 balloons
- index cards
- plastic bag
- sand
- red and green clay

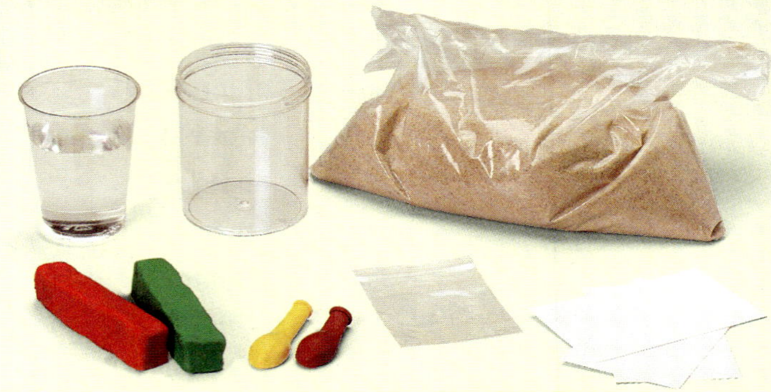

2 How New Cells Are Made
A12–13

PURPOSE Observe the growth of cells in an onion root tip and use the observations to infer how cells produce exact copies of themselves.

PREPARATION TIPS If you do not have access to a Microslide Viewer and microslides, use the cell pictures on page A18. Have students cover up the captions as they work.

TIME 20–30 minutes

EXPECTED RESULTS Students should be able to identify and place in order five stages of cell division.

MATERIALS
- Microslide Viewer
- Microslide of onion root tip

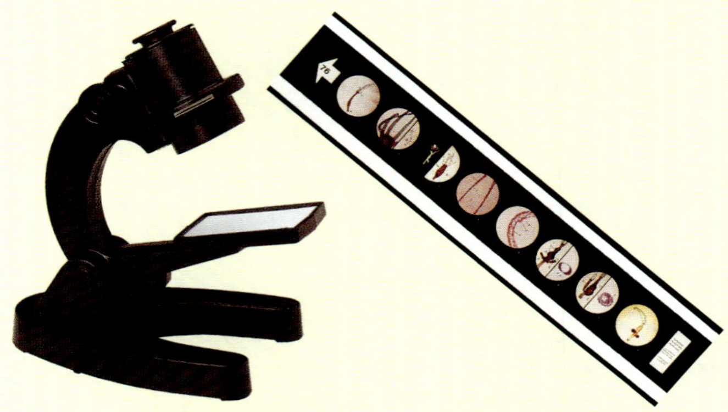

3 Seed Color in Corn
A22–23

PURPOSE Observe an ear of corn produced when a corn plant with white kernels was crossed with a corn plant with purple kernels.

PREPARATION TIPS Check a biological supply firm for purple-and-white corn. If unavailable, use an ear of yellow-and-white corn.

TIME 30–40 minutes

EXPECTED RESULTS Students should find about three purple kernels for every white kernel. Results may vary slightly. A greater quantity of data will yield more accurate results.

MATERIALS
- purple-and-white corn
- paper and pencil

Harcourt Science Activity Videos

Preview the Investigations The Activity Video for this unit shows students doing the Investigations. You may wish to view the video for classroom management ideas. The video can also be used to model the Investigation process for students.

Equipment Kits

Underlined items above are provided in the equipment kits (available separately).

. . . with Trade Books

These books provide in-depth information about cells and genetics.

Antoni Van Leeuwenhoek: First to See Microscopic Life by Lisa Yount (Enslow, 1994) tells of the scientist's many discoveries examining microscopic life. *NSTA Trade Book*

Cell Wars by Frank Balkwill (Carolrhoda, 1993) explains how cells fight off diseases and viruses in the human body.

The Cell Works by P. Bauerle and Norbert Landa (Barron's, 1997) explores a cell's nucleus and the double-helix structure of DNA; how new cells are made; and how they form organisms, based on their gene patterns.

Trade book titles are current at time of publication but may go out of print without notice.

. . . with the Take-Home Book

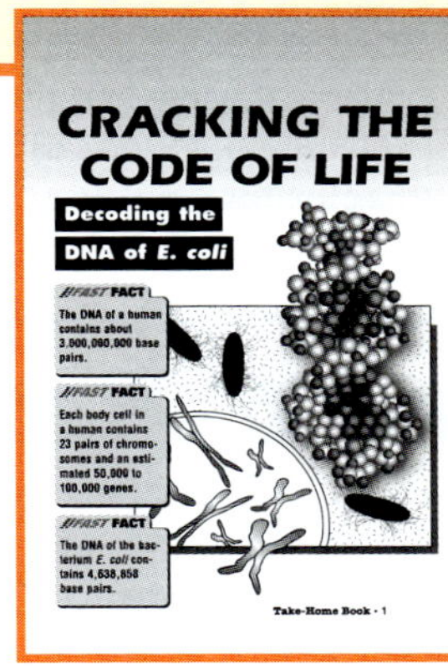

Cracking the Code of Life

CONTENTS

. . . with More Hands-On Activities

Investigation Challenge

Activities for Home or School

The cross-curricular Links that appear at the end of each lesson in the student edition and throughout the teacher's edition also provide activities.

. . . with Technology

Students can learn more about cells, genetics, and heredity through these components and special Internet sites. Look for Technology Links at the end of each lesson.

CNN Turner Learning

Harcourt Science Newsroom Video
Bioengineeered Plants

Harcourt Science Explorations CD-ROM
Now You See Them, Now You Don't

SCILINKS — THE WORLD'S A CLICK AWAY

These Internet Links are provided by the National Science Teachers Association.
www.scilinks.org/harcourt

Visit The Learning Site for related links, activities, and resources: **www.harcourtschool.com**

WELCOME TO THE LEARNING SITE

Assessment Options ✓ Chapter 1

Formal Assessment

- **Chapter Review and Test Preparation**, PE pp. A36–37
- **Chapter Test**, pp. AG1–5

Ongoing Assessment

- ✓ **Questions**, PE pp. A6–11, A14–21, A24–31
- **Lesson Review**, PE pp. A11, A21, A31
- **Informal Assessment**
  - **Performance**, TE p. A5
  - **Portfolio**, TE p. A13
- **Observation Checklist**, p. AGxiv

Student Self-Assessment

- **Self-Assessment–Investigate**, p. AGxvi
- **Self-Assessment–Learn About**, p. AGxvii
- **Project Summary Sheet**, p. AGxix

Performance Assessment

- **Chapter Review and Test Preparation**, PE p. A37
- **Chapter Performance Task**, pp. AG6–7
- **Project Evaluation Checklist**, p. AGxviii

Portfolio Assessment

My Work

- **Science Experiences Record**, p. AGxxii
- **Guide to My Science Portfolio**, p. AGxxiii
- **Portfolio Evaluation**, p. AGxxiv

Chapter Test AG1

Cells, Genetics, and Heredity

Part I Vocabulary 1 point each

Match each term in Column B with its meaning in Column A.

	Column A	Column B
Q	1. Storage space in a cell	A cell membrane
A	2. Controls what enters and leaves the cell	B cell wall
E	3. Jellylike substance in a cell	C chloroplast
J	4. Structures that release stored energy	D chromosomes
C	5. Structure that contains chlorophyll	E cytoplasm
B	6. Outer layer that protects a plant cell	F DNA
M	7. Surrounds the nucleus and holds it together	G dominant
L	8. Control center of the cell	H gene
F	9. Chemical that directs life activities	I meiosis
H	10. Segment of DNA that controls traits of offspring	J mitochondria
D	11. Strands that copy themselves and split in cell division	K mitosis
K	12. Cell division that produces body cells	L nucleus
I	13. Cell division that produces reproductive cells	M nuclear membrane
P	14. The joining of male and female reproductive cells	N Punnett square
G	15. Type of trait that gives an organism its appearance	O recessive
O	16. Type of trait that is masked or covered	P sexual reproduction
N	17. Chart of possible offspring combinations	Q vacuole

Chapter Test AG2

Part II Science Concepts and Understanding 3 points each

Use the following drawings to answer Questions 18–19.

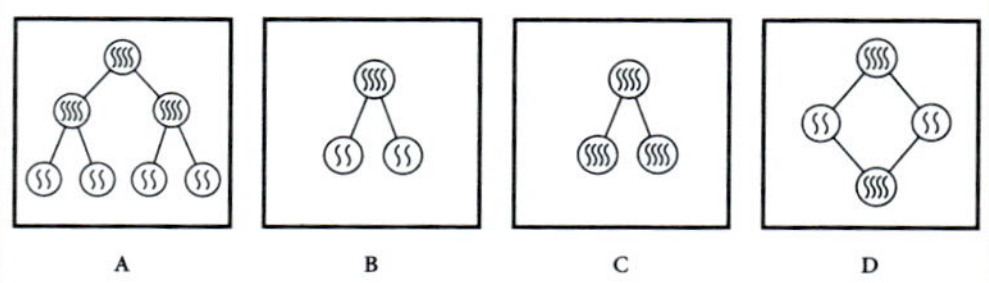

A 18. Which of the drawings represents meiosis?

C 19. Which of the drawings represents mitosis?

Use the following diagram to answer Questions 20–24. Use the letters of the terms below to complete the diagram.

A cell wall B cell membrane C chromosomes D nuclear membrane E nucleus

A 20.
B 21.
E 22.
D 23.
C 24.

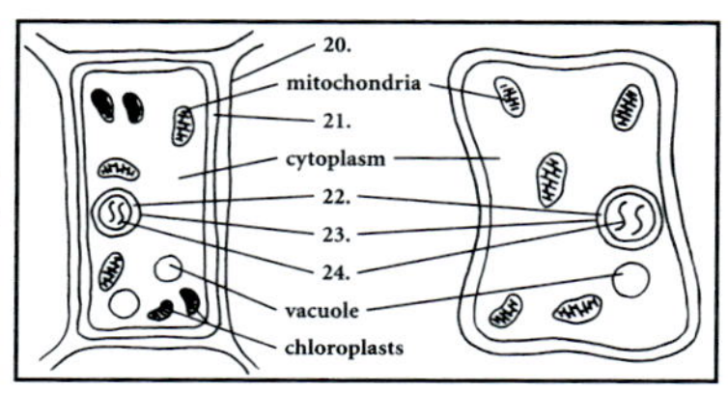

Chapter Test AG3

Write the letter of the best choice.

C 25. If a cell is very active, it needs more —
A chromosomes
B vacuoles
C mitochondria
D cytoplasm

H 26. Which is **NOT** a part of the cell theory?
F New cells come only from existing cells.
G All living things are made up of cells.
H All living things must have a nucleus in each cell.
J All life activities take place in cells.

A 27. What is meant by "DNA is the blueprint"?
A Every cell in an organism contains all the information about the organism in its DNA.
B Every cell in an organism has information for only that cell type.
C DNA controls the genes only when the chromosomes send information.
D Understanding DNA is understanding the universe.

H 28. Why didn't scientists before Robert Hooke develop the cell theory?
F Aristotle said there was nothing smaller to see, so no one looked.
G Cells didn't exist before Hooke's time.
H Cells couldn't be seen with the naked eye.
J Only plant cells could be seen.

B 29. Meiosis is the process —
A of elimination
B of forming reproductive cells
C of cell division to produce new body cells
D of making food from light

H 30. A yellow plant and a green plant produced offspring that were all green. The yellow color —
F is a dominant trait
G will never show up again
H is a recessive trait
J can be either dominant or recessive

Chapter Test AG4

Part III Critical Thinking 6 points each

Answer the following on a separate sheet of paper.

31. Why is the nucleus important in cell activities and reproduction?
Possible answer: It contains DNA, which is the blueprint for the organism. It gives information needed for repair and reproduction.

32. Explain why cells are specialized.
Possible answer: It is more efficient to have cells perform only certain functions; this way organisms can become very complex.

33. Why aren't all brothers and sisters exactly alike?
Possible answer: Each sibling inherits genetic material from each parent. The combination of genes is different except in the case of identical twins.

Part IV Process Skills Application

Compare **your school to a single cell. Match the school parts in Column B to the cell parts in Column A.**

		Column A	Column B
2	B	34. nucleus	A storeroom
2	D	35. cell membrane	B principal
2	C	36. mitochondria	C boiler room
2	A	37. vacuoles	D fence
2	E	38. nuclear membrane	E principal's office

4 39. Explain your answers to Questions 34–38.
Possible answer: In the nucleus, DNA "decides" what is needed for the cell, similar to a principal in the office deciding what the school needs. The nuclear membrane encloses the nucleus, just as the principal's office forms the principal's immediate surroundings. The cell membrane contains or surrounds the materials of the entire cell, like the fence that surrounds the school yard. The mitochondria produce energy for the cell, just as the boiler room produces heat energy for the school. The vacuoles store food, water, and waste for the cell, as the storeroom stores all the materials needed by the school.

Chapter Test AG5

4 40. Suppose that when two dogs breed, 75 percent of the offspring have long tails and 25 percent have short tails. Also suppose that long tails are dominant and short tails are recessive. **Use the numbers** provided to complete the following Punnett square.

LL = long tail
ll = short tail

	L	l
L	LL	Ll
l	Ll	ll

2 41. In Question 40, what kind of tail does each parent have?
Both parents have long tails, though they each carry the recessive gene for short tails.

4 42. Write 1, 2, 3, or 4 to show the correct **time relationships** in which the stages of meiosis occur.

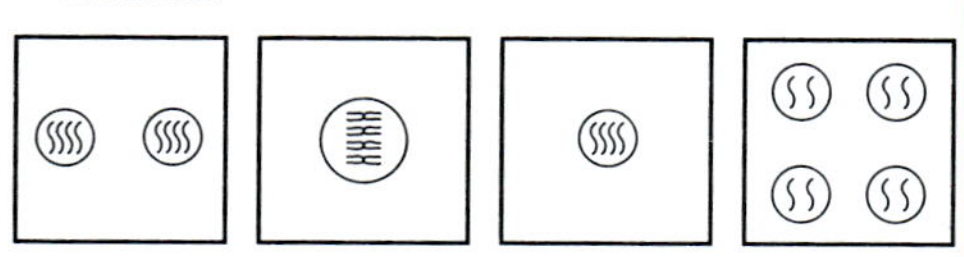

3 2 1 4

2 43. Describe how you could **make a model** of mitosis, and list the materials you would use. Include the following three mitotic stages: parent cells, doubling of chromosomes in the cells, and daughter cells.
Answers should include materials (such as chenille sticks and clay), the name of each stage (parent, doubling, daughter), and a description of the appearance of each cell.

Performance Assessment AG6

PERFORMANCE TASK

Discover a New Species

Materials: pencil or pen

Suppose that you have discovered a new organism. Upon close examination of it, you observe that some of the organisms have a horn and some do not. You set up three test crosses to breed the organism so you can identify which characteristic is dominant and get the following results.

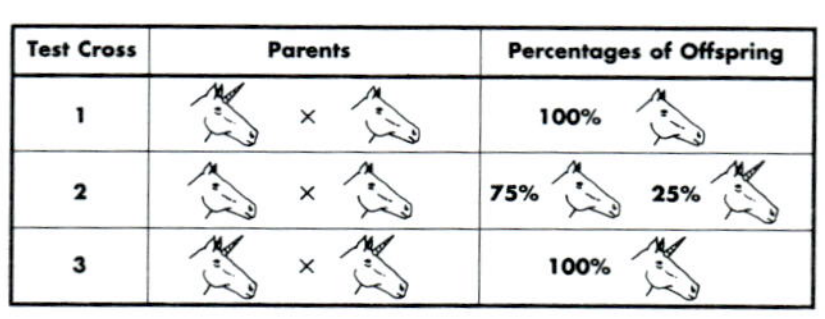

Test Cross	Parents	Percentages of Offspring
1	×	100%
2	×	75% 25%
3	×	100%

1. Complete a Punnett square for each of the test crosses above. Use HH for dominant and hh for recessive.

Test Cross 1

	H	H
h	Hh	Hh
h	Hh	Hh

Test Cross 2

	H	h
H	HH	Hh
h	Hh	hh

Test Cross 3

	h	h
h	hh	hh
h	hh	hh

2. Which characteristic is dominant? How do you know?

3. Cross an offspring from Test Cross 1 with an offspring from Test Cross 3. Complete the Punnett square.

4. In Question 3, what percentage of the offspring would have a horn?

Assessment Continued

Performance Assessment AG7

Discover a New Species

PERFORMANCE TASK

Materials Performance Task sheet, pencil or pen

Time 25–30 minutes

Suggested Grouping individuals or pairs

Science Processes infer, compare, classify/order, model

Preparation Hints none

Introduce the Task Explain to students that they will be interpreting and mapping the genetics of a new species. They will need to identify a trait as dominant or recessive. Then they will cross a pair of offspring.

Promote Discussion When students finish, have them compare their findings. Ask students whether one offspring from each test cross would be enough to tell whether a characteristic is dominant or recessive. Conclude with students that many offspring are necessary to come up with valid percentages. However, even if only one offspring has a recessive trait, we can assume that both parents carry at least one recessive gene.

Scoring Rubric

Performance Indicators

_____ Completes test cross Punnett squares correctly: Test Cross 1 crosses HH with hh, Test Cross 2 crosses Hh with Hh, and Test Cross 3 crosses hh with hh.

_____ Identifies dominant characteristic as no horn because none of the offspring in Test Cross 1 have horns.

_____ Completes cross of offspring correctly: Hh with hh.

_____ Calculates percentage of offspring with horns, in Question 4, to be 50 percent.

Observations and Rubric Score

3 2 1 0

Workbook Support

Investigate Log WB2

The Structure of Cells

Materials

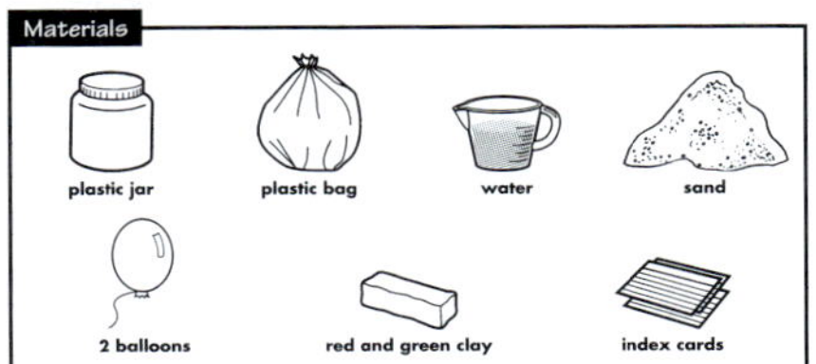

Activity Procedure

1. **Observe** the pictures of the magnified plant cells on page A5 of your science book. Find the parts that all the cells have in common.
2. Now use your observations to write a plan for how you will **make a model** of a typical plant cell. Decide how you can best show the cell, using the materials you have.
3. Show your plan to your teacher. When your teacher approves, make your model.
4. **Observe** your model. Write a description of each cell part you have modeled, and tell what you think the part's function might be. Use a separate index card for each cell part.

Investigate Log WB3

Draw Conclusions

1. Why was it useful to **observe** several plant cells before making your model?
Plant cells may differ; to find out which cell structures are typical, you need to observe several cells.

2. What can you **infer** about the function of the thick layer that surrounds the cell?
The thick layer supports the cell, keeps it separate from other cells, and gives the cell its shape.

3. What can you **infer** about the function of the thin layer that lines the cell?
The thin layer keeps the cell's materials inside the cell.

4. **Scientists at Work** Scientists often **make models** to help them make inferences. How did making a model help you make inferences about the parts of a cell?
The kinds of materials needed to represent parts of the cell can help suggest their functions; a three-dimensional model gives a better idea of what a cell is like.

Investigate Further • A fungus looks much like a plant, but it is almost never green in color. Use a Microslide Viewer or microscope with prepared slides and look at three more plant cells. What differences do you see? How would you use these observations to change the model you made in this activity?
Answers will depend on the plant cells examined.

Hypothesize what you might expect a fungus cell to look like. Then **design an experiment** to test your hypothesis.

Process Skills Practice WB4

Make a Model

Models are familiar or easy-to-see representations of processes or structures that are complicated or difficult to see. Sometimes you make your own model, and sometimes you can learn from a model that someone else has made.

Think About Making a Model

Organisms grow when their cells divide. Each body cell splits into two identical cells, and then each of these cells splits again. You could sit at a microscope and watch cells divide, but that's often not possible. To get an idea of how fast the number of cells in an organism increases, you can use beans to model the dividing cells.

Start with 1 bean. Next to it, make a pile of 2 beans to show how the first cell splits. Then make a third pile containing 2 beans for each of the beans in the second pile. Continue making new piles until the newest pile contains at least 100 beans.

1. How many piles of beans do you have? 8
2. How many beans are in the last pile? 128 beans
3. Assuming the beans are cells, how many divisions did it take to get at least 100 cells?
7 divisions
4. If each cell divides every 3 minutes, how long would it take to get 100 cells?
3 minutes × 7 divisions = 21 minutes
5. What did the model tell you about the number of cells present after each division?
It showed that the number of cells after each division doubled.

Reading Skills Practice WB5

Compare and Contrast

Read the selection below. Then fill in the chart to show the ways that red and white blood cells are alike and different.

Red Cell, White Cell

Blood cells are one type of cell found in all animals. Human beings have two kinds of blood cells: red and white. We need the right balance of both to stay healthy.

Millions of red blood cells cluster in each drop of blood. Hemoglobin, a bright red protein chemical that includes the element iron, gives blood its red color. The hemoglobin picks up oxygen molecules from the lungs and then releases the oxygen to the cells throughout the body. As the hemoglobin molecules then travel through the body, they carry away carbon dioxide from the tissues. The bones produce new blood cells constantly; 120 days is the average life cycle for a red blood cell.

White blood cells keep the body healthy. When they find germs within the body, they manufacture antibodies to devour the bacteria. A single drop of blood may contain between 7,000 and 25,000 white blood cells. The number of white blood cells increases when the body is fighting disease. An extremely high white blood cell count can signal a cancer in the blood called leukemia. People with leukemia may have up to 50,000 white blood cells in just one drop of blood. Some white blood cells have a fairly short life, living only a few days to a few weeks.

Characteristic	Red Blood Cells	White Blood Cells
Part of	a single drop of blood	a single drop of blood
Located in all of these	animals	animals
Number per drop of blood	millions	7,000–25,000
Function	deliver oxygen, remove carbon dioxide	fight disease
Life span	120 days	a few days to a few weeks or longer

Lesson Concept Review/Assessment WB6

How Do Plant and Animal Cells Differ?

Lesson Concept

All living things are made up of cells. Animal cells and plant cells have the same parts, except for cell walls and chloroplasts. Plant cells have them, but animal cells do not.

Vocabulary

cell membrane (A8), **cell wall** (A8), **cytoplasm** (A8), **mitochondria** (A8), **chloroplast** (A9), **vacuole** (A9), **nucleus** (A9), **chromosome** (A9), **nuclear membrane** (A9), **DNA** (A10)

Answer all questions in one or more complete sentences.

1. What are the three main parts of the cell theory? Cells are the basic building blocks of all life; all life activities take place in cells; new cells are produced from existing cells.
2. Compare a plant cell and an animal cell. Both cells have a cell membrane, cytoplasm, mitochondria, vacuoles, nuclear membrane, nucleus, and chromosomes. A plant cell also has a cell wall and chloroplasts.

What is the purpose of each of the following structures?

3. cell membrane It holds cell material inside and controls what goes in and out of the cell.
4. mitochondria They release energy stored in sugar to carry out cell activities.
5. vacuoles They are spaces within the cell that store food, water, and wastes.
6. chromosomes They are structures that contain the information necessary to determine all the characteristics of the organism.
7. nucleus The nucleus is the control center of the cell. It sends out instructions in the form of chemicals through pores in the nuclear membrane.
8. What is DNA? DNA is a complex chemical that is part of chromosomes. It contains information about every part of the organism and directs the life activities of a cell.

Workbook Support Continued

Investigate Log WB7

How New Cells Are Made

Materials

Microslide Viewer

Microslide of onion root tip

Activity Procedure

1. You will **observe** one instant in the life of the onion root tip. At that instant, different cells were preserved at different stages of cell division. Place the slide on the Microslide Viewer. Adjust the focus so that you can **observe** the cells.
2. **Record** your observations of cell division by drawing at least four stages of division.
3. **Show the sequence** of the process of cell division by numbering your drawings from 1 to 4.

Investigate Log WB8

Draw Conclusions

1. How could you tell that the cells you **observed** were dividing? The shape and contents of the cells differed; in some cells it looked as if a new cell wall was forming.
2. **Write the sequence** of the stages of cell division. 1. The nucleus disappears and rodlike structures become visible; 2. the rods line up in the center of the cell; 3. the rods separate and go to opposite ends of the cell; 4. a new cell wall forms; 5. the rods form a nucleus again.

   Describe each of the cells you drew.

   Stage 1: Accept answers that contain information about the stages.

   Stage 2: ______

   Stage 3: ______

   Stage 4: ______

   Stage 5: ______
3. **Scientists at Work** Scientists sometimes need to **find the sequence** of events to better understand a process. What clues might a scientist studying the division of onion cells look for in order to figure out the sequence? A scientist might look at cells that are not dividing and determine what a cell "at rest" looks like. Next, the scientist might look for cells that do not differ much from the cell at rest. He or she would continue until the sequence of events was established.

Investigate Further Look at other slides of plant cells. Try to find any areas other than the root tip where plant cells are dividing. If you have access to a microscope, use one to observe prepared slides of an onion root tip. See page R3 for tips on using a microscope. Answers will vary depending on the slides the students are using.

Process Skills Practice WB9

Show the Sequence and Compare

One way to order data is to show the sequence in which the events occur. Ordering data this way makes it easier to understand the parts of a process. When you compare data, you identify characteristics of objects or processes that are similar and different.

Think About Ordering and Comparing

The period of time from the beginning of one cell division to the beginning of the next cell division is called the cell cycle. This is the time it takes for a cell to divide and form two cells. In most growing plant and animal cells, the cell cycle is between ten hours and twenty hours. In this yeast cell, the cell cycle is two hours. During that time, the cell takes in nutrients and makes all the material needed for the new cell. If nutrients aren't available, the cell won't divide. A cell slows down and its cell cycle gets longer if the temperature gets too low.

At noon this yeast cell was put into a dish in a lab and its cell cycle began. At 2 P.M. the dish had no more nutrients in it. At 4 P.M. the lab technician added nutrients to the dish. At 8 P.M. the heat went out in the building and the lab got cold. The yeast cells slowed down, and their cell cycle became twice as long.

Using the information in the paragraphs above, match the pictures of the yeast cells with the times at which they occurred. Then write the number of yeast cells present at each time on the line.

1 noon
2 2 P.M.
2 4 P.M.
4 6 P.M.
8 8 P.M.
8 10 P.M.
16 midnight

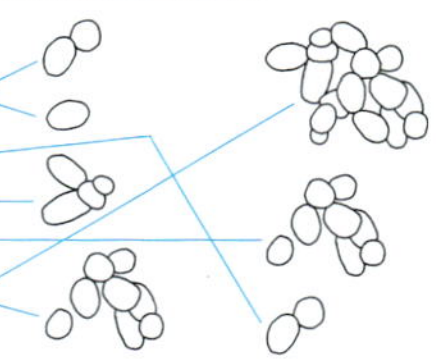

1. How many new yeast cells were produced between 2 P.M. and 4 P.M.? none
2. How many new yeast cells were produced between 8 P.M. and 10 P.M.? none
3. How are the results obtained between 2 P.M. and 4 P.M. similar to the results obtained between 8 P.M. and 10 P.M.? How are they different? During both periods of time, no new yeast cells were produced. Between 2 P.M. and 4 P.M. it was due to lack of nutrients. Between 8 P.M. and 10 P.M. it was due to the cold temperature.

Reading Skills Practice WB10

Arrange Events in Sequence

Read the selection below. When you have finished, number the steps in the correct order. Circle the word clues that help you know the order of events.

Cell Reproduction Through Meiosis

3 Next, the cell divides.

4 After this, the chromosome copies split apart.

1 First, each chromosome of the original cell is duplicated.

5 Finally, the two cells divide again.

2 Then, the two members of each pair of chromosomes separate and move to opposite ends of the cells.

Lesson Concept Review/Assessment WB11

How Do Cells Reproduce?

Lesson Concept

Cells reproduce by mitosis or meiosis. Meiosis occurs in reproductive cells. The cells produced by meiosis contain half the number of chromosomes as the parent cells. Mitosis occurs in all other body cells. The cells produced by mitosis contain the same number of chromosomes as the parent cells.

Vocabulary

genes (A15) **mitosis** (A16) **meiosis** (A17)

Answer all questions in one or more complete sentences.

1. Describe how a new organism develops. A new organism starts out as one cell that keeps dividing. Not all the cells are the same. For example, some cells become muscle cells and some become bone cells.
2. Why don't you just keep growing and growing? Your DNA sends instructions that change the rate of cell division in your body.
3. What is a gene and what does it do? A gene is a section of an organism's DNA that controls specific cell activities or characteristics.
4. How do body cells divide? They divide by mitosis.
5. How do reproductive cells divide? They divide by meiosis.
6. Why must reproductive cells undergo meiosis instead of mitosis? Reproductive cells must undergo meiosis so that when the sperm and egg cells join, the new cell will have the correct number of chromosomes.
7. Why is meiosis important in producing variation in living things? The new cell produced when two reproductive cells combine contains genes from both parents. This means there are many possible combinations of genes in the offspring.

Investigate Log WB12

Seed Color in Corn

Materials

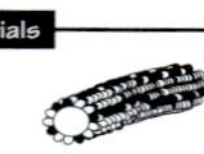
purple-and-white corn

paper and pencil

Activity Procedure

1. Take one ear of purple-and-white corn. **Observe** the corn, and **estimate** the number of purple kernels and the number of white kernels.
2. **Count** the purple kernels and white kernels. **Record** the two numbers in the chart below.
3. **Find the ratio** of purple kernels to white kernels. How many purple kernels are there for every white kernel?
4. Share your results with the class.
5. **Find the class totals** for purple kernels and white kernels.
6. Now **find the class ratio** of purple kernels to white kernels.

Number of Kernels of Corn	
Purple	**White**

Ratio ______ : ______

Workbook Support Continued

Investigate Log WB13

Draw Conclusions

1. What **ratio** of colors did you find for your ear of corn? The ratio of purple kernels to white kernels should approximate 3:1.

2. How did your results compare with the results of the whole class? The ratio for the class should be closer to 3:1 than most individual ratios.

3. **Scientists at Work** Scientists make **inferences** from their **observations**. They often **use numbers** to support their inferences. The larger the number of observations they study, the more likely their inferences are to be correct. What might you infer about the ratio of purple kernels to white kernels? The ratio varies for different ears of corn but is usually close to 3:1. Students may realize that the larger the sample is, the more accurate the results are likely to be.

   Should it change for each ear of corn? Yes; students should realize that the ratio will vary from ear to ear.

   How could you **use numbers** from your class and other classes to support your **inference**? Larger samples likely will support students' inferences.

Investigate Further Get an ear of yellow-and-white corn. **Find the ratio** of yellow kernels to white kernels. Is it the same as or different from the ratio of purple kernels to white kernels? The same; the ratio should be 3 yellow to 1 white.

Process Skills Practice WB14

Use Numbers and Infer

Scientists use numbers to help them order and compare their results. This helps them make reasonable inferences from their experiments.

Think About Using Numbers and Inferring

When scientists have many numbers to look at, they put the numbers in a table. This makes them easier to analyze and compare. Read the following paragraph. Put the family names and numbers in the table below and answer the questions.

Mr. and Mrs. Story both had brown eyes. They had four children. All four had brown eyes. Mr. and Mrs. Johnson also had brown eyes. They had four children. Three of their children had brown eyes, and one had blue eyes. Mr. and Mrs. Anderson both had blue eyes. All four of their children had blue eyes. Mr. Fleck had blue eyes, and Mrs. Fleck had brown eyes. All four of their children had brown eyes. Mr. Bailey had blue eyes, and Mrs. Bailey had brown eyes. Two of their children had brown eyes, and the other two had blue eyes. Mr. and Mrs. Bridgman both had brown eyes. Their four children all had brown eyes, too.

Family Name	Number of Parents		Number of Children	
	with brown eyes	with blue eyes	with brown eyes	with blue eyes
Story	2	0	4	0
Johnson	2	0	3	1
Anderson	0	2	0	4
Fleck	1	1	4	0
Bailey	1	1	2	2
Bridgman	2	0	4	0
Totals	8	4	17	7

1. What was the total number of parents? 12 How many of the parents had brown eyes? 8 How many had blue eyes? 4 What percent of the parents had brown eyes? $\frac{8}{12}$ = 66.7%
2. What is the total number of children? 24 How many of the children had brown eyes? 17 How many of the children had blue eyes? 7 What percent of the children had brown eyes? $\frac{17}{24}$ = 70.8%
3. What can you infer from this data? The percent of people with brown eyes increased in the children's generation. This may mean that it is more likely for people to have brown eyes than blue eyes.

Reading Skills Practice WB15

Using Graphic Sources

Use the Punnett square below to determine whether dimples would be present in children with the following genetic inheritance. (Dimples are dominant traits).

	D	d
D	DD	Dd
d	Dd	dd

Would dimples be present in a combination of the upper left-hand square? yes

Would the combination in the lower right-hand square produce dimples? no

Lesson Concept Review/Assessment WB16

How Are Traits Inherited?

Lesson Concept

Gregor Mendel's experiments with peas showed that traits are passed on from parents to offspring by factors that we now call genes. Dominant traits show up whenever a dominant gene is present. Recessive traits only show up when both genes are recessive.

Vocabulary

sexual reproduction (A24) **recessive** (A27)
dominant (A26) **Punnett square** (A29)

Answer all questions in one or more complete sentences when possible.

1. What was Gregor Mendel's contribution to science? He showed that traits were passed on from parents to offspring.
2. Give three reasons why Mendel chose pea plants for his experiments. Pea plants were easy to get. They grew rapidly. They showed traits that were easy to tell apart. It was easy to cross different plants. Left to themselves, the plants self-pollinated.
3. What type of reproduction allows children to inherit traits from both parents? Explain. Sexual reproduction allows children to inherit traits from both parents, because each new cell gets half its genes from each parent.
4. What are Mendel's factors called today, and where are they found? They are called genes, and they are found on the chromosomes.
5. Explain dominant and recessive traits, and give two examples of each. A dominant trait is one that will be shown every time its gene is present. Examples: tall pea plants, brown eyes, dimples. A recessive trait is one that will be shown only if both its genes are present. Examples: short pea plants, blue eyes, attached earlobes.
6. Right-handedness (R) is dominant, and left-handedness (r) is recessive. Complete the following Punnett square, and describe the offspring.
   Results: Half of the offspring are left-handed.

	r	r
R	Rr	Rr
r	rr	rr

Vocabulary Review WB17

Recognize Vocabulary

Place the letter of the word in the word list on the blank next to the phrase that best describes it.

A cell membrane	**G** vacuoles	**M** mitosis
B cell wall	**H** nucleus	**N** meiosis
C cytoplasm	**I** chromosomes	**O** sexual reproduction
D Punnett square	**J** nuclear membrane	**P** dominant
E mitochondria	**K** DNA	**Q** recessive
F chloroplast	**L** genes	

E 1. bean-shaped organelles that release stored energy
J 2. surrounds the nucleus
L 3. sections of DNA in the chromosomes
H 4. control center of the cell
K 5. complex chemical that contains information about the organism
Q 6. trait that is shown only when both parents have a gene for it
B 7. thick outer layer of plant cell
N 8. process that forms new reproductive cells
O 9. joining a male and female reproductive cell
C 10. colorless, jellylike substance inside cells
D 11. checkerboard-like diagram that shows gene combinations
F 12. structure in plant cells that allows them to make their own food
P 13. trait that is shown even if only one parent has a gene for it
A 14. holds cell material inside and controls flow of material in and out
I 15. structures in the nucleus that determine an organism's characteristics
M 16. cell division to produce new body cells
G 17. spaces that store food, water, and waste in cells

Writing Practice WB18

Write About Inherited Traits

Informative Writing–Explanation

Interview people from three generations of your own or a friend's family. Note examples of inherited traits. Use the table below to organize your notes. Summarize your conclusions. Then write a paragraph explaining what your results say about dominant and recessive traits.

Trait	You or a Friend	Parent (or Aunt/Uncle)	Grandparent (or Great-Aunt/ Great-Uncle)
Eye color			
Hair color			
Hair texture (curly or straight)			
Height (short, average, tall)			
Earlobes (attached or detached)			
Cleft chin?			
Dimples?			
Able to roll tongue?			

Conclusions

1. ______
2. ______
3. ______

CHAPTER 1

Cells, Genetics, and Heredity

Generate Science Questions

Invite students to read and discuss the Fast Facts. Ask students if, after reading the facts about plant and animal cells, they have any questions for which they would like to find answers. Lead the class in generating a list of questions, and record them on the board or on chart paper.

Ask students what kinds of investigations they would design to answer the questions they have. Explain that the chapter will help them answer their questions, but they may want to conduct additional investigations on their own.

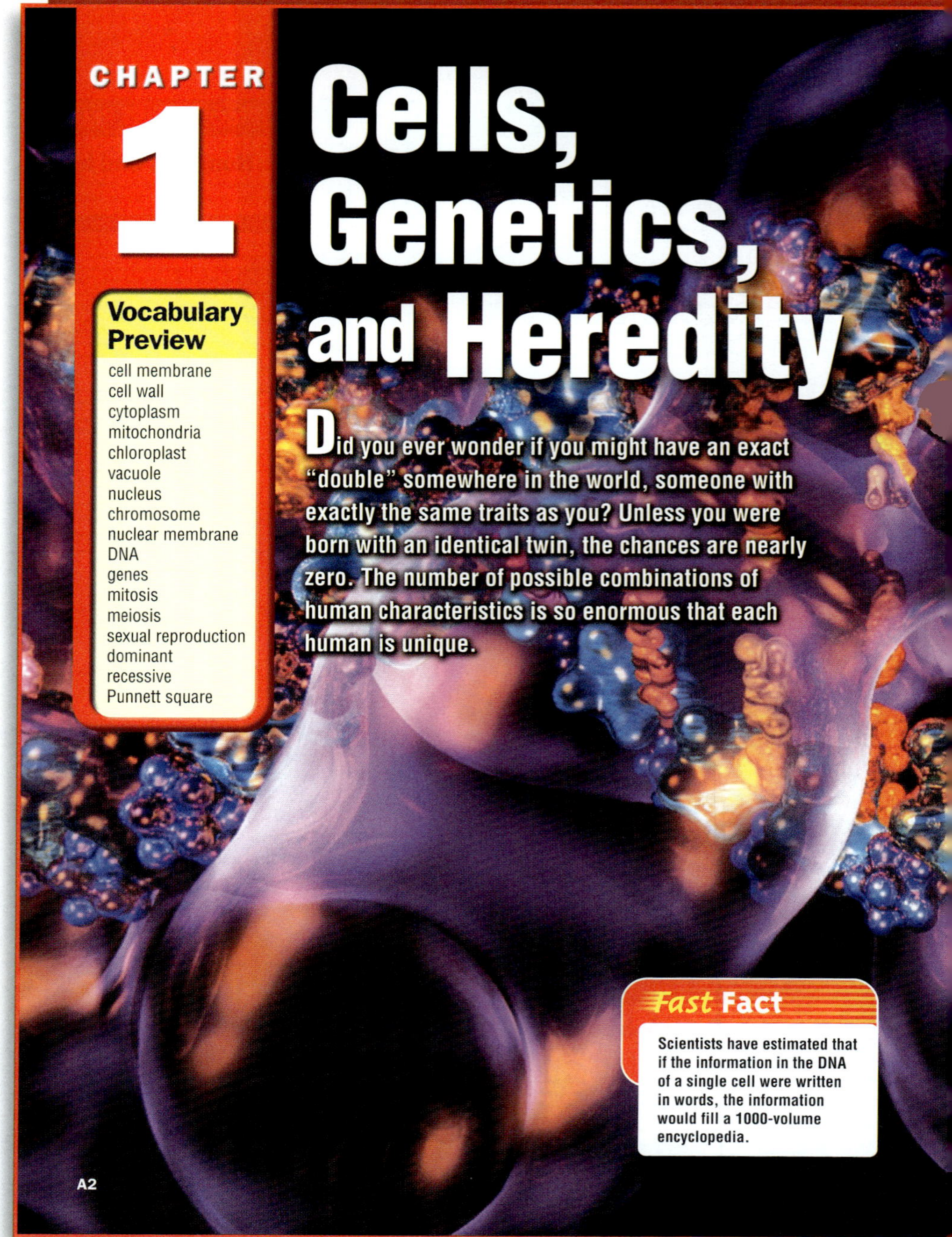

Reading Skills Checklist

Strategies for developing the following reading skills are provided in this chapter.

- ☑ use prefixes and suffixes to determine word meanings *p. A32*
- ☐ use context
- ☑ identify supporting facts and details *pp. A6, A8, A10, A14, A16, A17, A20, A24, A25, A26, A28, A30*
- ☑ arrange events in sequence *p. A16*
- ☑ identify the main idea *pp. A6, A8, A10, A14, A16, A17, A20, A24, A25, A26, A28, A30*
- ☐ summarize and paraphrase
- ☑ identify cause and effect *p. A26*
- ☑ predict probable future actions and outcomes *p. A29*
- ☑ use graphic sources for information *p. A26*
- ☑ draw logical conclusions *pp. A7, A15*
- ☑ compare and contrast *pp. A8, A29*
- ☐ distinguish fact and opinion
- ☑ make generalizations *p. A7*
- ☑ use reference sources *pp. A9, A10*

School-Home Connection

Distribute copies of the School-Home Connection, p. TR1. Have students take the page home to share with their families as you begin the chapter.

Follow Up Have volunteers compare the results of the activity they did at home. Additional School-Home Connections are provided by the **Activities for Home or School** (p. A35) and the **Take-Home Book** (p. A2d).

Teaching Resources, p. TR1

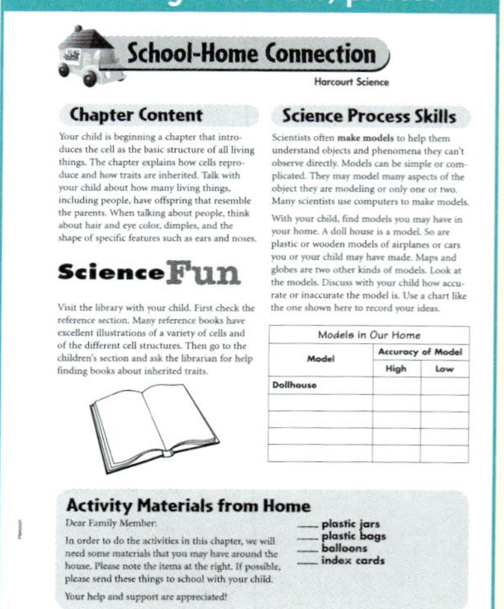

An artist's idea of what DNA looks like

Fast Fact

Gregor Mendel studied the inherited traits of pea plants for a period of eight years. During this time he grew over 30,000 plants.

Fast Fact

It's impossible to guess from an organism's size the number of chromosomes its cells have. A goldfish has 94 chromosomes, while humans have only 46.

Amoeba

Chromosomes per Cell in Various Organisms

Organism	Chromosomes
mosquito	6
housefly	12
pea	14
onion	16
carrot	18
starfish	36
human	46
amoeba	50
dog	78
goldfish	94

A3

Prereading Strategies

K-W-L Chart

Have students make a four-column chart headed What I Know, What I Want to Know, What I Learned, and "+" for What I'll Learn More About. Ask them to look at the chapter title and predict what the chapter will be about. Then have them fill in the first two columns. Students should fill in the third column as they study the chapter. At the end of the chapter, they can fill in the last column with additional things they would like to learn.

K-W-L Chart			
What I Know	What I Want to Know	What I Learned	+

Preview the Chapter

Have students scan the chapter, identify the lesson titles and major headings, and use them to make an outline of the chapter. They can write questions that occur to them as they make their outline.

Vocabulary Preview

Have students make a three-column chart for the words in the vocabulary list. In the first column, students should write each word; in the second column, the definition. The third column should tell how they remember the definition.

Science Journal		
Word	Definition	How I Remember

BACKGROUND

Webliography

Keyword heredity

www.harcourtschool.com

GO ONLINE

Human Genome Project

Begun in (approximately) 1990, the Human Genome Project is an international effort to map out the more than 3 billion DNA base pairs (50,000 to 100,000 genes) thought to make up the human genome. Although variation between individuals may seem considerable to us, variation between individual genomes is estimated at well below 1 percent. A complete draft of the genome was announced in 2000.

The ultimate goal of the Human Genome Project is to allow scientists to link particular human diseases with specific genes. Already, genes associated with cystic fibrosis and muscular dystrophy have been identified.

Fast Facts/Math Links

Find Possible Outcomes Challenge students to determine how many volumes the DNA of another one of the organism's cells would fill. 1000. All of an organism's cells have the same DNA.

Divide Challenge students to determine about how many pea plants Mendel grew each year, assuming he grew an equal number each year of his research. 30,000 ÷ 8 = about 3750 plants/yr

Find the Average Have students find the average number of chromosomes for the animals listed in the table. Note: An amoeba is not an animal. 6 + 12 + 36 + 46 + 78 + 94 = 272 ÷ 6 = 45.3

LESSON 1

How Do Plant and Animal Cells Differ?

Objectives

- Recognize that all living things are made up of cells.
- Compare the functions and structures of plant and animal cells.
- Describe how the nucleus of a cell directs its functions.

Motivate

More Fast Facts

Spark interest with these facts and questions.

- DNA is only 50 trillionths of an inch wide.

  Why do you think some people call DNA "the blueprint of life"? The DNA cell contains information and instructions about how the cell is built.

- The epidermis, or outer skin layer, sheds dead skin cells so quickly that your entire epidermis is replaced every four weeks.

  Why are we covered with a layer of dead skin cells? to protect the inner layers of live cells from the environment

Access Prior Knowledge

- What is a cell? What kinds of things does a cell do?

When Minutes Count . . .

Whole-Class Investigate Option Lead students through the planning and construction of a single, collective model. Based on students' observations, draw a diagram of a cell on the board. Have students vote on the materials to use for various cell parts, and invite volunteers to help with the construction of the model.

Using the Activity Video The video can be used to preview and model the investigate process with students. After viewing the video, students can conduct the investigation in small groups.

LESSON 1

How Do Plant and Animal Cells Differ?

In this lesson, you can . . .

- INVESTIGATE the parts that make up a plant cell.
- LEARN ABOUT the parts of plant and animal cells and how the cell parts work.
- LINK to math, writing, technology, and other areas.

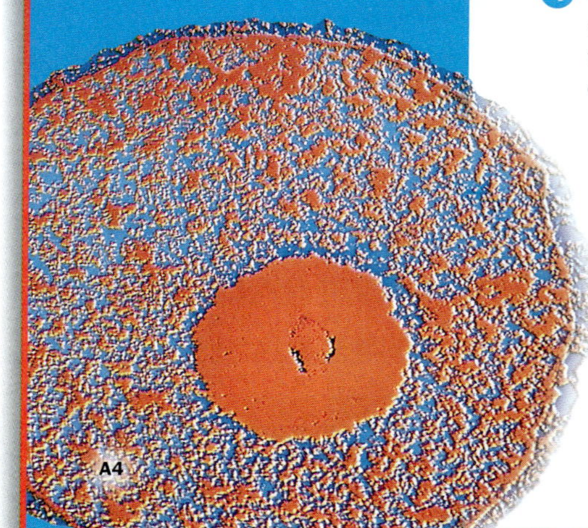

◀ All living things are made up of cells similar to the one shown here.

A4

INVESTIGATE

The Structure of Cells

Activity Purpose The parts of a cell are much too small to see without magnification. In this investigation you will **observe** photographs of plant cells as seen through a microscope. Then you will use your observations to **make a model** of a typical plant cell.

Materials

- plastic jar
- water
- 2 balloons
- index cards
- plastic bag
- sand
- red and green clay

Activity Procedure

1. **Observe** the pictures of the magnified plant cells. Find the parts that all the cells have in common.
2. Now use your observations to write a plan for how you will **make a model** of a typical plant cell. Decide how you can best show the cell, using the materials you have.
3. Show your plan to your teacher. When your teacher approves, make your model. (Picture A)
4. **Observe** your model. Write a description of each cell part you have modeled, and tell what you think the part's function might be. Use a separate index card for each cell part.

BACKGROUND

Cell Walls Unlike an animal cell, a typical plant cell includes an outer, rigid wall of cellulose. A plant's cell walls give it stiffness and enable it to grow, up to a height of 100 meters in the case of the giant redwoods. Cell walls remain intact for many years, even after the cell inside has died. Cell walls give lumber, cork, and natural rope their strength and shape.

Webliography GO ONLINE

Keyword cells

www.harcourtschool.com

For student self-assessment, see Assessment Guide p. AGxvi.

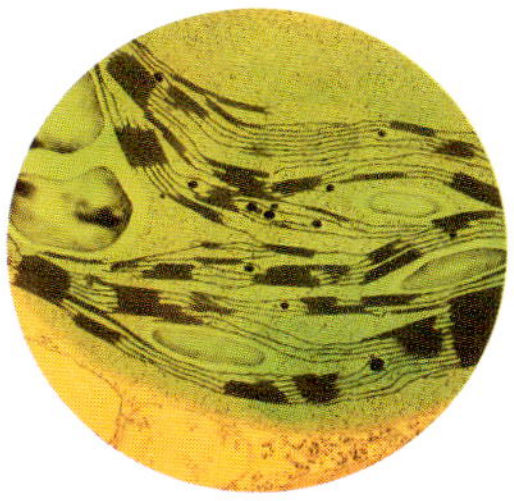
Plant cell 1

Plant cell 2

Plant cell 3

Picture A

Draw Conclusions

1. Why was it useful to **observe** several plant cells before making your model?
2. What can you **infer** about the function of the thick layer that surrounds the cell?
3. What can you **infer** about the function of the thin layer that lines the cell?
4. **Scientists at Work** Scientists often **make models** to help them make inferences. How did making a model help you make inferences about the parts of a cell?

Investigate Further

- Use a Microslide viewer or microscope with prepared slides and look at three more plant cells. What differences do you see? How would you use these observations to change the model you made in this activity?
- A fungus looks much like a plant, but it is almost never green in color. **Hypothesize** what you might expect a fungus cell to look like. Then **design an experiment** to test your hypothesis.

Process Skill Tip

Cells and their parts are too small to handle. When you **use a model,** you can handle the parts, move them around, and **infer** how they might work together.

A5

Process Skill Tip

Infer and Use a Model Invite students to look at the plant cell diagram on the Process Skill Tip Transparency. Point out that a diagram is a type of model—a simplified drawing of a complex object. This diagram can be used to show cell processes. Remind students that photosynthesis takes place within the chloroplasts of plant cells. H_2O and CO_2 combine in the presence of light to produce sugar and oxygen. Have students draw or use tokens on the transparency to infer and show how these materials move into and out of the cell during photosynthesis.

Process Skill Tip Transparency A1-1

Infer and Use a Model

Cell

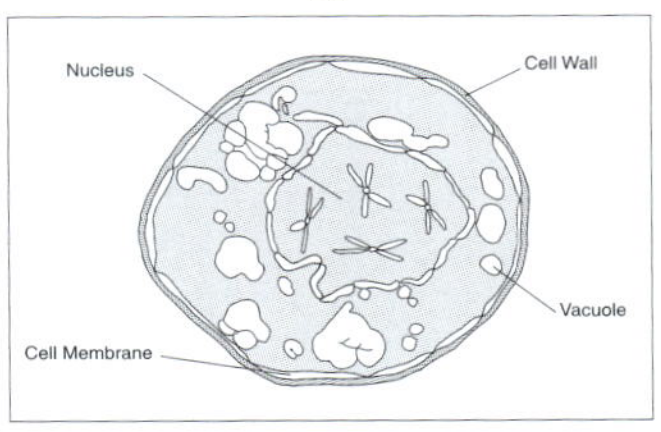

- Infer and model how the materials necessary for photosynthesis move into and out of the cell during this process. Some materials simply diffuse across the porous membrane. Other, larger materials may need to be moved across the boundary through pores.

INVESTIGATE

Time 30–40 min **Preparation Tips** See p. A2c.

Grouping individuals

Process Skill Tip use a model, infer

Other Process Skills observe, make a model

Expected Results Students' models should be three-dimensional and contain a cell wall, a cell membrane, a nucleus, and various organelles.

Teaching Tips

As students bring their model plans to you for approval, check for parts representing the cell wall, the cell membrane, the nucleus, and organelles.

Draw Conclusions

1. Possible answer: Observing several plant cells can help you determine which features are common to all plant cells.
2. The thick layer supports the cell and gives it its shape.
3. The thin layer keeps the cell's materials inside the cell.
4. Lead students to understand that their models helped them to better visualize how the parts of the plant cell fit together, which helped them infer how the parts worked and operated.

Investigate Further Students should suggest changes that would improve the accuracy of their models. Students can form a hypothesis stating what they expect a fungus cell to look like. Then students should design an experiment to test their hypothesis.

Informal Assessment

Performance Observe students to see that they are constructing their models based on what they see in the photographs and on their clearly labeled model plans.

INVESTIGATE LOG

Students can use **Workbook pp. WB2–3** to record the results of the investigation.

PROCESS SKILLS PRACTICE

To practice and apply process skills, see **Workbook p. WB4.**

LEARN ABOUT

1 Before Reading

SET A PURPOSE

FIND OUT Remind students that these statements will guide them in what to look for as they read.

VOCABULARY If students are doing the activity described in the Vocabulary Preview on page A3, remind them to add the definitions for these words to the chart.

Graphic Organizer for Chapter Concepts
Display the transparency (shown below) or have students use Workbook page WB1. Point out that by the end of the lesson, students should be able to fill in the parts of a cell on the transparency.

2 Guide the Learning

SCIENCE IDEAS

Help students focus on the supporting details for this **main idea:** **All living things are made up of cells.**

- **Are all different kinds of living things made up of cells?** Yes, that's what makes them living things.
- **Are all cells alike?** No, there are many different kinds of cells, even within one organism. Different cells perform different functions.
- **Are all organisms made up of millions of cells?** No, some organisms consist of just one cell.
- **How does an organism grow?** An organism grows larger by producing new cells.

VOCABULARY CARDS AND ACTIVITIES

Students can use the Vocabulary Cards to make their own graphic organizers or to add to an ongoing file of science terms. The Vocabulary Cards and a variety of strategies and activities are provided beginning on p. TR85 in the **Teaching Resources** book.

Ongoing Assessment

Use the ✓ questions in the pupil edition.

LEARN ABOUT

How Plant and Animal Cells Differ

FIND OUT

- what the cell theory is
- the similarities and differences between plant and animal cells

VOCABULARY

cell membrane
cell wall
cytoplasm
mitochondria
chloroplast
vacuole
nucleus
chromosome
nuclear membrane
DNA

The Discovery of Cells

Take a close look at a living thing, such as a plant leaf. No matter how hard you try, you won't be able to see its cells. Most cells are so small that thousands could fit on the period at the end of this sentence. Nobody even knew about cells until they could be magnified enough to be seen.

People built the first microscopes about 350 years ago. Microscopes magnified tiny things, opening up a whole new world to the scientists of the time. One of the first people to study the tiny building blocks that make up all living things was an English scientist named Robert Hooke. People still call these structures *cells*, the name that Hooke gave them more than 300 years ago.

✓ **Before the invention of the microscope, why couldn't scientists learn what living things were made of?** They could not see cells with the naked eye.

In 1665, using a microscope he built himself, Robert Hooke observed a thin slice of cork. He saw something never seen before. Hooke drew a picture of the hundreds of "tiny boxes" that he saw through his microscope. The boxes reminded him of the rooms in a monastery. Hooke gave them the same name—*cells*. ►

▼ Hooke's drawing of the cork cells he saw through the microscope

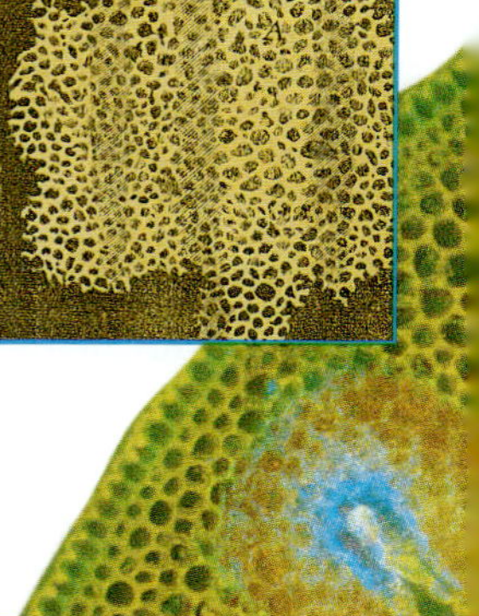

A6

Graphic Organizer for Chapter Concepts

Transparency A1 • Workbook, p. WB1

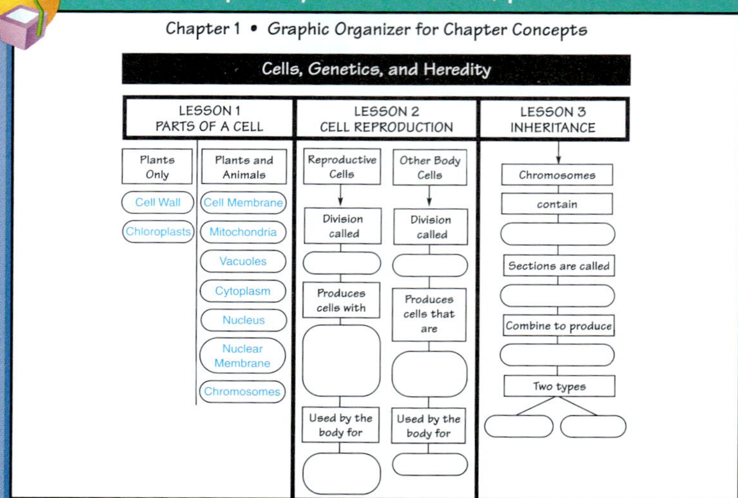

The Cell Theory

Since microscopes were invented in the 1600s, they have become much more powerful. As a result, scientists have discovered a lot about what cells are made of and what they do. By the end of the 1800s, scientists had developed a theory—the cell theory—that summarized much of what they had learned. The cell theory has three main parts:

Cells are the basic building blocks of all living things. All living things are made up of cells. Larger organisms most often have more cells and more kinds of cells than smaller organisms do. A human being, for example, has about 10 trillion cells of hundreds of different kinds. The photographs on this page show some of the kinds of cells in plants and animals.

All life activities take place in cells. Each kind of cell carries out the activities for which it is adapted. In animals movement happens because muscle cells contract, or shorten.

New cells are produced by existing cells. New cells come from other cells, not from anything else. Because organisms are made up of cells, the only way an organism can grow larger is by producing new cells. Your body grows because new cells are produced by existing cells. New cells also replace damaged or worn-out cells. Millions of cells in your body die every second, but new cells are produced to take their place. Your skin, for example, is continuously producing new cells as old skin cells die and fall away.

✔ **What are the three parts of the cell theory?** 1. All living things are made up of cells; 2. all the life activities of living things take place in cells; 3. new cells come only from existing cells.

▼ A cheek cell is flat and tough to protect the tissue beneath it.

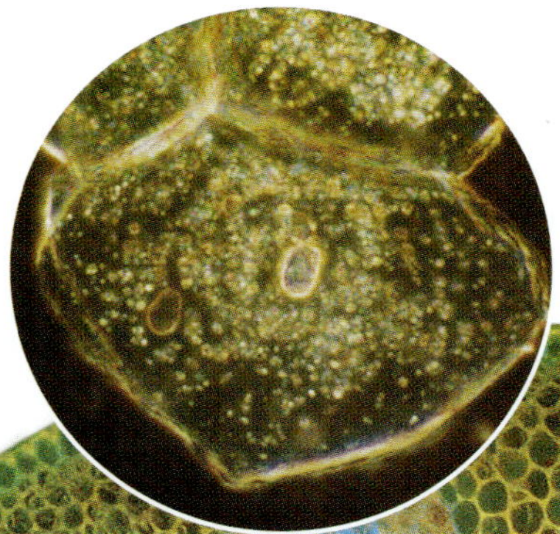

▼ This cross section of a stem shows different kinds of cells grouped together according to their function. Which cells do you think protect the stem?

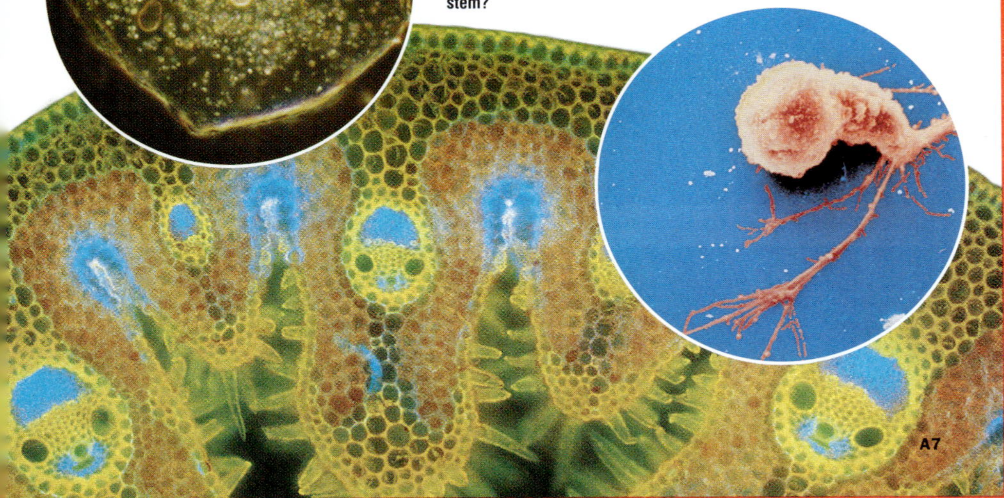

▼ A nerve cell has long, thin, wirelike extensions along which nerve signals can travel.

A7

VISUAL LEARNING

Have students compare the cells on this page and **make generalizations**. Are they all the same?

- ▶ In what ways are the cells alike? They are all too small to be seen without the use of a microscope. They are self-contained compartments that hold various materials.
- ▶ In what ways are the cells different? They have different colors and shapes. Some appear to be larger than others.
- ▶ Why don't all the cells look exactly alike? They all perform different functions and so have different shapes and sizes.

USE PROCESS SKILLS

Observe and Infer Have students observe the cells pictured on this page.

- ▶ Observe the nerve cell picture. Knowing that the nervous system quickly transmits signals throughout the body, what can you infer about the advantage of nerve cells having long extensions? Why wouldn't they travel just as quickly through short extensions? Perhaps the extensions allow messages to travel over greater distances within the body. The impulses can travel more quickly along these branches.

Critical Thinking Lead students to **draw logical conclusions** concerning the density of stem cells.

- ▶ Look at the cross section of a stem and observe how closely packed the cells are around the edge. What can you infer about the number of cells needed to form the outer surface of a stem? A lot of closely packed cells are needed to protect the stem.

BACKGROUND

Webliography GO ONLINE

Keyword cells

www.harcourtschool.com

Electron Microscopes enable us to see much smaller objects than we would be able to see through a traditional optical microscope. An SEM (Scanning Electron Microscope) works by sweeping the surface of an object with a beam of electrons. The object then emits its own electrons, which are collected and counted by an electronic device. The result is a three-dimensional image of the object's surface. An SEM can magnify an object up to 60,000 times.

A TEM (Transmission Electron Microscope) works by forcing electrons through a specimen to produce an image. Denser areas allow fewer electrons to pass; as a result, these areas appear darker on the image. TEM images are two-dimensional. They are frequently used to study cross sections of cells. The magnification power of the TEM is even greater than that of the SEM.

THE QUESTIONS KIDS ASK!

Why does cork have cells even though it is not living?

Address Misconceptions The cork was once part of a living thing—a tree. The "cells" you see in cork are actually only the remaining cell walls, not the other parts of the cell. Invite students to use a Microslide Viewer or a regular optical microscope to look at slices of cork.

Guide the Learning continued

SCIENCE IDEAS

Help students focus on supporting details for this **main idea:** Animal cells and plant cells are both similar and different.

- How are you similar to a flower or a bird? The cells of your body have parts in common with the cells of a flower and the cells of a bird.
- How can you identify the ways in which animal cells and plant cells are similar and different? One way is to compare and contrast the two types with magnified images or diagrams.

DEVELOP SCIENCE VOCABULARY

cell membrane, p. A8—Demonstrate that the cell membrane is flexible. Place a rubber band on a desk, and put some small objects inside it. Move the rubber band around to show that it can change shape while still enclosing its contents.

cell wall, p. A8—Point out that the cell wall is rigid and does not change shape. To demonstrate the rigidity of a cell wall, place a wire square around the rubber band (cell membrane).

cytoplasm, p. A8—Cytoplasm is a jellylike substance that supports the organelles. Mix a few raisins in a container of yogurt. Each object is suspended in the yogurt just as organelles are suspended in cytoplasm.

mitochondria, p. A8—Mitochondria are organelles that release energy to the cell so it can function. Ask students to picture their house as a cell and to identify the "mitochondria," or its energy supply.

The Parts of a Cell

You learned that cells have different functions and different shapes and sizes. However, all cells in all living things are alike in some ways.

All cells are covered with a thin layer called the **cell membrane**. This membrane holds the cell material inside and controls what enters and leaves the cell. The membrane lets food, oxygen, and water flow into the cell when they are needed, and it lets the cell's wastes flow out.

Plant cells have a **cell wall**—a stiff outer layer that protects the cell and gives it shape. Cell walls remain long after the rest of the cell dies. Cell walls are what Hooke saw when he observed cork through a microscope. Animal cells do not have cell walls.

Inside the cell membrane in both plant and animal cells is a colorless, jellylike substance called **cytoplasm** (SYT•oh•plaz•uhm). The cytoplasm holds the cell's *organelles*, tiny structures that carry out cell activities.

Each kind of organelle has a specific function. Bean-shaped organelles called **mitochondria** (myt•oh•KAHN•dree•uh) release the energy

THE INSIDE STORY

Comparing Animal and Plant Cells

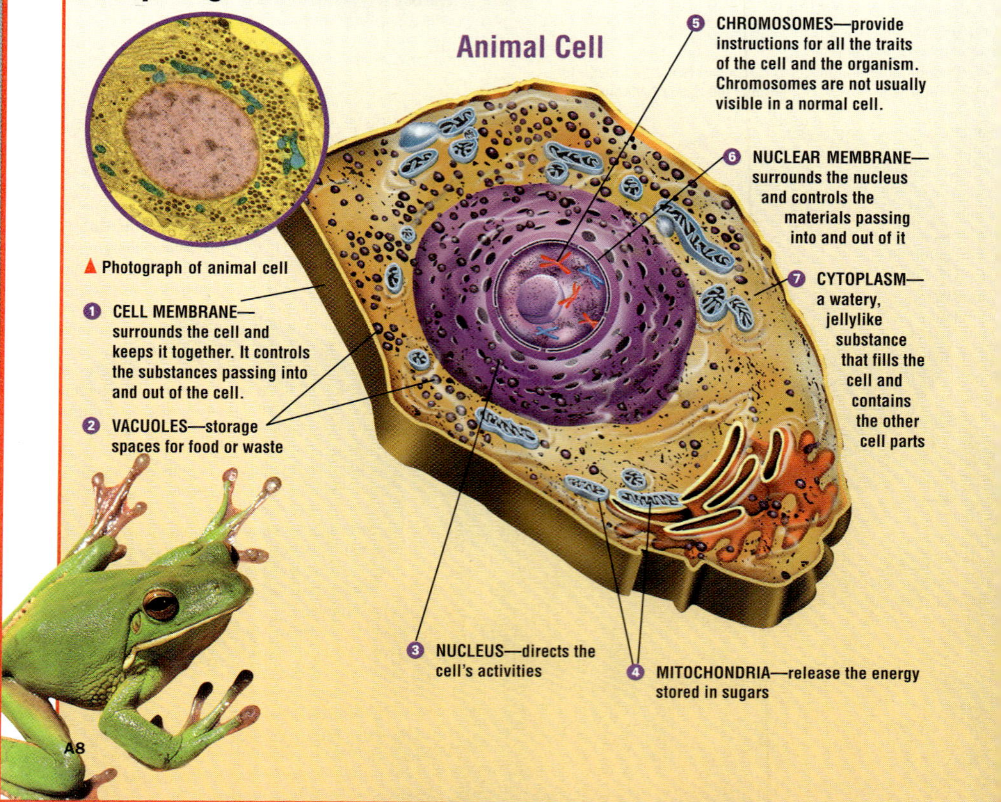

Reading Mini-Lesson

Compare and Contrast Words and phrases that show similarities include *as, also,* and *both;* and differences include *but, unlike, in contrast,* and *however.* Write these words on the board and display the transparency. Have students compare and contrast the characteristics of plant and animal cells found on pages A8 and A9 and write a paragraph using these words and phrases.

Reading Mini-Lesson Transparency A1-1

Compare and Contrast

Characteristic	Plants	Animals
cell wall	encloses the cell membrane and gives the cell a stiff shape	animal cells do not have a cell wall
chloroplasts	contain chlorophyll, which absorbs energy from sunlight, allowing the plant cells to make food	animal cells do not have chloroplasts
vacuoles	storage for food or waste	storage for food or waste
cell membrane	surrounds the cell and keeps it together; controls the substances passing into and out of the cell	surrounds the cell and keeps it together; controls the substances passing into and out of the cell

In plants, cell walls enclose the cell membrane and give the cell a stiff shape, while chloroplasts contain chlorophyll, which allows plants to make food through photosynthesis. Animals have neither cells walls nor chloroplasts. In plants and animals, vacuoles provide storage for food or waste. Cell membranes surround plant and animal cells and help keep them together.

READING SKILLS PRACTICE

To practice and apply this reading skill, see **Workbook p. WB5.**

stored in sugars to carry out the cell's activities. Each cell in your body contains between ten and several hundred mitochondria.

Plant cells contain a kind of organelle called a **chloroplast** (KLAUR•oh•plast) that enables the plant to make its own food. (Animals cannot make their own food. Their cells do not contain chloroplasts.) A **vacuole** (VAK•yoo•ohl) is a space in a cell enclosed by a membrane. Food, water, and wastes are stored in vacuoles.

Most cells have an organelle called the **nucleus**. The nucleus is the control center of the cell and directs all the cell's activities. It contains chromosomes. A **chromosome** (KROH•muh•sohm) is a structure that contains the information necessary to determine characteristics of the organism. The **nuclear membrane** surrounds the nucleus and holds it together.

✓ **What structures do plant cells have that animal cells do not have?** cell walls and chloroplasts

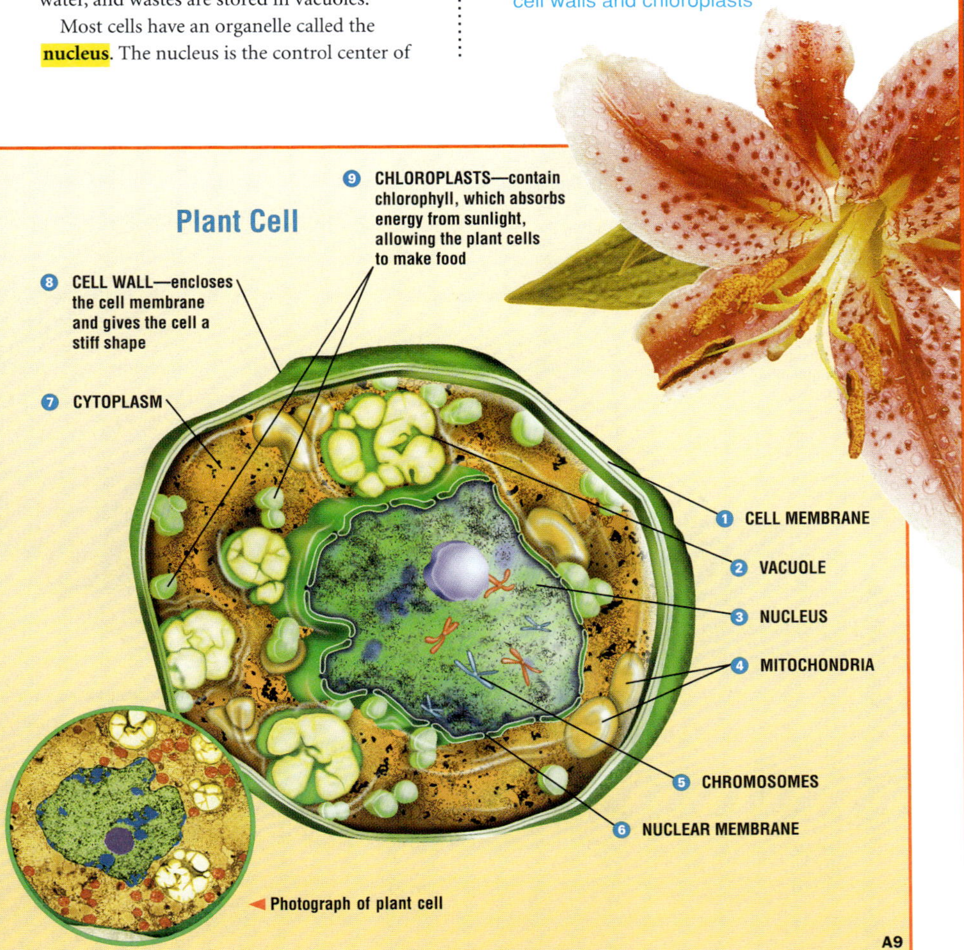

◄ Photograph of plant cell

A9

DEVELOP SCIENCE VOCABULARY

chloroplast, p. A9—Chloroplasts are found in the leaves (and in some stems) of plants and contain chlorophyll, a green pigment that plants use to make their own food. Have students explain the difference in color of some seasonal leaves.

vacuole, p. A9—Some vacuoles absorb waste products until they are full. At that point, they fuse with the cell membrane and expel the waste outside of the cell where it is carried away by the bloodstream. Invite students to identify a "waste vacuole" in their classroom "cell." rubbish bin

nucleus and **nuclear membrane**, p. A9—The nucleus acts as the manager of the cell, directing all cell activities from its "office" within the nuclear membrane. Ask students where the nucleus of their school is. the principal's office

chromosome, p. A9—A chromosome is the condensed form DNA strands take as they prepare for cell division. It contains DNA, which is often referred to as the cell's "blueprint." Ask students to explain a blueprint's role in construction.

VISUAL LEARNING

Have students compare the diagrams, photographs, and captions on these pages.

- ▶ **What parts are shared by both plant and animal cells?** cell membrane, cytoplasm, mitochondria, vacuoles, nucleus, nuclear membrane, chromosomes
- ▶ **What is the function of the cell wall?** to provide protection and shape

Health Link

Genetic Disorder Sickle-cell anemia is a genetic disorder that a person inherits from his or her parents. Sickle-cell anemia causes a distortion of the red blood cells into sickle (curved) shapes, slowing the blood flow and decreasing the amount of oxygen that reaches the cells of the body. Have students **use reference sources** to learn more about sickle-cell anemia and other genetic disorders such as muscular dystrophy, cystic fibrosis, Tay-Sachs disease, and Down syndrome.

ESL ACTIVITY

Challenge students to work in pairs to identify several "cells" and their **vacuoles** and **nucleus**. For example, have them identify the vacuole and nucleus of their town or city, of their home, and of a store or shopping mall.

Guide the Learning continued

SCIENCE IDEAS

Help students focus on supporting details for this **main idea:** A cell's functions are controlled by the cell's nucleus.

- ▶ How are chromosomes and DNA related? Chromosomes are made of DNA.
- ▶ What sort of instructions might the nucleus send to other parts of the cell? It might tell various cell parts to perform a specific action, such as to contract or to collect waste products.

USE PROCESS SKILLS

Compare and Infer Have students compare the nuclear membrane pictured here with the cell membrane pictured on page A9.

- ▶ Based on the structure of the membranes, what can you infer about the function of the membranes? They both have pores, so they both must allow certain materials to pass through while keeping others out.

DEVELOP SCIENCE VOCABULARY

DNA, p. A10—DNA is the genetic material contained within the nucleus of each cell. It is a double strand of chemicals that determine the characteristics of a cell. Ask students to **use reference sources** such as books or journals to research DNA and find out how to draw a diagram of a DNA double-helix on the board.

The Nucleus of a Cell

When you read about the cell theory, you learned that cells perform all the activities of a living organism. Just as a car needs a driver to control it, a cell needs instructions to work properly.

The nucleus is the control center of the cell. It contains chromosomes, tiny structures that look like rods. Chromosomes are made of proteins and **DNA**, a complex chemical that contains information about every part of an organism. DNA directs the life activities of the cell.

Some people call DNA "the blueprint of life." A blueprint for a house contains the information and instructions about how the house is built. In a similar way, the DNA in a cell contains the information and instructions about how the cell is built.

The photograph below shows the nucleus of a cell magnified many thousands of times. The nuclear membrane surrounds the nucleus and separates it from the rest of the cell. The nucleus sends out its instructions—in the form of chemicals—through tiny openings in the nuclear membrane. The chemicals are instructions for specific functions of the cell.

✔ **Why is the nucleus considered the control center of the cell?**

The nucleus contains the chromosomes and DNA for the cell. The DNA provides instructions for the cell to carry out all of its activities.

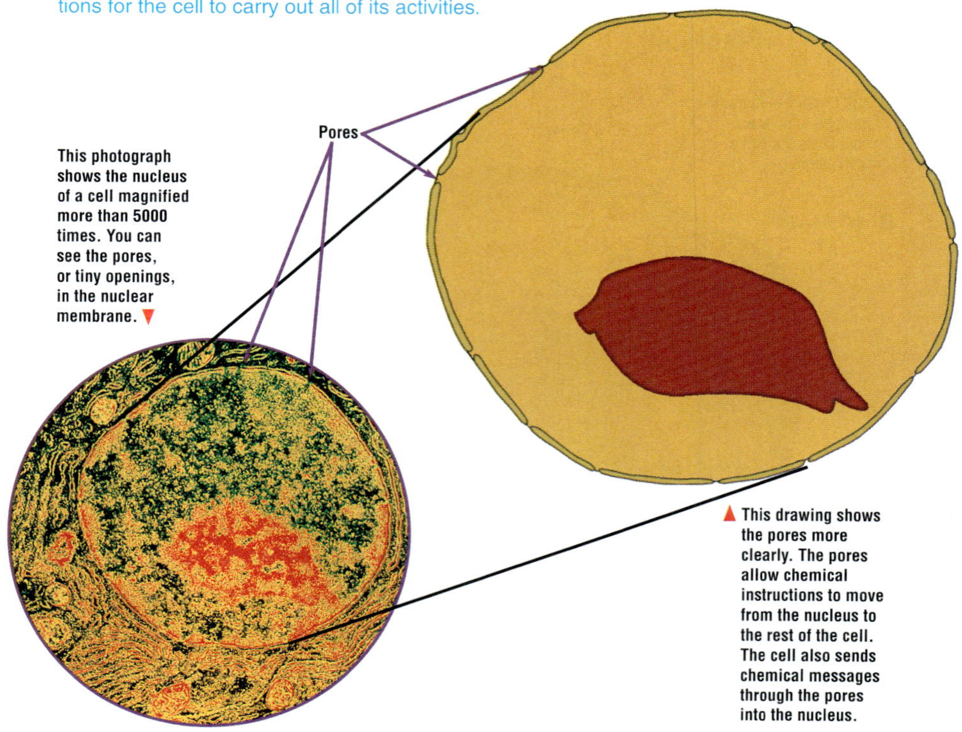

This photograph shows the nucleus of a cell magnified more than 5000 times. You can see the pores, or tiny openings, in the nuclear membrane. ▼

▲ This drawing shows the pores more clearly. The pores allow chemical instructions to move from the rest of the cell to the rest of the cell. The cell also sends chemical messages through the pores into the nucleus.

A10

Reaching All Learners

Kinesthetic Learners These students may find it easier to understand the nucleus by acting out its function. Students can link hands to form a circle that represents the nuclear membrane. Two of the students can then unlink their hands to demonstrate a "membrane pore." A student giving directions inside the circle can represent the DNA.

INVESTIGATION CHALLENGE

Hands-On Activities: More About Cells

Use these ideas for hands-on activities to extend students' understanding of key lesson concepts.

Have students use skin scrapings to make microscope slides for viewing. Ask them to use toothpicks to gently scrape cells from the inside of their cheek, the top of their tongue, underneath their fingernail, and the bottom of their foot. Have students draw pictures of the magnified cells.

Challenge students to make models to demonstrate semipermeable membranes. Provide materials such as nylon stockings, metal screening, colanders, sand, rice, beans, and water.

For student self-assessment, see Assessment Guide p. AGxvii.

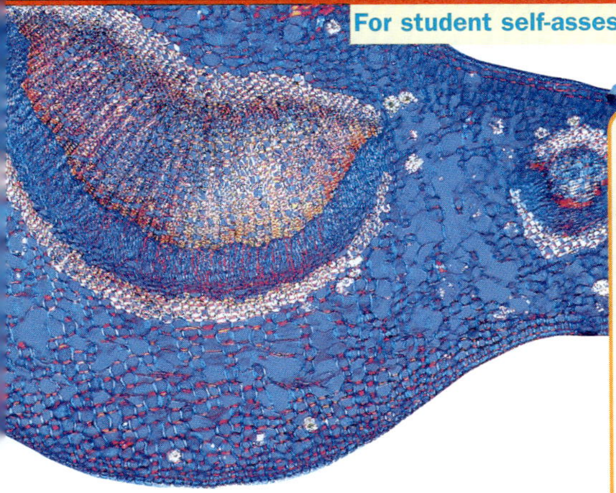

▲ A magnification of a plant cross-section shows the many different kinds of cells that help the plant survive.

Summary

According to the cell theory, all living things are made up of cells. All activities of living things take place in cells. All new cells come from existing cells. Plant and animal cells both have a cell membrane, cytoplasm, a nucleus, mitochondria, and vacuoles. Plant cells also have a cell wall and chloroplasts. The nucleus of a cell controls the cell's activities.

Review

1. How did cells get their name?
2. How does an organism grow larger?
3. What are the three parts of the cell theory?
4. **Critical Thinking** Which of your cells would you expect to have more mitochondria: your muscle cells or your bone cells? Explain your answer.
5. **Test Prep** Which of the following has a cell wall?
   A a nerve cell
   B a leaf cell
   C a cheek cell
   D a bone cell

LINKS

MATH LINK

Solve Problems A stack of 10 sheets of paper is about 1 mm thick. The thickness of a red blood cell is about $\frac{1}{20}$ the thickness of a single sheet of paper. A micron is 0.001 mm, so there are 1000 microns in a millimeter. How thick is a red blood cell in microns?

WRITING LINK

Informative Writing—Compare and Contrast Write an explanation that would help a fourth-grade student compare and contrast animal and plant cells.

SOCIAL STUDIES LINK

History of Science The cell theory resulted from the work of many scientists over a period of 200 years. Find out about the contributions of Matthias Schleiden, Antoni van Leeuwenhoek, Theodor Schwann, and Rudolf Virchow to the cell theory.

ART LINK

A Very Small City Think of ways to compare a cell and its parts with a small factory or a village. Imagine how you could represent each cell part. Mitochondria, for example, could be electric generators. Make a drawing or painting of your version of a cell.

TECHNOLOGY LINK

GO ONLINE

Learn more about plant and animal cells by visiting this Internet site.
www.scilinks.org/harcourt

SCILINKS

3 Wrap Up and Assess

SUMMARY

Graphic Organizer for Chapter Concepts After reading the summary, have students complete the Lesson 1 section of the graphic organizer (shown on page A6).

REVIEW

1. They reminded Hooke, their discoverer, of the cells in a monastery.
2. New cells are produced from existing cells within the organism.
3. Cells are the building blocks of all living things; all life activities take place in cells; new cells are produced by existing cells.
4. Muscle cells have more mitochondria because they require a lot of energy to contract and move.
5. **B,** a leaf cell

TEST PREP Test-Taking Tips

Item type multiple choice with question stem

To help students answer Question 5, share the following strategy with them. **Read the question carefully to find key words, such as *cell wall*, that are clues to the correct answer. Try to recall some facts that will help narrow the choices. Eliminate any answers that are obviously wrong. Then select the best answer to the question.**

LESSON CONCEPT REVIEW/ASSESSMENT

Students can use **Workbook p. WB5** to review the lesson concepts.

LINKS TIPS

Links Objective Extend understanding of cells.

Math Link Help students write equations to show the different relationships: thickness of 10 sheets = 1 mm; thickness of 1 cell = $\frac{1}{20}$ sheet; and 1 mm = 1000 microns. Thus, 1 sheet = $\frac{1}{10}$ mm; 1 cell = $\frac{1}{20} \times \frac{1}{10}$, or $\frac{1}{200}$ mm; 1 cell = $\frac{1}{200} \times 1000$, or 5 microns [1 mm = 1000 microns] 1 cell is 5 microns thick.

Social Studies Link Students should include the following: Matthias Schleiden, along with Theodor Schwann, formulated the cell theory; Antoni van Leeuwenhoek ground his own lenses which he used to see red blood cells; Rudolf Virchow determined that all living cells arise from preexisting cells.

HEALTH HANDBOOK

To learn how cells work together to form tissues, organs, and systems in humans, you may want to use the content and activities in the Health Handbook, pp. R24–R37 (T55–T61).

LESSON

2 How Do Cells Reproduce?

Objectives

- Compare the processes of mitosis and meiosis.
- Recognize that different combinations of parental DNA produce variation in new organisms.

Motivate

More Fast Facts

Spark interest with these facts and questions.

- The bodies of adult humans produce about 25 million new cells each second.

  Why does the body need so many new cells? to replace those that are damaged or no longer functioning

- Scientists can grow human skin in laboratory dishes.

  Why would scientists want to grow human skin? The new skin can be used to replace a person's skin that has been damaged by a fire or other accident.

Access Prior Knowledge

- What happens to the cells in your body when you grow? Do the cells get larger and larger, or does your body make new cells?
- Why do children look similar to their parents? Why don't brothers and sisters look exactly alike?

When Minutes Count . . .

Whole-Class Investigate Option Place the materials for observing an onion-root tip in a classroom center. Have students visit the center as partners or individuals. Then have them regroup to identify the five stages of cell division during a teacher-directed discussion.

Using the Activity Video The video can be used to preview and model the investigate process with students. After viewing the video, students can conduct the investigation in small groups.

LESSON 2

How Do Cells Reproduce?

In this lesson, you can . . .

- INVESTIGATE how exact copies are made of cells.
- LEARN ABOUT two different ways in which cells reproduce.
- LINK to math, writing, technology, and other areas.

A cell splitting to form two new cells

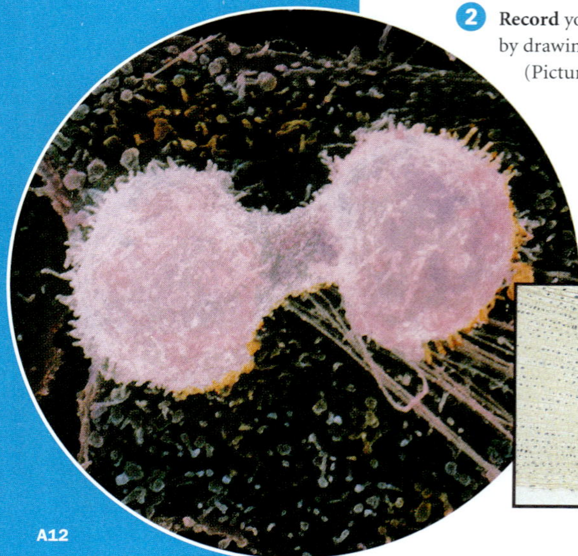

INVESTIGATE

How New Cells Are Made

Activity Purpose According to the cell theory, new cells can come only from existing cells. In this investigation you will **observe** the growth of cells in an onion root tip. You will use your observations to **infer** how cells produce exact copies of themselves.

Materials

- Microslide viewer
- Microslide of onion root tip

Activity Procedure

1. You will **observe** one instant in the life of the onion root tip. At that instant, different cells were preserved at different stages of cell division. Place the slide on the microviewer. Adjust the focus so that you can **observe** the cells. (Picture A)
2. **Record** your observations of cell division by drawing at least four stages of division. (Picture B)

This is a slice of the root tip of an onion. What you see is a single layer of cells that are no longer alive. A dye has been added so the structures inside the cells can be seen. Find the areas where new cells are being produced.

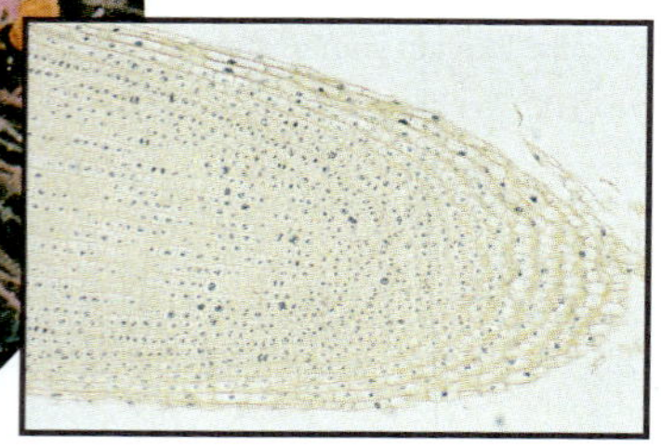

BACKGROUND

Webliography
Keyword cell division
www.harcourtschool.com

How Cells Divide There are five phases to the mitotic cycle—the process by which cells are made from existing cells. During **interphase,** each chromosome makes an exact duplicate of itself, and the pair of strands stay connected. In **prophase,** the chromosomes condense into shorter and thicker, more visible strands, the nuclear membrane breaks down, and the spindle starts to form. In **metaphase,** the chromosomes line up in the middle of the spindle. During **anaphase,** the two copies of each chromosome split apart and move to opposite ends of the cell. And in **telophase,** the final phase, new nuclear membranes form around the two sets of chromosomes; and the cell splits in half to form two cells, each with its own complete set of DNA.

For student self-assessment, see Assessment Guide p. AGxvi.

Picture A

Picture B

3 **Show the sequence** of the process of cell division by numbering your drawings from 1 to 4.

Draw Conclusions

1. How could you tell that the cells you **observed** were dividing?
2. **Write the sequence** of the stages of cell division. Describe each of the cells you drew.
3. **Scientists at Work** Scientists sometimes need to **find the sequence** of events to better understand a process. What clues might a scientist studying the division of onion cells look for in order to figure out the sequence?

Investigate Further Look at other slides of plant cells. Try to find any areas other than the root tip where plant cells are dividing. If you have access to a microscope, use one to observe prepared slides of an onion root tip. See page R3 for tips on using a microscope.

Process Skill Tip

To **show the sequence** of a group of events or items, first **compare** them carefully. Their differences should give you clues to the correct order.

A13

Process Skill Tip

Compare and Use Time/Space Relationships Students may have difficulty arranging sequential images that do not appear in order. Have students practice this skill by comparing and then arranging the images in the Process Skill Tip Transparency in proper time order.

Process Skill Tip Transparency A1-2

Compare and Use Space/Time Relationships

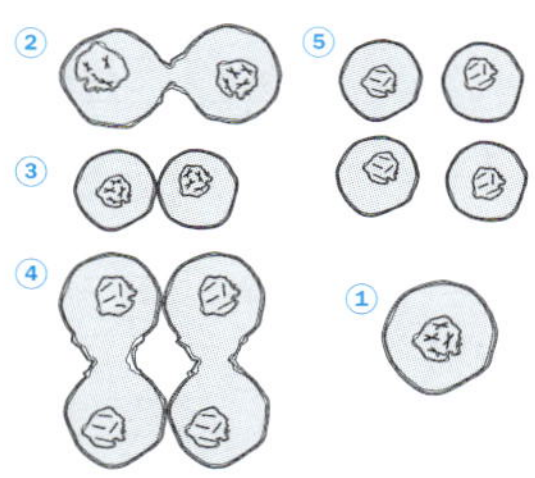

• Use these images to show the sequence of cell reproduction.

INVESTIGATE

Time 20–30 min **Preparation Tips** See p. A2c.

Grouping individuals

Process Skill Tip compare, use time/space relationships

Other Process Skills observe, infer, record data

Expected Results Students should be able to identify and place in order five stages of cell division.

Teaching Tips

As students begin to view the slide, help them to differentiate between cells at rest and cells that are undergoing some sort of internal transformation.

Draw Conclusions

1. Possible answers: The cells were different shapes, and they seemed to have different structures inside them.
2. First the cell's nuclear membrane disappears and dark rods form inside the cell. Then the rods line up in the center of the cell. Next the rods separate, half to each side of the cell. Then the cell splits and a new cell wall forms around each half to form two cells. Finally a nuclear membrane reforms inside each cell.
3. Scientists compare the appearance of a "resting" cell to the appearance of cells at various stages of division.

Investigate Further Have students look at additional slides of plant cells and identify other areas of a plant in which cell division occurs. Division is likely to be in the tips of stems.

Informal Assessment

Portfolio Students may want to keep their drawings of cell division in their portfolios.

INVESTIGATE LOG

Students can use **Workbook pp. WB7–8** to record the results of the investigation.

PROCESS SKILLS PRACTICE

To practice and apply process skills, see **Workbook p. WB9.**

LEARN ABOUT

1 Before Reading

SET A PURPOSE

FIND OUT Point out that these statements will guide students in what to look for as they read.

VOCABULARY If students are doing the activity described in the Vocabulary Preview on page A3, remind them to add the definitions for these words to the chart.

Graphic Organizer for Chapter Concepts Display the transparency (shown below) or have students use Workbook page WB1. Point out that by the end of the lesson, students should be able to fill in the answers under Cell Reproduction.

2 Guide the Learning

SCIENCE IDEAS

Help students identify supporting facts and details for this **main idea: Existing cells divide to produce new cells.**

- **How do living things grow larger?** Their existing cells divide to produce new cells.
- **When a body cell divides, how many new cells does it produce?** Each cell divides to form two new cells.

USE PROCESS SKILLS

Compare and Infer Have students compare the images of cells to the image of the fully formed frog.

- **What can you infer about what happened to the frog cells as they continued to divide and multiply?** They became different kinds of cells, such as eye cells and leg cells.
- **If a stage in frog development had a million cells, did its cells divide a million times?** Students may not understand the nature of exponential growth. Guide them to realize that if the cells divided only 20 times, the frog would already contain over 1 million cells. (2^{20} = 1,048,576)

Ongoing Assessment

Use the ✔ questions in the pupil edition.

LEARN ABOUT

How Cells Reproduce

FIND OUT

- how cells make copies of themselves
- how cells divide to produce reproductive cells

VOCABULARY

genes
mitosis
meiosis

How Do Plants and Animals Grow?

In the investigation, you observed onion cells producing new cells. Each dividing cell is too small to have a large effect on the onion plant. But many cells in the onion root divide at the same time. The cells they produce also divide, producing still more cells. Together, all of this activity has a large effect on the onion plant—it grows.

Most living things begin life as a single cell. The series of photographs below shows how a single cell can grow and multiply until it has produced a fully formed frog. As you can see in the photographs, the single egg cell splits into two cells. Those two cells then split, producing a total of four cells and then eight cells. The cells keep dividing this way, and by the time a tadpole hatches, it is made up of millions of cells.

You began life as a single cell, which divided to form new cells. As your cells divided, you grew inside your mother's body. By the time you were born, you were made up of billions of cells. They were not all the same. Some were bone cells, some were muscle cells, and some were skin cells.

One cell

The small pictures show three stages in the reproduction of cells in a frog embryo. The embryo starts as a single cell. The cell divides and forms new cells until a tadpole hatches. The tadpole can eventually grow to become a frog.

Two cells

Eight cells

Cells magnified about 50 times

Graphic Organizer for Chapter Concepts

Transparency A1 • Workbook, p. WB1

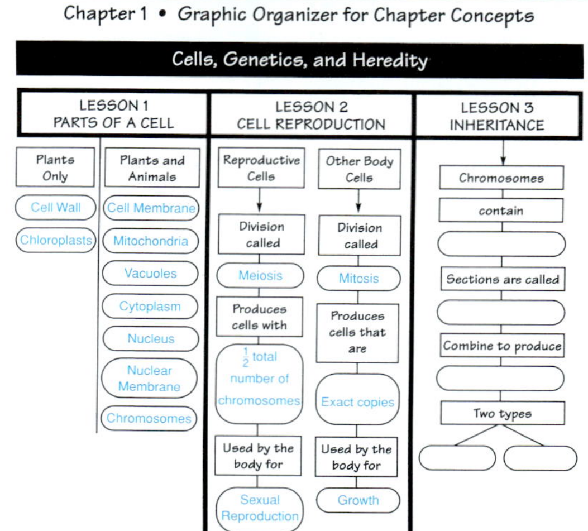

After you were born, you kept growing because your cells continued to divide. You'll keep growing bigger and taller as your cells keep dividing. But will you grow bigger and bigger every year of your life?

Most living things do not grow at the same rate throughout their lives. Growth often starts slowly, gets faster, and then slows down or stops. The graph shows the typical growth rate of a plant. Many living things have a similar pattern of growth.

Throughout your life your body cells continue to divide and replace cells that are damaged or worn out. However, you will probably stop growing larger sometime before you are 21 years old. The rate of cell division in your body will change, according to the instructions sent by your DNA.

✔ **When do plants grow quickly, and when do they grow slowly?** fast—between days 20 and 60. Slow—before day 20 and after day 60

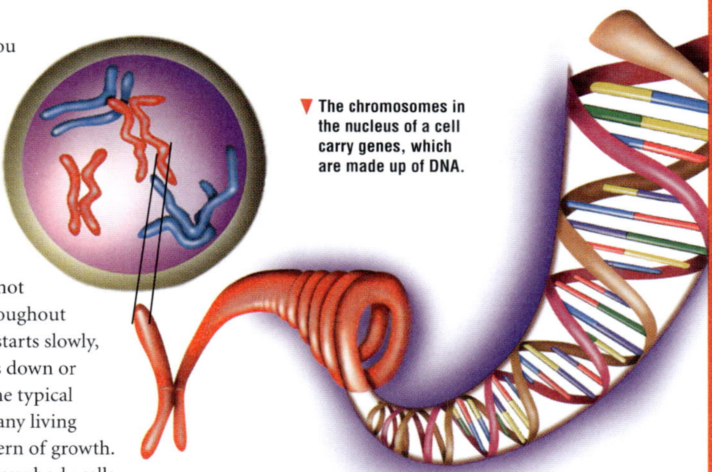

▼ The chromosomes in the nucleus of a cell carry genes, which are made up of DNA.

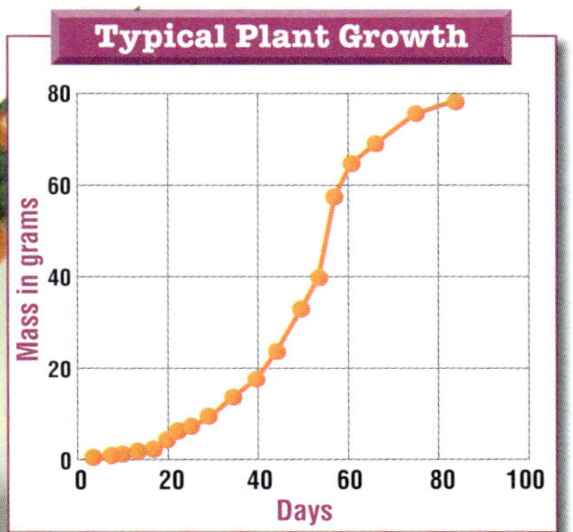

▲ This graph shows the typical growth rate of a plant.

DNA, the Blueprint of Life

The DNA in each cell controls the entire development of an organism. It is the "master plan" for every one of the organism's cells.

The nucleus of both plant and animal cells contains chromosomes. Chromosomes occur in pairs. Chromosomes contain segments of DNA called **genes**. Genes control specific cell activities and characteristics of every organism.

A human has 23 pairs of chromosomes, or 46 chromosomes in all. There are about 50,000 genes on these chromosomes.

When body cells divide, the new cells have the same DNA as the original cell. Most of an organism's body cells contains the same complete DNA. The DNA code that instructs the human brain to develop is also found in the cells of a person's liver. However, DNA codes are used only when they are needed. The DNA that instructs brain cells to grow is switched off in every cell except brain cells.

✔ **Where is an organism's DNA located?** An organism's DNA is in the chromosomes in the nucleus of every cell in the organism.

A15

ANALYZE AND INTERPRET DATA

Guide students as needed in reading the graph. Point out that the *x*-axis (horizontal axis) shows the passage of time and the *y*-axis (vertical axis) shows mass, which is similar to weight.

► **Is the plant getting older or younger as you move along the *x*-axis, from left to right? What is happening to size as you move up the *y*-axis?** From left to right along the *x*-axis the plant is getting older. From bottom to top along the *y*-axis, the plant is getting larger.

► **Within the first 20 days of development, is the plant growing very much? How do you know?** There is very little change in mass, which means that the plant is not growing very much.

► **What happens to the line graph between days 20 and 60? What does this mean?** There is a sharp rise in the line, which means that the plant is growing rapidly.

DEVELOP SCIENCE VOCABULARY

gene, p. A15—A gene is a discrete section of DNA that is often associated with a specific characteristic or a group of characteristics. For example, we say that baldness is linked to a specific gene. Ask students to **draw logical conclusions** in order to name some genes that they inherited from their parents, such as a gene for brown hair or a gene for long legs.

BACKGROUND

Webliography GO ONLINE

Keyword cell division

www.harcourtschool.com

Cell Differentiation Although all daughter cells produced from a parent cell contain the same genetic information, each may develop into a different type of cell. Cell differentiation involves the activation and deactivation of different genes in the cell's DNA. Because different genes code for different cell functions, cells with one set of genes activated will appear different and function quite differently from cells with another set of genes activated.

In some cases, the unused genes are permanently deactivated as the cell is differentiating. This process is called **determination.** Once a cell has undergone determination, it cannot change its function. So, for example, a light-sensitive retinal cell cannot change to become a contracting muscle cell, even though it contains the needed genetic information.

ESL ACTIVITY

Ask students to look through illustrated books or magazines to find pictures of family groups. Ask them to discuss characteristics of the people in the picture that might have been passed down from parent to child through their **genes**.

Guide the Learning continued

SCIENCE IDEAS

Help students focus on supporting details for this **main idea: All cells contain the same DNA.**

- **Why do all cells contain the complete DNA of the organism even if they don't use it all?** The first cell of an organism contains all the DNA. As it divides, it gives new cells an exact copy of its DNA, not just a fragment of it.
- **If all cells contain the same DNA, how can there be different kinds of cells?** Each cell uses only the DNA that relates to its specific functions.

USE PROCESS SKILLS

Observe and Infer Have students observe the chromosome pictures on page A16 and infer how the chromosomes separate into two cells.

- **How are the four chromosomes (each made up of two copies) divided between the two new cells?** The originals separate from their copies, and the two sets move to opposite ends of the cell to form two new cells, each containing four chromosomes.

Critical Thinking Lead students to understand why the order of steps in mitosis is important.

- **What would happen if the chromosomes lined up differently and one cell got both copies of two chromosomes and the other cell got both copies of the other two chromosomes?** The cells would not have identical copies of DNA and would not be able to function properly, nor could they pass complete copies of the organism's DNA on to new cells as they divided.

Mitosis

Take a look at your hand. Knowing how small each cell is, you can't even imagine how many cells are in just one fingertip. You began life as a single cell, so the cells in your body must have divided many times.

When a cell divides, some parts simply split into two. But the nucleus of the dividing cell cannot simply split into two parts, because all of the chromosomes from the original cell are needed in each new cell. If the chromosomes were simply divided into two groups, each new cell would have only half of the DNA. If cells continued to divide, the amount of DNA would keep decreasing. But a complete set of chromosomes is needed for every body cell.

The diagram shows what happens to a cell's chromosomes during cell division. Before division each of the chromosomes is copied exactly. The two identical sets of chromosomes are sorted out so that each new cell gets a complete set. The process of cell division that produces new body cells with complete sets of chromosomes is called **mitosis** (my•TOH•sis).

The cell in the diagram is from an organism that has two pairs of chromosomes, or four in all. Before mitosis, the chromosomes are copied so the cell has eight chromosomes. The chromosomes and their copies are joined at their centers and line up along the middle of the cell. They then split apart. One set of chromosomes goes to each side of the cell. Two new cells form.

Each new cell is an exact copy of the original cell. On page A18 you can observe photographs of the stages of mitosis.

✔ **If a cell that has 20 chromosomes undergoes mitosis, how many chromosomes will each of the new cells have? Explain.** 20, in mitosis the new cells have the same number of chromosomes as the parent cell.

Mitosis

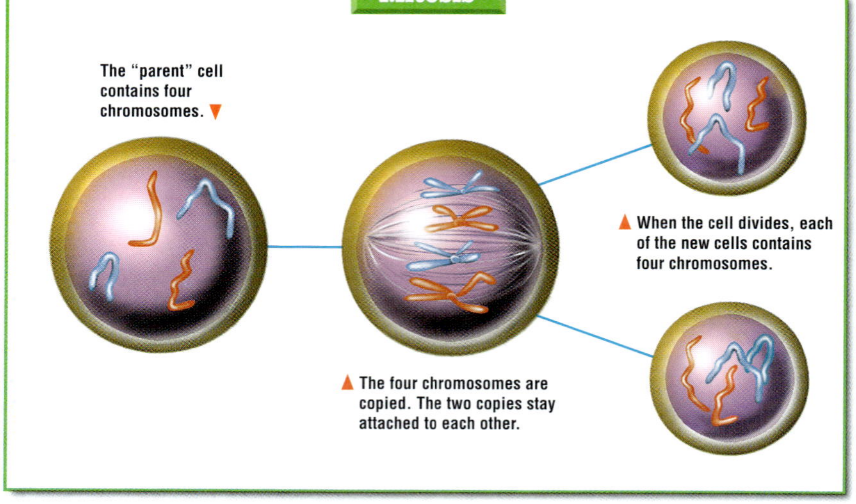

A16

Reading Mini-Lesson

Arrange Events in Sequence Signal words such as *first, next, then,* and *finally* indicate the sequence of events in a process. Display the transparency and have students number the steps in the correct order, identifying word clues that help them sequence events.

READING SKILLS PRACTICE

To practice and apply this reading skill, see **Workbook p. WB10.**

Reading Mini-Lesson Transparency A1-2

Arrange Events in Sequence

4 Finally, the single-stranded chromosomes become long and thin again.

2 Then, spindles form and the middle of each double-stranded chromosome lines up along the spindle's center.

1 First, the long, thin, double-stranded chromosomes shorten and thicken.

3 After all this, the two strands of each chromosome separate and the single-stranded chromosomes move along the spindle to opposite ends of the cell.

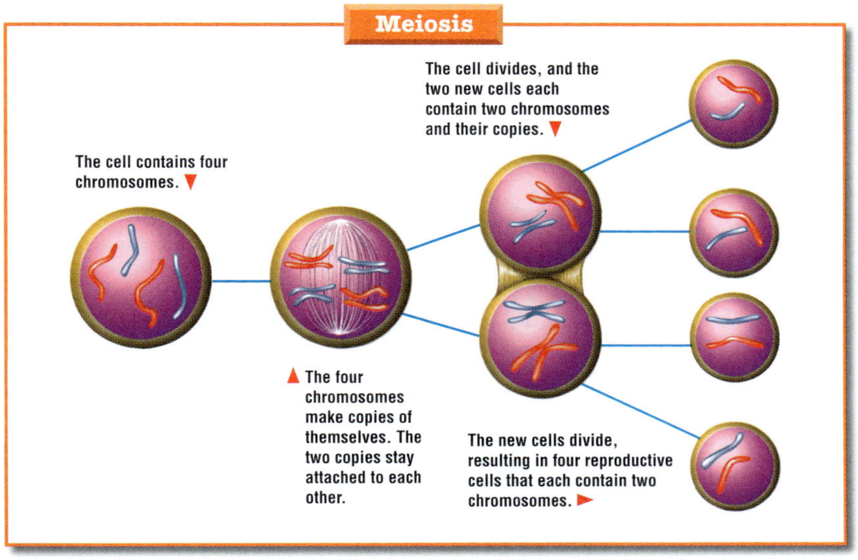

Meiosis

In mitosis exact copies of cells are made. This is how you grow and how dead cells in your body are replaced. Another kind of cell division takes place to form reproductive cells such as eggs and sperm.

The process that forms reproductive cells is called **meiosis** (my•OH•sis). In meiosis each new cell contains only half the number of an organism's chromosomes. As a result, when two reproductive cells join, they form a cell with the correct total number of chromosomes. If sperm and egg cells were formed by mitosis instead, each would have a complete set of chromosomes. When two reproductive cells joined, the resulting cell would have twice as many chromosomes as it should.

The diagram shows what happens to the chromosomes in a cell during meiosis. The cell in the diagram is from an organism that has two pairs of chromosomes, or four in all.

Meiosis actually includes two cell divisions. Before the first, each of the chromosomes in the original cell is copied. The new copies of the chromosomes remain attached to the originals. With their copies attached, one chromosome of each pair moves to opposite ends of the cell. Then the cell divides. Each new cell has half the chromosomes from the original cell. Each chromosome is still attached to its new copy.

The two cells then start a new division. In each cell, each chromosome and its copy split apart. Each of the four new cells that are produced has just two chromosomes—half the original number.

On page A19 you can observe photographs of an actual cell going through meiosis.

✔ **In what way does a cell formed by meiosis differ from a cell formed by mitosis?**

A cell formed by meiosis has half the number of chromosomes of the original cell. A cell formed by mitosis has the full number of chromosomes.

A17

DEVELOP SCIENCE VOCABULARY

mitosis, p. A16—This cell division produces new cells used for maintenance or growth of the organism. Discuss how different animals and plants can be identified by the number of chromosomes in their cells. For example, humans have 23 pairs (46 chromosomes) and peas have 7 pairs (14 chromosomes). Invite students to find the number of chromosomes in a cell in their favorite animal.

meiosis, p. A17—This cell division results in cells that mature into sex cells used in the formation of a new organism. It takes place in the cells of the testes and ovaries in most animals, and in the anthers and ovaries of flowers. Invite students to examine the parts of a flower and identify where meiosis would take place.

SCIENCE IDEAS

Help students identify supporting facts and details for this **main idea: Meiosis is a form of cell division that produces cells with only half the amount of genetic information contained in the parent cell.**

- **What is the function of cells produced by meiosis?** They are reproductive cells used to produce new organisms.
- **Why do reproductive cells contain only half of the DNA an organism needs?** The other half is supplied by the other parent.

Math Link

Solve Problems Challenge students to figure out the following problem that explores exponential growth.

You are at the beach one sunny morning, placing seashells in a large pail according to the following schedule: You begin by adding one shell. Then, once every minute, you double the number of shells that are currently in the pail. At noon, your pail is full. At what time was your pail half full? 11:59 A.M.

ESL ACTIVITY

The terms **meiosis** and **mitosis** may be confusing because they look and sound similar to each other. Have students work in pairs to practice pronouncing both terms. Model the pronunciation, emphasizing the *t* sound in the word *mitosis*. Then ask students to work together to define both terms. Challenge them to develop a mnemonic, or device to aid their memories, that will help them to remember the difference between the two terms.

Guide the Learning continued

USE PROCESS SKILLS

Infer Based on what students read and observe about the pictures on pages A18 and A19, have them infer the answers to the following questions.

- ▶ **In which parts of the onion plant is growth occurring most rapidly? Why do you suppose that is?** The root tip and the tip of the stem; the root tip grows down into the soil so that it can obtain nutrients. The stem grows tall so that the plant can capture sunlight. Extensive roots and tall stems help the plant to survive, so that's where the plant concentrates its growth.
- ▶ **Do you think the cells within the pollen grains of the lily anther have a complete set of chromosomes? Why or why not?** No. They are reproductive cells, so they contain only half of the total number of chromosomes found elsewhere in the lily.

Stages in Cell Division

The photographs on these pages show cells dividing.

The cell on the left is from the tip of an onion root. It is the same kind of cell that you observed in the investigation in this lesson. The root tip and the tip of the stalk are where most of the growth of an onion occurs.

The cell on the right is a male reproductive cell from a lily.

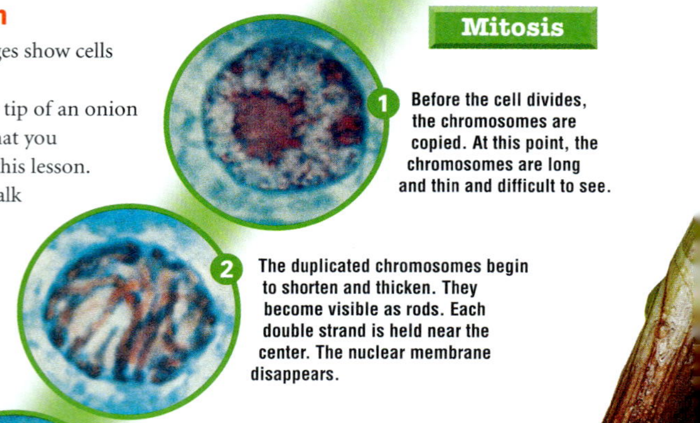
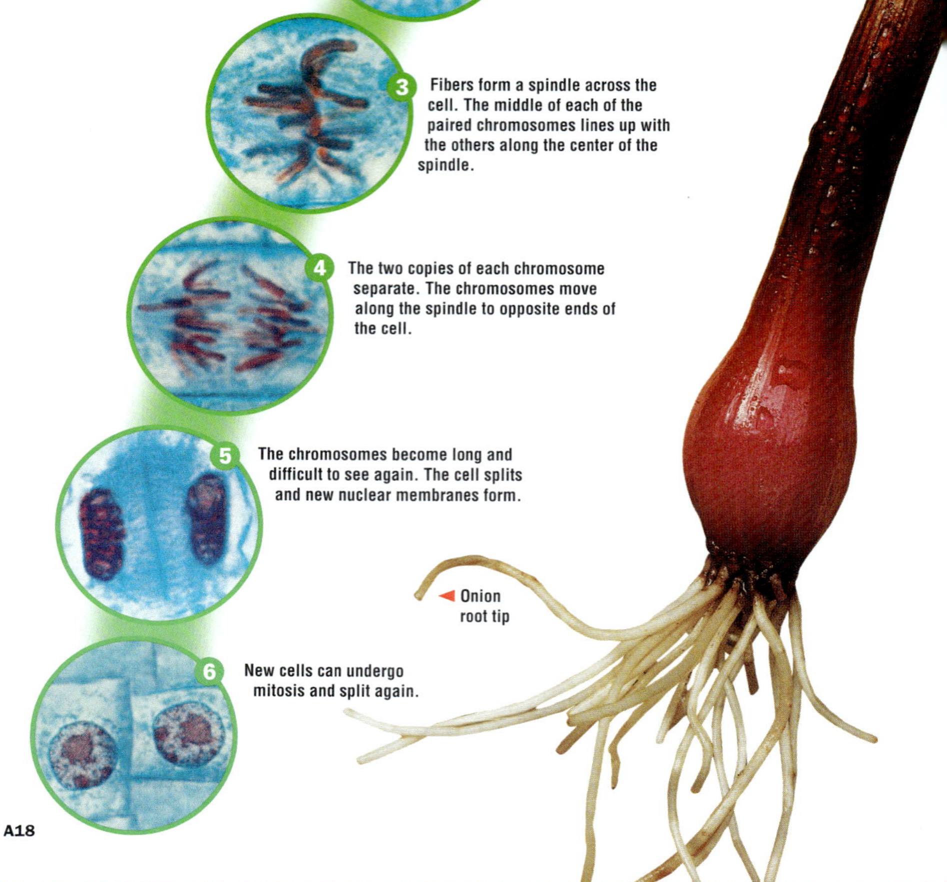

A18

THE QUESTIONS KIDS ASK!

How do cells know when to divide?

Address Misconceptions Students should understand that cells do not think or decide. The genetic material inside them along with signals sent from other parts of the organism and from the environment determine when they divide, specialize, and so on. Have students make models or drawings of mitosis.

BACKGROUND

Crossover in Meiosis In the paired chromosomes during meiosis, sometimes there is "crossover" or exchange of some of the DNA on each copy. This exchange increases the "mixing" of DNA that occurs in meiosis and adds further diversity to the DNA in each strand. Thus, in meiosis, the number of chromosomes in the final cell is $\frac{1}{2}$ the number in the original cell; and the strand of DNA may be a hybrid of the original two chromosomes.

Webliography GO ONLINE

Keyword cell division

www.harcourtschool.com

Meiosis

VISUAL LEARNING

Have students compare the photographs and captions on pages A18 and A19.

- **How are the stages of mitosis and meiosis alike?** In both cases, DNA is replicated so that the parent cell temporarily contains two identical copies of its DNA.
- **How are they different?** In mitosis, both new cells contain a complete set of the parental DNA. In meiosis, half of the original DNA goes to one new cell while the other half goes to the other new cell. As a result, each new cell has only one member of each pair of chromosomes.
- **Why can't you see single strands of DNA becoming double strands?** At the time when they are copying, they are too small to be seen. It is not until after they are copied and they condense that we can see them at this magnification.
- **When does the nuclear membrane reform around the DNA of the new cells?** after the cell has finished dividing

Cultural Connection

Domestication While sexual reproduction produces great variety in offspring, selective breeding can be used to develop those traits that are particularly favorable, including docility.

Evidence suggests that Europeans and Asians began domesticating animals more than 11,000 years ago. Sheep were perhaps the first domesticated animals—goats, swine, and cattle soon followed. These animals were integrated into the lives, and often homes, of ancient people as they provided a steady, reliable food source.

Guide the Learning continued

SCIENCE IDEAS

Help students focus on supporting details for this **main idea: An organism receives half of its genes from its mother and half from its father.**

- ▶ How does a new organism get a full set of chromosomes? A reproductive cell from the mother with half of the organism's chromosomes combines with a reproductive cell from the father with the other half of the organism's chromosomes.
- ▶ How does the fact that an organism receives genetic information from two parents affect variation of traits? Because an organism receives $\frac{1}{2}$ of its genetic information from each parent, two quite different organisms genetically, there will be a great variety of possible combinations.

USE PROCESS SKILLS

Compare and Infer Have students compare the traits of the puppies with those of their parents.

- ▶ Choose one puppy to examine. Which of its traits do you think were inherited from its mother? Possible answer: coloring
- ▶ Which of the puppy's traits do you think were inherited from its father? Possible answers: size and shape of ears, length of fur
- ▶ Does the puppy have any traits that don't appear to come from either parent? How do you think this is possible? Yes. Students may not know how the puppy could end up with these strange traits. Tell them that they will learn

Variation in Organisms

As a result of meiosis, sperm and egg cells each have half the number of chromosomes of an organism's body cells. The cell that is formed when female and male reproductive cells join has a complete set of chromosomes. Half are from the mother, and half are from the father. As a result, a new organism receives half its genes from its mother and half from its father.

Since the new organism has genes from both parents, it will have a new combination of traits, or characteristics—some from its mother and some from its father. The variation, or difference, in traits among members of plant or animal species is mostly due to the combining of cells that have gone through meiosis.

When you observe members of your own species, you can see a large variation in traits. Because people have 46 chromosomes and about 50,000 genes, there is an enormous number of possible combinations. The color of

▲ Offspring often resemble the parents. But they can also be quite different.

◀ Which of the puppies' traits were inherited from the mother dog (black and white), and which were inherited from the father dog (mostly white)?

A20

Reaching All Learners

Visually Impaired Allow time for visually impaired students to more closely examine the cell images in this lesson by using a hand lens. Help them as needed to identify structures that are not easily recognized, such as fuzzy genetic material.

INVESTIGATION CHALLENGE

Hands-On Activity: More About Variation

Use this idea for a hands-on activity to extend students' understanding of key lesson concepts.

Give students two jars full of different-colored beads. Tell them that each bead color represents a different gene. Challenge each student to pick and record ten pairs of beads (for example, red/blue or green/red). After each student has made the "genetic code" for his or her "organism" (ten pairs of colors in the order selected), have them compare and contrast their organisms with those of their classmates. Chances are there will be no two "genetically" identical organisms in the class.

For student self-assessment, see Assessment Guide p. AGxvii.

people's hair and eyes, their fingerprint pattern, their blood type—all of these traits result from the combination of genes that a person has inherited. In the next lesson, you will find out that children can even inherit traits that are not apparent in either of their parents.

✓ **How does meiosis produce variation?**
by randomly distributing the two members of each pair of chromosomes into new cells

Summary

When an organism grows, its cells divide over and over again by mitosis. During mitosis two new cells are formed, each identical to the original cell. Reproductive cells that each have half as many chromosomes as the original cell are formed by meiosis. When two reproductive cells join, the resulting cells have a complete set of chromosomes—half from the mother and half from the father.

Review

1. Why are chromosomes copied before mitosis?
2. How does a new cell formed by mitosis compare with the original cell?
3. Why is there variation in new organisms formed from the joining of two reproductive cells?
4. **Critical Thinking** In which cells would you expect mitosis to occur more often—the cells of an adult person's skin or the cells of an adult person's teeth? Explain.
5. **Test Prep** Suppose an organism has 16 chromosomes in each of its body cells. How many chromosomes are there in one of its reproductive cells formed following meiosis?
   A 4
   B 8
   C 16
   D 32

LINKS

MATH LINK

Use Numeric Patterns Some cells divide once every 20 minutes. At this rate, how many cells would there be 1 hour, 2 hours, and 3 hours after a single cell began mitosis?

WRITING LINK

Informative Writing—Description Suppose that you could shrink to the size of a cell and observe mitosis. Write a story for a classmate to describe what you would observe during the process, and what you might see as the result of mitosis.

HEALTH LINK

Genes and Disease Some diseases are caused by faulty genes. Sickle cell anemia, hemophilia, and cystic fibrosis, for example, are each caused by a single gene. Find out how these diseases are inherited and how they are treated.

LANGUAGE ARTS LINK

Word Origins Both *mitosis* and *meiosis* are words that came from the Greek language. Find out what the Greek words mean, and write a paragraph to explain how they relate to these two processes.

TECHNOLOGY LINK

To learn more about how genes and heredity lead to variation, watch *Bioengineered Plants* on the **Harcourt Science Newsroom Video.**

CNN Turner Learning

A21

3 Wrap Up and Assess

SUMMARY

Graphic Organizer for Chapter Concepts After reading the summary, have students complete the Lesson 2 section of the graphic organizer (shown on page A14).

REVIEW

1. so that each of the resulting cells will have a complete set of chromosomes
2. The cells are genetically identical.
3. There are many different possible combinations of parental genes because each parent's chromosomes were randomly separated during meiosis.
4. skin cells, as they die and are replaced almost constantly
5. **B,** 8

TEST PREP Test-Taking Tips

Item type multiple choice with question stem

To help students answer Question 5, share the following strategy with them. **Read the question carefully to find key words, such as *reproductive cells* and *meiosis,* that are clues to the correct answer. To help narrow the choices, try to recall how meiosis differs from mitosis. Eliminate any answers that are obviously wrong. Then select the best answer from the ones remaining.**

LESSON CONCEPT REVIEW/ASSESSMENT

Students can use **Workbook p. WB11** to review the lesson concepts.

LINKS TIPS

Links Objective Extend understanding of cell reproduction.

Math Link Students might want to make a chart. You may wish to introduce the use of exponents in calculating compounding growth rates. In the number 2^9, the *2* represents the multiplication factor, or base (the cells are doubling). The *9* represents the number of times the base is used as a factor. For example, $2^3 = 8$, $2^6 = 64$, $2^9 = 512$.

Writing Link Students may handle this assignment as a piece of fiction or treat it journalistically as nonfiction. Whatever approach they choose, their descriptions should be accurate. Writing models are provided in the **Teaching Resources** book.

LESSON

3 How Are Traits Inherited?

Objectives

- ▶ **Recognize that a plant's or animal's traits are inherited from its parents.**
- ▶ **Identify genes as the inherited factors that determine traits.**
- ▶ **Conclude that traits of dominant genes appear in the offspring regardless of whether there is a recessive gene or not, and those of recessive genes appear only when both genes for a trait are recessive.**

Motivate

More Fast Facts

Spark interest with these facts and questions.

- Men are 10 times more likely to be color-blind than women.

  Color-blindness is an inherited trait. What do you think that means? It is passed down through generations.

- About 500 million metric tons of corn are produced each year around the world.

  How many different varieties of corn can you name? Answers may include yellow, bread and butter, silver queen, blue, red, Indian, field, or feed corn.

Access Prior Knowledge

- ▶ What is a trait? Can you name some of your own physical traits?

When Minutes Count . . .

Whole-Class Investigate Option Divide the kernels from a single cob among the class and have students count the various colored kernels. Keep in mind that a smaller sample may not yield the exact 3:1 color ratio you are looking for.

Using the Activity Video The video can be used to preview and model the investigate process with students. After viewing the video, students can conduct the investigation in small groups.

LESSON 3

How Are Traits Inherited?

In this lesson, you can . . .

INVESTIGATE how the color of corn kernels is inherited.

LEARN ABOUT how traits are passed from parents to offspring.

LINK to math, writing, social studies, and technology.

◀ Offspring resemble their parents in many ways. This is because of the way meiosis allows for both parents to contribute to the genes of the offspring.

A22

INVESTIGATE

Seed Color in Corn

Activity Purpose Have you ever seen ears of corn hanging on doors in the fall? Why does the corn have kernels that are different colors? In this investigation you will **observe** an ear of corn produced when a corn plant with white kernels was crossed with a corn plant with purple kernels.

Materials

- purple-and-white corn
- paper and pencil

Activity Procedure

1. Take one ear of purple-and-white corn. **Observe** the corn, and **estimate** the number of purple kernels and the number of white kernels.
2. Copy the chart form on the next page. **Count** the purple kernels and white kernels. **Record** the two numbers. (Picture A)
3. **Find the ratio** of purple kernels to white kernels. How many purple kernels are there for every white kernel?
4. Share your results with the class.
5. **Find the class totals** for purple kernels and white kernels.

BACKGROUND

Gregor Johann Mendel (1822–1884) An Austrian monk, Mendel carried out experiments in variation and heredity in his monastery's experimental gardens. Most of his work was done between 1856 and 1863, during which time he grew thousands of pea plants and examined them for characteristics. Statistical analysis of Mendel's data supported his inferences about the mechanics of inheritance. These ideas later became what are known as Mendel's Laws: the law of segregation (the two forms of each pair of genes segregate randomly during meiosis); and the law of independent assortment (the inheritance of one trait is independent of the inheritance of any other trait).

Webliography GO ONLINE

Keyword heredity

www.harcourtschool.com

For student self-assessment, see Assessment Guide p. AGxvi.

Picture A

Number of Kernels of Corn	
Purple	White
|||	

6 Now **find the class ratio** of purple kernels to white kernels.

Draw Conclusions

1. What **ratio** of colors did you find for your ear of corn?
2. How did your results compare with the results of the whole class?
3. **Scientists at Work** Scientists make **inferences** from their **observations**. They often **use numbers** to support their inferences. The larger the number of observations they study, the more likely their inferences are to be correct. What might you infer about the ratio of purple kernels to white kernels? Should it change for each ear of corn? How could you **use numbers** from your class and other classes to support your **inference**?

Investigate Further Get an ear of yellow-and-white corn. **Find the ratio** of yellow to white kernels. Is it the same as or different from the ratio of purple to white kernels?

Process Skill Tip

Scientists **use numbers** when they estimate, count, and find ratios. These numbers help them support or change **inferences** they have made.

A23

INVESTIGATE

Time 30–40 min **Preparation Tips** See p. A2c.

Grouping individuals or groups

Process Skill Tip use numbers, infer

Other Process Skills observe, record data

Expected Results Students should find about three purple kernels for every white kernel. Results may vary slightly. A greater quantity of data will yield more accurate results.

Teaching Tips

- If students have trouble keeping track of which kernels they have already counted, suggest that they use their pencil tip to poke a small hole in each kernel they have counted. If the kernels are too hard to penetrate, they can mark them with a felt-tip pen.
- Another method is for students to count 2 or 3 rows, find an average, and estimate totals. They could then compare the estimate with an actual count.
- Note that this ratio is a part-to-part ratio, not a part-to-whole ratio. Thus the fractional part of white corn to the total is $\frac{1}{4}$, or 25%.

Draw Conclusions

1. Possible answer: approximately 3 purple:1 white
2. The class results should have been closer to a 3:1 ratio than the students' individual counts.
3. Students should infer that the ratio is about 3:1. No. Pool all the numbers and come up with an overall ratio. Point out that scientists sometimes pool their findings, too, so that the scientists can have a greater total amount of data on which to draw their inferences.

Investigate Further Ratios should be similar to what students encountered in the investigation.

Process Skill Tip

Use Numbers and Infer Expose the first section of the Process Skill Tip Transparency. Explain that these two students were picked randomly from Mr. Howard's classroom. Ask students to infer the ratio of boys to girls in the classroom. Then expose the second section and repeat the question. Note that this group was also picked randomly. Continue in this fashion through the fifth section, which reveals that the actual ratio of boys to girls is 1:1. Ask students to describe the effect of a larger sample size on the accuracy of their inferences.

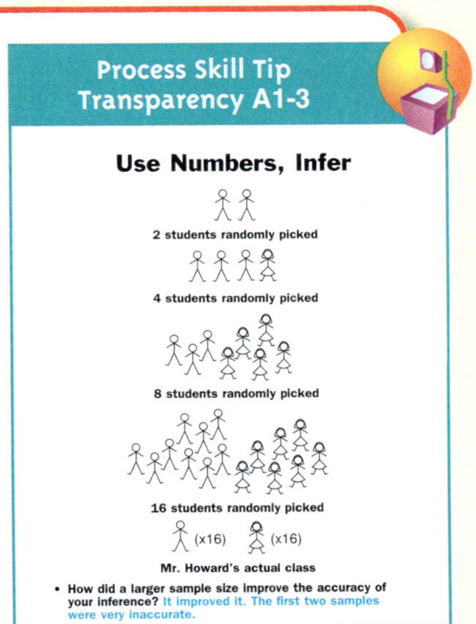

INVESTIGATE LOG

Students can use **Workbook pp. WB12–13** to record the results of the investigation.

PROCESS SKILLS PRACTICE

To practice and apply process skills, see **Workbook p. WB14.**

LEARN ABOUT

1 Before Reading

SET A PURPOSE

FIND OUT Remind students that these statements will guide them in what to look for as they read.

VOCABULARY If students are doing the activity described in the Vocabulary Preview on page A3, remind them to add the definitions for these words to the chart.

Graphic Organizer for Chapter Concepts Display the transparency (shown below) or have students use Workbook page WB1. Point out that by the end of the lesson, students should be able to fill in the answers concerning inheritance.

2 Guide the Learning

SCIENCE IDEAS

Help students focus on supporting details for this **main idea: Traits are passed from parent to offspring.**

- What were some of the traits of Mendel's pea plants that were passed from parent to offspring? plant height, seed color
- What did Mendel notice about the offspring of his pea plants? The plants produced offspring that looked like themselves.
- Do human offspring look exactly like their parents? Explain your ideas. No. They may share some characteristics, but they are not identical.
- Do girls look only like their mothers and boys only like their fathers? Explain your ideas. No. Girls may show their fathers' traits, just as boys may exhibit some of their mothers' traits.

DEVELOP SCIENCE VOCABULARY

sexual reproduction, p. A24—Students may be confused by the idea that an organism can have both male and female parts. This is because most organisms they are familiar with, including humans, are either male or female. Assure students that this phenomenon is normal in the plant world. If you can obtain a pea plant or another self-pollinating plant, open one of its flowers and identify the pistil and the stamen(s) for them.

LEARN ABOUT

FIND OUT

- how living things inherit traits from both parents
- what controls the traits that organisms show
- how organisms can carry traits that they do not show

VOCABULARY

sexual reproduction
dominant
recessive
Punnett square

How Traits Are Inherited

Mendel's Pea Plants

When a baby is born, family friends may point out how the baby looks like its parents. They might say, "She has her father's eyes and her mother's nose" or "He looks just like his grandfather." Offspring often look like their parents—and sometimes their grandparents. Tall parents often have children who will become tall, for example, and plants with red flowers usually produce other red-flowered plants.

An Austrian monk named Gregor Mendel was the first person to show how traits are passed from parents to offspring. In the mid-1800s, Mendel carefully observed the pea plants he grew in his garden. Pea plants have many different traits. For example, some are tall and others are short; some have green seeds and others have yellow seeds. Mendel found that pea plants usually produced offspring like themselves. Tall plants produced tall plants, and short plants produced short plants. But what would happen if tall and short plants were crossbred, or bred with each other? Mendel chose seven traits that were easy to tell apart. After his experiments, he was able to describe how inheritance works.

In many species of plants, pollen from one plant is carried to another plant. Pollen contains male reproductive cells. When the reproductive cells of the plants join, they form a seed that will produce a new plant. Reproduction by the joining of a male and a female reproductive cell is called **sexual reproduction**. The pea plant reproduces by sexual reproduction, but its male and female parts are completely enclosed by the petals of its flowers. The male and female reproductive cells that join to form a seed come from the same plant.

◄ Mendel experimented with pea plants for eight years. He studied the results of crossbreeding plants that had different traits. Mendel published the results of his experiments in 1865, but they were ignored for a long time.

Graphic Organizer for Chapter Concepts

Transparency A1 • Workbook, p. WB1

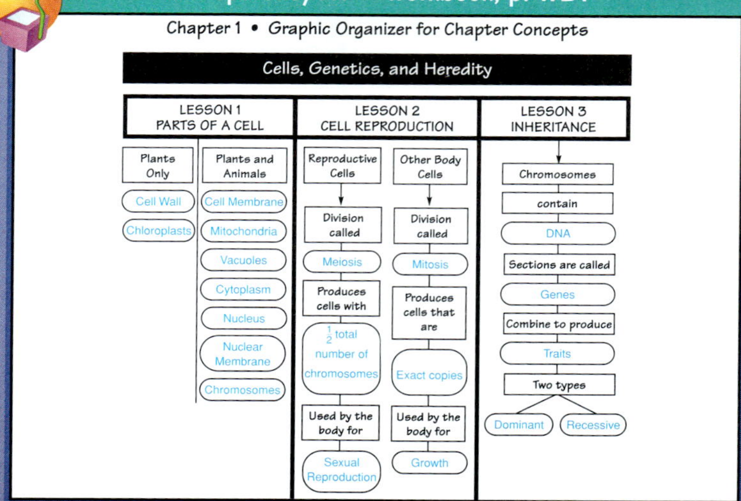

Mendel took pollen from a plant that had purple flowers and pollinated a plant that had white flowers. All the offspring had purple flowers. When he bred tall and short plants with each other, all the offspring were tall. The offspring of plants with opposite traits showed only one of the traits. It seemed that the other trait had disappeared.

Mendel allowed the offspring of his experimental plants to self-pollinate. That is, he allowed the male and female reproductive cells of the same plant to join and produce a seed. Mendel planted the seeds they produced. The traits that had disappeared now returned. Some new plants had purple flowers and others had white flowers. Some new plants were tall and others were short.

Mendel kept careful counts of his results. When he counted the new generation from self-pollinated plants, Mendel found that there were about three plants with purple flowers for every plant with white flowers, a ratio of 3 to 1. And there were about three tall plants for every short plant. In each case, the ratio was about the same—3 to 1.

Mendel repeated his experiment with other traits. The results were always the same—the new plants showed only one trait. But in the second generation, the missing trait returned in about one of every four offspring. Mendel realized that the missing trait had been present all along, even though it did not show.

✔ **Why did Mendel choose pairs of traits that were easy to tell apart?** He needed to identify plants that had the traits and to make accurate counts of those plants.

Mendel chose to experiment with pea plants for several reasons. Pea plants were easy to get. They grow rapidly and show traits that are easy to tell apart. It is easy to cross different pea plants. But left to themselves, they self-pollinate with little chance of any accidental pollination between plants. ▶

▲ When pea plants that have purple flowers are crossed with pea plants that have white flowers, all the offspring have purple flowers.

◀ The seeds of the pea plant also show traits. They can be green or yellow, round or wrinkled.

▲ The pods of pea plants are either yellow or green. When Mendel crossed plants that had different colored pods, all the offspring had green pods.

A25

SCIENCE IDEAS

Help students focus on supporting details for this **main idea: When plants with two different traits were crossbred, only one of the traits showed up in the offspring.**

- Which traits did Mendel cross? plant height (tall and short) and flower color (purple and white)
- What were the results of his crossing? All the offspring of the first cross were tall. All the offspring of the other cross had purple flowers.
- What happened to the other traits? They seemed to have disappeared.
- What happened in the second generation of pea plants? The missing trait showed up in about one-fourth of the plants.
- What was the part-to-part ratio of tall plants to short plants (or purple flowers to white flowers) in the second generation? about 3:1

VISUAL LEARNING

Have students look at the pea plant pictures.

- How many different pea plant traits are shown in the pictures? four: flower color, seed color, seed texture, and pod color
- What do you notice about the color of the flowers on the tall pea plant? They are all the same color. A pea plant has either purple flowers or white flowers.
- How many pods were produced by the one pea plant? Why was this an advantage to Mendel? Many. Multiple flowers, seeds, and pods yielded more data. The more data he had, the more certain he was of the accuracy of his inferences.

BACKGROUND

Webliography

Keyword heredity

www.harcourtschool.com

GO ONLINE

Cross-Pollination Mendel's pea plants were self-pollinating; that is, pollen from the stamen of a flower fertilized the pistil of the same flower to produce a seed. In order to cross a particular trait, like seed color, Mendel had to transfer pollen from one plant to the pistil of a different plant. In this way, the offspring plants inherited one seed-color gene from a green-seed plant, and another gene from a yellow-seed plant.

Although several plants reproduce by means of self-pollination, others use cross-pollination, which enhances the genetic variety of the offspring. Genetically diverse offspring are advantageous because they are less likely to suffer massive die-offs. Seeds produced by cross-pollination are also generally of a higher quality than those produced by self-pollination.

Ongoing Assessment

Use the ✔ questions in the pupil edition.

Guide the Learning continued

SCIENCE IDEAS

Help students focus on supporting details for this **main idea: Each parent carries two factors for a trait: one trait is dominant and one is recessive.**

- ▶ When a parent produces a reproductive cell, how many factors for a trait does it donate to the cell? one
- ▶ Where does the offspring's other factor come from? the other parent
- ▶ **Critical Thinking** If an organism received a factor for rough skin from one parent and a factor for smooth skin from the other, and the organism itself had smooth skin, which factor would you say is dominant? Which is recessive? dominant factor: smooth; recessive factor: rough

ANALYZE AND INTERPRET DATA

Guide students to **identify cause and effect** to explain why each generation appears as it does.

- ▶ **Can you tell what two factors the parent plants have just by looking at them? Explain.** The short plant has two factors for shortness. The tall plant could have either two tallness factors or one of each.
- ▶ **Explain why all the first-generation plants are tall, even though one parent was short.** The tallness factor is dominant.
- ▶ **Explain why one of the second-generation plants is short?** The first generation all had one recessive factor. In one out of four offspring, two recessive factors produced a short plant.

Dominant and Recessive

When Mendel did his experiments, little was known about how traits are inherited. The most common belief of the time was that a cross of two different traits would result in a blend. If a tall plant was crossed with a short plant, for example, people expected the result to be a plant of medium height. Mendel proved that idea to be wrong.

Mendel suggested that each plant carries two factors for each trait that it shows. When a parent plant makes reproductive cells, only one factor is passed on to the offspring. Because the offspring receives a factor from each parent, it carries two factors for the trait. In a pea plant the factors for height are a factor for tallness and a factor for shortness.

Each of the factors a plant carries can be the same, or they can differ. If a pea plant has two factors for tallness, it will be tall. But if the factors are different, one factor may be dominant. The **dominant** (DAHM•uh•nuhnt) factor determines the trait that is shown. A trait that is caused by the presence of a dominant factor is a

Mendel's Discovery

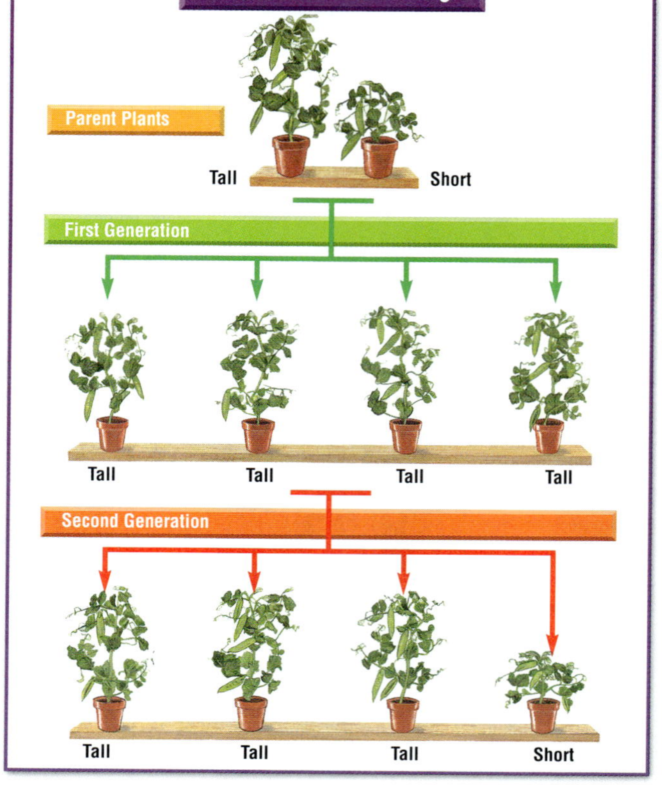

◀ When Mendel crossed a tall pea plant with a short pea plant, all the offspring were tall. The factor for tallness in the plant is dominant. In the second generation, one-fourth of the offspring showed the recessive factor for shortness.

A26

Reading Mini-Lesson

Use Graphic Sources for Information Point out that tables, graphs, and diagrams can convey certain types of information more efficiently than words. Display the transparency. Ask students what information they can get from the upper left-hand box and the lower right-hand box of the Punnett square on page A29.

Reading Mini-Lesson Transparency A1-3

Use Graphic Sources for Information

Mixed Green and Pure Yellow Pea Pods

	G	g
g	Gg	gg
g	Gg	gg

Upper left-hand square The upper left-hand box shows that one of the offspring of this cross would be a hybrid green pea pod.

Lower right-hand square The lower right-hand box shows that one of the offspring of this cross would be a pure yellow pea pod.

READING SKILLS PRACTICE

To practice and apply this reading skill, see **Workbook p. WB15.**

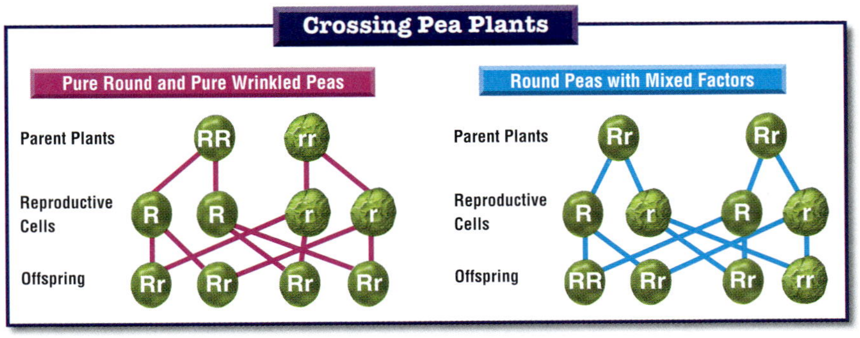

▲ This diagram shows how factors are inherited when a pea plant that has round seeds is crossed with a pea plant that has wrinkled seeds. One plant has two factors for round seeds (R), and the other plant has two factors for wrinkled seeds (r). The factors separate when reproductive cells are formed. The bottom row shows the possible combinations of factors in the offspring.

▲ This diagram shows the results of crossing two of the first generation plants from the diagram shown at the left. How many of the offspring have the factor for wrinkled seeds? How many of the offspring have wrinkled seeds?

dominant trait. One dominant trait in pea plants is tallness.

When the factors of a pair are different and one is dominant, the other is **recessive** (rih•SES•iv). The trait that results from the combination of two recessive factors is a recessive trait. A plant will show a recessive trait only if both of its factors for that trait are recessive. For example, a short pea plant must have two factors for shortness. A pea plant with one factor for tallness and one factor for shortness will be tall, because the factor for tallness is dominant.

Using the idea of dominant and recessive factors, Mendel could explain why a recessive trait showed in one-fourth of the offspring of the second generation. Mendel began his experiments with what he called truebreeding seeds. The seeds for tall plants would always produce tall plants, and the seeds for short plants would always produce short plants. Truebreeding tall plants have two factors for tallness. Truebreeding short plants have two factors for shortness.

The diagram above on the left shows what happens when a plant that has truebreeding round seeds is crossbred with a plant that has truebreeding wrinkled seeds. A capital letter stands for a dominant factor, in this case an *R* for *round seeds*. The matching lowercase letter stands for the recessive factor, in this case, for wrinkled seeds. All of the possible combinations of factors in the offspring are the same. Each plant has a factor for round seeds and a factor for wrinkled seeds. All the plants' seeds are round because the factor for round seeds is dominant.

The diagram on the right shows what happens when two of the offspring plants are crossed. This time, there are three possible combinations, shown in the third row. The first plant has two factors for round seeds, so it has round seeds. The next two plants have factors for both round and wrinkled seeds. They have round seeds because round seeds are dominant. The final plant has two factors for wrinkled seeds, so it has wrinkled seeds. One-fourth of the plants show the recessive factor. Mendel's explanation supports the results he obtained in his experiments.

✔ **If the offspring of two short pea plants are crossed, will they produce any tall pea plants? Explain.** No; they have only factors for shortness.

A27

USE PROCESS SKILLS

Compare Have students compare the two diagrams at the top of this page.

▶ **How do the parent cells in each diagram differ?** The cells on the left are true-breeding; that is, they contain two factors for the same trait. The cells on the right are not true-breeding—they have one factor for round seeds and one for wrinkled seeds. These plants are mixed.

▶ **What is different about the reproductive cells produced by the true-breeding plants and those produced by the mixed plants?** The factors in cells produced by each true-breeding plant are the same. Each mixed plant produces cells with two different factors, one R and one r.

▶ **What is different about the offspring produced in each diagram?** The offspring of the true-breeding plants are all the same. They all have round seeds and carry one dominant and one recessive factor. The offspring of the mixed plants vary. Three have round seeds; one has wrinkled; and they carry different combinations of factors.

DEVELOP SCIENCE VOCABULARY

dominant, p. A26—Ask students to describe other uses they have heard of for the words *dominant* or *dominate*. Note that the verb is often used in sports to express the superiority of one team.

recessive, p. A27—Ask students to think of other uses they have heard of for the words *recessive* or *recede*. For example, the verb form is used to describe how water moves away from shore.

BACKGROUND

Exceptions to Dominance

Although in many cases of inheritance one form of a gene is expressed and the other is not, intermediate characteristics are sometimes seen. In the four o'clock plant, for example, the factor R (red flowers) is not entirely dominant over the factor r (white flowers); as a result, the Rr pair is expressed as a pink flower. The two factors blend to produce a color somewhere between red and white.

Webliography GO ONLINE

Keyword heredity

www.harcourtschool.com

Guide the Learning continued

SCIENCE IDEAS

Help students focus on supporting details for this **main idea: Traits are passed down through the generations of a family.**

- **How are the traits passed down?** Parents pass the traits on to their children. These children later become parents themselves and pass their traits on to their offspring, who eventually pass them to the next generation.
- **Why is a dominant trait more likely than a recessive trait to be seen throughout generations of a family?** Because even if an offspring has only one factor for a dominant trait, he or she will show that trait. To show a recessive trait, an offspring must have two factors for it.
- **What physical traits do animals share with their parents?** Answers may include eye color, hair color, length of hair, and size.

USE PROCESS SKILLS

Observe and Investigate Have students observe the various facial traits of the person on this page. Then ask them to create a chart in which they can record the presence or absence of these traits on their own faces.

- **Are your ear lobes attached or detached? Does your chin have a cleft? Do you have dimples? What color are your eyes? Do you have a widow's peak?** Answers will vary. Have students record their findings in the chart.
- **Can you roll your tongue?** Answers will vary. Again, have students record their findings.

Inheritance of Traits

Mendel's explanation of inheritance doesn't apply only to pea plants. Most plants and animals have two parents and inherit some traits according to the principles that Mendel discovered.

You can see on the faces of your friends some traits that are passed from parents to offspring. Having dimples, for example, is an inherited trait. When such traits are dominant, they are likely to appear in many generations of a family.

You can often see dominant traits in the parents, children, and grandchildren of the same family.

The illustration shows some of the inherited traits that can be found on a person's face. There are other traits that show on a person's face as well. Eye color is not the result of a single dominant trait. But brown eyes are generally dominant and blue eyes are generally recessive.

✓ **Name three dominant traits and three recessive traits in humans.**

This face shows some of the traits that people inherit from their parents. Detached earlobes, dimples, and cleft chins are dominant traits. A widow's peak and attached earlobes are recessive traits. ▼

Dominant: dimples, cleft chin, detached earlobes, brown eyes; recessive: widow's peak, attached earlobes, blue eyes

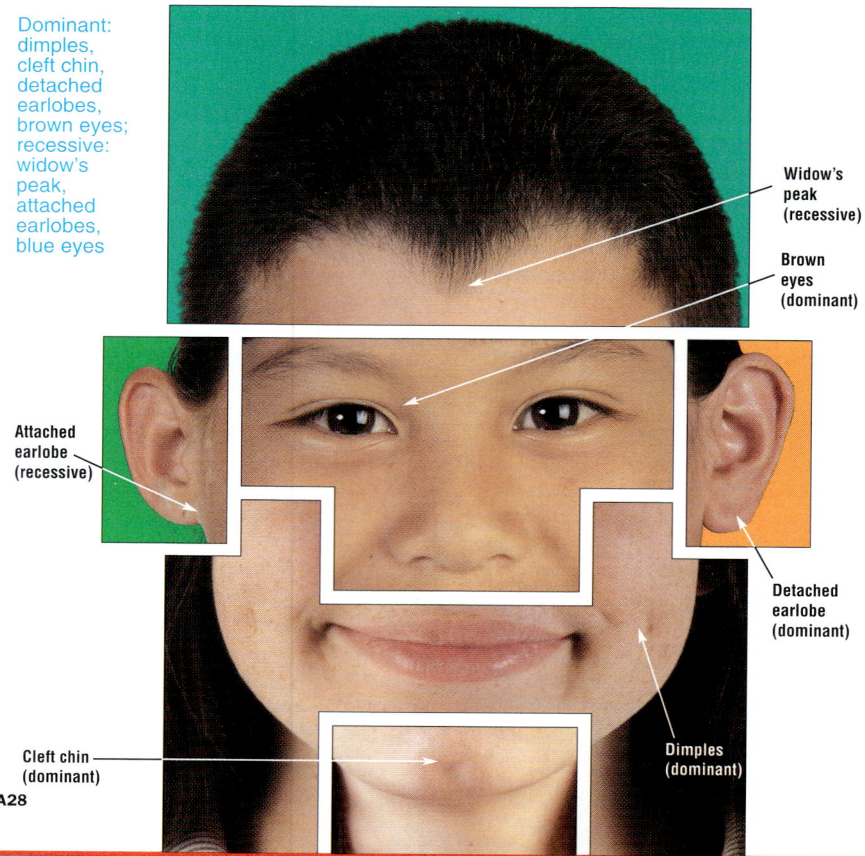

A28

Reaching All Learners

Visually Impaired Students who have trouble seeing facial traits can use their fingers to feel them. If this activity would be appropriate for your students, you may wish to provide close supervision. Provide time for students to check for detached earlobes, cleft chins, and dimples on themselves or a friend.

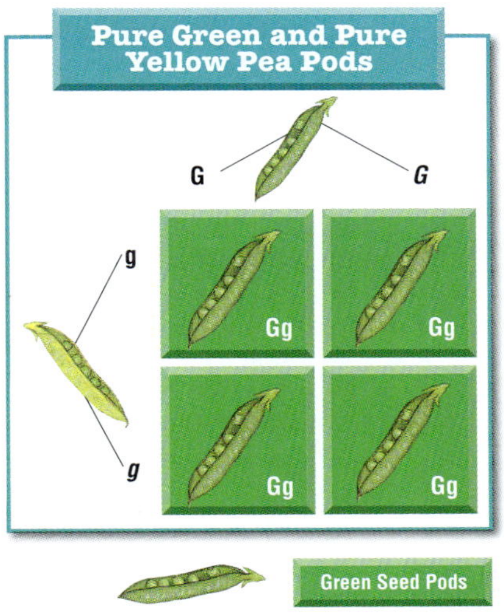

▲ When Mendel crossed plants that had green pods with plants that had yellow pods, all of the offspring had green pods.

▲ The results are very different if one of the green-pod plants that were offspring from the Punnett square at the left is crossed with a yellow-pod plant.

Punnett Squares

An easy way to find all the possible combinations of factors in the offspring of two parents is to use a checkerboard-type diagram called a **Punnett square**. The Punnett squares on this page show combinations of factors for the color of pea pods. Pea pods can be either green or yellow. The capital letter *G* stands for the dominant color, green. The small letter *g* stands for the recessive color, yellow.

The factors present in the parent plants are shown at the top and on the left side of a Punnett square. The factors in each parent are separated so that each factor has a single column or row of the square. The possible combination of factors in each box of the square is found simply by placing the column factor next to the row factor. Both Punnett squares on this page show the possible results of crossing a plant that has green pods with a plant that has yellow pods. Why are the results different?

The Punnett square at the top left of the page shows a cross between a plant with two factors for green pods and a plant with two factors for yellow pods. All the offspring have one factor for green pods and one factor for yellow pods. Yet they all have green pods because green is the dominant color for pods.

Suppose one of these offspring plants is crossed with a plant that has yellow pods. The plant with green pods has one green and one yellow factor. The plant with yellow pods must have two yellow factors, since yellow is recessive. The Punnett square at the top right shows the possible results.

✔ **If you have a pea plant that has green pods, how could you find out if it has a hidden factor for yellow pods?**

You could cross it with a plant that has yellow pods; if any of the offspring have yellow pods, the first plant would have a hidden factor for yellow pods.

A29

ANALYZE AND INTERPRET DATA

Have students **compare and contrast** the two Punnett square diagrams pictured at the top of page A29. Help them see that the factors of one parent are shown at the top of the square and that the factors of the other are shown to the left of the square.

- ▶ **What is different about the parents in each Punnett square?** In the square on the left, each parent is true-breeding. In the square on the right, a mixed parent is crossed with a true-breeding parent.
- ▶ **How do you determine which factors each offspring has?** In each Punnett square, you combine the factor from one parent (at the top of the column) with the factor from the other parent (to the left of the row).
- ▶ **If factors came in groups of three rather than in pairs, how many boxes would the Punnett squares contain?** nine

DEVELOP SCIENCE VOCABULARY

Punnett square, p. A29—The Punnett square was devised by the English biologist Reginald Punnett as a shorthand method for determining possible factor combinations in offspring and the proportions of the combinations. Help students remember the correct spelling of *Punnett*. Just as 2 sides of the square display a set of double letters, so the word contains 2 pairs of double letters.

Math Link

Compute Probability Albinism (bb) is a congenital condition characterized by an inability to form pigment. Animals with this condition typically have pale skin, whitish hair, and pink eyes. Without pigment to protect them from ultraviolet rays, such animals are particularly sensitive to the sun. Albinism is transmitted from parent to offspring as a recessively inherited trait. Have students **predict probable future actions and outcomes** by finding the probability of two Bb parents producing an albino offspring. 1:4

ESL ACTIVITY

Help students understand why the **Punnett square** is an appropriate graphic for depicting possible factor combinations. Remind them that a square is a figure with 4 equal sides and 4 right angles. Point out that each pair of parents brings 4 factors (two each) to the reproductive mix and that there are 4 possible combinations of genetic factors.

Guide the Learning continued

SCIENCE IDEAS

Help students focus on supporting details for this **main idea:** **Factors, or genes, are located on chromosomes.**

- ▶ According to the chromosome theory, how are genes passed from parent to offspring? Genes are located on each parent's chromosomes, half of which are passed on to offspring. Like chromosomes, genes come in pairs. Therefore, half of a parent's genes go to its offspring.
- ▶ How is the process of meiosis similar to what Mendel described in his factors? Both indicate that a parent has paired genes that split, and that only one gene of each pair is passed on to offspring by that parent.
- ▶ Do you think that there is a gene for body size? Body size is probably determined by a number of different genes.

VISUAL LEARNING

Have students observe the drawing on this page.

- ▶ What is the boy able to do that the girl cannot? roll his tongue
- ▶ What do you think determines the boy's ability to perform this trick? his genes
- ▶ On the basis of the diagram, how many genes do you think are carried on each chromosome? Actual numbers vary. It is certain that more than one gene is carried on each chromosome.
- ▶ What do the banded drawings of the genes on this page tell you? that genes are found on larger structures—chromosomes

The Role of Genes

Mendel discovered the importance of factors in inheritance. At the time, however, nobody knew exactly what Mendel's factors were. In fact, Mendel's work was ignored until about 1900.

After the discovery of the process of mitosis, scientists hypothesized that chromosomes were responsible for heredity. In the early 1900s scientists observed that the behavior of chromosomes during meiosis seemed to be similar to the action of Mendel's factors. This led to the development of the chromosome theory, which states that factors, which we now know are genes, are located on the chromosomes. Because each organism has pairs of chromosomes, with one chromosome coming from each parent, genes also occur in pairs. As you learned, the number of chromosomes in a reproductive cell is halved during meiosis. In this way, the genes in a pair are separated, and each reproductive cell that is formed receives one of the genes.

Mendel was able to discover how heredity works in pea plants because the traits he studied were easy to tell apart. Not all traits are so distinct, however, and heredity does not always follow Mendel's laws. Many traits are controlled by more than one gene pair. Skin color and height in humans, for example, result from several different genes.

✔ **What are Mendel's factors and where are they located?** They are genes, located on the chromosomes.

A30

INVESTIGATION CHALLENGE

Hands-On Activity: Modeling Heredity

Use this idea for a hands-on activity to extend students' understanding of key lesson concepts.

Provide students with 2 paper bags and 2 marbles in each. Red marbles represent a recessive gene for short tails (t), and blue marbles represent a dominant gene for long tails (T). Place 1 blue and 1 red marble in a bag (long-tailed parent). Place 2 red marbles in the other bag (short-tailed parent). Have students make a Punnett square for this crossing (Tt and tt). Ask them to pick a marble from each bag and record the factors for offspring and the trait it would show. Have students replace the marbles, and repeat for a total of 10 trials. Challenge students to compare these actual results (varied) to the predicted results ($\frac{1}{2}$ short, $\frac{1}{2}$ long).

For student self-assessment, see Assessment Guide p. AGxvii.

◄ Having color is a dominant trait.

Summary

Gregor Mendel discovered that traits are inherited through factors from each parent. These factors were later found to be genes. Genes occur in pairs. When a reproductive cell is formed, it has a single gene from the pair, so offspring receive one gene from each parent. Traits of dominant genes are shown, and these hide recessive gene traits. Punnett squares can be used to see all possible ways genes from two parents can combine in the offspring.

Review

1. Why didn't Mendel have to pollinate the first generation pea plants in order to produce the second generation?
2. Round seeds are dominant, and wrinkled seeds are recessive. What genes does a pea plant with wrinkled seeds have?
3. How many genes out of a gene pair does a plant that is not self-pollinated get from each of its parents?
4. **Critical Thinking** Suppose that when two tall pea plants are crossed, all the offspring are tall. Could any of the offspring have a gene for shortness? Explain.
5. **Test Prep** A Punnett square is used to find —
   A gene combinations of offspring
   B the number of genes in offspring
   C the genes in true-breeding pea plants
   D the dominance of genes

LINKS

MATH LINK

Determine Probability When Mendel crossed pea plants that each had a tallness and a shortness gene, one-fourth of the offspring would have two genes for shortness. Write *T* (for the tallness gene) on one side of two different counters and *t* (for the shortness gene) on the other side. How could you model Mendel's results by tossing the counters? Why would your results become closer to Mendel's the more times you tossed the counters?

WRITING LINK

Informative Writing—Explanation Suppose Mendel did the investigation at the beginning of this lesson. Write a paragraph for your teacher telling how Mendel would explain the results.

SOCIAL STUDIES LINK

Traits in History Choose two figures in American history who were part of the same family—the family of President John Adams for example. Find portraits or photographs of family members in books. See if you can observe any traits they had in common.

TECHNOLOGY LINK

Find out more about genes and heredity by experimenting with *Traits—Now You See Them, Now You Don't* on the **Harcourt Science Explorations CD-ROM.**

A31

LINKS TIPS

Links Objective Extend understanding of genetics.

Math Link A tossed counter will indicate one gene (T) 50 percent of the time, and the other gene (t) the other 50 percent. Thus, it symbolizes the chance that a single gene has of ending up in an offspring. Students should find that about 50 percent of the time they toss both counters, they get Tt; 25 percent of the time they get TT; and 25 percent of the time they get tt. The more trials they run, the closer their results will be to this 2:1:1 ratio. Point out that only one-fourth of the tosses produce two shortness genes (tt). As the number of trials increases, the actual results approach the predicted results.

Writing Link Remind students that Mendel would not know some of the terms we use today ("cool"), so they should avoid using such words. They should conclude that Mendel would have described purple kernel color as a dominant trait. Writing models are provided in the **Teaching Resources** book.

3 Wrap Up and Assess

SUMMARY

Graphic Organizer for Chapter Concepts After reading the summary, have students complete the Lesson 3 section of the graphic organizer (shown on page A24).

REVIEW

1. They were self-pollinating.
2. It must have two genes for wrinkled seeds, because recessive traits are expressed only when both genes are recessive for that trait.
3. one
4. Yes. They may be carrying a shortness gene as the recessive gene in their pair.
5. **A,** gene combinations of offspring

TEST PREP Test-Taking Tips

Item type multiple choice with a sentence fragment

To help students answer Question 5, share the following strategy with them. **Read the question carefully. Try to recall some facts that will help narrow the choices. Draw what you remember a Punnett square to look like and analyze the information it contains. Then select the best answer to the question.**

LESSON CONCEPT REVIEW/ASSESSMENT

Students can use **Workbook p. WB16** to review the lesson concepts.

VOCABULARY REVIEW

Students can use **Workbook p. WB17** to review the chapter vocabulary.

Objectives

- Connect chapter concepts with the history of science.

1 Before Reading

SET A PURPOSE

Ask students to read the title of the article and predict what it will be about. Record predictions on the board, along with any questions students hope to have answered. After students have read the article, encourage them to research the answers to any questions that were not answered in the text.

2 Guide the Learning

SCIENCE IDEAS

Help students gather information from this nonfiction science article by asking these questions:

- **What are genes made of, and where are they located?** Genes are made of DNA molecules, and they are located on chromosomes.
- **What is meant by "gene mapping"?** Gene mapping involves locating the chromosome sites where specific genes—sequences of DNA—occur.

  Critical Thinking Lead students to infer a useful result of gene mapping.
- **Why might having a map of the human genome be useful?** If doctors knew where an abnormal gene was located, they might be able to alter it and make it normal.

SCIENCE THROUGH TIME

Understanding GENETICS

Mendel's Factors

The father of genetics was Gregor Mendel, an Austrian monk and scientist who extensively studied pea plants. Mendel saw that parent plants passed on what he called factors—now known as genes—to their offspring. Mendel published his findings about the laws of heredity in 1866, but it wasn't until the 1900s that scientists began to understand his work and recognized genetics as a science.

A New Century, A New Science

In the early 1900s, a number of scientists in different fields made genetic discoveries. Chromosomes were identified. Genes were found on chromosomes. Scientists determined that genes work by directing the production of enzymes. Enzymes are the proteins that influence chemical reactions in cells.

Linking It All Together

In the 1940s scientists identified a component of chromosomes, DNA, that carries genetic codes. In 1953 research by Rosalind Franklin, Maurice Wilkins, James Watson, and Francis Crick led to the discovery of the molecular structure of DNA—a double helix.

Mapping

Once the structure of DNA was known, scientists could begin mapping the precise locations of genetic codes. The full set of an organism's genes and traits is called its genome. The Human Genome Project, an effort to map the 3.2 billion DNA base pairs that make up human genes, was begun in 1988. A rough draft of the complete human genome was announced in June 2000. A final map is expected in 2003.

Discoveries in Genetics

1800 — 1900

1866 Mendel publishes his work on inheritance.

1943 The role of DNA in inheritance is first demonstrated.

1953 The double helix is discovered.

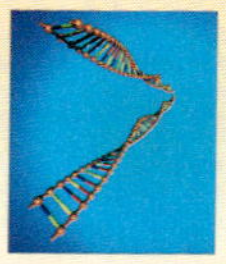

1970s The era of genetic engineering begins.

1980s The genes of *E. coli* bacteria and fruit flies are mapped.

1981 Chinese geneticists clone the first animal, a species of fish known as the golden carp.

A32

Reading Mini-Lesson

Use Prefixes and Suffixes to Determine Word Meanings List the words *genetics, gene, genome, genetic,* and *geneticist* on the board. Have students identify the common root and then discuss the meaning of each word. Ask how different suffixes affect the meaning of each word. Then write the words *generate, generation, regenerate, degenerate,* and *cogeneration* on the board. Have students find, if necessary, the meanings of these words and discuss them. Ask the class how the different prefixes and suffixes affect the meanings of these words.

Genetic Engineering

In the 1960s scientists discovered *plasmids*, small rings of free-floating DNA in cells. In the 1970s they discovered how to make unlimited copies of genes by inserting the genes into laboratory bacteria such as *E. coli*. This made possible the experimental recombining of selective traits in organisms. Since then, hundreds of genetically engineered bacteria and plants have been developed. The goal is to make useful products, such as vaccines and disease-resistant plants, and to find other ways to avoid or alter the course of diseases.

Controversial Frontiers

Genetic engineering led to the development of cloning—the process by which an offspring genetically identical to its parent is produced using the genes of only one parent. In February 1996, Scottish scientists successfully cloned a sheep. Since then mice, pigs, goats, cows, and other animals have been cloned. Genetic engineering and cloning are controversial. Some people worry that the technologies could be used for harmful rather than helpful purposes, or that they could have dangerous or unpredictable effects.

THINK ABOUT IT

1. Why was discovering the structure of DNA so important in the study of genetics?
2. Why are people concerned about genetic engineering and the use of cloning?

Dolly was the first mammal to be reproduced by cloning. ▶

1985 A gene that leads to human cancer is discovered.

1990 A four-year-old girl with a rare medical disorder becomes the first person to benefit from gene therapy.

1996 Dolly the cloned sheep is born and continues to develop normally.

2000 The first draft of the Human Genome Project is completed.

2000

CHAPTER CONCEPTS

This article focuses on how our understanding of heredity has changed through the years.

- ▶ What important observation did Mendel publish in 1865? Mendel observed that parents passed on "factors" to their offspring. This is why offspring look somewhat like their parents.
- ▶ What did these "factors" turn out to be? genes
- ▶ How are genes passed from parent to offspring? Through meiosis, sex cells containing half the parent's genes are formed. When two sex cells—one from the mother and one from the father—join, an offspring with a complete set of genes is formed. In cloning, a complete set of genes from one parent is used to produce the offspring.
- ▶ How has an understanding of bacterial DNA helped scientists make useful products? Scientists can make bacteria in the laboratory capable of doing different things by inserting different genes into them. New combinations of genes result in different traits. These bacteria can be engineered to produce vaccines and disease-resistant plants.

VISUAL LEARNING

Help students interpret the time line on this page.

- ▶ How long did it take scientists to determine the shape of the DNA molecule after they had discovered it? 10 years
- ▶ When was the first fish cloned? in 1981
- ▶ When was the first mammal cloned? in 1996, fifteen years later
- ▶ When was the first draft of the human genome completed? in 2000

3 Wrap Up and Assess

THINK ABOUT IT

1. Once scientists knew the structure, they could begin mapping the precise locations of genes.
2. Genetic engineering and cloning could be used for harmful rather than helpful purposes, or they could have dangerous results.

BACKGROUND

DNA and Human Genes

Mapping the human genome has been compared to putting together a 3.2-million-piece jigsaw puzzle. The number of individual human genes has been estimated at between 30,000 and 120,000, in combinations of between 2.8 and 3.5 billion base pairs, which form the double helix. The base pairs of DNA are referred to by the letters A, T, C, and G, standing for adenine, thymine, cytosine, and guanine, the chemical names of the "bases," or tag ends, of gene fragments. The "A" base pairs up with the "T" base, and the "C" base pairs up with the "G" base. These base pairs form the rungs of the spiral DNA ladder, and the pair sequences make up the genetic codes in the genes.

Webliography GO ONLINE

Keyword heredity

www.harcourtschool.com

PEOPLE IN SCIENCE

Objectives

- ► **Connect chapter concepts with the contributions of scientists.**

1 Before Reading

SET A PURPOSE

Geneticists such as Barbara McClintock study different aspects of heredity. Have students read to find out the important discoveries Dr. McClintock made about chromosomes.

2 Guide the Learning

THE WORK OF SCIENTISTS

Help students focus on the methods and processes scientists use and the impact of their work.

- ► How do you think Dr. McClintock's background in botany helped her in her genetic research? She understood the possible variations in plants and how they might occur.
- ► Can you think of any other famous scientists or inventors whose ideas were "ahead of their time"? Possible answers: Thomas Edison, Alexander Graham Bell

3 Wrap Up and Assess

THINK ABOUT IT

1. She found that chromosomes can "cross over," meaning that pieces of chromosome that are near each other can break off and recombine to make new genetic combinations.
2. Her ideas were too far ahead of their time. She was working just after Mendel's work had been rediscovered and had not been widely accepted. Her work built on Mendel's ideas.

PEOPLE IN SCIENCE

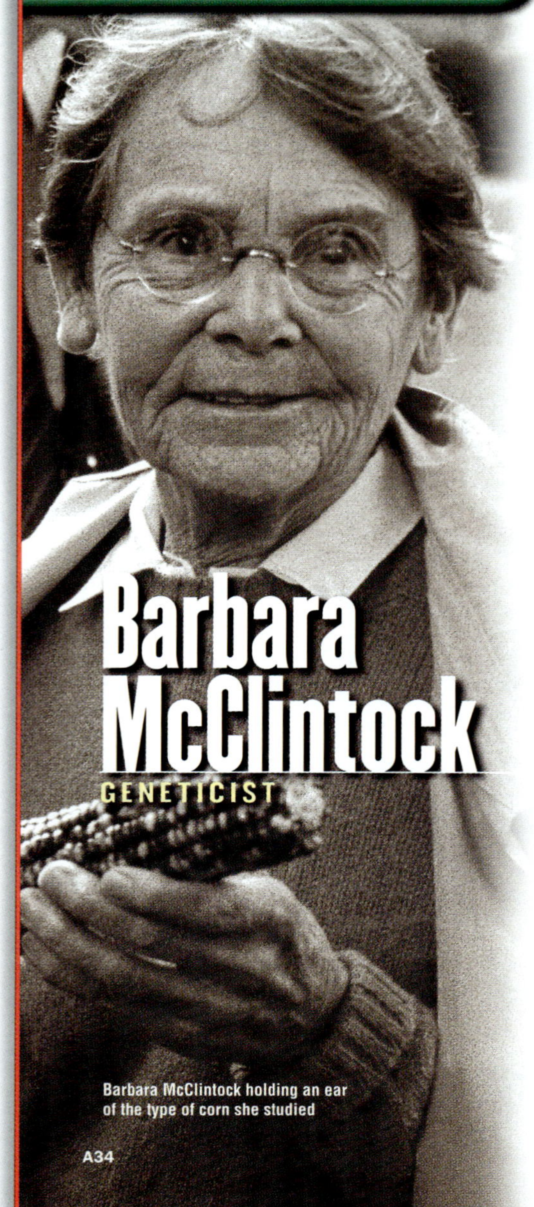

Barbara McClintock

GENETICIST

Barbara McClintock holding an ear of the type of corn she studied

Dr. Barbara McClintock is considered one of the greatest geneticists of the twentieth century. She made her mark by doing research that has helped us better understand heredity. Dr. McClintock earned many important honors for her work. She was elected to the National Academy of Sciences in 1944, and she received the Nobel Prize for her work in genetics in 1983.

Dr. McClintock earned a Ph.D. in botany (BAHT•uhn•ee)—the study of plants—at Cornell University in 1927. For the next 14 years, she stayed there doing genetic research on corn. After 1941, she worked alone at a lab on New York's Long Island, doing genetic research. At the time, scientists were just beginning to understand the connection between heredity and the cell structures that control it. From her research Dr. McClintock concluded that certain pieces of DNA can move from one place on a chromosome to a different place, or even to a place on a different chromosome in the same cell. This important idea was unusual and was not well received at the time. Much later, however, the work of other geneticists supported this idea.

Dr. McClintock's work was even more remarkable because it was done at a time when Mendel's work had just been rediscovered. For a long time Mendel's discoveries had been largely unknown and at the time Dr. McClintock was working, they had not yet been widely accepted. Yet Dr. McClintock had based much of her work on Mendel's ideas. As a result, many of her ideas were too far ahead of their time to be accepted by other scientists. As Mendel's work was studied, understood, and confirmed, however, Dr. McClintock's work also became more widely accepted. Even though her ideas were not accepted at first, she continued to do her research and to support her hypotheses with additional evidence. Now her work is an important part of our understanding of genetics.

THINK ABOUT IT

1. What important discovery did Dr. McClintock make?
2. Why were Dr. McClintock's ideas not accepted at first?

A34

BACKGROUND

Genetic Recombination and Gene Mapping Genetic recombination is any exchange of genetic material. Crossing-over is a form of genetic recombination that involves the exchange of segments between chromosomes as they line up during meiosis. Recombined genes appear as new traits in offspring and so can be quantified. Years ago scientists discovered that they can use observations of the physical traits of offspring to map their genes. Because clusters of adjacent genes break off and cross over together, genes that are far apart are more likely to end up recombined than those that are closer together. By noting the percentage of new (recombined) physical traits in offspring of breeding pairs, scientists can map out the relative distances between genes along a chromosome. If many offspring display a new recombined trait, chances are that the recombined genes are far apart along the chromosome.

Webliography GO ONLINE

Keyword biographies

www.harcourtschool.com

ACTIVITIES FOR HOME OR SCHOOL

SALTING POTATOES

How Does a Cell Membrane Work?

Materials

- potato cut into equal sized pieces
- ruler
- 2 bowls
- distilled water
- salt
- plastic wrap

Procedure

1. Measure and record the dimensions of your potato pieces. They should all be about the same size, so measure only one.
2. Mix 4 tablespoons of salt into one cup of water. Put the water into one of the bowls and add half of the potato pieces.
3. Put one cup of distilled water into a bowl. Add half of the potato pieces.
4. Cover the bowls with plastic wrap and put them in a cool place.
5. Let the pieces sit for one full day. Pour off the water and measure and record the dimensions of one potato piece from each bowl.

Draw Conclusions

How did the dimensions of the potatoes in each bowl change? What could explain the differences in size you saw? Infer what you think happened to the cells that resulted in this change.

BREEDING MANX CATS

How do cats pass genes to their offspring?

Materials

- 2 small cards marked *N* for *no tail*
- 2 small cards marked *n* for *tail*

Procedure

1. Among Manx cats, the lack of a tail is a dominant trait. NN is one possible gene combination for a Manx cat with no tail. Use the cards to show another possible combination.
2. Show the gene combinations of two tailless Manx cats that had a kitten with a tail. Draw a Punnett square, if necessary. What gene combination does the kitten have?
3. Suppose a tailless Manx cat and a Manx cat with a tail had one kitten. Could you predict definitely whether the kitten would have a tail? Why or why not?
4. If the cats from Step 3 had four kittens, how many would you expect to be tailless? Why?

Draw Conclusions

What do you need to know in order to determine the probability that a gene will be passed on from parents to offspring?

A35

School-Home Connection

These activities provide an excellent opportunity to assign hands-on activities for the students and their families. The materials are often easy to obtain and are safe to work with. Encourage students to complete at home those activities that you do not wish to do in the classroom. Remind them that communicating their results is always important, but especially so for home activities since you will not be there to witness them. Ask students to use diagrams, sketches, graphs, or even photographs to document their results.

ACTIVITIES FOR HOME OR SCHOOL

Salting Potatoes

Objectives

- **Explore how water moves across cell membranes.**

Suggested Time 10-minute observation sessions over six days

Hints

- Keep the potatoes in a cool, dry place for observation. Replace water that evaporates to keep the potatoes completely covered.

Draw Conclusions Students should find that water moves out of the potato pieces in salt water, resulting in the potato pieces' shriveling. This result occurs because the individual cells are losing water to the surrounding environment and, as a result, are shrinking.

Breeding Manx Cats

Objectives

- **Relate physical appearance to genetic makeup.**

Suggested Time 20–30 minutes

Procedure

1. Nn
2. Parents are both Nn; kitten is nn.
3. No; the tailless Manx might have a recessive gene for a tail, so the kitten could be Nn or nn.
4. You would expect that at least two out of four kittens would be tailless since the tailless Manx would have at least one dominant gene for no tail. If it had two dominant genes, all four kittens would be tailless.

Draw Conclusions You need to know whether a gene is dominant or recessive and what gene combination each parent carries.

ACTIVITIES FOR HOME OR SCHOOL

Reproducible copies of these activities are provided in the **Teaching Resources** book.

CHAPTER 1

Review and Test Preparation

VOCABULARY REVIEW (2 points each)

1. G, chromosomes
2. K, cell wall
3. H, nuclear membrane
4. E, chloroplast
5. A, cell membrane
6. F, nucleus
7. C, DNA
8. J, cytoplasm
9. D, mitochondria
10. L, vacuoles
11. B, sexual reproduction
12. I, Punnett square

CONNECT CONCEPTS (2 points each)

13. cells
14. mitosis
15. meiosis
16. body
17. eggs (or sperm)
18. sperm (or eggs)
19. offspring
20. genes
21. traits
22. dominant
23. recessive

CHECK UNDERSTANDING (3 points each)

24. C, Cells are microscopic.
25. J, having cell walls
26. A, nucleus
27. G, stays the same
28. A, is halved

CRITICAL THINKING (6 points each)

TEST PREP Test-Taking Tips

Item type open-ended format in which the student is asked to write a short answer

To help students answer Question 29, share the following strategy with them. **First think of what chloroplasts do (help turn light energy into food energy). Then think about where roots grow.**

29. Chloroplasts contain chlorophyll, which along with sunlight is necessary for photosynthesis. Since roots are not exposed to light and do not carry out photosynthesis, their cells do not need chloroplasts.

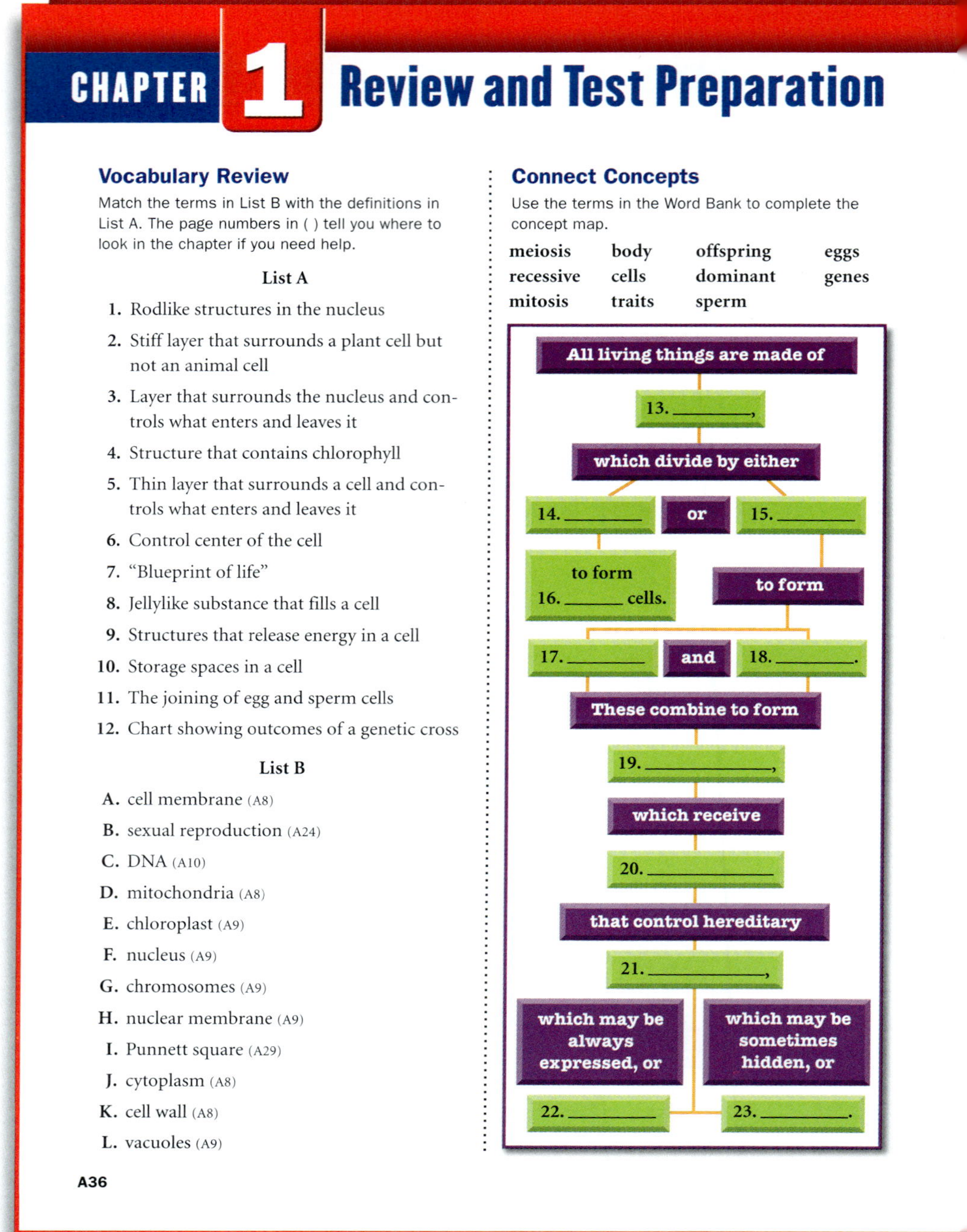

CHAPTER 1 Review and Test Preparation

Vocabulary Review

Match the terms in List B with the definitions in List A. The page numbers in ( ) tell you where to look in the chapter if you need help.

List A

1. Rodlike structures in the nucleus
2. Stiff layer that surrounds a plant cell but not an animal cell
3. Layer that surrounds the nucleus and controls what enters and leaves it
4. Structure that contains chlorophyll
5. Thin layer that surrounds a cell and controls what enters and leaves it
6. Control center of the cell
7. "Blueprint of life"
8. Jellylike substance that fills a cell
9. Structures that release energy in a cell
10. Storage spaces in a cell
11. The joining of egg and sperm cells
12. Chart showing outcomes of a genetic cross

List B

A. cell membrane (A8)
B. sexual reproduction (A24)
C. DNA (A10)
D. mitochondria (A8)
E. chloroplast (A9)
F. nucleus (A9)
G. chromosomes (A9)
H. nuclear membrane (A9)
I. Punnett square (A29)
J. cytoplasm (A8)
K. cell wall (A8)
L. vacuoles (A9)

Connect Concepts

Use the terms in the Word Bank to complete the concept map.

meiosis	body	offspring	eggs
recessive	cells	dominant	genes
mitosis	traits	sperm	

A36

CHAPTER TEST

See **Assessment Guide pp. AG1–5** for the Chapter Test. Assessment Options are provided on p. A2e.

TAKE-HOME BOOK

Use the **Take-Home Book** (described on p. A2d) to provide more chapter content and activities.

Check Understanding

Write the letter of the best choice.

24. Which statement is **NOT** part of the cell theory?
- **A** All living things are made of cells.
- **B** All life activities take place in cells.
- **C** Cells are microscopic.
- **D** New cells are produced by existing cells.

25. Plant cells differ from animal cells by —
- **F** having vacuoles
- **G** having no DNA
- **H** having a nuclear membrane
- **J** having cell walls

26. The control center of the cell is the —
- **A** nucleus
- **B** mitochondrion
- **C** cell membrane
- **D** cytoplasm

27. Mitosis is cell division in which the number of chromosomes —
- **F** is halved
- **G** stays the same
- **H** doubles
- **J** triples

28. Meiosis is cell division in which the number of chromosomes —
- **A** is halved
- **B** is doubled
- **C** remains the same
- **D** is reduced by two

Critical Thinking

29. Would you expect to find chloroplasts in the cells of a root? Explain.

30. Why can't reproductive cells of most animals be formed by the process of mitosis?

31. Suppose green pods are dominant and yellow pods are recessive. Two yellow-podded plants are crossed. What can you tell about the color of the pods of their offspring?

Process Skills Review

32. How might you **use a model** to describe the structures and functions of a plant cell to a younger brother or sister?

33. Make a diagram or a poster **showing the sequence** of the stages of mitosis in an animal cell. Why is it important for this process to take place in a certain sequence?

34. Suppose that in a certain kind of mouse, black fur is dominant over red fur. **Use numbers** to describe the ratio of the expected offspring of a cross between two mice that each have one recessive gene for fur color.

Performance Assessment

Dominant and Recessive

Suppose that in a certain kind of plant, red flowers are dominant and yellow flowers are recessive. A plant that has two genes for red flowers is crossed with a plant that has yellow flowers. Use a Punnett square to show the possible offspring of this cross. State the possible offspring as percentages of plants with red flowers and with yellow flowers. Also show the possible offspring of a cross that forms the second generation. Give the percentages of the possible offspring.

	R	R
r	Rr	Rr
r	Rr	Rr

	R	r
R	RR	Rr
r	Rr	rr

A37

30. Mitosis produces cells that have a complete set of chromosomes. If two such cells joined in reproduction, the resulting cell would have twice as many chromosomes as a normal cell.

31. They have yellow pods. Since having yellow pods is a recessive trait, the parent plants must have had genes for only yellow pods.

PROCESS SKILLS REVIEW
(7 points each)

32. Accept reasonable answers that include the main parts of the cell that would be shown in a model.

33. Diagrams should include the stages of mitosis indicated in the illustration on page A18. This sequence is important because each new cell must get a complete set of chromosomes.

34. Each mouse will have the genes Bb. The cross of Bb by Bb will result in a ratio of 3:1, or three black-haired mice for every one red-haired mouse.

Performance Assessment

Scoring Rubric

Performance Indicators

___ Draws accurate Punnett square for first cross. (See diagram on left.)

___ States percent of possible offspring as 100% red-flowered.

___ Draws accurate Punnett square for second cross. (See diagram on right.)

___ States percent of possible offspring as 75% red-flowered, 25% yellow-flowered.

Observations and Rubric Score

3 2 1 0

WRITING ABOUT SCIENCE

Have students write a composition that involves writing for a specific purpose. You may wish to use the prompt that is provided in the **Workbook.** The prompt is accompanied by a graphic organizer that will help students organize their ideas before writing. Models for writing are provided in the **Teaching Resources** book. You may wish to reproduce those for students or display them on a transparency.

WRITING PRACTICE

A chapter writing prompt and a prewriting activity are provided on **Workbook p. WB18.**

CHAPTER 2 Planning Guide Classification

LESSON	PACING	OBJECTIVES	MATERIALS
1 What Are the Kingdoms of Organisms? A40–49 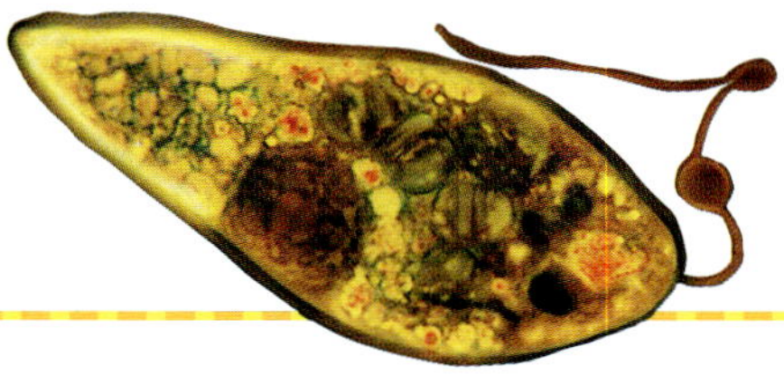	3 days	▶ Recognize that scientists classify animals in order to show ways in which they are related. ▶ Identify the characteristics scientists use to classify organisms. ▶ Conclude that all living things belong to one of five kingdoms. ▶ Recognize that classification systems change as additional information about organisms becomes available.	**Investigate** 1 bag of beans, poster board, clear adhesive tape or glue
2 How Are Kingdoms Subdivided? A50–57 	3 days	▶ Recognize that kingdoms are divided into smaller groups. ▶ Describe the two-part scientific name of an organism. ▶ Demonstrate how to use a dichoto-mous key to identify living things.	**Investigate** bean classifications from the Investigation in Lesson 1, paper and pencil
End of Chapter A58–63 **Science and Technology** **People in Science** **Activities for Home or School** **Chapter Review and Test Preparation**		▶ Evaluate the impact of research and technology on scientific thought, society, and the environment. ▶ Identify careers related to science. ▶ Connect chapter concepts with the contributions of scientists. ▶ Extend chapter concepts with hands-on investigations. ▶ Review chapter concepts.	

PROCESS SKILLS	VOCABULARY	RESOURCES AND TECHNOLOGY	REACHING ALL LEARNERS
Process Skill Tip classify **Other Process Skills** observe, compare	**classification** **Linnaean system** **Animalia** **Plantae** **Fungi** **Protista** **Monera** **adaptation**	**Workbook,** pp. WB19–24 **Vocabulary Cards,** pp. TR97–100 **Transparencies** A2, A2-1 **Harcourt Science Activity Video** **Internet Site**	**Kinesthetic Learners**, pp. A45, A46 **ESL Activity**, pp. A45, A47 **Investigation Challenge**, p. A48 **Advanced Learners**, p. A1g
Process Skill Tip gather data, record data **Other Process Skills** compare	**genus** **species** **dichotomous key**	**Workbook,** pp. WB25–31 **Transparencies** A2, A2-2 **Harcourt Science Activity Video** **Harcourt Science Newsroom Video**	**Visually Impaired**, p. A55 **ESL Activity**, p. A53 **Investigation Challenge**, p. A56
		Take-Home Book, pp. TH9–12 *The Tiny World of Protozoa* **Activities for Home or School,** p. TR66 **Chapter Test**, pp. AG8–12	**Performance Assessment**, p. AG13 **Portfolio Evaluation**, p. AGxxiv

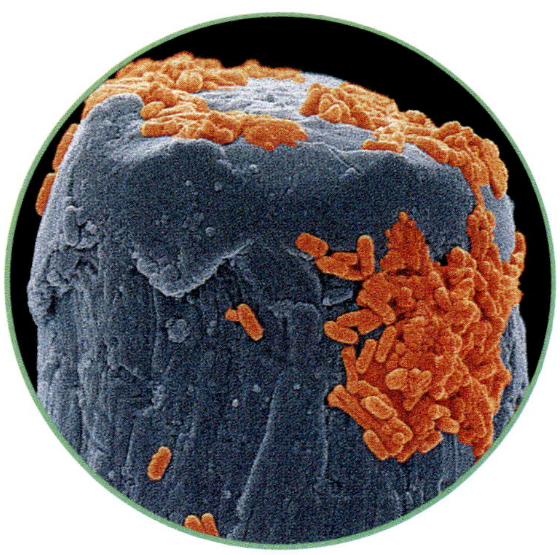

Prepare for Activities

Use this page to help you organize and prepare materials for the Investigations, which begin each lesson of this chapter.

LESSON INVESTIGATION	MATERIALS

1 Classifying Beans

A40–41

PURPOSE Classify beans and share the classification system.

PREPARATION TIPS Use 15 bean soup or other, similar dried bean soup mix for your bean mixture. The activity can also be done using bird seed, buttons, or bottle caps. Students may discern different ways of sorting the beans by using a hand lens.

TIME 30–40 minutes

EXPECTED RESULTS Students will sort their beans on the basis of many different criteria. Accept all groupings that students can justify.

MATERIALS

- 1 bag of beans
- poster board
- clear tape or glue

2 How to Develop a Key

A50–51

PURPOSE Develop a key to help others identify their beans by the names used in Lesson 1.

PREPARATION TIPS Be prepared to distribute students' classification systems completed in the Lesson 1 Investigate. An oversized sheet of paper will work best for student keys. Have the beans handy in case students need to take another look at them.

TIME 30–40 minutes

EXPECTED RESULTS Student keys should indicate a separate category for each type of bean.

MATERIALS

- bean classification from the Investigation in Lesson 1
- paper and pencil

Harcourt Science Activity Videos

Preview the Investigations The Activity Video for this unit shows students doing the Investigations. You may wish to view the video for classroom management ideas. The video can also be used to model the Investigation process for students.

Equipment Kits

Underlined items above are provided in the equipment kits (available separately).

. . . with Trade Books

These books provide in-depth information about the classification of organisms.

Pond Water Zoo: An Introduction to Microscopic Life by Peter Loewer (Atheneum, 1996) examines the microscopic life in a drop of pond water.

Richard Orr's Nature Cross-Sections by Moira Butterfield and Richard Orr (Dorling Kindersley, 1995) illustrates the variety of life in various habitats through cross-section views. *NSTA Trade Book*

Simon & Schuster Children's Guide to Insects & Spiders by Jinny Johnson (Simon & Schuster, 1996) describes family characteristics and habits of over 100 insects and arachnids. *NSTA Trade Book*

What Is a Reptile? by Robert Snedden (Sierra Club, 1994) answers numerous questions about reptiles with the help of many illustrations and photos. *NSTA Trade Book*

Trade book titles are current at time of publication but may go out of print without notice.

. . . with More Hands-On Activities

Investigation Challenge

Activities for Home or School

The cross-curricular Links that appear at the end of each lesson in the student edition and throughout the teacher's edition also provide activities.

. . . with the Take-Home Book

The Tiny World of Protozoa

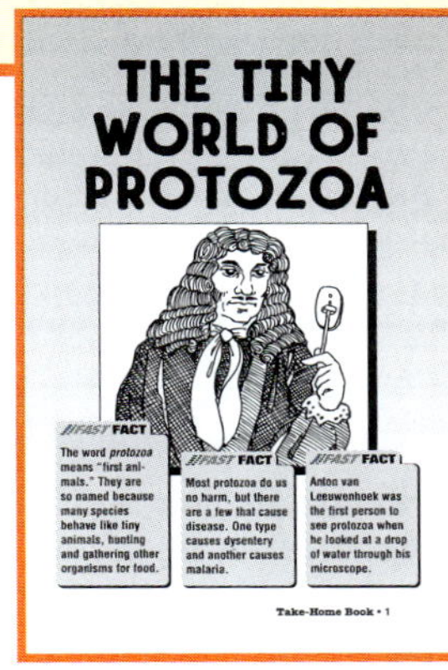

CONTENTS

. . . with Technology

Students can learn more about classification through these components and special Internet sites. Look for Technology Links at the end of each lesson.

CNN Turner Learning

Harcourt Science Newsroom Video
Dinosaur Rock

Visit The Learning Site for related links, activities, and resources: www.harcourtschool.com

Welcome to The Learning Site

Assessment Options ✓ Chapter 2

Formal Assessment

- **Chapter Review and Test Preparation,** PE pp. A62–63
- **Chapter Test,** pp. AG8–12

Ongoing Assessment

- ✓ **Questions**, PE pp. A42–49, A52–57
- **Lesson Review**, PE pp. A49, A57
- **Informal Assessment**
  - **Performance,** TE p. A41
  - **Portfolio,** TE p. A51
- **Observation Checklist,** AGxiv

Student Self-Assessment

- **Self-Assessment–Investigate**, p. AGxvi
- **Self-Assessment–Learn About**, p. AGxvii
- **Project Summary Sheet**, p. AGxix

Performance Assessment

- **Chapter Review and Test Preparation**, PE p. A63
- **Chapter Performance Task**, pp. AG13–14
- **Project Evaluation Checklist**, p. AGxviii

Portfolio Assessment

- **Science Experiences Record**, p. AGxxii
- **Guide to My Science Portfolio**, p. AGxxiii
- **Portfolio Evaluation,** p. AGxxiv

Chapter Test AG8

Classification

Part I Vocabulary 1 point each

Use the letters of the terms in the Word Banks to complete the sentences.

The current **1.** _A_, or grouping, system, is based on the **2.** _D_ system developed by Carolus Linnaeus. Like Aristotle, Linnaeus grouped all living things into two kingdoms. But Linnaeus also grouped living things of each kingdom into smaller and smaller groups based on their features. Linnaeus also gave each living thing a Latin name that consisted of two words. Today we continue to use Latin words to name living things. The first part of a Latin name is the **3.** _C_, which always begins with a capital letter. This word is the name given to a group of organisms that share major features and are closely related. The second part of the name, the **4.** _E_, identifies one specific organism and always begins with a lowercase letter. We can use a **5.** _B_ to identify members of a certain genus or species.

A classification
B dichotomous key
C genus
D Linnaean
E species

Today scientists group living things into five kingdoms instead of Linnaeus' two kingdoms. Scientists classify living things by their structures and features, by their DNA, and by how they get their food. They also classify living things by observing other **6.** _F_ that help them meet a particular need in their natural habitats. Members of the Kingdom **7.** _I_ are single-celled and lack a nucleus, yet they are adapted to almost every known environment. Members of the Kingdom **8.** _K_ have a nucleus and resemble animals, plants, or both. The members of the Kingdom **9.** _H_ have a nucleus and cell walls and are multicellular, but they cannot move or make their own food. Organisms grouped in the Kingdom **10.** _J_ are multicellular, make their own food, and cannot move. The Kingdom **11.** _G_ includes more than one million kinds of multicellular organisms. They take in food and can move.

F adaptations
G Animalia
H Fungi
I Monera
J Plantae
K Protista

Chapter Test AG9

Part II Science Concepts and Understanding

For Questions 12–17, write the correct term in each blank below. One term has already been filled in for you. 2 points each

Kingdom
12. Phylum
13. Class
14. Order
15. Family
16. Genus
17. Species

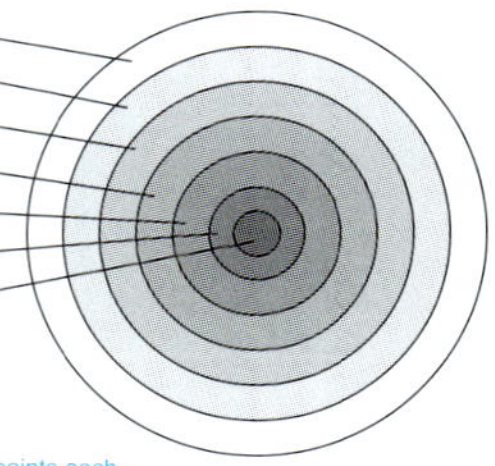

Write the letter of the best choice. 2 points each

_A_ **18.** If a living thing does **NOT** ____, it **MUST** be in the kingdom ____.
- A have a nucleus; Monera
- B have a backbone; Animalia
- C use chlorophyll to make food; Fungi
- D remain in one place; Protista

_H_ **19.** A classification system is good as long as it is —
- F approved
- G historical
- H useful
- J in Latin

_A_ **20.** The presence or absence of vascular tissue is used to classify —
- A types of plants
- B types of animals without a backbone
- C the more plantlike protists
- D monerans

_H_ **21.** In the scientific name *Epsaltria australis, Epsaltria* is the name of the —
- F kingdom
- G family
- H genus
- J species

Chapter Test AG10

_D_ **22.** *Hyla versicolor* and *Hyla chrysoscelis* **DO NOT** belong to the same —
- A kingdom
- B family
- C genus
- D species

For Questions 23–31, write *true* if the statement is true, or change the underlined term to make it true. 3 points each

_no_ **23.** All bacteria have a nucleus.
true **24.** Animalia is the only kingdom in which most members can move.
animals or plants **25.** All fungi are multicellular.
never **26.** Cold-blooded fish are sometimes classified by the temperature of their blood.
true **27.** All flowers are parts of plants.
true **28.** Latin is still used to name living things.
true **29.** All animals are in the kingdom Animalia.
true **30.** Scientists use the number and position of teeth in a mammal to help classify it.
can **31.** A living thing cannot have a skeleton without having a backbone.

Part III Critical Thinking 5 points each

Answer the following on a separate sheet of paper.

32. So far, between one million and two million kinds of living things have been discovered and named. Do you think more will be discovered? Why or why not? Possible answer: Yes, because we have not studied all areas of Earth extensively (such as the sea); because more kinds of living things are being found all the time.

33. *Felis domesticus* is a domesticated cat, or house cat. Why do you think all domesticated cats are classified in the same genus and species, even though they don't all look alike? Possible answer: because what is used to classify animals (presence of backbone, size, shape, position of skeleton, body covering, and reproduction) is the same for all house cats

34. Why do you think it would be difficult to classify an organism by using a dictionary? Possible answer: because in order to classify a living thing, you would have to read every description

Chapter Test AG11

Part IV Process Skills Application

Use the following pictures to answer Questions 35–38. Write the letter or letters to *classify* the organisms for each question. Note that the drawings are not to scale.

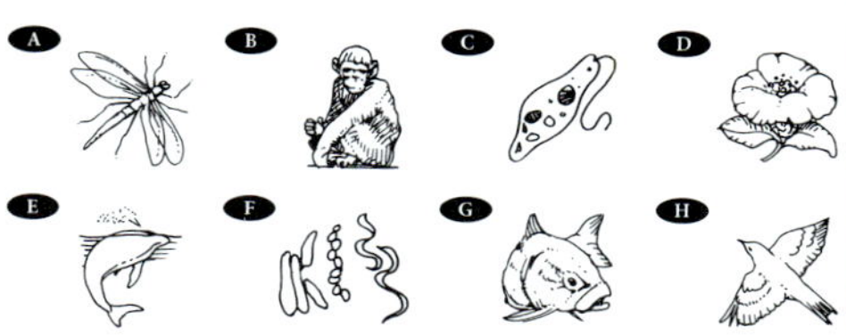

3 **35.** Which organisms are in the Kingdom Animalia?
A, B, E, G, and H

3 **36.** Which organisms are unicellular?
C and F

3 **37.** Which organisms have cells that have a nucleus? How do you know?
All but F have nuclei. F shows bacteria, which are Monerans and have no nucleus.

4 **38.** Which organisms have similar adaptations? **Observe** each drawing carefully. Then write the letters of the living things and name the adaptation.
Accept answers with at least two groups of living things with similar adaptations. Answers may include E and G—ways to move under water; A and H—wings for flight; B and E—feed young with milk; C and D—can make their own food.

Chapter Test AG12

For Questions 39–42, circle the living thing that fits each description. Then *classify* the organism by naming the kingdom it belongs to. 3 points each

39. moves, lives both in water and on land, has protective body covering

Kingdom: _Animalia_

40. has chlorophyll, does not move, is multicellular

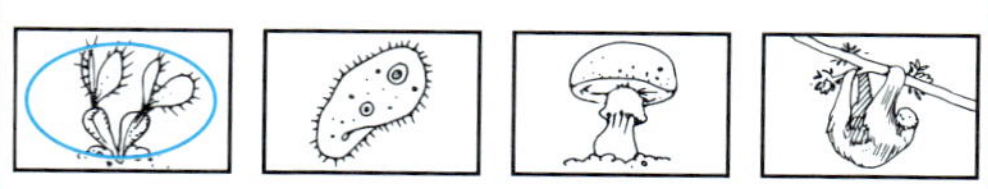

Kingdom: _Plantae_

41. cannot produce own food or move, has cell walls and a nucleus, is mostly multicellular

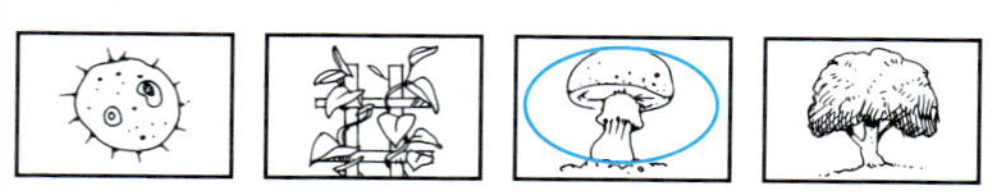

Kingdom: _Fungi_

42. is single-celled, has no nucleus, has a cell wall, is found everywhere

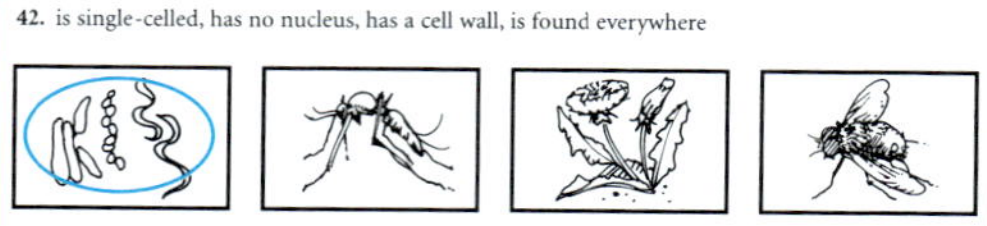

Kingdom: _Monera_

Assessment Continued

Performance Assessment AG13

A Dichotomous Key?

PERFORMANCE TASK

Materials

markers or colored pencils | drawing paper

1. Look at the classification key at the right. Is it a dichotomous key? Why or why not?

____

____

2. Suppose you were to make a dichotomous key to classify both living and nonliving things that can fly. What are some questions you would ask?

____

____

3. Design your own dichotomous key for both living and nonliving things that can fly. Start by writing the key below onto another sheet of paper. Add at least two more questions to classify things that fly.

Classifying Organisms into Kingdoms

1. a. Can move. Go to 2.
   b. Cannot move. Go to 3.
2. a. Can make its own food. Go to 4.
   b. Cannot make its own food. *Animalia.*
3. a. Can make its own food. *Plantae.*
   b. Cannot make its own food. *Fungi.*
4. a. Has nucleus. *Protista.*
   b. Does not have nucleus. *Monera.*

Classifying Things That Can Fly

1. a. It is alive. Go to 2.
   b. It is not alive. Go to 3.
2. a. It has feathers. Go to 4.
   b. It does not have feathers. Go to 5.
3. a. It is engine powered. Go to 6.
   b. It is not engine powered. Go to 7.

Performance Assessment AG14

PERFORMANCE TASK

A Dichotomous Key?

Materials Performance Task sheet, markers or colored pencils, drawing paper

Time 20–30 minutes

Suggested Grouping individuals or pairs

Science Processes communicate, classify/order

Preparation Hints Copy the dichotomous key at the top of the page onto the board.

Introduce the Task Tell students that they will consider a classification key and decide whether it is a dichotomous key. Guide students through the steps of the classification key by asking them for the name of a living thing (for example, insects, trees, flowering plants). Then ask whether those types of living things can move, can make their own food, or have a nucleus. Tell students that they will be making their own classification keys for things that fly.

Promote Discussion When students finish, have them view others' work. Ask for volunteers to explain their dichotomous keys and tell how they chose questions for their dichotomous keys.

Scoring Rubric

Performance Indicators

_____ Demonstrates the understanding that the provided classification key is a dichotomous key because it asks questions that have only two answers.

_____ Writes questions that have only two answers for classifying things that fly.

_____ Lists only yes-or-no type questions for classifying things that fly.

_____ Draws a dichotomous key that begins at high level, uses only yes-or-no type questions, and ends at low level, classifying things that fly.

Observations and Rubric Score

3 2 1 0

Workbook Support

Investigate Log WB20

Classifying Beans

Materials

- 1 bag of beans
- poster board
- clear adhesive tape or glue

Activity Procedure

1. Pour out the beans from your bag, and **observe** them carefully. In this investigation the beans represent newly discovered organisms.
2. **Classify,** or sort, the beans into separate groups of similar beans. Use all the characteristics you can observe, and place all beans with the same characteristics in the same group.
3. Draw a picture of one bean from each of your groups, give it an identifying name, and write a short description of it.
4. Make a poster showing how you grouped the beans, the names you gave them, and a description of each bean type. Tape or glue a few beans of each type next to your descriptions.

Investigate Log WB21

Draw Conclusions

1. **Compare** your results with the results of other members of the class. How many different ways did all of you find to classify the beans? Answers will depend on class and group size.
2. What characteristics did you use to **classify** the beans? Accept all reasonable answers. Possible answers: color, shape, texture, size

   What characteristics did some of your classmates use? Accept all reasonable answers. Possible answers: color, shape, texture, size
3. Now that you have seen other classification systems, how might you change your own? Explain your answer. Accept all reasonable answers that reflect a logical reason for changing the system or a definite improvement.
4. **Scientists at Work** Scientists **classify** living things to show how they are similar and how they are different. Why do you think it's important for scientists to agree on a set of characteristics for classifying living things? If they didn't, there would be numerous disagreements about how to classify organisms.

Investigate Further Choose a different set of objects such as marbles, buttons, or rocks. **Develop a classification system** for the objects. What similarities does this system have to the system you found for the beans? What differences are there? Accept reasonable answers. Look for similarities and differences between the systems.

Process Skills Practice WB22

Classify

Classifying is arranging things into groups according to traits or characteristics that they share. Classification is based on careful observation. Often there is more than one way to arrange objects.

Think About Classifying

You are helping with a neighborhood yard sale. A neighbor brings in the following items of clothing.

1. woman's blue coat
2. man's red flannel shirt
3. man's heavy brown jacket
4. woman's white silk shirt
5. boy's blue jeans
6. child's red snowsuit
7. boy's white sneakers
8. woman's black shoes
9. child's white T-shirt
10. child's brown rubber boots

Design for the items three different classification systems that you think would be most useful to the customers. Name the systems. Identify the characteristics of the groups within each system. List the numbers of the items that belong in each group.

System 1: The two most useful classification systems would be by user—child, man, woman—and by type of item—shirts, shoes, jeans, coats, and so on. Items could also be classified by color.

System 2:

System 3:

How would classifying items help customers shop? Once the items were classified and signs were put up, shoppers would know where to look for what they wanted.

Reading Skills Practice WB23

Summarize and Paraphrase a Selection

Read the selection below about the platypus. Then write a summary sentence for each paragraph and paraphrase the entire selection.

The platypus presented a real classification challenge to scientists. Europeans first discovered this Australian animal in the late eighteenth century. Although it is warm-blooded, the platypus lays eggs as reptiles do. Webbed and clawed feet, a ducklike bill, and a small, furry body make the platypus resemble a mix-and-match puzzle.

Platypuses hunt for food at the bottom of rivers and lakes. A platypus can swim rapidly because of its streamlined body design and webbed feet. A double layer of dense yellowish or brown fur protects the platypus's skin and keeps the animal warm in the water. The platypus shuts skin flaps over its ears and eyes when diving underwater. Electrosensors on its bill guide the platypus to insect larvae, worms, small snails, and crayfish.

A platypus egg is leathery, like a lizard's or snake's egg. If squeezed, it will not break. Platypus eggs are about one-tenth the size of chicken eggs. The mother platypus lays one to three eggs in a tunnel; she incubates them between eight and ten days.

The spurs on the hind legs of the adult male platypus are another unique feature of this species. These spurs, made of the same material as human fingernails, carry venom. The venom can seriously wound large animals and kill small ones. The platypus uses the spurs as a defense or a weapon.

Summary sentences
The platypus is an animal that was difficult to classify.
The platypus is adapted so that it can hunt for food underwater.
A mother platypus lays eggs similar to a reptile's eggs.
Male platypuses have venom in spurs on their hind legs.

The platypus is an interesting and unusual mammal. It has fur like a mammal, a bill like a duck, and webbed feet. Platypuses are able to swim rapidly. They look for food at the bottom of lakes and rivers. The duckbill, which has electrosensors, helps them locate food. Platypus mothers lay eggs that are like a reptile's. Males have venomous spurs for weapons and defense.

Lesson Concept Review/Assessment WB24

What Are the Kingdoms of Organisms?

Lesson Concept

All organisms are classified into one of five kingdoms, depending on their cell structure and how they get food.

Vocabulary

classification (A42)	**Plantae** (A44)	**Monera** (A44)
Linnaean system (A43)	**Fungi** (A44)	**adaptation** (A45)
Animalia (A44)	**Protista** (A44)	

Answer all questions in one or more complete sentences.

1. What is classification? It is arranging things into groups with similar characteristics.
2. Why did Aristotle's classification system become less useful? As more organisms were discovered, there were things that did not fit into any of Aristotle's groups.
3. Who developed the modern system of classification? The Swedish scientist Carl von Linné (Carolus Linnaeus) developed the modern system.
4. What determines an organism's classification? Organisms are classified by their structures and features, by similarities in their DNA, and by the way they get food.
5. What are the five kingdoms of the modern system? They are Animalia, Plantae, Fungi, Protista, and Monera.
6. What are the two major divisions of the animal kingdom? They are animals with backbones and those without backbones.
7. List three adaptations or characteristics that are used to classify plants. They are presence of vascular tissue, method of reproduction, and structure of stems, roots, and leaves.
8. How are fungi different from plants? Fungi do not make their own food.
9. Name two types of organisms in the kingdom Monera. Two types of organisms are bacteria and cyanobacteria.
10. How are protists different from monerans? The cells of monerans don't have nuclei while those of protists do.

Investigate Log WB25

How to Develop a Key

Materials

- bean classifications from the investigation in Lesson 1
- paper and pencil

Activity Procedure

1. Use your classification system from the investigation on pages A40–A41.
2. Make a simple flowchart or other diagram that would allow others to correctly identify their beans by the same name you gave to yours. Don't rely on pictures only!
3. Test your identification key by asking another student to use it to identify one or more beans.

Workbook Support Continued

Investigate Log WB26

Draw Conclusions

1. **Compare** your key with those of your classmates. How many kinds of keys were made? Are some keys easier to use than others? Explain your answer.
   Accept all answers that focus on why one key may be easier to use than another.

2. Why might a key be a more useful tool than a set of pictures of organisms?
   It is often difficult to see important characteristics in photographs. Sometimes size is not clear. If you needed picture cards for all organisms you would need 1,500,000 cards for animals alone.

3. **Scientists at Work** When scientists classify items or make keys for identification, they **gather** and **record data** carefully and precisely. Why might scientists want to avoid using imprecise terms such as *small, big, heavy,* and *light* in a key? It is important for everyone who uses a key to use it in the same way. Terms that are imprecise may mean different things to different people.

Investigate Further Choose another group of objects or organisms. **Develop a classification key** to help people identify the objects or organisms.

Process Skills Practice WB27

Gather and Record Data

To organize your observations you can gather and record data in such a way that it is easy to make inferences and draw conclusions. Sometimes the best way to do this is by making graphs or diagrams.

Think About Gathering and Recording Data

An ear of North American corn from 7000 years ago was only 2.5 centimeters (1 in.) long. Since then, corn has changed greatly. There are many different varieties with many different uses. Some corn is picked when the kernels are soft. This is what humans eat. Some of the kernels are yellow and some are white. Of the yellow kernels, some are large and some are small. Some of the corn is picked when the kernels are hard. Some of this hard corn is yellow, and some of it is a variety of other colors. The hard yellow corn may be used for animal feed or for making corn oil.

Complete the key below to organize the information about the different kinds of corn. Each step should give a choice between two characteristics.

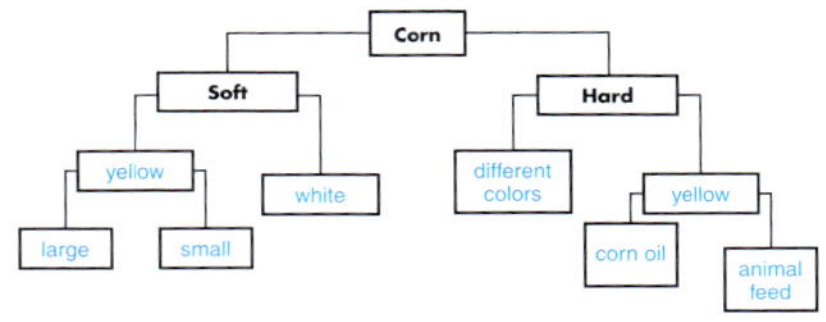

1. What other traits could you have used as the first step in your key?
   Answers will vary. Some might suggest starting with color.

2. If your first step had been different, would your key have been different? Explain.
   If the starting point had been different, the entire system would change. The type and number of groupings would change.

3. How would organizing the data in such a key help a farmer choose the kind of seed he wanted? A farmer could follow the key to the seed with the desired characteristics.

Reading Skills Practice WB28

Distinguish Fact and Opinion

Read the selection about entomologists. Then decide if the statements in the table are based on fact or opinion.

The research entomologists at the National Museum of Natural History spend their days collecting, examining, and classifying insects. According to these "bug scientists," between 20 million and 80 million plant and insect species exist. However, fewer than 2 million have been classified. I think that studying insects would be an interesting job.

When entomologists travel to other countries, they study insects within the animals' ecosystems. Researchers can sometimes apply the principles they learn about insects to human life. For example, by studying the genetic inheritance of butterflies, one researcher solved the problem of incompatible blood types between human mothers and their babies.

I believe that the United States needs to supply funding for more research into the insect world. Insects pollinate plants, but they also often damage crops. One scientist suggests that greater knowledge of insects could save farmers billions of dollars a year. It is important to know more about these tiny animals.

Statement	Fact or Opinion?
I think that studying insects would be an interesting job.	Opinion
According to the researchers, between 20 million and 80 million plant and insect species exist.	Fact
I believe that the United States needs to supply funding for more research into the insect world.	Opinion
Researchers can sometimes apply the principles they learn about insects to human life.	Fact

Lesson Concept Review/Assessment WB29

How Are Kingdoms Subdivided?

Lesson Concept

Each kingdom is divided into lower levels of classification. There are seven levels in the classification system of living things.

Vocabulary

genus (A53) **species** (A53) **dichotomous key** (A56)

Answer all questions in one or more complete sentences.

1. List the levels in the Linnaean classification system in order from largest to smallest.
   They are kingdom, phylum, class, order, family, genus, and species.

2. Which category contains the most types of organisms? The least? Kingdoms contain the most types of organisms. Species contain the least.

3. What do you know about the kingdom, phylum, and class of two organisms that are in the same order? The organisms are in the same kingdom, phylum, and class.

4. In which division would the organisms be the most similar? Species would be the most similar. Usually there is only one type of organism in each species.

5. Describe an organism's scientific name. The name, which is in Latin, consists of two parts. The first part is the genus, a group of closely related organisms. The second part is the species, which usually specifies just one type of organism. Both names are written in italics or underlined. The genus is always capitalized.

6. Describe how a dichotomous key works. The key consists of pairs of statements. Only one statement is true for the organism you are trying to identify. The statements refer you to other statements. You keep going through the key until you identify your organism.

Vocabulary Review WB30

Recognize Vocabulary

Fill in the blanks with the letters of the word that best fits the definition. The word that appears in the vertical box fits the definition of the sentence below the puzzle.

classification
genus
Plantae
dichotomous key
Monera
Animalia
adaptation
Protista
Linnaean system
species
Fungi

1. L I N N A E A N S Y S T E M
2. D I C H O T O M O U S K E Y
3. A N I M A L I A
4. S P E C I E S
5. P L A N T A E
6. C L A S S I F I C A T I O N
7. P R O T I S T A
8. F U N G I
9. M O N E R A
10. G E N U S

An adaptation helps an organism meet a need in its natural habitat.

1. name of a method of classification
2. guide to identifying an organism based on choices between statements
3. kingdom that includes organisms with backbones
4. smallest division in the modern system of classification
5. kingdom of organisms that have cell walls and chloroplasts
6. arrangement of organisms into groups that have similar traits
7. kingdom of mostly one-celled organisms with both plant and animal characteristics
8. kingdom whose organisms have cells similar to those of plants but do not make their own food
9. kingdom containing bacteria and cyanobacteria
10. first part of an organism's scientific name

Writing Practice WB31

Compare and Contrast Characteristics

Informative Writing–Compare and Contrast

Write a handout for science students, explaining how to tell the difference between animal-like and plantlike protists. Use the Venn diagram below to help you organize your writing.

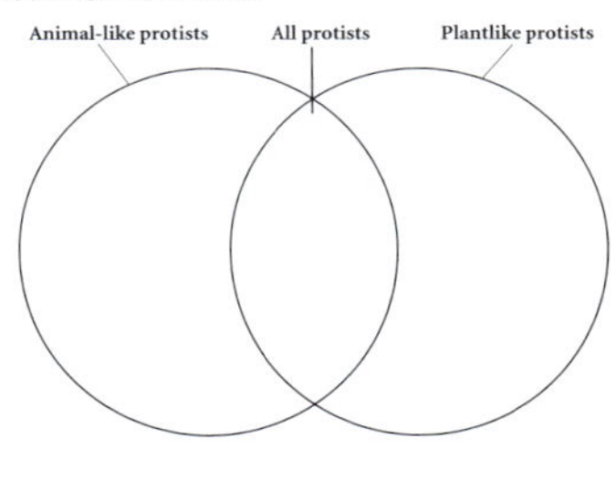

CHAPTER 2

Classification

Generate Science Questions

Invite students to read and discuss the Fast Facts. After they read the facts about classification, ask students if they have any questions they would like to have answered. Lead the class in generating a list of questions, and record them on the board or on chart paper.

Ask students what kinds of investigations they would design to answer the questions they have. Explain that the chapter will help them answer their questions, but they may want to conduct additional investigations on their own.

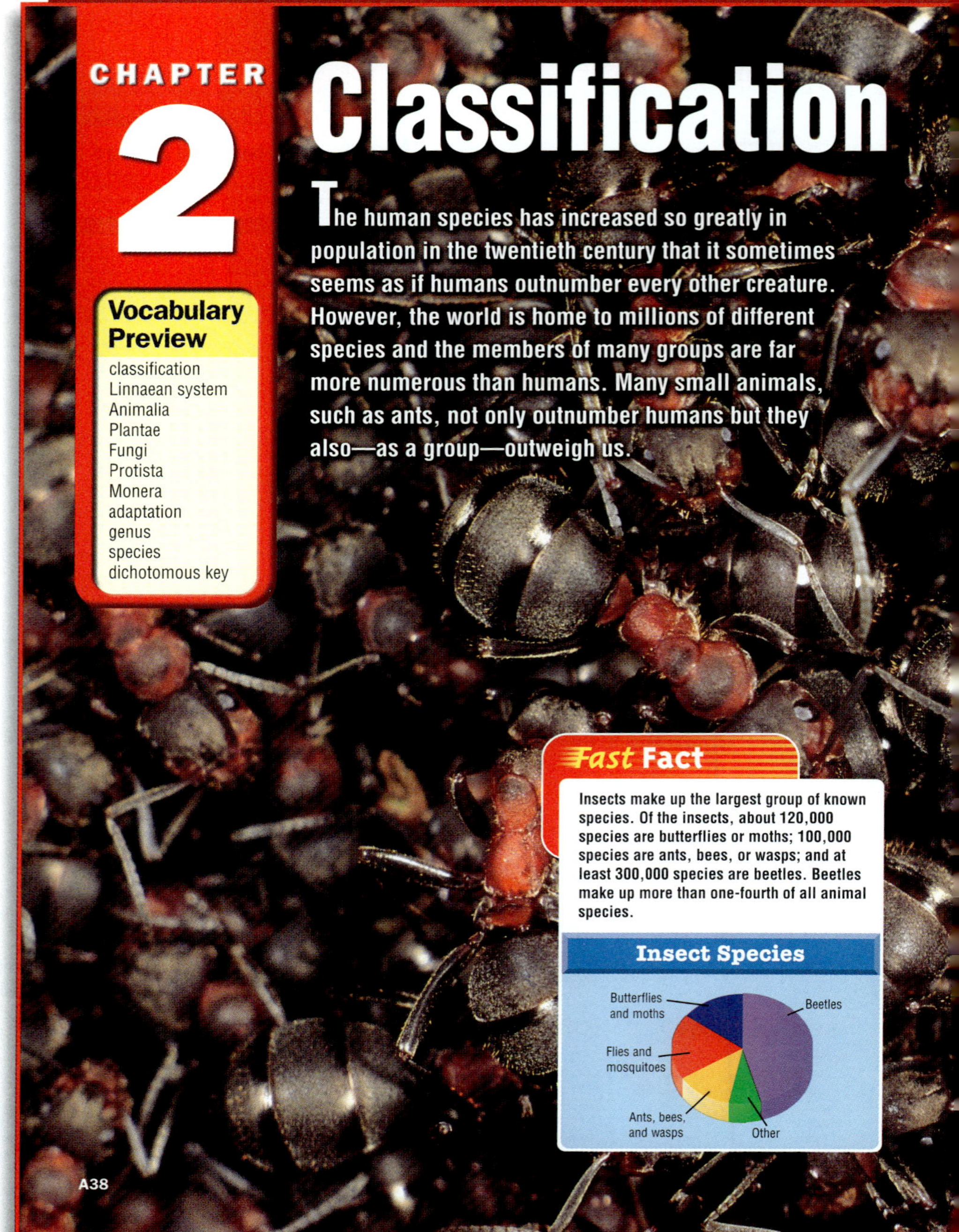

CHAPTER 2

Classification

The human species has increased so greatly in population in the twentieth century that it sometimes seems as if humans outnumber every other creature. However, the world is home to millions of different species and the members of many groups are far more numerous than humans. Many small animals, such as ants, not only outnumber humans but they also—as a group—outweigh us.

Vocabulary Preview

classification
Linnaean system
Animalia
Plantae
Fungi
Protista
Monera
adaptation
genus
species
dichotomous key

Fast Fact

Insects make up the largest group of known species. Of the insects, about 120,000 species are butterflies or moths; 100,000 species are ants, bees, or wasps; and at least 300,000 species are beetles. Beetles make up more than one-fourth of all animal species.

Insect Species

Butterflies and moths
Beetles
Flies and mosquitoes
Ants, bees, and wasps
Other

A38

Reading Skills Checklist

Strategies for developing the following reading skills are provided in this chapter.

- ❏ use prefixes and suffixes to determine word meanings
- ☑ use context to determine/confirm word meaning *p. A58*
- ☑ identify supporting facts and details *pp. A42, A45, A46, A47, A48, A52, A54, A55*
- ❏ arrange events in sequence
- ☑ identify the main idea *pp. A42, A45, A46, A47, A48, A52, A54, A55*
- ☑ summarize and paraphrase a selection *p. A46*
- ❏ identify cause and effect
- ❏ predict probable future actions and outcomes
- ☑ use graphic sources for information *pp. A43, A48, A53*
- ☑ draw logical conclusions *pp. A42, A55*
- ❏ compare and contrast
- ☑ distinguish fact and opinion *p. A55*
- ❏ make generalizations
- ☑ use reference sources *pp. A47, A49*

School-Home Connection

Distribute copies of the School-Home Connection, p. TR3. Have students take the page home to share with their families as you begin the chapter.

Follow Up Have volunteers compare the results of the activity they did at home. Additional School-Home Connections are provided by the **Activities for Home or School** (p. A61) and the **Take-Home Book** (p. A38d).

Teaching Resources, p. TR3

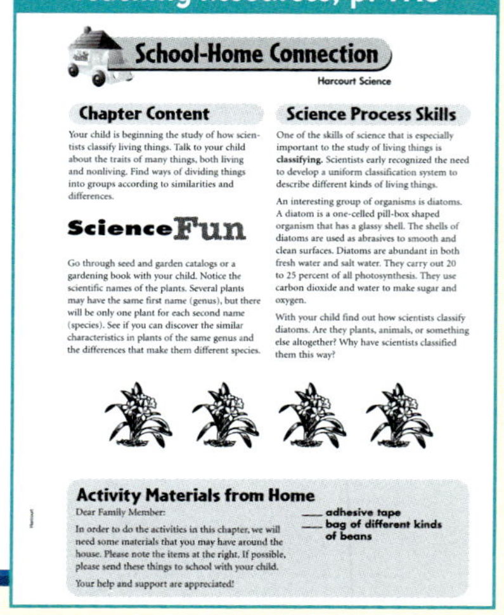

School-Home Connection

Harcourt Science

Chapter Content

Your child is beginning the study of how scientists classify living things. Talk to your child about the traits of many things, both living and nonliving. Find ways of dividing things into groups according to similarities and differences.

ScienceFun

Go through seed and garden catalogs or a gardening book with your child. Notice the scientific names of the plants. Several plants may have the same first name (genus), but there will be only one plant for each second name (species). See if you can discover the similar characteristics in plants of the same genus and the differences that make them different species.

Science Process Skills

One of the skills of science that is especially important to the study of living things is **classifying.** Scientists early recognized the need to develop a uniform classification system to describe different kinds of living things.

An interesting group of organisms is diatoms. A diatom is a one-celled pill-box shaped organism that has a glassy shell. The shells of diatoms are used as abrasives to smooth and clean surfaces. Diatoms are abundant in both fresh water and salt water. They carry out 20 to 25 percent of all photosynthesis. They use carbon dioxide and water to make sugar and oxygen.

With your child find out how scientists classify diatoms. Are they plants, animals, or something else altogether? Why have scientists classified them this way?

Activity Materials from Home

Dear Family Member:

In order to do the activities in this chapter, we will need some materials that you may have around the house. Please note the items at the right. If possible, please send these things to school with your child.

Your help and support are appreciated!

___ adhesive tape
___ bag of different kinds of beans

Fast Fact

Some fungi are predators. The fungus *Arthrobotrys dactyloides* uses loops of threads to trap roundworms. The fungus then digests the roundworm.

Fast Fact

Bacteria are the hardiest of all living things. Some survive temperatures below freezing for years. Others can live in the boiling water of hot springs. Still others can survive in very hot acid. Bacteria have been found inside granite rocks, in the depths of the ocean, and in the stratosphere.

A39

Prereading Strategies

Preview the Chapter

Have students scan the chapter to identify the lesson titles and major headings and then use them to make an outline of the chapter. Encourage students to write questions that may occur to them as they make their outlines.

Preview the Illustrations

Have students scan the chapter, looking at the illustrations and reading the captions. Ask them to generate a list of questions based on what they see in the pictures.

Vocabulary Preview

Have students make a three-column chart for the words in the vocabulary list. In the first column, students should write each word, and in the second column, they should write what they think the word means. In the third column, have students write the definitions as they read through the chapter.

Vocabulary Chart

Word	What I Think It Means	Definition

BACKGROUND

Webliography

Keyword classification

www.harcourtschool.com

GO ONLINE

Kingdoms Up until relatively recent times, scientists recognized only two kingdoms of organisms—plants and animals. All organisms were forced, however awkwardly, into one of these two divisions, even though many showed traits of both kingdoms.

In ancient times, the division of organisms into those that moved and those that were stationary seemed perfectly logical. These traits were ones early taxonomists could observe easily. With optical instruments, however, taxonomists began to expand their classification system. They realized, for instance, that while mushrooms did not move, they also did not photosynthesize their own food and, therefore, were not producers like true plants. Similarly, they saw that some microorganisms were capable of both movement and photosynthesis. Thus were the kingdoms of Monera, Protista, and Fungi born.

Fast Facts/Math Links

Estimate Challenge students to estimate how many worms this fungus could kill in one year if it took 27 hours to kill each worm. 27 hours = 1.125 days; 365 days/yr ÷ 1.125 days/worm = about 324 worms

Find the Range Challenge students to find the most restrictive temperature range bacteria could live in based on the information given. The highest temperature still below freezing is 31°F (⁻1°C); water boils at 212°F (100°C). The shortest range would be 31°F to 212°F (⁻1°C to 100°C).

LESSON 1

What Are the Kingdoms of Organisms?

Objectives

- Recognize that scientists classify animals in order to show ways in which they are related.
- Identify the characteristics scientists use to classify organisms.
- Conclude that all living things belong to one of five kingdoms.
- Recognize that classification systems change as additional information about organisms becomes available.

Motivate

More Fast Facts

Spark interest with these facts and questions.

- The giant puffball fungus measures up to 2 m (6.6 ft) in circumference and weighs up to 20 kg (44 lb). In just one day, a giant puffball can release 7 billion spores.

  What common fungi do you sometimes get on top of your pizza? mushrooms

- The tiny, single-celled diatom is one of the smallest, simplest plantlike organisms. It would take up to 50 diatoms to cover the period at the end of this sentence.

  Diatoms are considered part of the kingdom Protista. How many kingdoms are there? five

Access Prior Knowledge

- What is meant by *classification*? Can you provide examples of things that are classified?

When Minutes Count . . .

Whole-Class Investigate Option Divide a single bag of beans among the class and have students sort them. Encourage students to consider all proposed methods before deciding how to sort the beans.

Using the Activity Video The video can be used to preview and model the investigate process with students. After viewing the video, students can conduct the investigation in small groups.

LESSON 1

What Are the Kingdoms of Organisms?

In this lesson, you can . . .

- INVESTIGATE how to classify objects.
- LEARN ABOUT the kingdoms into which scientists classify living things.
- LINK to math, writing, health, and technology.

A40

INVESTIGATE

Classifying Beans

Activity Purpose To talk to one another and be understood, we need to agree on the meanings of words. To talk about living things, we need to know that everyone is talking about the same organism. Classifying gives people who understand the classification system a way to define organisms. In this activity you will **classify** beans and share your classification system.

Materials

- 1 bag of beans
- poster board
- clear adhesive tape or glue

Activity Procedure

1. Pour out the beans from your bag, and **observe** them carefully. In this investigation the beans represent newly discovered organisms. (Picture A)
2. **Classify,** or sort, the beans into separate groups of similar beans. Use all the characteristics you can observe, and place all beans with the same characteristics in the same group. (Picture B)

BACKGROUND

Webliography GO ONLINE

Keyword classification

www.harcourtschool.com

Carolus Linnaeus (1707–1778), a Swedish naturalist, was the first person to look at physical characteristics when classifying organisms. In grouping flowering plants, for example, he looked for similarities in the structure of the sexual flower parts (pistil and stamen). In classifying animals, he looked at anatomy, including internal systems.

Linnaeus was also the first to develop the idea of identifying organisms by genus and species (e.g., *Canis lupus*). This system is still in use.

You may wish to share this information with students.

For student self-assessment, see Assessment Guide p. AGxvi.

3. Draw a picture of one bean from each of your groups, give it an identifying name, and write a short description of it.

4. Make a poster showing how you grouped the beans, the names you gave them, and a description of each bean type. Tape or glue a few beans of each type next to your descriptions.

Picture A

Picture B

Draw Conclusions

1. **Compare** your results with the results of other members of the class. How many different ways did all of you find to classify the beans?
2. What characteristics did you use to **classify** the beans? What characteristics did some of your classmates use?
3. Now that you have seen other classification systems, how might you change your own? Explain your answer.
4. **Scientists at Work** Scientists **classify** living things to show how they are similar and how they are different. Why do you think it's important for scientists to agree on a set of characteristics for classifying living things?

Investigate Further Choose a different set of objects such as marbles, buttons, or rocks. **Develop a classification system** for the objects. What similarities does this system have to the system you found for the beans? What differences are there?

Process Skill Tip

To **classify** objects, you first need to observe and describe them carefully. By doing this, you can identify similarities and differences between the objects. Then you can tell which ones belong in the same group.

A41

INVESTIGATE

Time 30–40 min **Preparation Tips** See p. A38c.

Grouping individuals or groups

Process Skill Tip classify

Other Process Skills observe, compare

Expected Results Students will sort their beans based on different criteria. Accept all groupings that students can justify.

Teaching Tips

- Suggest that students observe each bean carefully and evaluate its shape, size, color, texture, and density before deciding how to classify it.
- Be sure to collect the students' classification systems for further use in Lesson 2.

Draw Conclusions

1. probably more than one
2. Answers may include color, shape, size, texture, and density.
3. Answers should reflect how students would fix apparent weaknesses in their classifications.
4. Lead students to conclude that if scientists sorted organisms by different criteria (color, shape, or size), they would form different groups. Only when scientists sort by the same criteria are groups comparable.

Investigate Further Answers will vary. The greater the variety within each category of item (marbles, buttons, rocks), the more diverse and interesting the students' classification schemes.

Process Skill Tip

Classify Display the Process Skill Tip Transparency. Invite students to observe the watches and sort them as analog or digital. Then have them compare their classification results with those of their classmates.

Challenge students to come up with new classification criteria, and have them compare their systems with the others.

Process Skill Tip Transparency A2-1

Classify

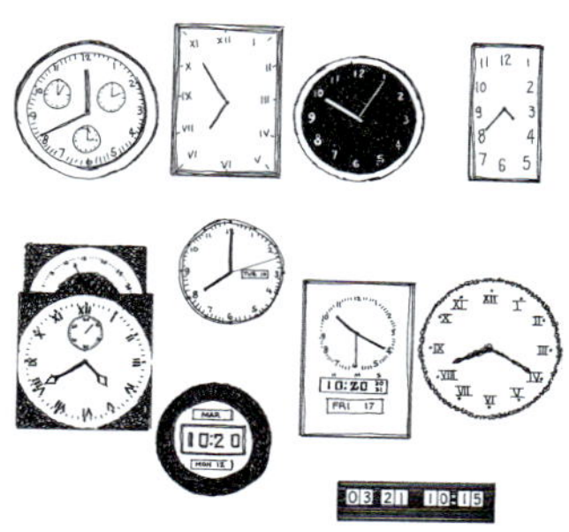

• How can you sort these watches? Number type, digital and analog, background color, shape, second hand or no second hand, time and date or time only.

Informal Assessment

Performance Read students' bean descriptions to see if they include similarities and differences. See if they can explain the reasoning behind their classifications.

INVESTIGATE LOG

Students can use **Workbook pp. WB20–21** to record the results of the investigation.

PROCESS SKILLS PRACTICE

To practice and apply process skills, see **Workbook p. WB22.**

LEARN ABOUT

1 Before Reading

SET A PURPOSE

FIND OUT Remind students that these statements will help guide them in what to look for as they read.

VOCABULARY If students are doing the activity described in the Vocabulary Preview on page A39, remind them to add the definitions for these words to the chart.

Graphic Organizer for Chapter Concepts Display the transparency (shown below) or have students use Workbook page WB19. Point out that by the end of the lesson, students should be able to list all five of the kingdoms on the transparency.

2 Guide the Learning

SCIENCE IDEAS

Help students focus on supporting details for this **main idea: Scientists classify organisms on the basis of their similarities with other organisms.**

- **Why are apples, oranges, pears, and grapes all found in the same section of the grocery store?** They are all fruits.
- **What do golden retrievers, poodles, and terriers have in common? What group do they form?** They are all dogs.

Critical Thinking Help students **draw logical conclusions** about the value of classification.

- **What can an organism's classification tell you about it?** Encourage students to provide examples that illustrate how classification groups organisms with like characteristics.

VOCABULARY CARDS AND ACTIVITIES

Students can use the Vocabulary Cards to make their own graphic organizers or to add to an ongoing file of science terms. The Vocabulary Cards and a variety of strategies and activities are provided beginning on p. TR85 in the **Teaching Resources** book.

Ongoing Assessment

Use the ✔ questions in the pupil edition.

LEARN ABOUT

Classification of Living Things

FIND OUT

- why living things are classified
- how living things are classified into kingdoms

VOCABULARY

classification
Linnaean system
Animalia
Plantae
Fungi
Protista
Monera
adaptation

Why Living Things Are Classified

If you go to a grocery store to buy apples, you can find them by remembering that apples are fruits and looking for them where fruits are displayed. The store arranges items in groups so that customers know where to find them. The arrangement of things in groups of similar items is called **classification**.

Scientists classify living things to describe what they are like and how they relate to other organisms. So far, between 1 and 2 million kinds of living things have been discovered and named. However, scientists estimate that there may be more than 10 million different kinds of living things. Without a system of classifying, scientists could not share information about all these organisms.

There is no one correct classification system for living things. A system is correct as long as it is useful. For scientists the best classification system is one that allows them to identify organisms correctly and to interpret how living things are related. However, even a good

The History of Classification

This time line describes some of the ways in which people have classified living things over the last 2,500 years.

Aristotle 350 B.C.

Aristotle's classification system groups organisms first as plants or animals. It then groups animals by how they move—fliers, swimmers, or walkers. It groups plants by size—as herbs, shrubs, or trees.

B.C. | A.D.

A.D. 300 – A.D. 1500s

A system used by herbalists classifies plants by how they are used medically and whether or not they are poisonous.

A42

Graphic Organizer for Chapter Concepts

Transparency A2 • Workbook, p. WB19

Classification

Kingdoms	Subdivisions of Kingdoms
1. Animalia	1.
2. Plantae	2.
3. Fungi	3.
4. Protista	4.
5. Monera	5.
	6.

system may need to be improved. Scientists sometimes need to change their classification systems because of new information they have learned. The time line on these pages summarizes the history of changes in classification systems.

One of the first systems for classifying living things was developed about 350 B.C. by the Greek philosopher Aristotle. At the time, people recognized only about 1000 different kinds of living things.

As scientists learned more, Aristotle's system became less useful. World explorers returned from their travels with thousands of plants and animals Europeans had never seen before. The invention of the microscope led to the discovery of tiny forms of life. The number of known living things increased greatly, and the newly discovered ones did not all fit well into Aristotle's system.

In the 1700s, the Swedish scientist Carl von Linné developed a new classification system for living things. First, he grouped all living things into two major groups that he called kingdoms —the plant kingdom and the animal kingdom. Next, he organized the members of each kingdom into smaller groups based on their features. Then, he divided these groups into smaller groups. This system, called the **Linnaean system**, was useful for storing and finding information about living things. It became the basis for our present system of classification.

Von Linné also invented a system for naming living things. Scholars of the 1700s communicated in Latin, and von Linné gave every organism that was known at the time a Latin name that consisted of two words. He gave himself a Latin name as well. He has become known as Carolus Linnaeus.

✔ **Why do scientists classify living things?** so that they have a common system that allows them to identify organisms and to see relationships between different kinds of organisms

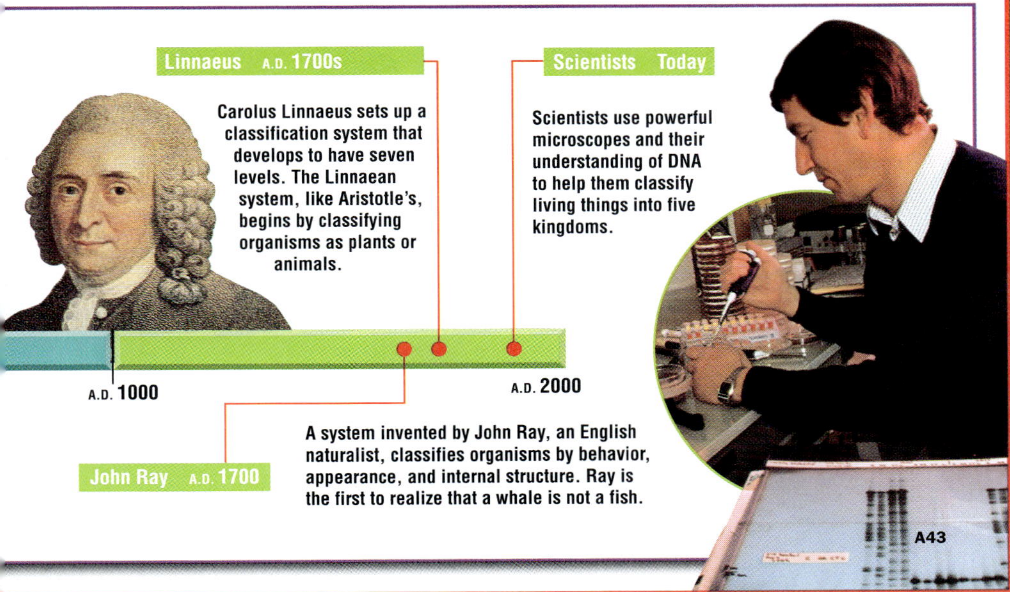

A43

VISUAL LEARNING

Have students look at the time line on pages A42–A43 and **use this graphic source for information**.

▶ What do you notice about the rate at which the classification system changed? It stayed the same for a long time but has changed much in the past 400 years.

▶ Why has the classification system changed so much in the past 400 years? Advances in scientific instruments and technology have allowed scientists to observe things never seen before, thus causing changes in the classification system.

DEVELOP SCIENCE VOCABULARY

classification, p. A42—Write the word *classification* on the board and circle the first five letters *class*. Explain that a class is a group of related organisms or objects. Ask students to make generalizations about how the students in your class are related. They are about the same age, they follow the same daily schedule at school, and they are learning the same subjects.

Linnaean system, p. A43—Ask students who invented the Linnaean system. Linnaeus Explain that *Linnaean* is an adjective that describes the system. Like many adjectives, it is a modified noun with the *-an* suffix. Challenge students to name a system that might come from Europe or one that came from Mars. European system; Martian system

BACKGROUND

Webliography

Keyword classification

www.harcourtschool.com

GO ONLINE

Aristotle (384–322 B.C.) was a Greek philosopher and scientist who made contributions to biology, logic, physics, astronomy, ethics, and philosophy.

Perhaps because his father was a physician, Aristotle approached philosophy from a biological perspective. He was the first person to divide the world of living creatures into plants and animals. He then subdivided the organisms he had placed in these two categories on the basis of general shape and size (shrubs in one group, trees in another) and where the organism resided (air, land, or water).

Scientists abandoned Aristotle's system when they found that it did not show natural relationships. For example, under Aristotle's system, whales, starfish, and plankton would be grouped together simply because they all lived in the sea, even though the organisms were clearly unrelated.

Language Arts Link

Latin Linnaeus used Latin words to describe the genus and species of organisms because Latin was the language used by the scientific community of his day. Originating in ancient Rome, Latin became the principal language of most of Western Europe as Roman rule expanded. Until the eighteenth century it was the chief language of academics and diplomats. Many Latin words, including those used to describe genus and species, are still used by scientists today. Provide a Latin I textbook, and encourage students to note common word endings they see used in this chapter.

Guide the Learning continued

DEVELOP SCIENCE VOCABULARY

Animalia, p. A44—*Animalia* comes from the Greek word meaning "having breath." Ask students why they think this classification name made sense to scientists thousands of years ago. They saw animals as "breathing" and the other form of life they identified—plants—as "not breathing."

Plantae, p. A44—*Plantae* comes from the Latin word meaning "sprout," or "shoot." Ask students to consider the large variety of organisms in this kingdom (from clover to giant redwoods) and to explain why they are all classified as plants.

Fungi, p. A44—Allow students to handle a variety of edible raw mushrooms. Then ask them to explain why the Greeks might have used their word for *sponge* to describe this organism. (Be sure students wash their hands after handling any biological material.)

Protista, p. A44—Explain that *proto-* means "primitive." Protists are some of the simplest organisms on Earth. If possible, show students magnified slides of pond water and have them look for amoebas, paramecia, and other protists.

Monera, p. A44—All monerans are single-celled. *Monera* comes from the Greek word for "single." Show students a carton of yogurt. Explain that yogurt is made by the action of bacteria (*Lactobacillus*).

How Living Things Are Classified

Although we still use the Linnaean system, some of the reasons for placing an organism in a particular group have changed. Also some of the groups themselves have changed. For example, Linnaeus started with only two kingdoms. In the present system of classification, there are five kingdoms. Some scientists have suggested that there are six or seven kingdoms.

An organism's structure and features are important characteristics for classifying. For example, fish are classified in part by the shape and position of their fins, and mammals are classified in part by the structure, number, and placement of their teeth. However, modern classification is based on other characteristics as well.

There are differences in the cells of different organisms. There are also differences in the ways organisms get their food. Scientists use these two characteristics—cell structure and ways of getting food—to place organisms in kingdoms. Scientists have discovered that many organisms are very different from most animals and plants. They do not, therefore, belong in either the kingdom **Animalia** (an•uh•MAYL•yuh) or the kingdom **Plantae** (PLAN•tee). These organisms have been placed in one of three other kingdoms.

The members of the kingdom **Fungi** (FUHN•jy) have cells similar to those of plants and often look like plants. However fungi do not make their own food. They take the nutrients that they need from the environment.

The members of the kingdom **Protista** (proh•TEES•tuh) can have traits of plants, animals, or fungi. Most protists are one-celled. Their cell contains one or more nuclei and most of the structures found in plant and animal cells. Some protists make their own food. Others take in food from the environment, as animals and fungi do. Still others get food in both ways.

▲ Scientists use DNA analysis to help them classify organisms. The patterns produced by the DNA of organisms are compared. Similarities and differences are noted and the relationship of the organisms is inferred from the comparison.

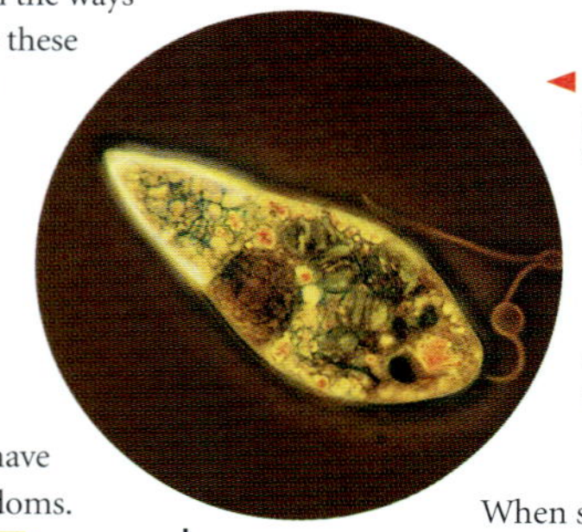

◄ *Euglena* is a type of protist. It has chloroplasts and can make its own food, as plants do. It can also take in nutrients from its surroundings, as fungi do. *Euglena* can even move, as animals do, by lashing a whiplike structure back and forth.

When scientists first examined bacteria, they found cells unlike those of members of any of the other kingdoms. Bacteria have cell walls as plant cells do, but the structure of the walls is different.The cells of bacteria do not have a nucleus. These organisms either make their own food or take in nutrients from their environment. As a result of their differences from other organisms, bacteria are placed in kingdom **Monera** (moh•NAIR•uh).

✔ **What traits do scientists use to classify living things into kingdoms?** cell structure and the method of getting food

A44

THE QUESTIONS KIDS ASK!

How could a fungus live on rotten, dead stuff?

Address Misconceptions Although fungi look like plants, they do not have chlorophyll and, thus, cannot make their own food. They need to take nutrients from the environment. In fact, one of the reasons the dead stuff *is* decaying is that many fungi are acting as decomposers and are using nutrients. Allow students to observe bread mold growing and to notice its intricate structure through a hand lens. Keep the mold sealed while observations are made.

BACKGROUND

Webliography

Keyword classification

www.harcourtschool.com

GO ONLINE

Eubacteria and Archaebacteria

Although many scientists still follow the five-kingdom system, others divide the Monerans into two kingdoms. These scientists have concluded that the archaebacteria can no longer be classified with the true bacteria.

Archaebacteria were discovered in 1982 in a deep-sea thermal vent. These bacteria derived energy not from sunlight but from the oxidation of chemicals coming out of the vents. Their genetic makeup was found to be different from that of any other organism.

Because of these differences, some scientists divide the kingdom Monera into two kingdoms: kingdom Eubacteria and kingdom Archaebacteria.

Plants

The kingdom Plantae includes a great variety of organisms, from tiny mosses to giant redwood trees. How do scientists separate plants into related groups?

Scientists look at the different adaptations of plants to group them. An **adaptation** is a structure or feature of an organism that helps it meet a particular need in its natural habitat.

One adaptation of some plants is the presence of tissues that move materials such as water and nutrients through the plant. The presence or absence of this tissue is used to divide plants into two major groups. The stems of lilies, for example, can transport water and sugar. Mosses have no system to transport these materials. Because of this difference, lilies and mosses are placed in separate groups.

The way plants reproduce is also used to place them in groups. Some plants, such as ferns, produce spores to reproduce, while others produce seeds. Some plants, such as pine trees, produce seeds that have no outer covering. Others, such as apple trees, produce seeds that are covered by a fruit. Some plants, such as orchids, have flowers that contain the reproductive organs of the plant. Others have no such structure. Scientists use all of these features to help them classify plants.

Scientists also look at the structures of the roots, stems, and leaves of plants. They look at the size, shape, and division of the leaves of plants. They may look at the bark of a tree or the shape of a flower to help determine how the plant is related to others.

Scientists have classified all the kinds of living plants into 12 groups called divisions. These large divisions are divided into smaller groups to help scientists organize and understand the organisms that make up the kingdom Plantae.

✔ **What characteristics do scientists use to classify plants?**

adaptations such as presence or absence of vascular tissue; means of reproduction; structures of stems, roots, and leaves

Equisetum is a plant commonly called a horsetail or scouring rush. These tough, strong plants got their name because they were used in frontier times to scour, or scrub, pots and pans. ►

◄ Orchids are one of many different groups of flowering plants. Worldwide, there are about 24,000 different kinds of orchids. Many of them grow in rain forests.

◄ The sundew plant is one of about 350 kinds of plants that trap small animals such as insects. These insect-eating plants usually grow in poor soil. They make their own food but need nutrients that they get from their prey.

A45

USE PROCESS SKILLS

Observe and Infer Have students look at the plant pictures on this page.

- ► **Why do you suppose the petals on the orchid have such beautiful colors?** to attract insects that will help it cross-pollinate with other orchids
- ► **What can you infer about the sundew plant?** It must give off an odor or have a taste that attracts insects.

SCIENCE IDEAS

Help students focus on supporting details for this **main idea: Plants may be divided into groups on the basis of the structure of their roots, stems, and leaves.**

- ► **How do you think roots can vary?** Some plants have shallow roots, and others have roots that go many meters down into the soil.
- ► **How do you think stems can vary?** Some are fleshy (like those of grasses), and others are woody (like those of trees). Some plants have one stem that grows straight up. Others have several stems that branch out.

DEVELOP SCIENCE VOCABULARY

adaptation, p. A45—Adaptations are features that enable an organism to survive in its particular environment. Ask students to explain how an arctic hare and an arctic wolf are both adapted to their cold, snowy environments. coloration, fur, fat

ESL ACTIVITY

Explain that the terms ***Animalia***, ***Plantae***, ***Fungi***, ***Protista***, and ***Monera*** are Latin. Review the fact that Latin was the language of scientists in the 1700s, when Linnaeus created his system for naming living things. Have students use a dictionary or encyclopedia to find out if their first language is related to Latin.

Reaching All Learners

Kinesthetic Learners Have a group of visual learners model the classification system by using a set of seven nesting boxes. Have students label the largest box *Kingdom,* the next *Phylum,* then *Class, Order, Family, Genus,* and finally, *Species.* Encourage kinesthetic learners to nest the boxes saying the names of the classification labels as they do this. Then have them take the boxes apart.

Guide the Learning continued

SCIENCE IDEAS

Help students focus on supporting details for this **main idea: Animals may be divided into groups on the basis of whether or not they have a backbone.**

- How are animals without a backbone supported? They have shells (oysters), or outer casings (beetles), or other supportive structures.
- **Critical Thinking** Why do you think having a backbone can be an advantage? Encourage students to provide specific examples. For example, a backbone provides support, allows flexibility of movement, or protects a highly developed nerve cord.

USE PROCESS SKILLS

Observe and Infer Have students observe the animal pictures on page A46.

- What do mollusks have in common? All are softbodied. Many have an external hard shell.
- Why would the sea urchin need spines? The sea urchin cannot move quickly to escape predators. It must defend itself without moving.
- Why do you suppose the water bear lives in wet sand? The watery surroundings supply tiny food particles. In the dry sand, the water bears would dry out and lack a constant supply of food.
- Why is the entrance to the meerkats' den on top of a mound? What do you suppose the meerkats are looking at so attentively? The mound elevates the meerkats and enables them to see an approaching predator.

▲ A Spanish dancer is a kind of mollusk. Mollusks also include oysters, squid, and garden snails.

▲ A sea urchin has an internal skeleton but no backbone. Most sea urchins have a body covered with sharp spines that protect them from animals that would eat them.

▲ This animal, called a water bear, lives between the grains of sand on a wet beach or shoreline. It is one of many animals that have been discovered through the use of the microscope. The animal pictured here has been magnified more than 1000 times.

Animals

If you ask a friend to name an animal, you're likely to get a correct answer. Your friend might name a pet, such as a cat or a dog; a farm animal, such as a cow or a pig; or a "zoo animal" such as a giraffe or a bear. Yet all of these animals are members of just one group—mammals. There are about 4500 different kinds of mammals in the world. But there are more than a million different kinds of animals.

Scientists use the presence of a backbone to define one large group of animals. Animals with a backbone are organized into groups based on their body coverings, how they reproduce, how they breathe, and other characteristics. There are five large groups of animals with a backbone.

For animals without a backbone, scientists look at whether or not there is a skeleton, what it is made of, how it is shaped, and whether it is inside the body or outside the body tissue. For example, sea urchins and clams both have skeletons made of limestone. However, the clam's skeleton, or shell, is on the outside of its body and the sea urchin's skeleton is on the inside of its body. There are many more animals without a backbone than there are animals with a backbone.

✓ **What characteristics are used to classify an animal?** presence or absence of a backbone; size, shape, position of skeleton; body covering; method of reproduction

◀ Meerkats are one example of animals with a backbone. The meerkat is a mammal, one of the five large groups of animals with a backbone.

A46

Reaching All Learners

Kinesthetic Learners Have students physically act out the part of animals with a backbone and then animals without a backbone. Then have them write stories about what it would be like to be an invertebrate.

Reading Mini-Lesson

Summarize and Paraphrase a Selection Have students reread the first four paragraphs of the selection of p. A44 titled "How Living Things Are Classified." Display the transparency and have students write the main points in their own words. Then have them write three or four sentences to summarize.

Reading Mini-Lesson Transparency A2-1

Summarize and Paraphrase a Selection

Certain aspects of the Linnaean system of classification have changed over the years.

Characteristics such as structure and features are still used for classification, but other characteristics are important too.

Certain organisms are classified into kingdoms based on their cell structure and the way that they get food.

Some organisms are classified in the kingdom Fungi because they get food from their surroundings, instead of making their own.

There are many characteristics used to classify an organism. Scientists can consider an organism's structure, features, cell structure, and method for getting food in order to place it in a kingdom. Many organisms do not fit into kingdom Animalia or kingdom Plantae, and so they are classified in one of the other three kingdoms. Fungi is a kingdom that contains organisms that may look like plants but which get food from the environment.

READING SKILLS PRACTICE

To practice and apply this reading skill, see **Workbook p. WB23.**

Fungi and Protists

Organisms in the kingdom Fungi are all around you. Yeast—which makes bread rise—and bread mold are two types of fungus.

Scientists classify fungi by their shape, size, way of reproducing, whether they are one cell or many, how and where they live, and other similar characteristics. One-celled fungi like yeast are unusual. Most fungi, like the mushroom and the bracket fungi common in forests, are made up of more than one cell.

Many organisms that make up the kingdom Protista can have characteristics of plants, animals, or fungi. Most of these organisms are single-celled.

Scientists classify protists by whether they are more animal-like, plantlike, or funguslike. They group the plantlike protists mainly by their color and the types of materials they use to store food. Scientists group animal-like protists by shape, size, what the organism uses to move around, and other characteristics. They group funguslike protists mainly by their shape and reproductive structures.

Protists may be unfamiliar to you because you can't see them without magnification. But protists are all around you. One gram of soil—much less than a teaspoonful—can contain more than 150,000 protists.

✓ **What characteristics do scientists use to classify protists?** whether they are plantlike, animal-like, or funguslike; plantlike: color and food storage; animal-like: shape, size, movement structures; funguslike: shape and reproductive structures

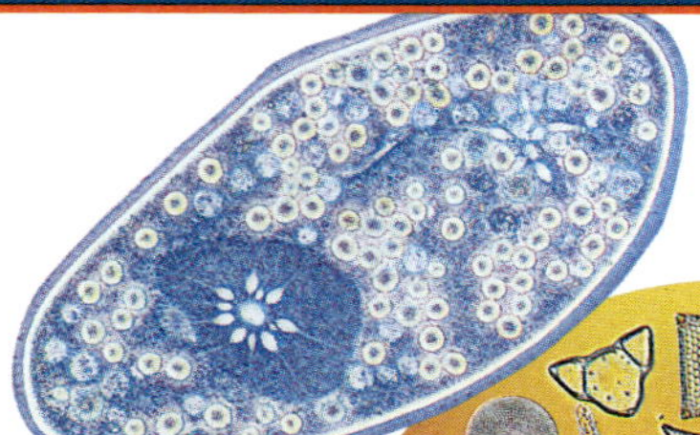

▲ The protist *Paramecium* swims by moving the tiny hairs on its surface.

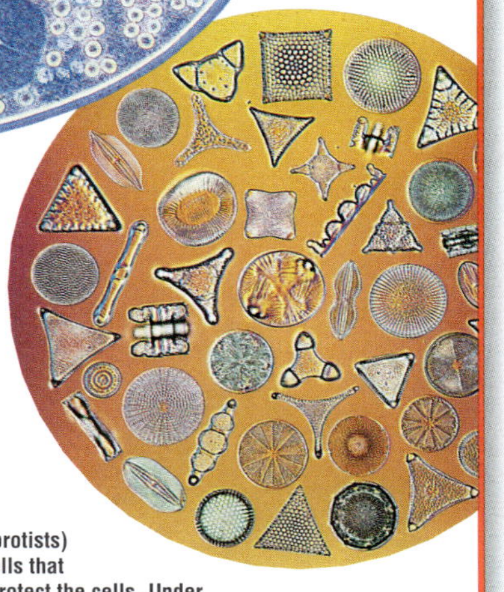

Diatoms (plantlike protists) and foraminifera (animal-like protists) both have shells that support and protect the cells. Under magnification you can see the beautiful shapes and colors of the diatom shells. ▲

Many kinds of fungi live in the forest. Some of the mushrooms, such as the morel shown here, are considered tasty treats. However, many of the mushrooms that grow in the wild are extremely poisonous. ►

When bracket fungi grow on dead trees like this one, they take in nutrients from the tree. ►

A47

VISUAL LEARNING

Have students look at the pictures of fungi and protists on page A47.

- ► What do you notice about the shape of the diatoms? Most are symmetrical, like a crystal.
- ► What structures can you identify in the paramecium? The main nucleus is clearly visible. (The paramecium may have several smaller nuclei, as well.) Vacuoles may also be seen. Vacuoles are used in excretion and storage.
- ► Can you tell whether or not a mushroom is poisonous just by looking at it? Explain your answer. No. Poisonous mushrooms cannot be distinguished from nonpoisonous ones by sight except by experts on fungi. So one should never eat a mushroom unless it has been grown for humans.

SCIENCE IDEAS

Help students focus on supporting details for this **main idea:** **Some protists are animallike, others are more plantlike, and still others are funguslike.**

- ► In what ways are some protists animallike? They move around to capture and eat food.
- ► In what ways are some protists plantlike? They make their own food by photosynthesis.
- ► What would be the advantage of being able to move like an animal and photosynthesize like a plant? You would be able to make your own food and to capture prey.
- ► In what ways are protists different from both plants and animals? They do not have specialized cells that are organized into tissues and organs.

Health Link

Malaria is actually several different diseases caused by several different species of protist. Several days after entering the host organism, the parasitic protists invade the host's red blood cells. As the parasites grow, the red blood cells burst. This bursting is what causes the high fever associated with malaria.

The invading protists are transmitted by the female *Anopheles* mosquito, which bites an infected individual and then transfers some of the infected blood into her next victim.

Although considerable progress has been made in the fight against malaria, it still kills up to 2 million people each year, usually in developing countries. Have students **use reference sources,** such as an encyclopedia, to find current means of treatment and prevention of malaria.

ESL ACTIVITY

The plural ending *-i* in the term ***fungi*** may confuse some ESL students. Explain that most nouns are made plural by adding *-s* or *-es*. In fact, *funguses* is an acceptable plural form of *fungus* though it is not commonly used. This same irregular plural form is also used for the noun *nucleus* to form its plural *nuclei.* Have students with intermediate levels of fluency draw simple pictures of mushrooms, use the labels *fungus* and *fungi,* and discuss the terms.

Guide the Learning continued

SCIENCE IDEAS

Help students focus on supporting details for this **main idea:** The kingdom Monera consists of different types of bacteria.

- Do all bacteria look alike? Explain your answer. No, they come in different shapes.
- What are the shapes of bacteria? spirals, spheres, rods
- What determines the shape of a bacterial cell? its cell wall
- Can you see bacteria? Explain your answer. You can't see a single bacterium, but you can see bacteria when they clump together as a colony.

ANALYZE AND INTERPRET DATA

Guide students in interpreting the chart at the top of page A48 and **using this graphic source for information.**

- What do bacteria have in common with plants? They have cell walls.
- Which kingdom shows the greatest diversity of organisms? kingdom Animalia
- Why do you think Plantae and Animalia were the first two kingdoms identified? Most plants and animals can be seen with the naked eye. Protists and monerans are mostly single-celled and can be seen only with a microscope.
- Why do you think fungi were mistaken for plants for so long? Like plants, they grow slowly and do not move.

The Kingdoms of Living Things

Kingdom	Description	Kinds Known
Plantae	multicelled, cell walls, nucleus, make their own food, cannot move	about 300,000
Animalia	multicelled, no cell walls, nucleus, eat food, most can move	more than 2,000,000
Fungi	mostly multicelled, cell walls, nucleus, take in food, cannot move	about 100,000
Protista	mostly single-celled, some have cell walls, nucleus, take in food or make their own food, many can move	about 60,000
Monera	single-celled, cell wall, no nucleus, take in food or make their own food, some can move	about 5000

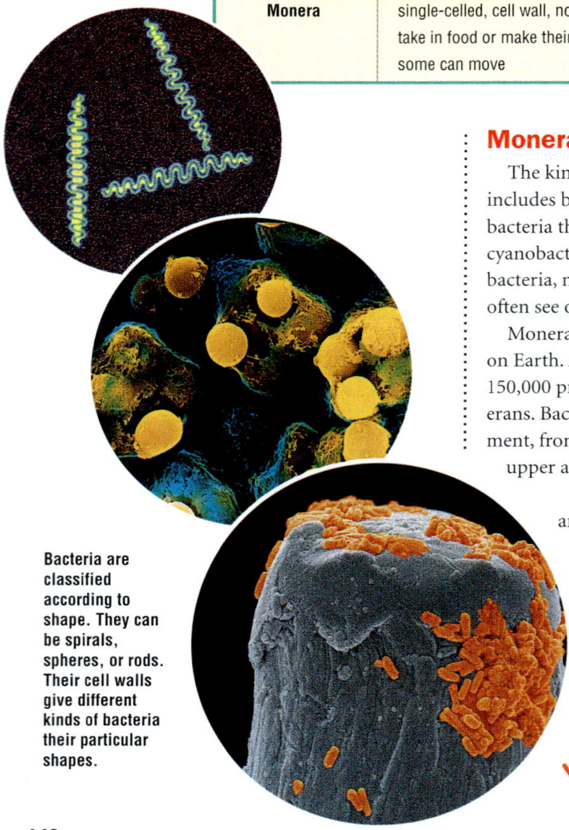

Bacteria are classified according to shape. They can be spirals, spheres, or rods. Their cell walls give different kinds of bacteria their particular shapes.

Monerans

The kingdom Monera consists of bacteria. It includes both bacteria that cause disease and bacteria that do not cause disease, including cyanobacteria. Cyanobacteria, or blue-green bacteria, make up the green scum that you can often see on the surface of ponds.

Monerans are the most numerous organisms on Earth. A gram of soil that might contain 150,000 protists could hold over 2 billion monerans. Bacteria exist in almost every environment, from the depths of the oceans to the upper atmosphere.

Scientists classify bacteria by the size and shape of the cell. They also consider how the cells are arranged in relation to each other.

All of life on Earth belongs to one of the five kingdoms—Plantae, Animalia, Protista, Fungi, and Monera. The chart on this page defines each of the kingdoms by its most important features.

✓ **How do monerans differ from all other organisms?**
Their cells do not have nuclei.

A48

INVESTIGATION CHALLENGE

Hands-On Activity: Modeling Microscopic Organisms

Use this idea for a hands-on activity to extend students' understanding of key lesson concepts.

Invite students to use modeling clay, construction paper, tissue paper, wire, glue, glitter, and other materials to design models of protists and monerans. You may provide other pictures for students to examine in addition to those found in their books.

Challenge students to duplicate each of the bacterial shapes described in the text: spirals, rods, and spheres. Protist models should include any flagella or cilia possessed by the organism. Have students compare these models to those of plant cells.

For student self-assessment, see Assessment Guide p. AGxvii.

Summary

Scientists classify living things to organize information about them and to show relationships between them. The characteristics they use to classify organisms include the structure and number of their cells; the ways they get food; and their color, shape, size, mobility, and DNA. All living things are placed in one of five kingdoms—Plantae, Animalia, Protista, Fungi, or Monera.

Review

1. What is the purpose of classification?
2. Which two characteristics are the most important for classifying organisms into kingdoms?
3. What characteristics does an organism that belongs to the kingdom Plantae have?
4. If you discovered a new multicelled organism that could move, in which kingdom might you place it? Explain your answer.
5. **Critical Thinking** What scientific invention caused scientists to decide that there should be more than two kingdoms? Explain your answer.
6. **Test Prep** Which kingdom is made up of organisms that have cells without cell walls?
   A Plantae
   B Fungi
   C Animalia
   D Monera

LINKS

MATH LINK

Estimate A gram of soil can hold 150,000 protists or 2,000,000,000 monerans. About how many times as great as the number of protists is the number of monerans?

WRITING LINK

Informative Writing—Description Suppose that you could tour the microscopic world of organisms. Write a journal entry for your teacher to describe the kinds of things you might see on your journey. Be sure to tell which kingdom each of the living things you see belongs to.

HEALTH LINK

Organisms and Disease Some protists, fungi, and monerans cause human diseases. Malaria, for example, is caused by protists. Find out about one organism that causes a disease. Research the disease and how modern medicine treats it.

TECHNOLOGY LINK

GO ONLINE

Visit the Harcourt Learning site for related links, activities, and resources. **www.harcourtschool.com**

Welcome to The Learning Site

LINKS TIPS

Links Objective Extend understanding of the kingdoms of organisms.

Math Link Some students may be intimidated by the large numbers in this division problem. Before dividing, have them divide the divisor and the dividend by the same power of 10, or have them cross off equal numbers of zeros from both figures. (200,000 divided by 15). Then help them determine the placement of the first digit in the quotient and the number of digits in the quotient. about 13,000 times as great

Writing Link Students should choose an environment to explore and then **use reference sources** to find out what sort of microscopic organisms would normally be found there. Encourage them to illustrate their entries. Writing models are provided in the **Teaching Resources** book.

3 Wrap Up and Assess

SUMMARY

Graphic Organizer for Chapter Concepts After reading the summary, have students complete the Lesson 1 section of the graphic organizer (shown on page A42).

REVIEW

1. to organize and show relationships among different organisms
2. cell structure and the way the organism gets food
3. It is multicellular and has cell walls, a nucleus, and chloroplasts, which it uses to make its own food. It cannot move.
4. probably in the animal kingdom because, although protists and monerans can move, they are mostly single-celled
5. the microscope, which allowed scientists to see organisms they couldn't see with the naked eye and to study the structure of cells in more detail
6. C, Animalia

TEST PREP Test-Taking Tips

Item type multiple choice with question stem

To help students answer Question 6, share the following strategy with them. **Read the question carefully. Try to recall some facts that will help narrow the choices, such as which organisms can move freely and which are fixed in one place. Then select the best answer to the question.**

LESSON CONCEPT REVIEW/ASSESSMENT

Students can use **Workbook p. WB24** to review the lesson concepts.

LESSON 2

How Are Kingdoms Subdivided?

Objectives

- Recognize that kingdoms are divided into smaller groups.
- Describe the two-part scientific name of an organism.
- Demonstrate how to use a dichotomous key to identify living things.

Motivate

More Fast Facts

Spark interest with these facts and questions.

- Koalas almost never need to drink; they get most of their water from the eucalyptus leaves they eat. The word *koala* is an aboriginal word meaning "no-drink animal."

  How can koalas exist without drinking? They are adapted to living with very little liquid.

- Newborn kangaroos are about the size of a paper clip and weigh no more than one gram.

  How do the tiny baby kangaroos survive? They live in a pouch on their mother's abdomen, where they feed and grow until they are big enough to come out into the world.

Access Prior Knowledge

- What do you, a dog or cat, and a cow have in common?
- Are insects animals? What characteristics do insects have?
- How can you identify a tree?

When Minutes Count . . .

Whole-Class Investigate Option Have the class develop a single key as a group exercise. Have them run trials with each type of bean to confirm that their key works.

Using the Activity Video The video can be used to preview and model the investigate process with students. After viewing the video, students can conduct the investigation in small groups.

LESSON 2

How Are Kingdoms Subdivided?

In this lesson, you can . . .

INVESTIGATE how to develop a system for identifying beans.

LEARN ABOUT the seven levels of classification and how to use a dichotomous key.

LINK to math, writing, language arts, and technology.

INVESTIGATE

How to Develop a Key

Activity Purpose In Lesson 1 you developed a classification system for beans. In this investigation you will develop a key to help others identify their beans by the same names you used.

Materials

- bean classifications from the investigation in Lesson 1
- paper and pencil

Activity Procedure

1. Use your classification system from the investigation on pages A40–A41. (Picture A)
2. Make a simple flowchart or other diagram that would allow others to correctly identify their beans by the same name you gave to yours. Don't rely on pictures only! (Picture B)
3. Test your identification key by asking another student to use it to identify one or more beans.

Shells on a beach show the diversity of living things.

A50

BACKGROUND

Webliography

Keyword keys

www.harcourtschool.com

Dichotomous Key A key is a series of step-by-step instructions that allows a person to identify an object on the basis of its characteristics. Each step may offer several paths from which to choose. When multiple options are given at each step, identification can become complicated.

A dichotomous key is generally easier to use. At each step, this key asks the reader to decide which of two statements best fits the organism he or she is trying to identify. The reader then moves on to the next pair of statements until the organism is identified. Many dichotomous keys contain simple yes-or-no questions relating to a particular feature or trait (for example, whether the organism has a backbone).

For student self-assessment, see Assessment Guide p. AGxvi.

Picture A

Picture B

Draw Conclusions

1. **Compare** your key with those of classmates. How many kinds of keys were made? Are some keys easier to use than others? Explain your answer.
2. Why might a key be a more useful tool than a set of pictures of organisms?
3. **Scientists at Work** When scientists classify items or make keys for identification, they **gather** and **record data** carefully and precisely. Why might scientists want to avoid using imprecise terms such as *small, big, heavy,* and *light* in a key?

Investigate Further Choose another group of objects or organisms. **Develop a classification key** to help people identify the objects or organisms.

Process Skill Tip

To **gather** and **record data,** you first need a set of careful observations. Then determine which of your observations are the most important for someone who is doing the same kind of work.

This is one way to make a key to help people identify animals. ►

INVESTIGATE

Time 30–40 min **Preparation Tips** See p. A38c.

Grouping individuals or groups

Process Skill Tip gather and record data

Other Process Skills compare

Expected Results Student keys should indicate a separate category for each type of bean.

Teaching Tips

Employ questioning strategies to clarify students' ideas about how to identify each type of bean. For example, ask "What characteristics are important in identifying your group?"

Draw Conclusions

1. Students will probably produce more than one type of key. The easiest ones to use will probably be dichotomous.
2. The key asks the reader to look for one characteristic. A picture may depict more than one characteristic at a time.
3. Relative terms have no meaning in a key because the person trying to identify an organism may have only a single specimen of the species with nothing to compare to.

Investigate Further Students may choose to make a key for animals in the schoolyard or plants in a park. If you are unable to take your class outside, have students make a key for inanimate objects, such as pens or notebooks.

Informal Assessment

Portfolio Students may want to keep their dichotomous keys in their science portfolios.

Process Skill Tip

Gather and Record Data Display the Process Skill Tip Transparency. Invite students to observe the two insects and to make careful notes about the characteristics of each.

Then challenge students to complete the chart to record the data. For instance, the first characteristic may be body shape. Other characteristics could include presence of dots, type of antennae, and wing shape.

Process Skill Tip Transparency A2-2

Gather and Record Data

Specimen A Specimen B

	Specimen A	Specimen B
Shape of body	thin	fatter oval
Type of Antenna	simple straight	feathery
Wing Shape	wide	narrow
Wing Decoration	spots and bands	light and dark fields, some spots
Body Decoration	wide, close, light bands	narrower, distant, light bands

- Gather and record data to compare the two insects.

INVESTIGATE LOG

Students can use **Workbook pp. WB25–26** to record the results of the investigation.

PROCESS SKILLS PRACTICE

To practice and apply process skills, see **Workbook p. WB27.**

LEARN ABOUT

1 Before Reading

SET A PURPOSE

FIND OUT Remind students that these statements will help guide them in what to look for as they read.

VOCABULARY If students are doing the activity described in the Vocabulary Preview on page A39, remind them to add the definitions for these words to the chart.

Graphic Organizer for Chapter Concepts Display the transparency (shown below) or have students use Workbook page WB19. Point out that by the end of the lesson, students should be able to list how kingdoms are subdivided.

2 Guide the Learning

SCIENCE IDEAS

Help students focus on supporting details for this **main idea: Scientists use seven main levels of classification, in which each group on one level is made up of smaller, related groups on the level below.**

Reinforce students' understanding of levels of classification by reminding them of the discussion in Lesson 1 about all fruits appearing in the same section of the grocery store.

- **Name some specific types of a given fruit that would be grouped together within the fruit section.** apples (Red or Golden Delicious, Rome, McIntosh, or Granny Smith) or grapes (red, black, green, or Concord)
- **What is the highest level of scientific classification?** kingdom
- **What levels are beneath kingdom?** phylum, class, order, family, genus, and species
- **Which two organisms are more closely related: two species within one genus or two species within one family?** two species within one genus
- **Can a species belong to more than one order? Explain your answer.** No. An order can contain more than one species, but a single species can belong to only one order.

Ongoing Assessment

Use the ✓ questions in the pupil edition.

LEARN ABOUT

How Kingdoms Are Subdivided

FIND OUT

- how kingdoms are subdivided into phylum, class, order, family, genus, and species
- how organisms are named
- how to use a dichotomous key

VOCABULARY

genus
species
dichotomous key

Grouping Living Things

When you mail a letter to a friend, you can be pretty sure that it will be delivered to the right person. The mail carrier won't just ask around to find out if anyone knows where your friend might be. Instead, the postal service will use the same classification system you used when you wrote your friend's name and address on the envelope. The categories included in an address are the name of the person to whom the envelope is addressed, apartment number, building (or house) number, street, city, state, and sometimes country.

A member of one category of the postal classification system must be part of the next larger category. For example, a street is in a city, and a city is in a state. An address in Canada, for example, must not include a street in the United States.

The Linnaean classification system groups living things in a similar way. Linnaeus described several levels of classification. A total of seven levels is now used. The largest group is a kingdom. The other levels, in order from largest to smallest, are phylum, class, order, family, genus, and species. A group on one level is made up of smaller, related groups on the level below it. A kingdom is made up of several related phyla, and each phylum is made up of several related classes. The species in one genus are more closely related to each other than they are to species of any other genus. Families in one order are more closely related to each other than they are to families in other orders. You'll find out more on pages A54 and A55.

Just as an apartment in Atlanta is also in Georgia, an organism that belongs to an order also belongs to the class that contains that order. And just as a resident of the United States cannot live on a street in Mexico, a member of one kingdom cannot belong to a family that is part of another kingdom.

Each part of the postal classification system—building number, street, city, state, and country—is a part of all of the levels above it. ▼

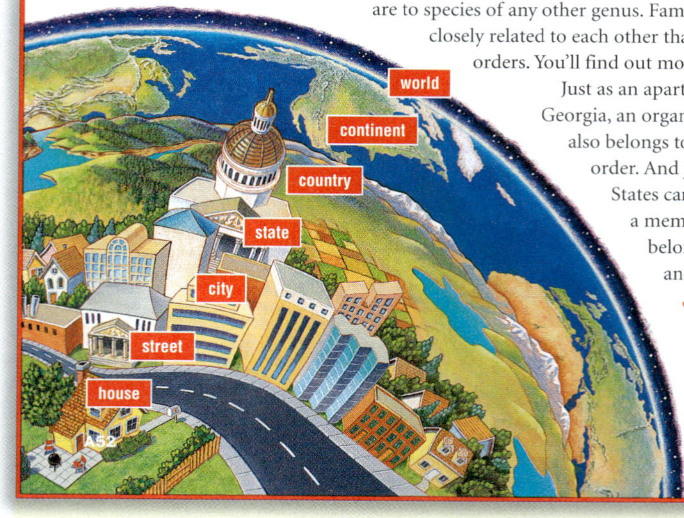

✓ **In which of Linnaeus's levels would you expect all the members to be the most similar to each other? Why?** Species; this level is not divided into smaller groups.

Graphic Organizer for Chapter Concepts

Transparency A2 • Workbook, p. WB19

Classification

Kingdoms	Subdivisions of Kingdoms
1. Animalia	1. Phylum
2. Plantae	2. Class
3. Fungi	3. Order
4. Protista	4. Family
5. Monera	5. Genus
	6. Species

▲ These two animals show why scientists need a single system for naming organisms. In different parts of the country, each animal is called a gopher.

Number	Phylum	Includes
1,000,000	Arthropoda	insects, spiders, crabs
150,000	Mollusca	oysters, snails, squid
54,000	Chordata	fish, mammals, birds
15,000	Annelida	earthworms, leeches
7000	Echinodermata	sea stars, sea urchins

Number of Species in Some Phyla of the Animal Kingdom

▲ Keeping track of all the species in the animal kingdom would be impossible without a naming system. The scientific names of the animals provide information about how animals are related between various species.

Naming Living Things

What is a gopher? The answer depends on the person you ask. In the Midwest, people call a striped ground squirrel a gopher; in the Southeast, people know a burrowing tortoise as a gopher. Two different organisms may be called by the same name, or one organism may be called by different names. For example, the animal called a mountain lion in some parts of the country is called a puma, cougar, or panther in other parts of the country.

Identifying an organism can be difficult when several organisms are called by the same name, or when one organism is called by several names. To avoid this confusion, scientists use a system of scientific names for living things. The system was set up by Linnaeus, who gave each organism a two-part name.

The first part of an organism's name is its genus name. A **genus** (JEE•nuhs) is a group of organisms that share major characteristics and are therefore closely related. For example, cats that purr are part of the genus *Felis*, which is Latin for "cat." A genus name always begins with a capital letter and is either underlined or in italics.

The second part of a scientific name identifies one specific organism in the genus. For example, the house cat is *Felis domesticus*, and the mountain lion is *Felis concolor*. Each different kind of organism is a **species** (SPEE•sheez). A species name always begins with a lowercase letter and is also underlined or in italics.

If all scientists use an organism's scientific name, they can be sure they are all referring to the same organism when they talk or write about it. The scientific name also contains information about how the organism is related to other organisms.

✔ **Why do scientists use scientific names for organisms?** to avoid confusion and to be sure they are referring to the same organism

A53

ANALYZE AND INTERPRET DATA

Guide students in interpreting the information in the **graphic source** at the top of the page.

- ▶ **Which phylum contains the most species?** Arthropoda contains the most species.
- ▶ **How many times as many kinds of arthropods are there as kinds of mollusks?** about 7 times
- ▶ **Do all animals that live in the ocean belong in the same phylum? Explain.** No. Sea stars and sea urchins are in a different phylum from oysters.

Students could use a graphing program such as ***Data ToolKit*** to convert the table to a circle graph.

USE PROCESS SKILLS

Compare and Infer Have students compare the two animals at the top of page A53.

- ▶ **Describe both the physical and the behavioral characteristics of these two animals.** One looks like a tortoise; the other, a ground squirrel or chipmunk. Both burrow into the ground.

DEVELOP SCIENCE VOCABULARY

genus, p. A53—The word *genus* comes from the Latin noun that means "kind." Organisms within one genus are clearly recognizable as being of one kind. For example, the genus *Canis* includes all dogs, wolves, jackals, and coyotes. Ask students to guess which animals belong to the genus *Panthera*. tiger, lion, jaguar, leopard

species, p. A53—Species names often describe what the organism looks like or where it is typically found. Ask: Why are house cats called *Felis domesticus*? *Domesticus* means "house."

Language Arts Link

Share with students the following mnemonic for remembering the levels of classification in the appropriate order:

King **P**hillip **C**ame **O**ver **F**or **G**reat **S**paghetti.

Have students develop other mnemonics.

ESL ACTIVITY

Invite students to develop a two-part naming system for objects they use. For example, they might make up "**genus**" names for categories such as objects they play with, eat, listen to, or wish they had. Then they could add "**species**" names for specific types of objects within each genus.

Guide the Learning continued

VISUAL LEARNING

Have students look at the pictures of organisms on pages A54–A55.

- At which classification level do cats most resemble the other members of the group? the species level
- What is the highest level at which the resemblance of the cats to the other animals at that level is still obvious? the family level
- At what level does the whale first appear with the cat? the class level
- What does a cat have in common with a gila monster? Both are vertebrates.

SCIENCE IDEAS

Help students focus on supporting details for this **main idea: Members of a lower classification level resemble each other in more ways than do members of a higher classification level.**

- What characteristics do cats share with all other animals? They are multicellular, are able to move, and cannot produce their own food.
- What characteristics do cats share with all other mammals? Besides all of the characteristics above, they produce milk for their young, are covered in fur, and are warm-blooded.
- What characteristics do cats share with other cats? All of the above, plus they have retractable claws, erect ears, and long whiskers. They also possess keen night vision and exceptional senses of hearing and smell.

THE INSIDE STORY

The Seven Levels of Classification

There are seven levels in the classification system of living things. At each level, an organism belongs to a group of living things with shared characteristics. The chart on these pages shows the classification of *Felis domesticus*, the common house cat. As you can see from the chart, members of a group at a lower level of classification resemble each other in more ways than members of a group at a higher level do. Members of a species share at least one characteristic that no other animals have.

You can get a lot of information from the classification of an organism. The group that it belongs to at each level of classification lets you know some of the characteristics of the organism. For example, the house cat belongs to the order of carnivores, so you know that it eats meat. It is in the class of mammals, so you know that it has fur, and can produce milk for its young.

✓ **Are all mammals carnivores? How can you tell?**

No; carnivores are only one of the orders that make up the class of mammals.

A54

Cultural Connection

Cats The ancient Egyptians were the first to domesticate cats, bringing the animals into their homes as long ago as 2500 B.C. Cats were prized for their ability to hunt rodents and reptiles in the house and in the fields. The Egyptians were so respectful of cats that they mummified them upon death, just as they did the pharaohs.

Level	Name	Description
Kingdom	Animalia	The kingdom *Animalia*, the animals, is divided into phyla. The phyla include chordates, mollusks, worms, jellyfish, sponges, echinoderms, and arthropods such as *Libellula pulchella*—the ten-spot dragonfly.
Phylum	Chordata	The phylum *Chordata* includes all vertebrates and some animals that have a cartilage rod instead of a backbone. Chordates include classes of mammals, fishes, birds, amphibians, and reptiles such as *Heloderma suspectum*—the Gila (HEE•luh) monster.
Class	Mammalia	The class *Mammalia*, animals that have fur or hair and can produce milk for their young, is made up of many orders. The orders include carnivores, primates, elephants, rodents, bats, rabbits, and cetaceans such as *Delphinapterus leucas*—the beluga whale.
Order	Carnivora	The order *Carnivora*, meat-eating animals, includes families of cats, dogs, weasels, seals, and bears such as *Ursus horribilis*—the grizzly bear.
Family	Felidae	The family *Felidae*, the cat family, consists of the genus *Felis* and the genus *Panthera*—cats that roar, such as *Panthera tigris*, the tiger.
Genus	*Felis*	*Felis domesticus* belongs to the genus *Felis*, purring cats, which also includes such cats as *Felis pardalis*, the ocelot.
Species	*Felis domesticus*	House cats belong to the species *Felis domesticus*.

A55

USE PROCESS SKILLS

Compare and Infer Have students look at the pictures and captions on pages A54–A55.

- ▶ In what ways are polar bears and tigers alike? In what ways are they different? They both eat meat and produce milk for their young. The polar bear is not a member of the cat family; therefore, it cannot roar and is not as agile.
- ▶ What do all the members of the order Carnivora have in common? They all eat meat and they are all mammals.

Critical Thinking Lead students to **draw a logical conclusion** about classification.

- ▶ The dragonfly eats mosquitoes. Why, then, isn't it considered a member of Carnivora? Carnivora is a sublevel of mammals. All animals that belong to the order Carnivora are therefore mammals. Dragonflies are not mammals, so they cannot be members of Carnivora.

SCIENCE IDEAS

Help students focus on supporting details for this **main idea: An organism's classification can tell you a lot about its characteristics.**

- ▶ On the basis of the cat's classification, what do you know about how it meets its energy needs? It is an animal; therefore, it must consume plants or other animals to meet its energy needs.
- ▶ What do you know about what it eats? Because it is a carnivore, it eats meat.
- ▶ What do you know about internal structure? It is a vertebrate; therefore, it has a backbone.

Reading Mini-Lesson

Distinguish Fact and Opinion Remind students that facts can be proved, and opinions cannot. Display the transparency. Ask students to determine whether the statement is a fact or an opinion.

READING SKILLS PRACTICE

To practice and apply this reading skill, see **Workbook p. WB28.**

Reading Mini-Lesson Transparency A2-2

Distinguish Fact and Opinion

Statement	Fact/Opinion
Scientists classify organisms by looking at their cell structures and how they get food.	Fact
The kingdom Fungi contains organisms that are very interesting.	Opinion
The Linnaean system is the best way to classify living things.	Opinion
The first part of an organism's name is its genus name.	Fact

Reaching All Learners

Visually Impaired On the front board, enlarge the picture map shown on pages A54–A55, using posters of each organism. If you cannot find a poster of a gila monster or a dragonfly, substitute another organism in that classification level.

Guide the Learning continued

ANALYZE AND INTERPRET DATA

Guide students in interpreting the key on this page.

- ▶ After reading Step 1 in the key, what kind of animal do you think we are trying to key? one that lives in water
- ▶ What does Step 1 distinguish for you? whether your animal is a fish or a whale
- ▶ If you were sent from Step 1 to Step 3, could your animal be a whale? No, only fish appear in Step 3.
- ▶ If you had started the key at Step 5, would you have correctly identified your animal? Probably not. If your animal were a whale with a tail half the length of its body, you would have incorrectly identified it as a thresher shark.
- ▶ What do you know about sharks from Step 3? They all have gill slits behind the mouth.
- ▶ Why aren't the phrases in Step 5 dichotomous? Step 5 does not lead to further steps. It simply identifies two kinds of organisms.

DEVELOP SCIENCE VOCABULARY

dichotomous key, p. A56—Write the word *dichotomous* on the board and circle *di* and *tom.* Explain that *di-* means "two" and *-tomy* means "cutting." *Dichotomous* comes from the Greek word meaning "cut in two." Point out to students how at each step in the key, one group of organisms is cut into two subgroups.

Using a Key

Have you ever hiked through the woods and noticed an animal that you had never seen before? Even people who study nature cannot recognize every living thing they encounter. When bird-watchers and others interested in nature go hiking or camping, they often carry guides that help them identify species of living things they observe. One type of guide is a field guide—a book with pictures or photographs of local animals or plants.

Another guide that can be used to identify an organism based on its characteristics is a **dichotomous key** (dy•KAHT•uh•muhs kee). It consists of a series of choices that lead to the correct name of the organism you want to identify. *Dichotomous* means "divided into two parts." Each step of the key presents a pair of phrases describing an organism's characteristics. Only one of the phrases applies to the organism you are identifying, and that choice directs you to another pair of phrases and another choice. The process continues until you reach the name of the organism.

The dichotomous key on this page can be used to identify certain species of animals. Follow the steps in the key.

✔ **What are the names of the animals in the photographs labeled *A*, *B*, and *C*?** *A,* narwhal; *B,* thresher shark; *C,* hammerhead shark

Dichotomous Key

Step		
1	a. tail fins are horizontal—whale	go to 2
	b. tail fins are vertical—fish	go to 3
2	a. has teeth or tusk—toothed whale	go to 4
	b. has no teeth	baleen whale
3	a. has gill slits behind mouth—shark	go to 5
	b. has no gill slits	nonshark fish
4	a. black with white underside	killer whale
	b. tusk, gray with dark spots	narwhal
5	a. head is hammer shaped	hammerhead shark
	b. tail is half the body length	thresher shark

INVESTIGATION CHALLENGE

Hands-On Activity: Key Cards

Use this idea for a hands-on activity to extend students' understanding of key lesson concepts.

Invite the class to come up with 10 different animals. Write the names of the animals on the board. Then make a list of six pairs of opposite characteristics, like "has teeth/does not have teeth" or "has hair/does not have hair," that apply to the animals named.

Now give each student or group of students a stack of index cards. Have students write the name of each animal on a separate card. Have them write each characteristic (from each pair) on a separate card.

Then challenge them to design a dichotomous key using their cards. Each step (pair of characteristics) should lead to another step or to the identification of an organism. (Students may need to come up with new characteristic pairs as they go, in order to distinguish between animals.)

For student self-assessment, see Assessment Guide p. AGxvii.

Summary

There are seven levels of classification. A kingdom is divided into smaller groups: phylum, class, order, family, genus, and species. The scientific name of an organism has two parts—its genus and species names. A dichotomous key is one type of guide used to identify living things. It presents a series of choices that lead the user to the correct identification of a living thing.

Review

1. What two parts make up an organism's scientific name?
2. What are the advantages of using an organism's scientific name?
3. Which organisms share more characteristics, the members of a species or the members of an order? Explain your answer.
4. **Critical Thinking** Can two species be members of the same family but not the same genus? Can they be members of the same genus but not the same family?
5. **Test Prep** The scientific name of a red maple tree is correctly written as —
   - A *Acer rubrum*
   - B Acer Rubrum
   - C *Acer Rubrum*
   - D *Acer* rubrum

LINKS

MATH LINK

Compute Percents Of an estimated 1,400,000 described species, about 62% are arthropods. Of the arthropods, about 86% are insects. About how many species of insects are there?

WRITING LINK

Expressive Writing—Poem or Song Lyrics Write a poem or song that would help your classmates remember the seven divisions of classification and their proper order from kingdom down to species.

LANGUAGE ARTS LINK

Naming a Species The first person to describe a new species has the privilege of naming it. People often name species after themselves or their friends. For example, *Rhea darwinii* is a bird named after Charles Darwin. Some names are expressive: *horrida*, *tormentor*, and *abominator* are each a kind of mosquito. Some groups have been given funny names: a genus of squash bug is named *Polychisme* (pronounced "Polly kiss me"). Invent a new organism, draw or describe it, and make up a scientific name for it.

TECHNOLOGY LINK

To learn more about classification, watch *Dinosaur Rock* on the **Harcourt Science Newsroom Video**.

A57

3 Wrap Up and Assess

SUMMARY

Graphic Organizer for Chapter Concepts After reading the summary, have students complete the Lesson 2 section of the graphic organizer (shown on page A52).

REVIEW

1. its genus and its species
2. Each organism has only one scientific name. It can have more than one common name.
3. The members of a species share more characteristics. Each sublevel of classification indicates closer relationships among the members, and species is a lower classification level than order.
4. Yes. No, if two species belong to the same genus, they also belong to all the same levels that appear above genus (family, order, class, phylum, and kingdom).
5. A, *Acer rubrum*

TEST PREP Test-Taking Tips

Item type multiple choice with a sentence fragment

To help students answer Question 5, share the following strategy with them. **Read the question carefully. Try to recall some facts that will help narrow the choices, such as how the names are capitalized and italicized. Eliminate any answers that are obviously wrong. Then select the best answer.**

LESSON CONCEPT REVIEW/ASSESSMENT

Students can use **Workbook p. WB29** to review the lesson concepts.

VOCABULARY REVIEW

Students can use **Workbook p. WB30** to review the chapter vocabulary.

LINKS TIPS

Links Objective Extend understanding of how kingdoms are subdivided.

Math Link Remind students that the problem requires two multiplications. Because the question asks *about* how many species there are, students should round the factors. 1,400,000 × 0.6 = 840,000; 840,000 × 0.9 = 756,000, or about 760,000

Language Arts Link The scientific names students suggest can reflect the physical appearance of their organism, its geographic region, or its behavior. Encourage students to alter words that are in common use to make them sound more scientific.

Objectives

- Evaluate the impact of research and technology on scientific thought, society, and the environment.
- Identify careers related to science.

1 Before Reading

SET A PURPOSE

Ask students to examine the title and pictures for the article and predict what it will be about. Record their predictions on the board along with any questions they hope to have answered. After reading the article, encourage students to research the answers to any questions that were not answered in the text.

2 Guide the Learning

SCIENCE IDEAS

Help students gather information from this nonfiction science article by asking these questions:

- What two things are needed in order to see an object through a microscope? a light source to light up the object and a lens to focus the light and magnify the object
- What limits how much a microscope can magnify an object? the lenses
- How many times can an optical microscope magnify an image before the image gets blurry? about 1000 times
- How are scientists then able to see things smaller than an optical microscope can show? Rather than using visible light to form magnified images, they use electrons, X rays, ultraviolet light, or ions.
- How is an atomic force microscope like a record player? Students may know that a record is played by dragging a thin needle along the contours of the groove in the record's vinyl. Similarly, the atomic force microscope pulls a tiny needle across an object in order to read its contours.

SCIENCE AND TECHNOLOGY

Microscope Technology

Scientists are developing microscopes that allow them to look clearly at smaller and smaller objects.

Loriciferans (lor•ih•SIF•uh•renz) are a new phylum of animals that have been studied by using a scanning electron microscope. The small image on the left is a scanning electron micrograph. The large image above is a drawing made from a scanning electron microscope image.

How does a microscope work?

All microscopes need some form of light or energy to light up an object to be viewed. Another device in the microscope magnifies and focuses an image formed by the light. Most of the microscopes you may be familiar with are optical microscopes. They use visible light to light up objects. A glass lens or series of lenses focuses the light and magnifies the object.

A58

Reading Mini-Lesson

Use Context to Determine or Confirm Word Meaning

Remind students that they can often figure out or confirm word meanings by reading the sentences that precede, include, and follow words they are wondering about. Copy this chart onto the board. Ask volunteers to locate each term in the lesson. Have them read each term in context and fill in the second column of the chart, basing their guesses on what the context has told them the word might mean. Then have students complete the third column of the chart by listing each word's meaning as it appears in a dictionary.

Term	What It Might Mean	What It Means
optical		
image		

How small an object can microscopes make visible?

All microscopes are limited by the type of light or energy they use to enable scientists to view things. In theory, a microscope can show details in objects that are as small as the wavelength of the type of light or energy it uses. So, in theory, optical microscopes can show detail in objects that are as small as a single wavelength of visible light. But in reality, a microscope is limited by the lenses that are used in it. Because of this limitation, optical microscopes can't show very clearly anything that has been magnified more than about 1000 times. That can make a very tiny object appear large. But there are even smaller objects that an optical microscope can't help us see.

How do scientists see even smaller objects?

Improved microscopes use forms of light and energy such as electrons, X rays, ultraviolet light, and ions to enable scientists to view tiny objects. Electron microscopes use beams of electrons to light up objects. The wavelength of an electron is much smaller than the wavelength of visible light. So, a beam of electrons (rather than a beam of light) can produce images of much smaller objects. An electron microscope can show objects that measure less than 2 angstroms. One angstrom is one ten-billionth of a meter (about 4-billionths of an inch).

The atomic force microscope produces images by dragging a tiny needle across an object. The needle bumps up and down as the atoms on the tip of the needle are repelled by the atoms on the object. A computer then turns these movements into a three-dimensional color image. This microscope enables scientists to see the outlines of atoms.

Why is developing new types of microscopes important?

The microscope is one of the most important scientific tools. In medicine, microscopes are used to examine blood and tissue samples for diseases. Engineers use microscopes to examine and develop tiny computer chips and electronic devices. In the future, new microscopes could help scientists map genes and discover more about the structure of viruses. New treatments for diseases, smaller and faster computer chips, and new types of materials could result from the ability to see and work with smaller objects.

THINK ABOUT IT

1. Why do scientists use microscopes?
2. In what way have microscopes been improved over time?

CAREERS

MICROSCOPE TECHNICIAN

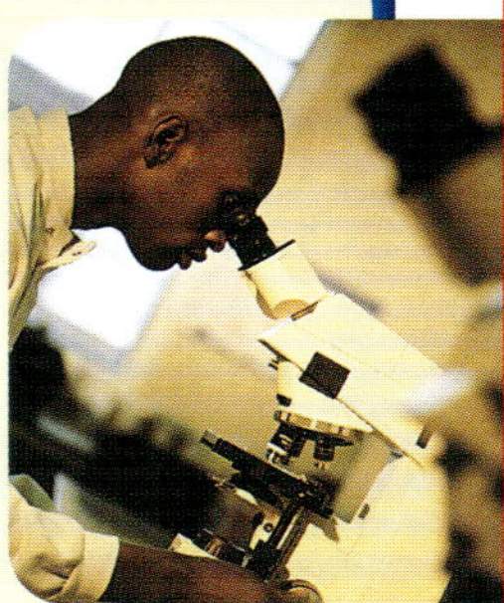

What They Do

Microscope technicians can help doctors determine the state of a patient's health by using microscopes to examine samples of the patient's blood and tissue. The cause of a specific disease can often be determined only by examining a patient's blood or tissue under a microscope.

Education and Training

Microscope technicians need a background in physics and optics. They also should be familiar with using computers. Microscope technicians first earn a high school diploma and then must take technical courses to qualify for a license.

WEB LINK
For Science and Technology updates, visit the Harcourt Internet site.
www.harcourtschool.com

BACKGROUND

Webliography
Keyword careers
www.harcourtschool.com

Microscopy is a field of investigation that is used to study objects that are too small to be seen by the naked human eye. Microscope technology has led to the forming of the field of microanalysis. Equipment in this field includes the electron microprobe analyzer (EMPA), the cathode luminescence (CDL), the energy dispersive spectroscope (EDS), and the Fourier transform infrared microspectrometer (FTIR). Using these tools, a microscopist explores the relationship of structure and properties of a wide variety of materials in order to more fully understand why a particular item behaves the way it does. This discipline is applied to many fields, including materials science, geology, biological science, physics, chemistry, and engineering.

DEVELOP SCIENCE VOCABULARY

microscope, p. A58—Write *micro* and *scope* on the board. Challenge students to define each word part. *Micro* means "small"; *scope* indicates "an instrument for observing." Lead students to conclude that a microscope is used to observe small objects. Then have students research definitions for *telescope*, *microchip*, or other words using the parts *micro* and *scope*.

angstrom, p. A59—This term comes from the name of the Swedish physicist known for his pioneering work on wavelengths of electromagnetic radiation. Explain that an angstrom is also a unit of wavelength. Challenge students to explain how a unit describing wavelength can also be a unit of length in using an electron microscope. If the object is very small, a wavelength measure might be an appropriate unit.

CHAPTER CONCEPTS

This article focuses on the development of more and more powerful microscopes.

- **How could an optical microscope be used to classify plants and animals?** A scientist could look at the cells of an organism in order to determine if they were plant or animal. Then the scientist would know how to classify the organism.
- **How could a more powerful microscope be used to help scientists classify organisms of the same genus?** Scientists could use the microscope to observe individual genes on an organism's chromosomes. Shared genes can indicate a relationship between two organisms.

3 Wrap Up and Assess

THINK ABOUT IT

1. They use them to study and work with very small objects that the unaided eye cannot see.
2. Scientists have invented microscopes that use forms of light or radiation that allow the viewing of smaller and smaller objects.

PEOPLE IN SCIENCE

Objectives

- ▶ **Connect chapter concepts with the contributions of scientists.**

1 Before Reading

SET A PURPOSE

How can a scientist make observations if he or she is blind? Have students read to find out how biologist Geerat Vermeij continues to explore the physical world.

2 Guide the Learning

THE WORK OF SCIENTISTS

Help students focus on the methods and processes scientists use and the impact of their work.

- ▶ What do you imagine a typical day of research is like for Dr. Vermeij?
- ▶ How might Dr. Vermeij's discoveries about mollusks help us to learn about other animals? Understanding how mollusks develop and how they are related to one another may help clarify the development and relationships of other animals.

3 Wrap Up and Assess

THINK ABOUT IT

1. He is an expert on the development of mollusks that have shells.
2. Because he is blind, Vermeij observes by touching the animals he studies, and he notices details about every ridge, twist, and turn in their shells.

PEOPLE IN SCIENCE

Geerat Vermeij

BIOLOGIST

"My fourth-grade teacher aroused in me a lasting curiosity about things unknown. She created a freedom for me to observe, an encouragement to wonder, and an environment in which to ask questions."

Dr. Geerat Vermeij (GIR•aht ver•MAY) is an expert on how mollusks develop. Mollusks include animals that have shells (such as clams and snails) as well as animals that do not (such as squid and octopods). Some mollusks live on land, but many live in water. Dr. Vermeij's detailed descriptions of every ridge, twist, and turn in the shells of certain mollusks have helped us understand how these animals change as they mature.

Dr. Vermeij has been blind since he was four years old. As a result, he has always relied on sounds, odors, shapes, and textures to observe and study the world around him. He says his fourth-grade teacher sparked his interest in science with her shell collection. He spent hours exploring the shells with his fingers. He observed them through their shapes and textures. Soon he imagined himself exploring the places where they were found and learning about the creatures that once lived inside them.

Geerat Vermeij decided he would go to college and become a scientist. However, he did not plan to stay indoors in a laboratory. "There is no substitute for making one's own observations in the wild," he says. That's what he has done. Braving venomous snakes, dangerous crocodiles, rough seas, and the edges of high cliffs, he has traveled around the world to observe the creatures he studies.

Dr. Vermeij's work has helped explain how a mollusk's shell protects its soft body and how these animals compete for survival. He continues to explore the world through touch and his other senses and to encourage others to do the same.

THINK ABOUT IT

1. What is Dr. Vermeij's area of study?
2. How might the fact that Dr. Vermeij is blind have resulted in his unique contribution to our knowledge of certain mollusks?

A60

BACKGROUND

Talking Computers are one of many electronic devices that have allowed sight-impaired scientists to carry on research. These computers have both speech recognition and speech synthesis functions. Thus far, the technology has allowed for the recognition of only a very limited vocabulary and, even then, usually from only one specific user. Engineers have been similarly challenged in developing machines that can synthesize spoken words effectively. In most cases, word fragments are spliced together to form whole words without regard for syllable emphasis, resulting in an unnatural, mechanical sound. While only in its rudimentary stages of development, the talking computer has already allowed blind persons and persons with other disabilities access to the technologies essential to working and doing research.

Webliography GO ONLINE
Keyword biographies
www.harcourtschool.com

ACTIVITIES FOR HOME OR SCHOOL

SORTING PASTA

How can you use a dichotomous key to categorize different kinds of pasta?

Materials

- 5 or 6 kinds of uncooked pasta
- transparent tape

Procedure

1. Begin the key by writing these statements:
   1a. The pasta is tube-shaped. Go to 2.
   1b. The pasta is not tube-shaped. Go to 3.
2. Sort your pasta into a pile that is tube-shaped and a pile that is not tube-shaped.

3. Add this pair of statements to your key:
   2a. The tube is straight.
   2b. The tube is curved.
4. If only one kind of tube-shaped pasta is straight, tape an example opposite Statement 2a. If only one kind of tube-shaped pasta is curved, tape an example opposite 2b.
5. Think of statements for 3a and 3b to divide the pasta that is not tube-shaped into two groups. Continue to divide the pasta into pairs of groups to complete your key.

Draw Conclusions

If two kinds of tube-shaped pasta are straight, how could you divide them into two groups? What would you write beside Statement 2a? What pair of statements would divide the straight, tube-shaped pasta into two groups?

SORTING FUNGI

How can you identify types of mold?

Materials

- several kinds of moldy food (such as bread, citrus fruit, cheese) in tightly sealed, clear, plastic bags
- hand lens

Procedure

1. Examine each kind of mold with the hand lens. **CAUTION Do not open bags to make your observations.** Mold is part of the Kingdom Fungi and includes many species. Use this dichotomous key to identify the kinds of mold.

1a.	Black, white, or gray fuzz	Go to 2.
1b.	Yellow or blue-green fuzz	Go to 3.
2a.	No rootlike threads and many dark balls (spores)	*Aspergillus flavus*
2b.	Many white or gray rootlike threads	Go to 4.
3a.	Yellow rootlike threads and many dark balls	*Aspergillus niger*
3b.	Blue-green or gray rootlike threads and few dark balls	*Penicillium*
4a.	Beige or white rootlike threads and few dark balls	*Mucor*
4b.	Gray or white rootlike threads and many dark balls	*Rhizopus*

Draw Conclusions

How did you differentiate among the types of mold? Why do you think mold grows on food?

A61

School-Home Connection

These activities provide an excellent opportunity to assign hands-on activities for the students and their families. The materials are often easy to obtain and are safe to work with. Encourage students to complete at home those activities that you do not wish to do in the classroom. Remind them that communicating their results is always important, but especially so for home activities since you will not be there to witness them. Ask students to use diagrams, sketches, graphs, or even photographs to document their results.

Sorting Pasta

Objectives

- **Use a dichotomous key to classify everyday objects.**

Suggested Time 20–30 minutes

Hints

- Collect many different pasta shapes for this activity, including elbow macaroni, bow ties, twists, small and large tubes, spaghetti, and flat noodles.
- Students can use pictures of different pasta types rather than actual pasta pieces, if desired.

Draw Conclusions Students could separate the straight pasta tubes by size using, for example, the following key:

Go to 4

4a The straight tube is less than 1 inch long.

4b The straight tube is more than 1 inch long.

Sorting Fungi

Objectives

- **Observe and classify common household molds.**

Suggested Time 20–30 minutes

Hints

- Place individual food items in self-sealing plastic bags several days before doing this activity. Make certain the bags are completely sealed before storing them in a warm place.
- Encourage students to handle the bags as little as possible so as not to crush the spore heads.

Safety Tip Many people are highly allergic to mold spores. Be certain that all moldy foods observed are completely sealed within the plastic bags.

Draw Conclusions Students should find that the color of fuzz, presence or color of rootlike threads, and number of balls (spores) are important characteristics for identifying a fungus. They should realize that mold is a fungus, so it cannot produce its own food and must absorb nutrients from the food it grows on.

ACTIVITIES FOR HOME OR SCHOOL

Reproducible copies of these activities are provided in the **Teaching Resources** book.

CHAPTER 2

Review and Test Preparation

VOCABULARY REVIEW (2 points each)

1. Protista
2. genus; species
3. classification
4. Animalia; Plantae
5. adaptation
6. Fungi
7. dichotomous key
8. Linnaean system
9. Monera

CONNECT CONCEPTS (2 points each)

The order of answers for Questions 10–14 may vary.

10. Animalia
11. Plantae
12. Protista
13. Monera
14. Fungi
15. kingdom
16. phylum
17. class
18. order
19. family
20. genus
21. species

CHECK UNDERSTANDING (5 points each)

22. **C,** protist
23. **G,** they have no nucleus
24. **C,** to avoid confusion about the identity of the organism
25. **H,** family
26. **A,** genus and species

CRITICAL THINKING (5 points each)

27. Answers may vary but could include the following: in the library; in sorting clothes; in sorting food, dishes, utensils, and other kitchen items; in sorting items in a medicine cabinet; in sorting sheets and towels; in keeping track of school work.
28. She would have to check the characteristics of her organism against the descriptions of all closely related species that have already been named and described.

CHAPTER 2 **Review and Test Preparation**

Vocabulary Review

Write the correct term to complete the sentences. The page numbers in ( ) tell you where to look in the chapter if you need help.

classification (A42)
Linnaean system (A43)
Animalia (A44)
Plantae (A44)
Fungi (A44)
Protista (A44)
Monera (A44)
adaptation (A45)
genus (A53)
species (A53)
dichotomous key (A56)

1. One-celled organisms that have one or more nuclei and are similar in structure to plant and animal cells are in the kingdom ____.
2. The first part of a scientific name is the ____ name, and the second part is the ____ name.
3. The arrangement of things in groups of similar items is called ____.
4. Animals belong to the kingdom ____, and plants belong to the kingdom ____.
5. An ____ is a trait that better suits an organism to its environment.
6. Organisms that have cells similar to plants but do not make their own food belong to the kingdom ____.
7. A series of choices that leads to the correct identification of an organism is called a ____.
8. The ____ is a classification system in which living things are divided into smaller and smaller groups.
9. The kingdom ____ includes bacteria and blue-green bacteria.

Connect Concepts

Complete the chart.

The five kingdoms are —

10. ________ 11. ________ 12. ________ 13. ________ 14. ________

The seven levels of the Linnaean system, in order from largest to smallest, are —

15. ________
16. ________
17. ________
18. ________
19. ________
20. ________
21. ________

A62

CHAPTER TEST

See **Assessment Guide pp. AG8–12** for the Chapter Test. Assessment Options are provided on p. A38e.

TAKE-HOME BOOK

Use the **Take-Home Book** (described on p. A38d) to provide more chapter content and activities.

Check Understanding

Write the letter of the best choice.

22. A certain one-celled organism has chloroplasts, a nucleus, and a whiplike flagellum. It is classified as —
    A a plant
    B an animal
    C a protist
    D a moneran
23. A characteristic that separates monerans from all other organisms is that —
    F they have a cell wall
    G they have no nucleus
    H they are single-celled
    J they cannot move
24. Scientists discussing a particular organism use its scientific name —
    A to establish how it is related to other organisms
    B to impress other scientists
    C to avoid confusion about the identity of the organism
    D to place it in a dichotomous key
25. Organisms that are members of the same order are not necessarily members of the same —
    F phylum
    G class
    H family
    J kingdom
26. An organism's scientific name is made up of the names of its —
    A genus and species
    B kingdom and species
    C kingdom, family, and species
    D family, genus, and species

Critical Thinking

27. Describe classifications you use in your everyday life.
28. A scientist believes she has discovered a new species. How can she check to make sure that her species has not been described before?
29. Not all scientists classify organisms in exactly the same way. In fact, some scientists use a seven-kingdom classification system rather than a five-kingdom system. Are they wrong? Explain.

Process Skills Review

30. What characteristics could you use to **classify** a group of buttons?
31. Under the classification system currently in use, how do scientists **organize data** about living things?

Performance Assessment

A Key for Diatoms

Study the photograph of diatoms on page A47. Choose one diatom to classify. Make a list of the characteristics of your diatom, based on what you observe in the photograph. Design a dichotomous key with at least four steps to identify the diatom you have selected.

TEST PREP Test-Taking Tips

Item type open-ended format in which the student is asked to write a short answer

To help students answer Question 29, share the following strategy with them. **Ask students to first think of the function of a classification system (to help organize and show relationships among organisms). Then ask them to think about how our understanding of the world around us changes all the time. Finally, have them decide if the terms *right* and *wrong* apply here.**

29. A classification system is not right or wrong, but is correct as long as it is useful and consistent.

PROCESS SKILLS REVIEW (7 points each)

30. Answers may vary but could include color, shape, size, texture, number of holes.

31. It groups living things into categories in seven levels according to their characteristics. Organisms in the same category have more similarities. The lower the level of the category that they share, the greater number of similarities between them.

Performance Assessment

Preparation You may wish to compose a dichotomous key yourself before grading students' keys. Steps might include "has sides/does not have sides (is circular)," "has three sides/has more than three sides."

Scoring Rubric
Performance Indicators

___ Lists more than two characteristics for the selected diatom based on observation, comparison, and categorization of the shape.

___ Organizes characteristics in a logical order from general characteristics to more specific.

___ Writes at least four appropriate questions with only a "yes" or "no" type answer at each level of classification.

___ Identifies a specific shape by using the key.

Observations and Rubric Score

3 2 1 0

WRITING ABOUT SCIENCE

Have students write a composition that involves writing for a specific purpose. You may wish to use the prompt that is provided in the **Workbook.** The prompt is accompanied by a graphic organizer that will help students organize their ideas before writing. Models for writing are provided in the **Teaching Resources** book. You may wish to reproduce those for students or display them on a transparency.

WRITING PRACTICE

A chapter writing prompt and a prewriting activity are provided on **Workbook p. WB31.**

UNIT A EXPEDITIONS

Objectives

- Classify animals and plants using a classification system.
- Compare the similarities and differences of plants.

Hints

A field trip to the Birmingham Zoo, the Shaw Nature Reserve, or a similar destination may not be feasible for your class. If so, consider visiting The Learning Site on the Internet. This site provides links to sites appropriate for student learning.

GO ONLINE www.harcourtschool.com

UNIT A EXPEDITIONS

There are many places where you can go to improve your understanding of living things. By visiting the places below, you can learn how plants and animals differ and how they are classified. You'll also have fun while you learn.

Birmingham Zoo

WHAT A collection of unique animals from around the world

WHERE Birmingham, Alabama

WHAT CAN YOU DO THERE? Visit the zoo, see the many exhibits, and learn about the animals living there.

Shaw Nature Reserve

WHAT A part of the Missouri Botanical Garden with 2500 acres of natural landscape and plant collections

WHERE St. Louis, Missouri

WHAT CAN YOU DO THERE? Tour the garden, see the wide variety of plants, and ask about the research conducted there.

GO ONLINE Plan Your Own Expeditions

If you can't visit the Birmingham Zoo or the Shaw Nature Reserve, visit a zoo or nature reserve near you. Or log on to The Learning Site at **www.harcourtschool.com** to visit these science sites and other places. There you can observe plants and animals.

A64

UNIT TEST

See **Assessment Guide pp. AG15–AG18** for the Unit Assessment. The Unit Assessment includes items for all the chapters in this Unit.

Living Things Grow and Respond

UNIT B

Unit Overview

Unit Theme Plants and animals exhibit predictable patterns of change in both structures and ways of life as they grow and develop. The changes they undergo help plants and animals adjust to their surroundings and to sudden changes in conditions around them.

UNIT B

Skills for Lifelong Learning

Science Process Skills

The Science Process Skills are important inquiry tools and are essential for investigating the natural world. Opportunities for developing process skills are provided throughout *Harcourt Science*. The pages shown in this chart indicate where the process skills receive special emphasis.

Science Process Skill	Skill Tip and Mini-Lesson	Skills Practice (Workbook)	Reinforcement
Observe	B27, B43, B51, B103, B113	WB48, WB58, WB63, WB89, WB94	B1j, B5, B9, B13, B33, B39, B72, B75, B77, B85
Compare	B33, B51, B103	WB53, WB63, WB89	B5, B10, B15, B30, B37, B43, B55, B67, B75, B79, B80, B107, B113
Classify/Order	B85	WB81	B51
Gather, Record, Display, or Interpret Data			B1j, B5, B27, B33, B51, B75, B85, B103
Use Numbers			
Communicate	B43, B75	WB58, WB76	B1j, B45, B117
Plan and Conduct Simple Investigations	B67	WB71	B1j, B5, B27, B33, B43, B51, B67, B103
Measure			B1j
Predict			B13, B27, B113
Infer	B33	WB53	B27, B43, B51, B67, B72, B85, B103
Draw Conclusions	B13	WB40	B1j, B5, B13, B15, B21, B27, B33, B43, B51, B61, B67, B75, B85, B90, B97, B103, B113, B125
Use Time/Space Relationships			B58
Formulate or Use Models			
Hypothesize	B27, B113	WB48, WB94	B1j, B5, B13, B33, B43, B51
Identify and Control Variables	B5	WB35	B15
Experiment			B13

Reading, Writing and Math Skills

Effective science instruction integrates science content and experiences with all areas of the elementary curriculum. *Harcourt Science* provides students with many opportunities to develop reading, writing, and math skills via meaningful activities and strategies built into every lesson in this unit.

Reading Skills Checklist

Strategies for developing the following reading skills are provided in this unit.

- **use prefixes and suffixes to determine word meanings** B15, B35, B53, B71
- **use context to determine/confirm word meaning** B54
- **identify supporting facts and details** B6, B7, B8, B14, B16, B28, B34, B36, B38, B44, B46, B48, B52, B54, B68, B70, B76, B78, B80, B81, B86, B88, B91, B92, B104, B106, B108, B110, B114, B116, B118, B121
- **arrange events in sequence** B30
- **identify the main idea** B6, B7, B8, B14, B16, B28, B34, B36, B38, B44, B46, B48, B52, B54, B68, B70, B76, B78, B80, B81, B82, B86, B88, B91, B92, B104, B106, B108, B110, B114, B116, B118
- **summarize and paraphrase a selection** B45, B47
- **identify cause and effect** B78
- **predict probable future actions and outcomes** B94
- **use graphic sources for information** B40, B120
- **draw logical conclusions** B70, B109
- **compare and contrast** B7, B8, B10, B15, B18, B48, B55, B90
- **distinguish fact and opinion** B8, B90
- **make generalizations** B6, B15, B87
- **use reference sources** B16, B39, B40, B56, B58, B73, B107, B122

Writing Links

Prompts that provide opportunities for students to express their ideas in writing are provided in every lesson in this unit.

Informative Writing—Compare and Contrast B11
Informative Writing—Description B17, B57
Informative Writing—Explanation B31, B83, B111
Informative Writing—How-to B41, B73
Expressive Writing—Poem B49
Narrative Writing—Story B93
Persuasive Writing—Opinion B121
Writing About Science B23, B63, B99, B127

Math Links

Opportunities to practice math skills, solve problems, and connect math and science are provided in every lesson.

Make a Graph B3, B16, B40
Write an Equation B3
Compute Percents B11
Compute Ratios B17
Choose an Appropriate Graph B25
Find the Average B25
Write a Ratio B31, B93
Use Formulas B41
Write a Fraction B49
Find Multiples B57
Organize Data B65
Analyze/Interpret Data B65, B101
Find Symmetry B72
Compare Graphs B73
Use Mental Math B83
Use Logical Reasoning B101
Write Percents B111
Solve Problems B121

UNIT B

How to Integrate Your Day

Use these topics to help you integrate science into your daily planning.

Curriculum Integration

Fine Arts

Seed Crafts B11
The Botanical Museum at Harvard University B46
Flowers in Art B49
Dried Flowers B57
Spider Webs B71
Sepia B93
Mermaids B106
Camille Saint-Saëns B117
Unicorns and Camelopards B121

Health/Physical Education

Vegetables B55
Schistosomiasis B81
Exercise for Your Pet B73

Math

Make a Graph B3, B16, B40
Write an Equation B3
Compute Percents B11
Compute Ratios B17
Choose an Appropriate Graph B25
Find the Average B25
Write a Ratio B31, B93
Use Formulas B41
Write a Fraction B49
Find Multiples B57
Organize Data B65
Analyze/Interpret Data B65, B101
Find Symmetry B72
Compare Graphs B73
Use Mental Math B83
Use Logical Reasoning B101
Write Percents B111
Solve Problems B121

Writing

Informative Writing—Compare and Contrast B11
Informative Writing—Description B17, B57
Informative Writing—Explanation B31, B83, B111
Informative Writing—How-to B41, B73
Expressive Writing—Poem B49
Narrative Writing—Story B93
Persuasive Writing—Opinion B121
Writing About Science B23, B63, B99, B127

Language Arts

Naming Plants B31
Debating Arthropods B90
Word Meanings B111

Social Studies

Cereal Grains B9
Calendars B17
Mosses in History B31
Trees B39
How Native Americans Used Conifers B41
Flowers as Emblems B49
Origins of Fruit B55
Dams and Reservoirs B83
Cultured Pearls B88

Using Technology

Harcourt Science **includes an array of technology products that**

- bring interactive experiences to the classroom.
- add depth to chapter concepts.
- meet individual needs.

Technology Links at the end of each lesson in the student's book help you integrate technology at the point of its most effective use.

Computer Software

Harcourt Science Explorations CD-ROM

These engaging CD-ROMs provide interactive science explorations. Activities enhance and expand key chapter concepts through simulations and investigations.

Data ToolKit

Data ToolKit is a tool-based program that enables students to organize, communicate, and interpret data they have collected during their investigations.

Videos

CNN Turner Learning

Harcourt Science Newsroom Videos

These videos bring exciting science news events to your classroom from the leaders in news broadcasting—CNN and Turner Learning. For each chapter in this unit, there is a **Harcourt Science Newsroom Video** segment that helps students see the relevance of science to their lives. The videos also add interesting depth to chapter concepts.

Internet

GO ONLINE

Smithsonian Institution®

With *Harcourt Science*, it's easy to make the Smithsonian Institution a part of the science classroom. Through a special agreement, you can visit this Internet site to see correlated links to the Web sites of Smithsonian museums, galleries, and the National Zoological Park. Resources include virtual tours, on-line exhibits, and hands-on investigations—all especially selected to complement and enrich *Harcourt Science.*

www.si.edu/harcourt/science

The Learning Site

Visit The Learning Site for a world of science resources, including science News Breaks, interactive learning games and activities, school-home resources, and an animated Science Glossary.

www.harcourtschool.com

NSTA SciLinks

These special Internet links are provided by the National Science Teachers Association in partnership with Harcourt. The SciLinks Web page connects students to selected Web sites that enhance and enrich *Harcourt Science* chapter content.

www.scilinks.org/harcourt

Other Technology Resources

Software

Steck-Vaughn Interactive Science Encyclopedia

This CD-ROM is a storehouse of articles, images, video clips, animations, and activities. Students can use this encyclopedia to gather additional information about chapter concepts and conduct investigations and simulations.

The Way Plants Grow

Queue (MAC CD/WIN CD)
Explore the life cycle of a variety of plants including sunflowers, grapes, dandelions, sweet potatoes, and rice. Students learn about the development of plants from seed to maturity and such concepts as pollination and photosynthesis.

Video

What Is A Plant

National Geographic, 1991. (15 minutes) Examines the different parts of a plant and the plant life cycle.

Classification of Living Things

Knowledge Unlimited, 1992. (22 minutes) Discusses the concept of kingdoms of living things, as well as bacteria, fungi, viruses, plants, vertebrates and invertebrates.

UNIT B

Reaching All Learners

Strategies for Multi-Age Classrooms

Develop Science Process Skills

- Pair older and younger or less-advanced students to develop the science process skill of *comparing* that they use for the investigation in Chapter 2, Lesson 2. The older student can narrate steps and observations while the younger student draws the experiment.

Build Science Concepts

- Have older, more-advanced or younger, less-advanced students work in a small group to make a visual presentation of a science process skill to share with other students. This provides reinforcement of the skill for all students.
- When students in different grade levels are working on similar science concepts, such as plant growth, you may want to group them to present a visual, kinesthetic, or tactile summary of the materials covered. Let the students decide which approach they want to use.

Strategies for ESL Students

Build Concepts

Graphic Organizer Use this graphic organizer to summarize and reinforce the most important concepts in the unit. Where possible, repeat the use of a graphic organizer in different lessons.

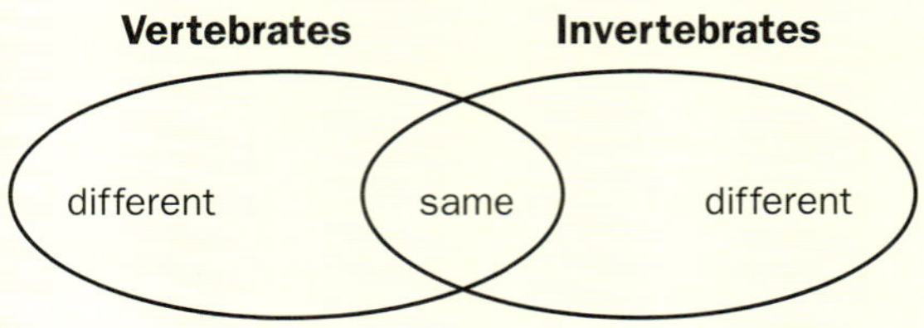

Chapter Graphic Organizer A Graphic Organizer for Chapter Concepts is provided for each chapter in this unit. ESL students can use this to summarize major concepts and reinforce chapter vocabulary. The Graphic Organizer for Chapter Concepts appears in the workbook and is also provided as an overhead transparency.

Develop Oral Language

Interviews ESL students can write questions they have about living things and then interview each other. Encourage volunteers to share their findings orally as well as in writing.

Illustrations Have pairs of students search through magazines for photos that show types of plants and animals, cut them out, and glue them on paper. Then have students write key words about the characteristics of each under the appropriate picture. Finally, have students share the pictures, reading aloud the key words that they wrote.

ESL Activities

Point-of-use activities for building concepts and developing oral language are provided in the lessons throughout the unit.

Strategies for Special Needs Students

Learning Difficulties

- After students read a particular section of their textbooks, have them represent the information through drawings, by acting out, or by another means.
- Students who are working below grade level will benefit from focusing on several questions before reading or rereading. On the board, write three to five questions that you want students to find answers to.
- Students with learning difficulties may also benefit from using the Graphic Organizer for Chapter Concepts to summarize key ideas. The Graphic Organizer appears in the workbook and is also provided as an overhead transparency.

Visually Impaired

- Provide hand lenses for visually impaired students to use to see details. For example, a hand lens would be helpful in examining the photographs of mosses used in the investigation in Chapter 2, Lesson 1.
- These students may benefit from listening to the Text on Tape audiocassettes.

Hearing Impaired

- Provide as many additional photographs and models as possible to reinforce the investigation of living things.

Strategies for Advanced Learners

The following strategies provide an opportunity for advanced learners to study the content of the unit in greater depth.

Chapter 1 Plant Growth and Responses

Ask students to investigate the meanings of various types of tropisms. Have them determine which tropisms can affect plants and which more typically affect animals. Then have students design an experiment to test how plants react to different stimuli. Have students record and analyze their data, draw a conclusion about each experiment, and report the results to the class.

Chapter 2 Types of Plants

Challenge students to identify various types of pollinators. Have students research the type of flowers each pollinator prefers. Students might also discover that the wind often acts as a pollinator. Ask students if any types of flowers typically have more than one pollinator. Also have them consider whether having more than one pollinator would be advantageous for the flower.

Chapter 3 Invertebrates

Ask students to investigate the general characteristics of earthworms and mealworms. Have students construct a chart outlining differences and similarities between the two species. Have students then write a creative story about an earthworm and a mealworm who were lifelong friends. Have them include the differences in the life cycles as part of the plot of the story.

Chapter 4 Vertebrates

Invite students to investigate the similarities and differences between turtles and tortoises. Have them compare the various types of shells, eating habits, behavior, reproduction habits, and so on. Then have students design a habitat for each, taking into consideration the differing needs of the two types of reptiles. Ask students to investigate to find any tortoises or turtles that are on the endangered or extinct species lists.

UNIT B

Unit Materials List

Quantities are indicated for a class of 30 students working individually or in groups of 5, depending on the nature of the activity. Where shared equipment is suggested, a smaller number of items is specified. Quantities are also listed for those materials included in the Materials Kit.

Nonconsumable Materials	Class Quantity Needed	Kit Quantity	Activity Page
balance	1*	1	B1
cup, measuring	6	6	B1, B4
goggles, safety	30	Safety Kit	B32
hand lens	6	6	B32, B42, B50, B74
knife, plastic	6	1 pkg of 24	B50
Microslide Viewer	6	6	B74
Microslide, hydra	6	6	B74
pictures, mosses	6 sets		B26
ruler, metric	6		B1
shell, clam	3*	3	B84
shell, sea urchin	3*	3	B84
shell, snail	3*	3	B84
shell, spirula	3*	3	B84
shell, starfish	3*	3	B84
sponge, natural	6 sm. pieces	3 med. pieces	B74

*Student groups should share these materials.

Consumable Materials	Class Quantity Needed	Kit Quantity	Activity Page
aluminum foil	1 roll	1 roll	B1
box, cardboard	6		B66
cellophane, 4 colors	6 sheets of each color	6 sheets of each color	B1
cone, pine	6	6	B32
cone, spruce	6	6	B32
conifer needles	12		B32
cups, plastic clear	36	1 pkg of 50	B4, B12
cups, foam	72	75	B1
diagram, fish	30	1 pkg of 50	B102
diagram, frog	30	1 pkg of 50	B102
diagram, grasshopper	30	1 pad of 50	B84
diagram, skeleton of reptile, bird, and mammal	30	See TR	B112
fern frond	6		B32
flower	6		B42
marker, permanent	6		B1, B4
paper towels	1 roll		B4, B12, B42, B50
pencil	6		B1
pencils, colored	30 sets		B112
planaria	6	Coupon	B74
plastic wrap	1 roll	1 roll	B1
seeds, bean	2 pkgs	2 pkgs	B1, B4
seeds, corn	1 pkg	1 pkg	B50
seeds, lima bean	1 pkg	1 pkg	B50
shoe box	30		B1, B4
soil, potting	3 bags	3 bags	B1
supplies, diorama	6 sets		B66
tape, masking	1 roll		B1, B4

Additional kit options are available. Contact your sales representative for details.

UNIT B LIFE SCIENCE

Living Things Grow and Respond

UNIT EXPERIMENT

Lighting Up Plants

Plants need light to grow and make food. The amount and type of light can affect plant growth. While you study this unit, you can conduct a long-term experiment related to plants and light. Here are some questions to think about. How exactly does light affect plant growth? For example, how does the color of light affect the growth of plants? Plan and conduct an experiment to find answers to these or other questions you have about light and plant growth. See pages x–xvii for help in designing your experiment.

B1

Informal Assessment

Research shows that Informal Assessment is an effective way to monitor and evaluate students. Performance assessment and teacher observation are particularly useful for assessing projects and other, similar tasks.

Performance Together with students, develop a rubric that will be used to evaluate their completed projects. Use *Developing Your Own Rubric* on page AGxi of the Assessment Guide. Challenge and guide students to produce their best work. Then use the rubric to assess their performance.

Classroom Observation You can gain insight into students' learning by observing how they use scientific processes, collect and record information, and select and use appropriate tools to plan and implement their project. While students are working on their projects, use the *Project Evaluation Checklist* on page AGxviii of the Assessment Guide for ongoing assessment.

UNIT B EXPERIMENT

Objectives

- **Promote scientific inquiry.**
- **Use a scientific method to plan and conduct a long-term investigation.**
- **Design an experiment to show how plants need light to grow and make food.**

A lesson plan for guiding students through the Unit Experiment is provided on pages B1i–B1j.

Science Fair Project Ideas

In addition to the Unit Experiment, you may wish to allow students to plan and carry out these science fair projects related to the chapters in Unit B. Students can use the framework on the Unit Experiment pages to help them plan and conduct experiments.

Chapter 1 Leaf Lean Ask students to design and conduct an experiment that will test phototropism. A simple test is to place bean plants in a windowsill and observe the direction the leaves lean.

Chapter 2 Seed Needs Ask students to compare the germination of seeds from gymnosperms and spores from ferns. Students can put a few samples of each type of seed and spore in moist paper towels to begin germination.

Chapter 3 Make a Butterfly Garden Students can study the habits and adaptations of butterflies native to their geographic region. Students can research what plants are needed to attract the butterflies to their butterfly garden.

Chapter 4 The Zoo Review Have students design areas within a zoo that would simulate habitats. Students can draw the zoo to scale.

UNIT B EXPERIMENT

Choose from these options to help students complete the unit experiment.

Option 1 Independent Inquiry Assign the experiment for students to complete independently. Students should use the prompt as a springboard for writing their own questions and designing their own experiment. They can write a hypothesis, design a procedure, select materials, and conduct the experiment. Students can refer to pages x–xxiv in the Pupil Edition for guidance.

Option 2 Guided Inquiry Suggest students use the Experiment Log (Workbook pp. WB375–378) to help them plan their experiment. The Experiment Log pages shown below appear in the Workbook, but without answers. Use the lesson plan that follows to guide students as they design the experiment.

Option 3 Structured Inquiry Have students complete the experiment by testing the hypothesis and using the procedure that have been provided for them. Display the overhead transparencies shown below (or provide photocopies from the transparency package), and have students copy the hypothesis, variables, and procedure into the Experiment Log (Workbook pp. WB375–378). Students then conduct the experiment and gather and record their data. Use the lesson plan on page B1 to guide students as they complete the experiment.

BACKGROUND

Light and Plant Growth
Light is the driving force in plant growth. Understanding the effects of different-colored light on plant growth could be helpful when choosing glass for greenhouse construction, developing artificial grow lights, or attempting to control growth in plants.

Although full-spectrum light, such as sunlight, provides for optimum growth in plants, each part of the color spectrum does affect plant growth. For example, plants stay short, stocky, and leafy when they receive plenty of blue light. Plants tend to stretch, flower, and fruit early when exposed primarily to red light.

Webliography GO ONLINE

Keyword plant adaptations
www.harcourtschool.com

LESSON PLAN

Resources

Experiment Log, Unit B, Workbook pp. WB375–378

Experiment Transparencies, Unit B

Time 30–45 minutes to plan and set up, 20 minutes each day for two weeks to observe and record data

Expected Results In most cases the bean plants exposed to full-spectrum light (beneath the clear cellophane) will show the most growth when compared with the bean plants exposed only to red, green, or blue light.

Suggested Materials

- **12 small foam cups**
- **4 shoe boxes**
- **potting soil**
- **12 beans**
- **balance**
- **pencil**
- **aluminum foil**
- **centimeter ruler**
- **4 colors of cellophane, each 45 cm long (clear, red, green, blue)**
- **marker**
- **measuring cup**
- **tap water**
- **tape**

Underlined items above are provided in the equipment kits (available separately).

Preparation Tips Space should be made available near a window where plants can receive an equal amount of sunlight. If necessary, a grow light or a mixture of fluorescent and incandescent light sources may work as well as sunlight.

As an alternative to beans, students can use radish seeds, rye grass seeds, or seeds of another fast-sprouting plant.

Transparency Exp B, p. 1 Workbook p. WB375

Lighting Up Plants

1 Observe and Ask Questions

Does the color of light to which a plant is exposed affect how it grows? For example, will plants exposed to full-spectrum light (natural sunlight) grow taller than plants exposed only to red, green, or blue light? Make a list of questions you have about plant growth and colored light. Then circle a question you want to investigate.

Will plants exposed to full-spectrum light grow taller than plants exposed only to red, green, or blue light?

2 Form a Hypothesis

Write a hypothesis. A hypothesis is a suggested answer to the question you are investigating. You must be able to test the hypothesis.

Plants exposed to full-spectrum light will grow taller than plants exposed only to red, green, or blue light.

3 Plan an Experiment

To plan your experiment, you must first identify the important variables. Complete the statements below.

Identify and Control Variables

The variable I will change is *the color of light I expose the plants to.*

The variable I will observe or measure is *how much the plants grow.*

The variables I will keep the same, or *control*, are *the type of plant, the amount of water, the amount of light (measured in hours), the type of soil, the temperature.*

1 Observe and Ask a Question

Have students discuss the prompt and then brainstorm a list of other questions they may have about the topic. Students can record their questions in the Experiment Log and then circle the question they are going to test.

2 Form a Hypothesis

Guide students in forming a testable hypothesis.

- **Look at the question you have chosen to investigate. What do you predict might be an answer to this question?** Possible answer: Yes, plants exposed to sunlight (full-spectrum light) will grow taller.
- **Use your answer to form a statement. Your statement should include all the things you are testing. This statement will be your hypothesis.** Possible hypothesis: Plants exposed to sunlight will grow taller than plants exposed only to red, green, or blue light.

3 Plan an Experiment

Identify and Control Variables Remind students that an experiment is a fair test of a hypothesis. Identifying and controlling the variables will ensure that the test is conducted fairly. Have students complete the statements in the Experiment Log.

Develop a Procedure and Gather Materials Have students write a detailed procedure or have them use the procedure that appears on the transparency.

4 Conduct the Experiment

Gather and Record Data Students can record their data in the Experiment Log.

Common Error Alert Point out that it is important to measure the same plant part in the same way each time they collect data. Students may find it helpful to mark the leaf that they are measuring.

Interpret Data Students should continue to measure the plants for two weeks. Suggest that students plot line graphs to show the change in size of the plants over time. A bar graph could also be used to compare sizes of the plants at the end of the experiment.

5 Draw Conclusions and Communicate Results

Have students answer the questions in the Experiment Log.

Scientists at Work Emphasize to students that there is no "right" answer for a science experiment. Hypotheses may not be supported by the data. If so, the hypothesis needs to be changed, *not* the data. It is always important to consider results honestly and objectively.

Investigate Further Encourage students to build upon this experiment to develop a science fair project or other investigation.

Transparency Exp B, p. 2 Workbook p. WB376

Develop a Procedure and Gather Materials

Write the steps you will follow to set up an experiment and collect data.

1. Use a pencil to make two small holes in the bottom of each of 12 foam cups.
2. Label three cups as follows: "Full Spectrum," "Red," "Green," "Blue."
3. Put potting soil in each cup until it is two-thirds full. Use the balance to be sure that you put equal amounts of soil in each cup.
4. Insert one bean into the soil of each cup. Add about a centimeter of potting soil to each cup to cover the bean.
5. Line four boxes with foil. Place the three cups labeled "Full Spectrum" in one box. Place each of the other three groups of cups into a separate box.
6. Add water to each cup until the soil is thoroughly moist.
7. Tape the appropriatly colored cellophane over each box to cover it.
8. Place the boxes on a windowsill or place where they will get an equal amount of sunlight.
9. Each day, add a small (and equal) amount of water to each cup. Check to see that each box is receiving the same amount of light throughout the day.
10. After the beans sprout, observe the new plants each day for two weeks. Use a ruler to measure the height of the plants. Record average heights for each box.

Use extra sheets of blank paper if you need to write down more steps.

Materials List Look carefully at all the steps of your procedure, and list all the materials you will use. Be sure that your teacher approves your plan and your materials list before you begin.

foam cups, balance, measuring cup, marker, pencil, tap water, potting soil, beans, centimenter ruler, shoe boxes, aluminum foil, tape, cellophane (clear, red, green, blue)

Transparency Exp B, p. 3 Workbook p. WB377

4 Conduct the Experiment

Gather and Record Data Follow your plan and collect data. Use the chart below or a chart you design to record your data. **Observe** carefully. **Record** your observations and be sure to note anything unusual or unexpected.

Plant Observations

	Average Height (mm)						
Color of Light	Day 1	Day 2	Day 3	Day 4	Day 5	Day 6	Day 7
Full Spectrum							
Red							
Green							
Blue							

	Average Height (mm)						
Color of Light	Day 8	Day 9	Day 10	Day 11	Day 12	Day 13	Day 14
Full Spectrum							
Red							
Green							
Blue							

Other Observations

Transparency Exp B, p. 4 Workbook p. WB378

Interpret Data

Make a graph of the data you have collected. Plot the graph on a sheet of graph paper, or use a software program.

5 Draw Conclusions and Communicate Results

Compare the **hypothesis** with the data and chart, and then answer these questions.

1. Was your hypothesis supported by the results of the experiment? Explain. Students should justify their responses with data and observations.
2. How would you revise the hypothesis? Explain. Students may revise their hypotheses to match the results of the experiment more closely.
3. What else did you **observe** during the experiment? Students should use their notes and be as specific as possible.

Prepare a presentation for your classmates to **communicate** what you have learned. Display your data table and graph.

Investigate Further Write another hypothesis that you might investigate.

Responses will vary depending on experiment outcomes and student interests.

CHAPTER 1 Planning Guide Plant Growth and Responses

LESSON	PACING	OBJECTIVES	MATERIALS
1 How Do Plants Meet Their Needs? B4–11 	**3 days**	▶ Describe how plants grow from seeds. ▶ List what plants need in order to grow. ▶ Compare phloem and xylem. ▶ Distinguish between vascular and non-vascular plants.	**Investigate** paper towels, 5 clear plastic cups, 20 bean seeds, masking tape, permanent marker, water, graduate or measuring cup, shoe box with lid
2 How Do Plants Respond to Their Environments? B12–17 	**2 days**	▶ Describe how plants respond to light, gravity, and touch. ▶ Explain how tropisms help plants survive. ▶ Compare long-day and short-day plants.	**Investigate** clear plastic cup, paper towels, water, sprouted bean seed (from previous investigation)
End of Chapter B18–23 **Science and Technology** **People in Science** **Activities for Home or School** **Chapter Review and Test Preparation**		▶ Evaluate the impact of research and technology on scientific thought, society, and the environment. ▶ Identify careers related to science. ▶ Connect chapter concepts with the contributions of scientists. ▶ Extend chapter concepts with hands-on investigations. ▶ Review chapter concepts.	

PROCESS SKILLS	VOCABULARY	RESOURCES AND TECHNOLOGY	REACHING ALL LEARNERS
Process Skill Tip identify and control variables **Other Process Skills** observe, compare, record data, draw conclusions, hypothesize, plan simple investigations	**vascular plant** **xylem** **phloem** **nonvascular plant**	**Workbook,** pp. WB32–37 **Vocabulary Cards,** pp. TR99–102 **Transparencies** B1, B1-1 **Harcourt Science Activity Video** **Harcourt Science Newsroom Video**	**Visual Learners**, p. B10 **ESL Activity**, p. B9 **Investigation Challenge**, p. B10 **Advanced Learners**, p. B1g
Process Skill Tip draw conclusions **Other Process Skills** observe, hypothesize, experiment, predict	**tropism** **phototropism** **gravitropism** **long-day plant** **short-day plant**	**Workbook,** pp. WB38–44 **Transparencies** B1, B1-2 **Harcourt Science Activity Video** **Internet Site**	**Kinesthetic Learners,** p. B15 **Investigation Challenge,** p. B16
		Take-Home Book, pp. TH15–18 *Manipulating Plants* **Activities for Home or School,** p. TR67 **Chapter Test**, pp. AG19–23	**Performance Assessment**, p. AG24 **Portfolio Evaluation**, p. AGxxiv

Prepare for Activities

Use this page to help you organize and prepare materials for the Investigations, which begin each lesson of this chapter.

LESSON INVESTIGATION	MATERIALS

1 Germinating Seeds

B4–5

PURPOSE Control variables in an experiment to find out under which conditions seeds germinate.

PREPARATION TIPS Try to determine appropriate locations for the cups ahead of time. You will need a place that is warm, but it should not be directly over a heater or in direct sunlight. You will also need a refrigerator that has enough space for the cups. Sprouted seeds should be saved for the next lesson's investigation.

TIME 20 minutes for setup; 10 minutes a day for observations until the seeds germinate, or about 5 days

EXPECTED RESULTS Students should conclude that seeds need water, air, and warmth in order to germinate.

Materials

- paper towels
- <u>5 clear plastic cups</u>
- <u>20 bean seeds</u>
- masking tape
- permanent marker
- water
- graduate or <u>measuring cup</u>
- shoe box with lid

2 Do All Roots Grow Down?

B12–13

PURPOSE Experiment to find out if the root still grows downward and the shoot still grows upward when the seed is planted upside down.

PREPARATION TIPS Prior to the activity, find a place to store the cups. You will need a room that can be kept dark all the time, an empty shelf in a cabinet that can be kept closed, or a large box that can be closed to seal out all light. If all students are doing the activity, be sure that they label their cups with their names.

TIME 5–20 minutes for setup; 10 minutes for observation 2–3 days later

EXPECTED RESULTS Students should conclude that roots grow downward and shoots grow upward.

Materials

- <u>clear plastic cup</u>
- paper towels
- water
- sprouted bean seed (from previous investigation)

Harcourt Science Activity Videos

Preview the Investigations The Activity Video for this unit shows students doing the Investigations. You may wish to view the video for classroom management ideas. The video can also be used to model the Investigation process for students.

Equipment Kits

<u>Underlined items</u> above are provided in the equipment kits (available separately).

. . . with Trade Books

These books provide in-depth information about plant adaptations and responses to their environments.

Inside Guides: Incredible Plants by Barbara Taylor (Dorling Kindersley, 1997) reveals amazing facts about plants and uses three-dimensional models to show how the parts of a plant operate.

Plant Survival: Adapting to a Hostile World by Brian Capon (Timber Press, 1994) illustrates fascinating mechanisms that plants use to survive.

Sand and Fog: Adventures in Southern Africa by Jim Brandenburg (Walker, 1994) depicts the specially-adapted lifeforms in the Namib Desert through the eyes of an award-winning nature photographer. *NSTA Trade Book*

***Smart Survivors* (World of Discovery)** by Sneed B. Collard (Northwood Press, 1994) describes "smart things" that plants and animals do to assure their survival.

Trade book titles are current at time of publication but may go out of print without notice.

. . . with More Hands-On Activities

Investigation Challenge

More About Seed Germination B10

Tropisms B16

Activities for Home or School

Traveling Water B21

Growth and Touch B21

The cross-curricular Links that appear at the end of each lesson in the student edition and throughout the teacher's edition also provide activities.

. . . with the Take-Home Book

Manipulating Plants

CONTENTS

. . . with Technology

Students can learn more about plant growth and responses through these components and special Internet sites. Look for Technology Links at the end of each lesson.

CNN Turner Le@rning

Harcourt Science Newsroom Video
Giant Plant Gene

SCILINKS THE WORLD'S A CLICK AWAY

These Internet Links are provided by the National Science Teachers Association.
www.scilinks.org/harcourt

Visit The Learning Site for related links, activities, and resources: **www.harcourtschool.com**

WELCOME TO THE LEARNING SITE

Assessment Options Chapter 1

Formal Assessment

- **Chapter Review and Test Preparation**, PE pp. B22–23
- **Chapter Test**, pp. AG19–23

Performance Assessment

- **Chapter Review and Test Preparation**, PE p. B23
- **Chapter Performance Task**, pp. AG24–25
- **Project Evaluation Checklist**, p. AGxviii

Ongoing Assessment

- ✔ **Questions**, PE pp. B6–11, B14–17
- **Lesson Review**, PE pp. B11, B17
- **Informal Assessment**
  - **Classroom Observation**, TE p. B13
  - **Portfolio**, TE p. B5
- **Observation Checklist**, AGxiv

Student Self-Assessment

- **Self-Assessment–Investigate**, p. AGxvi
- **Self-Assessment–Learn About**, p. AGxvii
- **Project Summary Sheet**, p. AGxix

My Work Portfolio Assessment

- **Science Experiences Record**, p. AGxxii
- **Guide to My Science Portfolio**, p. AGxxiii
- **Portfolio Evaluation**, p. AGxxiv

Chapter Test AG19

Plant Growth and Responses

Part I Vocabulary 1 point each

Use the letters of the terms in the Word Banks to complete the sentences.

A xylem B nonvascular plants C phloem D vascular plants

Plants that have tissues that carry water, nutrients, and food to their cells are 1. D. Water and nutrients are carried from the roots to the leaves by the 2. A. The 3. C carries food to all the parts of the plant, where the food is either used or stored. Some plants do not have these tissues. The cells of these plants, called 4. B, must absorb water and nutrients directly from the soil or air.

E short-day plants G tropism I phototropism
F gravitropism H long-day plants

Plants respond to their environments in a variety of ways. The turning of a plant toward or away from a stimulus is called a 5. G. The word comes from a Greek word meaning "to turn." Plants respond to a variety of stimuli. For example, the roots of all plants grow downward, in the direction of gravity; this response is called 6. F. Turning toward sunlight is called 7. I. Plants also respond to changes in the seasons. Many plants do this by responding to changes in the length of the day. 8. E, such as poinsettias, strawberries, and ragweed, need long periods of darkness before they can bloom. 9. H, on the other hand, bloom when the nights are short.

Chapter Test AG20

Part II Science Concepts and Understanding 3 points each

Use the following pictures to answer Questions 10–11.

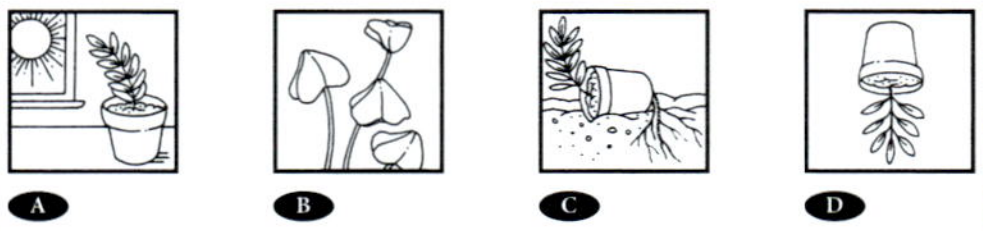

A 10. Which of the drawings shows phototropism?

C 11. Which of the drawings shows negative gravitropism?

Write the letter of the best choice.

B 12. Which of the following do seeds NOT need to germinate?
A water
B sunlight
C air
D warmth

G 13. If there is too much water in the soil, seeds —
F have to germinate faster
G won't get enough oxygen
H won't get enough carbon dioxide
J won't produce enough chlorophyll

B 14. When a seed germinates, it —
A causes disease
B grows roots
C is nonvascular
D dies

F 15. Which is NOT true of vascular plants?
F They are much smaller than nonvascular plants.
G They can grow to be quite large.
H Water and nutrients are absorbed through the root hairs.
J They have tissues that carry water and food.

Chapter Test AG21

Use the following pictures to answer Questions 16–17.

C 16. Which picture represents a nonvascular plant?

A 17. Which picture represents response to touch?

Write the letter of the best choice.

C 18. Food is stored in the roots of some plants during the winter. What happens to this food in the spring?
A The stems use the food to produce root hairs.
B The food is absorbed into the soil.
C The food moves through the phloem to the buds.
D The roots use the food for photosynthesis.

J 19. Where does plant growth NOT occur?
F at the tips of roots
G at the tips of shoots
H at the lateral buds
J at the tips of the leaves

A 20. The roots of a plant grow —
A from the tips
B from the place where they connect to the plant
C from the middle
D from the side

J 21. The process of photosynthesis requires three elements. Which of the following is NOT required?
F sunlight
G water
H carbon dioxide
J oxygen

Chapter Test AG22

B 22. When the Darwins experimented with seedlings and their responses to light, they were experimenting with —
A gravitropism in plants
B phototropism in plants
C seasonal changes in sunlight
D sensitivity of plants to touch

Part III Critical Thinking 8 points each

23. How are tropisms important to the survival of plants? Give two specific examples.
Possible answer: Unlike animals, plants cannot move from place to place. But they have to meet their needs for light, air, water, and nutrients. Tropisms help plants meet these needs. Examples include positive gravitropism to reach water, negative gravitropism and phototropism to reach light, and response to touch for support or protection.

24. Why is it a good idea to dust the leaves of plants in your home?
Possible answer: to keep the tiny pores (stomata) from getting clogged up. Carbon dioxide and oxygen pass into and out of those pores, and dust or dirt can impede that process.

25. Suppose you planted several grapefruit seeds in a pot. To make sure they would grow, you watered them every day, sometimes twice a day. But still, the seeds never sprouted. Explain why.
Accept reasonable answers such as: The seeds never sprouted because they were over-watered. They couldn't get enough oxygen, and they rotted.

Chapter Test AG23

Part IV Process Skills Application

9 26. Describe an experiment to test phototropism. How would you **control variables**? Which variable would you test?
Possible answer: You would need to control water, temperature, air, light, and nutrients. You would test the positions of the plant—for example, facing a window, using artificial light, and planted in a box. The control would be a plant in a normal position.

9 27. Suppose you placed two potted plants on a table several feet away from a sunny window. After a week, you noticed that one of the plants was bent toward the light, but the other one was not. What **conclusion** can you draw about these plants' need for direct sunlight?
Possible answer: The plant turning toward the sunlight is showing positive phototropism. The other plant is not showing any phototropism, negative or positive.

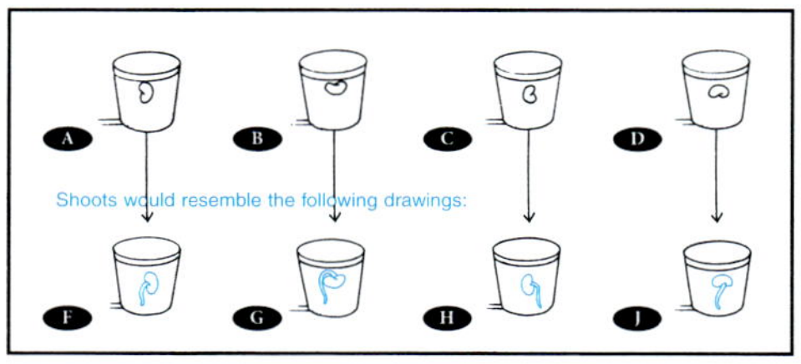

10 28. The diagram above shows an experiment with bean seeds. On the diagram, draw the way the shoots would look in cups F, G, H, and J. What **conclusion** can you draw from this experiment? What conditions need to be **controlled**?
A conclusion drawn from this experiment is that roots grow down and stems grow up. Amount of water, exposure to light, and temperature need to be controlled.

Assessment Continued

Performance Assessment AG24

PERFORMANCE TASK

Looking for Light

Materials

marker | two labels | two potted seedlings | light source | water

1. Put a label on each pot. Label one pot *Plant A* and the other *Plant B.*
2. Place both plants in a sunny window or near a light source.
3. Rotate Plant A one-half turn each day. Do not rotate Plant B.
4. Observe the plants each day and record your observations. Be sure to water the plants as needed.
5. After 10 days, write a paragraph to summarize your observations.
6. Draw pictures to show how the plants responded to the light.

Performance Assessment AG25

PERFORMANCE TASK

Looking for Light

Materials marker, two labels, two potted seedlings, light source, water

Time 25–30 minutes to set up; 10 minutes per day for 10 days

Suggested Grouping individuals, pairs, or groups of three

Science Processes identify and control variables, draw conclusions

Preparation Hints The seedlings in each group must be similar in size and type.

Introduce the Task Have students read the Performance Task sheet. Review phototropism with the class. Stress the importance of making observations at regular time intervals.

Promote Discussion Have students display their paragraphs on a bulletin board so that others can read them. Lead a class discussion about the similarities and differences in the students' experiences.

Scoring Rubric

Performance Indicators

_____ Records observations each day.

_____ Summarizes findings in a written paragraph that describes how the plants move to make sure they get the light they need.

_____ Identifies the variable (turning Plant A) and the control (not turning Plant B).

_____ Draws pictures that show how Plant B grows toward the light.

Observations and Rubric Score

3 2 1 0

Workbook Support

Investigate Log WB33

Germinating Seeds

Materials

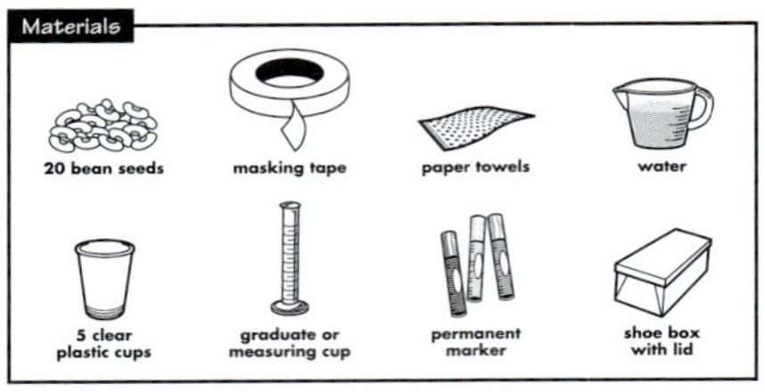

Activity Procedure

1. Fold a paper towel so that its width is the height of one of the cups. Use it to line the sides of the cup. Then fill the cup carefully with crumpled paper towels. Repeat this step for each cup.
2. Between the paper lining and the side of each cup, place four bean seeds. You should be able to see the beans through the plastic.
3. Using masking tape and a marker, label the cups *A, B, C, D,* and *E.* Follow the directions next to the letter for each cup:
   - **A** Slowly drip water into the cup until the paper towels are moist. Place the cup in the shoe box. Close the lid. Add water each day to keep the towels moist.
   - **B** Fill the cup with water.
   - **C** Add *no* water.
   - **D** Moisten the paper towels as for *A*, and keep them moist each day.
   - **E** Moisten the paper towels as for *A*, and keep them moist each day.
4. Place Cups *A, B, C,* and *D* in a warm place, but not directly over a heater and not in direct sunlight. Place Cup *E* in a refrigerator. **Observe** the cups daily to see how many beans are sprouting. **Record** your observations. **Compare** your results with those of your classmates.

Investigate Log WB34

Draw Conclusions

1. In which cups did you **observe** the seeds sprouting? The seeds in cups A and D should sprout. The seeds in cups B, C, and E should not sprout.
2. Based on the results of your experiment, what **conclusions can you draw** about what seeds need in order to germinate? Seeds need water, air, and warm temperatures to sprout.
3. **Scientists at Work** Scientists often **control variables** in an experiment to make sure their conclusions are accurate. Sometimes they need to refine an experiment to make sure they are testing only one variable at a time. What variables did you test in this experiment? How could you change the experiment to test only a single variable? Amount of air and water, temperature, amount of light. Possible answer: To test the proper amount of water, you could place seeds in three cups and keep all three cups in the same dark, warm place. One cup should be dry, one should be filled with water, and one should have moistened paper towels.

Investigate Further How might fertilizer affect germination? **Formulate** a hypothesis and **plan a simple investigation** to test your hypothesis. Students' hypotheses should state that fertilizer will speed up germination. In their plans for an investigation they should have a control without fertilizer.

Process Skills Practice WB35

Identify and Control Variables

This skill involves looking at the factors that will affect the outcome of an experiment. Once you have identified those variables, you must be sure that only one variable is changed in any one test in your experiment.

Think About Identifying and Controlling Variables

When you design an experiment, the first thing you do is decide exactly which factor you are testing. The factor you are changing is called the variable. Once you have decided on your variable, you must be very careful to keep all other conditions the same. Many experiments have given false results because a factor that should have been kept constant was not controlled.

1. Design and write the procedure for an experiment to test the effect the amount of water has on the growth of plants. Use three samples. State which factor is your variable. Then decide which factors you will have to control. Answers should include most of the following ideas: Plants should be as similar to one another as possible and planted at the same depth in the same soil in the same kinds of pots; all three plants should have the same light, air, and temperature; the amount of water should be different but controlled in each sample; adequate time should be allowed for results.
2. What was the variable you allowed to change? the amount of water
3. What factor did you expect to be affected by allowing this variable to change? plant growth
4. What variables needed to be controlled? light, air, temperature, type and amount of soil, depth of plants in soil, and type of pot
5. Why is it necessary to allow only one variable at a time for each test? Students should realize that to be able to draw valid conclusions from the results of an experiment, there should be only one variable tested at a time. In all other ways, samples should be treated the same.

Reading Skills Practice WB36

Distinguish Fact and Opinion

Read each fact below. Then write an opinion statement based on the fact, using the words above each box. Possible statements are given.

1. Although rain forests cover less than 7 percent of Earth's land surface, more than 50 percent of all known organisms live there.

   I believe...

   I believe that such a large number of species means we must carefully explore the rain forests to learn more about the diversity of life.

2. More than 500 species of birds live in 414 square kilometers (160 sq. mi.) of forest between Costa Rica and Panama.

   I think...

   I think that the rain forest must be noisy if many of those birds are songbirds.

3. Only a small portion of the plants and insects in the rain forest have been studied and classified.

   I believe...

   I believe that international teams of scientists should be paid to study the rain forests.

Lesson Concept Review/Assessment WB37

How Do Plants Meet Their Needs?

Lesson Concept

Most plants grow from seeds. All plants need water, sunlight, carbon dioxide, oxygen, and nutrients such as nitrogen, phosphorus, and sulfur. Vascular plants have tissues for transporting food, water, and nutrients. Nonvascular plants do not have those tissues and must absorb water and nutrients directly from the air and soil.

Vocabulary

vascular plants (B8) **xylem** (B8) **phloem** (B9) **nonvascular plants** (B10)

Read the following paragraph and then answer the questions.

It was nearly St. Patrick's Day, and the owner of a flower shop wanted some green flowers. She half-filled a bucket with water and then added green food coloring. Into the bucket she placed several bunches of white carnations. She put the bucket of flowers into the flower cooler and left it overnight. When she checked the flowers the next day, the white carnations were green.

1. Are carnations vascular or nonvascular plants? vascular plants
2. How did the color get into the flowers? The green food coloring moved up the xylem into the petals of the flowers.
3. How is the growth of vascular and nonvascular plants different? Why? Nonvascular plants do not grow as large as vascular plants do because they have no system to get nutrients from roots to leaves.
4. What five things are essential to plant growth? sunlight, water, carbon dioxide, oxygen, and nutrients
5. Are mosses and liverworts vascular or nonvascular plants? How are they attached to the soil? They are nonvascular plants. They attach to the soil by long hairlike cells called rhizoids.

Investigate Log WB38

Do All Roots Grow Down?

Materials

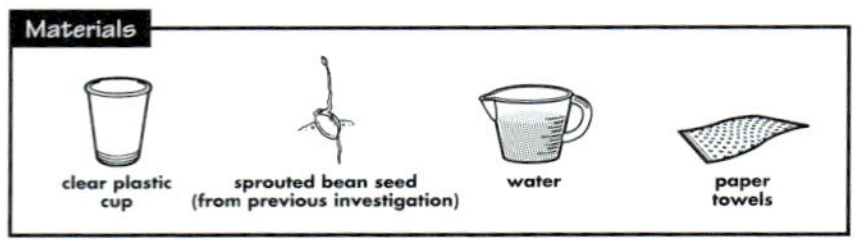

Activity Procedure

1. Line the inside of a plastic cup with several layers of paper towels. Moisten the paper towels.
2. Carefully place the sprouted bean seed between the paper towels and the side of the cup. Position it so that its root is pointing downward.
3. Place the cup in a cabinet or a room that is dark all the time.
4. After a day or two, turn the cup on its side. **Hypothesize** how this change might affect the growth of the sprout. Write your hypothesis. Keep the paper towels moist, and **observe** the results after another day or two.

Workbook Support Continued

Investigate Log WB39

Draw Conclusions

1. What happened to the seed's root after you turned the cup on its side? The root changed direction so that it again pointed downward.
2. How did your **observation** support or disprove your **hypothesis**? A reasonable explanation is that roots somehow respond to gravity and therefore always grow downward.
3. **Scientists at Work** Scientists have **experimented** to **observe** the different effect of gravity in outer space on germinating seeds. When wheat seeds germinated in an orbiting space shuttle, the roots and shoots grew in every direction.
   What can you **conclude** from this experiment? On Earth the response of plants to gravity is that roots grow downward and shoots grow upward. This conclusion is justified because roots and shoots do not grow in any one direction in orbit, where Earth's gravity is weaker and so does not have the same effect.

Investigate Further Try an **experiment** like the one you did to test this **hypothesis**: when you do not move the cup, the direction of the growth of shoots and roots depends on how the seed is positioned. Place four sprouted lima beans with roots pointing in different directions in a plastic cup lined with moist paper towels. Put the cup in a dark place. Keep the paper towels moist, but don't turn the cup.
If your hypothesis is correct, what would you **predict** will happen to each of the roots? Students should suggest that each root will grow downward.

Process Skills Practice WB40

Draw Conclusions

To be able to draw valid conclusions from an experiment, you must make and record observations that are accurate and complete. Then you can use your data to draw reasonable conclusions.

Think About Drawing Conclusions

Dante surveyed three different types of land for a biodiversity study: cultivated property, partly cultivated property, and wild land. He defined *cultivated* as a lawn that received a lot of care from its owner, *partly cultivated* as a lawn that received an average amount of care, and *wild* as an area that received no care or interference from human beings. He marked off an area 6 ft by 20 ft on each type of land. Then he wrote down the different kinds of plants he found, counting their number when possible. He used his data to make the following chart.

Cultivated Property	Partly Cultivated Property	Wild Land
Grass Clover Crab Grass 4 Dandelion 2 Sour Grass	22 Wild Mustard 11 Cheeseweed 11 Groundsel 4 Blue-Curls 3 Yellow Sweet Clover 2 Thistle 1 Red-Stem Filaree 1 Telegraph Weed "unidentified" grasses	Wild Oats 11 Wild Radish 11 Telegraph Weed 6 Milkweed 1 Lupine

1. What conclusion can you draw from Dante's data? Partly cultivated land is more biodiverse.
2. Is this the conclusion you expected when you first imagined surveying the three different types of land? Why or why not? Answers depend on the student, but most students would probably have expected more diversity in the wild land because human cultivation tends to limit diversity.
3. Why do you think Dante got the results he did? Possible answer: the partly cultivated land has more nutrients for the various species and not enough cultivation to eliminate many species. In the wild land, it is likely that some of the wild plants "strangled out" diversity by preventing some species from germinating there and by choking out young plants of rival species.

Reading Skills Practice WB41

Make Generalizations

Read the following paragraph about plant response to salt. Then write a generalization in the box below.

Scientists have recently found a gene that helps plants tolerate salt. Soil can become salty because of faulty irrigation practices that add sodium. In some areas of Australia, Pakistan, India, and countries of Southwest Asia, plants no longer grow where they once did because of increased salinity. There are also places in California and Arizona where plants no longer grow because of salty soil. Crops such as wheat, corn, soybeans, fruits, and vegetables do not produce as high a yield when grown in salty soil. Improving a plant's salt tolerance will increase its ability to grow in certain places.

Generalization

Improving a plant's salt tolerance will increase its ability to grow in certain places.

Lesson Concept Review/Assessment WB42

How Do Plants Respond to Their Environments?

Lesson Concept

A tropism is the turning of a plant away from or toward something in the environment. Gravitropism is a plant's response to gravity; phototropism is a plant's response to light. Some plants respond to being touched. Plants also adapt to the seasons by responding to changes in the length of the day.

Vocabulary

tropism (B14)	**phototropism** (B14)	**gravitropism** (B14)
long-day plant (B16)	**short-day plant** (B16)	

Read the following paragraph and then answer the questions.

There are two related plants called the Christmas cactus and the Easter cactus. In the northern part of the country, you can grow them as houseplants. In warm climates you can grow them outside. The Christmas cactus blooms in the winter around Christmastime, but the Easter cactus blooms in late March or early April.

1. What is different in each plant's environment between December and April? The number of hours of daylight and darkness is different.
2. Compare the conditions each plant needs for flowering. The Christmas cactus needs more hours of darkness to flower.
3. How might you make the Christmas cactus flower earlier? You could increase the daily hours of darkness by putting it in a dark place more hours of the day.
4. What kind of tropism did the plants in the investigation for Lesson 2 show? gravitropism
5. Leaves use their stomata to rid the plant of excess moisture. There are more stomata on the underside of leaves than on the top of the leaves. If you observe some trees on a humid day, you will see the leaves turned over so the bottom side faces up. What inference can you make to explain why this occurs? The leaves are turned over so their stomata are exposed more directly to sunlight. This way their moisture evaporates more quickly.

Vocabulary Review WB43

Recognize Vocabulary

Solve the puzzle, using the terms listed below.

vascular plants	**phototropism**	**xylem**
gravitropism	**phloem**	**long-day plant**
nonvascular plants	**short-day plant**	**tropism**

Across

4. plants without true roots
6. carries water and nutrients from roots to leaves
7. causes plant roots to grow down
8. turning of a plant as a reaction to something in its environment
9. plants that contain xylem and phloem

Down

1. turning of a plant in response to light
2. carries food from the leaves to other parts of the plant
3. plant needing less than twelve hours of daylight to flower
5. plant that flowers from mid-May to mid-July

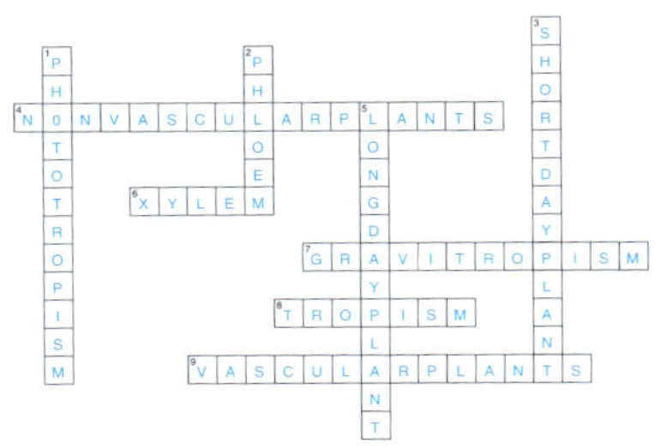

Writing Practice WB44

Write the Story of a Seed

Narrative Writing–First-Person Story

Imagine that you are a seed. Write the story of your growth and development as the narration for a video presentation. Use the story map below to help plan your writing. Make notes or sketches to indicate what the video should show for each stage of the story.

Character (Type of Seed):	Setting (Plant habitat):	Visuals
How does your life begin?		
How are you spread?		
How do you germinate?		
What does the fully developed plant look like?		

CHAPTER 1

Plant Growth and Responses

Generate Questions

Invite students to read and discuss the Fast Facts. Ask students if, after reading the facts about plant growth and responses, they have any questions for which they would like to find answers. Lead the class in brainstorming a list of questions, and record them on the board or on chart paper.

Ask students how they might go about finding answers to the questions they have. Suggest that the chapter will have investigations and reading material to help them answer their questions but that they may want to conduct other investigations or research on their own to find out more about how plants grow and respond to their environment.

CHAPTER 1

Plant Growth and Responses

Vocabulary Preview

- vascular plant
- xylem
- phloem
- nonvascular plant
- tropism
- phototropism
- gravitropism
- long-day plant
- short-day plant

You might not pay much attention to plants—but you couldn't stay alive without them. You eat plants, use furniture made of plants, and breathe oxygen produced by plants. A world without plants would also be a world without people.

Fast Fact

Several types of plants trap and "eat" insects. The pitcher plants of Malaysia, however, are large enough to trap birds or small mammals.

B2

Reading Skills Checklist

Strategies for developing the following reading skills are provided in this chapter.

- ☑ use prefixes and suffixes to determine word meanings *p. B15*
- ☐ use context to determine/confirm word meaning
- ☑ identify supporting facts and details *pp. B6, B7, B8, B14, B16*
- ☐ arrange events in sequence
- ☑ identify the main idea *pp. B6, B7, B8, B14, B16*
- ☐ summarize and paraphrase a selection
- ☐ identify cause and effect
- ☐ predict probable future actions and outcomes
- ☑ use graphic sources for information *p. B9*
- ☑ draw logical conclusions *p. B15*
- ☑ compare and contrast *pp. B7, B8, B10, B15, B18*
- ☑ distinguish fact and opinion *p. B8*
- ☑ make generalizations *pp. B6, B15*
- ☑ use reference sources *p. B16*

School-Home Connection

Distribute copies of the School-Home Connection, p. TR5. Have students take the page home to share with their families as you begin the chapter.

Follow Up Have volunteers compare the results of the activity they did at home. Additional School-Home Connections are provided by the **Activities for Home or School** (p. B21) and the **Take-Home Book** (p. B2d).

Teaching Resources, p. TR5

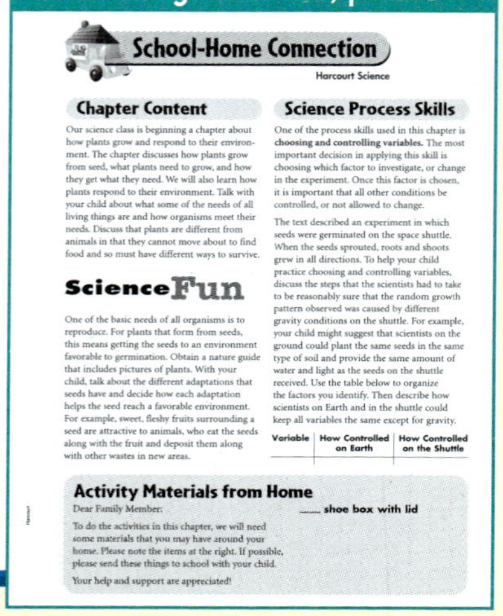

School-Home Connection

Harcourt Science

Chapter Content

Our science class is beginning a chapter about how plants grow and respond to their environment. The chapter discusses how plants grow from seed, what plants need to grow, and how they get what they need. We will also learn how plants respond to their environment. Talk with your child about what some of the needs of all living things are and how organisms meet their needs. Discuss that plants are different from animals in that they cannot move about to find food and so must have different ways to survive.

Science Fun

One of the basic needs of all organisms is to reproduce. For plants that form from seeds, this means getting the seeds to an environment favorable to germination. Obtain a nature guide that includes pictures of plants. With your child, talk about the different adaptations that seeds have and decide how each adaptation helps the seed reach a favorable environment. For example, sweet, fleshy fruits surrounding a seed are attractive to animals, who eat the seeds along with the fruit and deposit them along with other wastes in new areas.

Science Process Skills

One of the process skills used in this chapter is **choosing and controlling variables.** The most important decision in applying this skill is choosing which factor to investigate, or change in the experiment. Once this factor is chosen, it is important that all other conditions be controlled, or not allowed to change.

The text described an experiment in which seeds were germinated on the space shuttle. When the seeds sprouted, roots and shoots grew in all directions. To help your child practice choosing and controlling variables, discuss the steps that the scientists had to take to be reasonably sure that the random growth pattern observed was caused by different gravity conditions on the shuttle. For example, your child might suggest that scientists on the ground could plant the same seeds in the same type of soil and provide the same amount of water and light as the seeds on the shuttle received. Use the table below to organize the factors you identify. Then describe how scientists on Earth and in the shuttle could keep all variables the same except for gravity.

Variable	How Controlled on Earth	How Controlled on the Shuttle

Activity Materials from Home

Dear Family Member:

To do the activities in this chapter, we will need some materials that you may have around your home. Please note the items at the right. If possible, please send these things to school with your child.

Your help and support are appreciated!

____ shoe box with lid

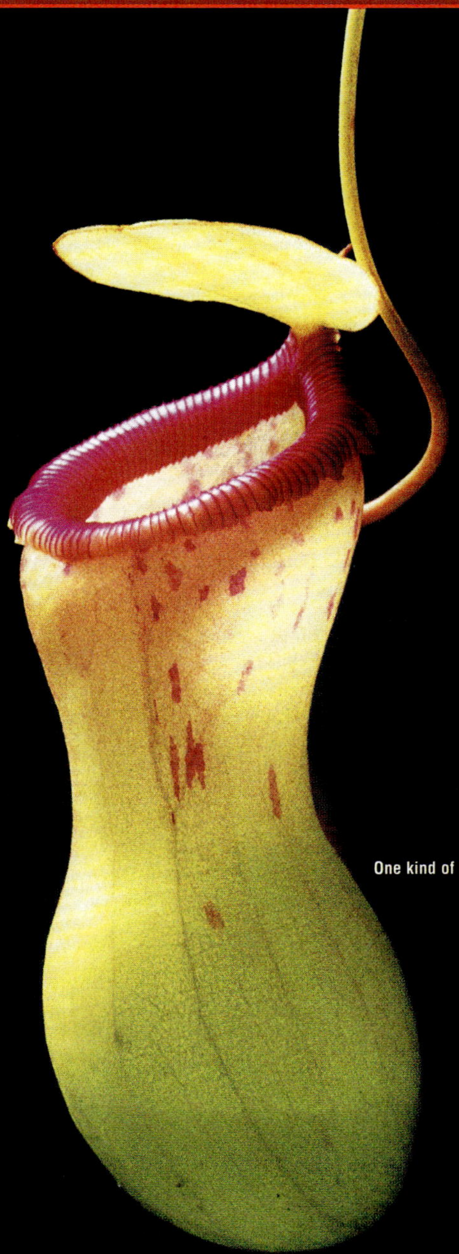

One kind of pitcher plant

Fast Fact

The tallest tree on record is not a giant redwood but an Australian eucalyptus 435 ft (133 m) tall. That's 160 ft (47 m) more than the tallest living redwood.

Tall Trees

Tree	Height (feet)	Height (meters)
Apple	20–30	(6–9)
Live oak	40–50	(12–15)
Paper birch	50–70	(15–21)
Elm	80–100	(24–30)
Ponderosa pine	100–180	(30–55)
Douglas fir	180–250	(55–76)
Redwood	200–275	(61–84)

Fast Fact

Farms provide us with more than food. In the middle of summer, one acre of corn produces enough oxygen each day for 130 people.

B3

Prereading Strategies

K-W-L Chart

Have students make a four-column chart headed What I Know, What I Want to Know, What I Learned, and "+" for What I'll Learn More About. Ask them to look at the chapter title and predict what the chapter will be about. Then have them fill in the first two columns. Students should fill in the third column as they study the chapter. At the end of the chapter, they can fill in the last column with additional things they would like to learn.

K-W-L Chart

What I Know	What I Want to Know	What I Learned	+

Preview the Chapter

Have students scan the chapter, identify the lesson titles and major headings, and use them to make an outline of the chapter. They can write questions that occur to them as they make their outline.

Vocabulary Preview

Have students make a three-column chart for the words in the vocabulary list. In the first column, students should write each word; in the second column, what they think the word means. In the third column, students should write the definitions as they read through the chapter.

Vocabulary Chart

Word	What I Think It Means	Definition

BACKGROUND

Webliography

Keyword plant adaptations

www.harcourtschool.com

GO ONLINE

Mimosa pudica, pictured on page B14, belongs to a genus that contains about 400 species. All grow in tropical parts of the Americas, Asia, and Africa. Many species in this genus are not sensitive only to touch. The leaves bend together and droop in response to chemical and electrical stimuli as well. Most species of mimosas have clusters of small yellow, orange, or purple flowers. The fruit is a pod. *Mimosa pudica* is a perennial shrub in its native setting. In North America it is grown as an annual.

Fast Facts/Math Links

Make a Graph Have students use the data provided in the Tall Trees table to make a bar graph. They may use two bars for each species to show the normal range of heights, or they may simply show the tallest trees in each species.

Write an Equation Challenge students to write an equation showing the number of acres of corn that would be needed to provide enough oxygen each day for all the students in your class. Now have them show the number of acres needed to supply your whole school with enough oxygen for a day.

$$a = \frac{\text{number of students}}{130}$$

LESSON 1

How Do Plants Meet Their Needs?

Objectives

- Describe how plants grow from seeds.
- List what plants need in order to grow.
- Compare phloem and xylem.
- Distinguish between vascular and nonvascular plants.

Motivate

More Fast Facts

Spark interest with these facts and questions.

- The lotus is a flowering plant that grows in water. When its seeds are ripe, they fall into the water and float away. Some lotus seeds germinate more than 200 years after they are produced.

  Why does it take some lotus seeds 200 years to germinate? The conditions are not suitable before that.

- The largest plants in the world are the giant sequoias in California. Some are more than 95 m (about 310 ft) tall.

  Over what distance does a giant sequoia move water? from the soil to the leaves, which may be more than 95 m (about 310 ft)

Access Prior Knowledge

- What needs do you think plants have? How do plants' needs differ from those of animals?
- Do you think all plants meet their needs in the same ways? Why or why not?

When Minutes Count . . .

Whole-Class Investigate Option Set up the investigation at a science center. Have small groups of students make observations daily and report back to the whole class.

Using the Activity Video The video can be used to preview and model the investigate process with students. After viewing the video, students can conduct the investigate in small groups.

LESSON 1

How Do Plants Meet Their Needs?

In this lesson, you can . . .

INVESTIGATE seed germination and seedling growth.

LEARN ABOUT plant growth, plant structures, and how plants meet their needs.

LINK to math, writing, art, and technology.

Banana leaves have slits that let water run through.

B4

INVESTIGATE

Germinating Seeds

Activity Purpose Most plants grow from seeds. Some seeds germinate—or sprout—quickly. Other seeds take years to germinate. No matter how long they take, seeds germinate only under certain conditions. In this investigation you will **control variables** in an experiment to find out what those conditions are.

Materials

- paper towels
- 5 clear plastic cups
- 20 bean seeds
- masking tape
- permanent marker
- water
- graduate or measuring cup
- shoe box with lid

Activity Procedure

1. Fold a paper towel so that its width is the height of one of the cups. Use it to line the sides of the cup. Then fill the cup carefully with crumpled paper towels. Repeat this step for each cup. (Picture A)
2. Between the paper lining and the side of each cup, place four bean seeds. You should be able to see the beans through the plastic.
3. Using masking tape and a marker, label the cups *A, B, C, D,* and *E.* Follow the directions next to the letter for each cup: (Picture B)

   A Slowly drip water into the cup until the paper towels are moist. Place the cup in the shoe box. Close the lid. Add water each day to keep the towels moist.

   B Fill the cup with water.

   C Add *no* water.

BACKGROUND

Webliography GO ONLINE
Keyword plant adaptations
www.harcourtschool.com

Types of Seeds A great diversity exists among seeds. Students may be familiar with acorns, maple-seed "helicopters," and pine nuts. They may have seen thistle, milkweed, or dandelion seeds floating through the air. In foods, students may have seen sunflower, watermelon, tomato, cucumber, pea, or other seeds. Finally, some students may have planted seeds in a garden or clay pot.

Special Adaptations of Vascular Plants **Epiphytes** are vascular plants that grow on the trunks or upper branches of trees. As the trees grow, the plants are raised up toward the sunlight with them. Epiphytes do not have roots in the ground; like nonvascular plants, they absorb water from air and rainwater. **Vines** are vascular plants that find their way out of the forest shadows by climbing trees. Some vines have tendrils that respond to touch by curling around branches or leaves, thus supporting the plant.

For student self-assessment, see Assessment Guide p. AGxvi.

Picture A

Picture B

D Moisten the paper towels as for A, and keep them moist each day.

E Moisten the paper towels as for A, and keep them moist each day.

4 Place Cups A, B, C, and D in a warm place, but not directly over a heater and not in direct sunlight. Place Cup E in a refrigerator. **Observe** the cups daily to see how many beans are sprouting. **Record** your observations. **Compare** your results with those of your classmates.

Draw Conclusions

1. In which cups did you **observe** the seeds sprouting?
2. Based on the results of your experiment, what **conclusions can you draw** about what seeds need in order to germinate?
3. **Scientists at Work** Scientists often **control variables** in an experiment to make sure their conclusions are accurate. Sometimes they need to refine an experiment to make sure they are testing only one variable at a time. What variables did you test in this experiment? How could you change the experiment to test only a single variable?

Investigate Further How might fertilizer affect germination? **Formulate** a hypothesis, and **plan a simple investigation** to test your hypothesis.

Process Skill Tip

A *variable* is a factor that could affect the results of an experiment. To make sure that their conclusions are accurate, scientists need to **identify and control variables**. To control variables, you need to set up experimental conditions to test only one thing at a time.

B5

Process Skill Tip

Identify and Control Variables

Students may have difficulty identifying the factors that might affect the outcome of an experiment. Display the Process Skill Tip Transparency on an overhead projector or draw similar plants on the board. Discuss the questions on the transparency.

Point out that each set of seedlings has a different combination of two factors—light and temperature. Explain that **variables** are factors that may affect the outcome of an experiment. Only one variable should be changed in any given test.

Process Skill Tip Transparency B1-1

Identify and Control Variables

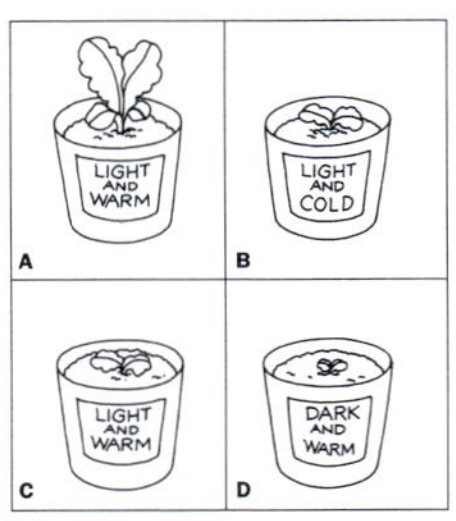

- What conditions affect plant growth? Which of these are being tested in A & B? C & D? Light, temperature, and water; temperature; light
- In A & B what is controlled? What varies? Should any of the other conditions vary? Light; temperature; no

INVESTIGATE

Time 20 min for setup; 10 min a day for observations until the seeds germinate, or about 5 days

Preparation Tips See page B2c.

Grouping pairs

Process Skill Tip identify and control variables

Other Process Skills observe, compare, record data, draw conclusions, hypothesize, plan simple investigations

Expected Results Students should conclude that seeds need water, air, and warmth in order to germinate.

Teaching Tips

Point out that scientists may observe by measuring or by comparing two or more objects (noting their similarities and differences). Sprouted seeds should be saved for the next lesson.

Draw Conclusions

1. The seeds in cups A and D should sprout. The seeds in cups B, C, and E should not sprout.
2. water, air, and warm temperatures
3. The variables tested were temperature and amount of air, water, and light. To test only the amount of water, you could place seeds in three cups. One cup should be dry, one should be filled with water, and one should contain moistened paper towels. Then you should put all three cups in the same dark, warm place.

Investigate Further Encourage students to formulate hypotheses about how fertilizer affects germination. Help students plan a simple investigation to test their hypotheses.

Informal Assessment

Portfolio Students may want to keep their recorded observations of the beans in their science portfolios.

INVESTIGATE LOG

Students can use **Workbook pp. WB33–34** to record the results of the investigation.

PROCESS SKILLS PRACTICE

To practice and apply process skills, see **Workbook p. WB35.**

LEARN ABOUT

1 Before Reading

SET A PURPOSE

FIND OUT Remind students that these statements will help them know what to look for as they read.

VOCABULARY If students are doing the activity described in the Vocabulary Preview on page B3, remind them to add the meanings of these words to the chart.

Graphic Organizer for Chapter Concepts Display the transparency (shown below) or have students use Workbook page WB32. Point out that by the end of the lesson, students should be able to fill in the plant needs and structures on the transparency.

2 Guide the Learning

SCIENCE IDEAS

Help students focus on the supporting details for this **main idea: Under proper conditions, a seed will germinate and begin to grow.**

- **What things does a seed need in order to germinate?** the right amounts of water, warmth, and oxygen
- **What happens to a seed when it absorbs water?** It swells.
- **What happens when a seed germinates?** It splits open. A tiny root grows downward, and a tiny shoot begins to grow upward. Leaves form at the tip of the shoot.

VISUAL LEARNING

Have students observe the photographs and read the captions on this page.

- **On the basis of what you read and observed, make a generalization about the location of growth in a plant.** Plants grow at the tips of roots and shoots.

VOCABULARY CARDS AND ACTIVITIES

Students can use the Vocabulary Cards to make their own graphic organizers or to add to an ongoing file of science terms. The Vocabulary Cards and a variety of strategies and activities are provided beginning on p. TR85 in the **Teaching Resources** book.

LEARN ABOUT

Adaptations of Plants

FIND OUT
- how plants grow from seeds
- what plants need in order to grow and how they meet their needs
- the differences between vascular and nonvascular plants

VOCABULARY
vascular plant
xylem
phloem
nonvascular plant

Most Plants Grow from Seeds

Most plants produce new plants from seeds. Walk through an overgrown field or an empty lot, or even down a city street, and you'll find seeds of many sizes, colors, and shapes. Some, such as the seeds of the dandelion, are like parachutes. Others are burs that stick to clothing and fur like Velcro. Look on trees, and you'll find many other kinds of seeds. Maybe you've seen the winged seeds on maples, the acorns on oaks, and the seeds in the cones on pine trees.

But how do seeds germinate? When conditions are right, seeds absorb water. The seeds swell. If there is not enough water in the soil, the seeds won't sprout. If there is too much, the seeds will not get enough oxygen, and they will rot.

Seeds also need warmth to germinate. In regions where winters are cold, seeds do not begin to sprout and grow until spring arrives. Some will sprout in the colder ground of early spring. Others need the sun-warmed ground of late spring or early summer.

When a seed germinates, it splits open and a small seedling appears. A tiny root begins to grow downward. Then a tiny shoot begins to grow upward. Soon leaves form at the tip of the shoot. In some plants, branches may grow from lateral, or side, buds. The tips of the side branches grow, and they, too, form leaves and more lateral buds. In this way, branches produce more branches.

✔ **What happens first when a seed germinates?** It splits open. A small root forms.

A shoot grows longer as the cells at its tip divide and grow longer.

Growth also occurs at the very tip of a root. The root pushes deeper into the soil as the cells at the root tip divide and grow longer.

A tree seedling grows only where plant cells are dividing rapidly. This is most often at the tip of the root and the tip of the shoot.

B6

Graphic Organizer for Chapter Concepts

Transparency B1 • Workbook, p. WB32

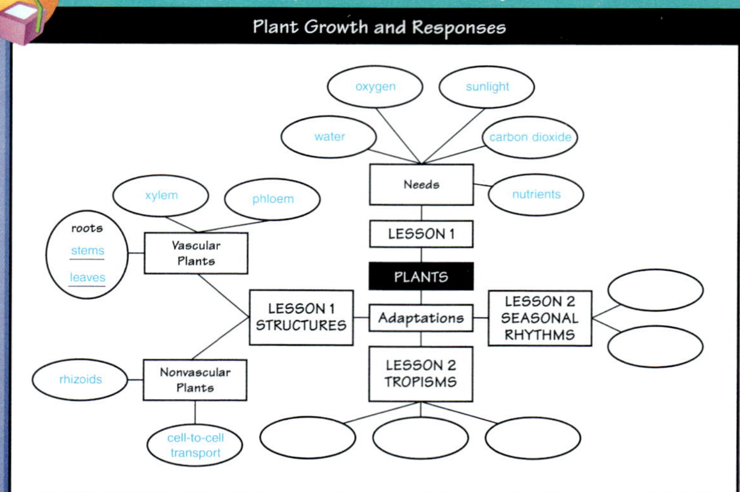

Plant Needs

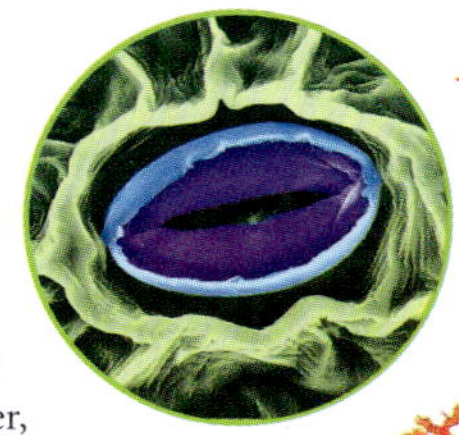

◀ Gases, such as oxygen and carbon dioxide, may enter or leave the leaf through the stomata. Water is lost through open stomata in a process called transpiration.

After a seed has germinated and begun to grow, the new plant needs certain things to stay healthy. It needs water, sunlight, carbon dioxide, oxygen, and nutrients. Plants use these things to carry on life processes within the cells and to make and transport the food they need.

Water makes up as much as 95 percent of some plants. All life processes inside plant cells take place in a watery solution. In addition, all the materials carried through a plant are transported in water. Plants also use water in producing food.

The process of photosynthesis, in which plants make food, requires water, sunlight, and carbon dioxide. Plants have chloroplasts, which are full of the green pigment chlorophyll. Chlorophyll captures the energy from sunlight. The energy is used to combine carbon dioxide and water to produce sugar and oxygen.

Plants also use oxygen to release the energy from food for life processes. When oxygen combines with the sugar, the process produces energy, carbon dioxide, and water.

Plants also need nutrients such as nitrogen, phosphorus, and sulfur. They are used to make important parts of plant cells. They also help carry out some of the processes of the cells.

✔ **What five essential materials do plants need?** water, sunlight, carbon dioxide, oxygen, and nutrients

◀ During photosynthesis, the chlorophyll in these chloroplasts captures the energy of the sun to make sugar:
sunlight + carbon dioxide + water → sugar + oxygen

Plant cells use sugar for energy for life processes:
sugar + oxygen → energy + carbon dioxide + water

◀ In many plants, tiny root hairs absorb water and nutrients from the soil.

B7

SCIENCE IDEAS

Help students focus on the supporting details for this **main idea:** **In order to grow and be healthy, plants need certain materials.**

- ▶ **What materials do plants need to survive?** water, sunlight, oxygen, carbon dioxide, and nutrients (nitrogen, phosphorus, and sulfur)
- ▶ **Why do plants need water?** All life processes inside plants occur in a watery solution. Nutrients and other materials are transported through plants in water. Water makes up as much as 95 percent of some plants.
- ▶ **What is photosynthesis? Why is it important?** The process by which green plants use sunlight to make food; this food provides plants with the energy they need for life processes.

VISUAL LEARNING

Have students **compare and contrast** the photographs and captions on page B7.

- ▶ **Compare the steps in photosynthesis with the steps plants go through to use sugar for energy. What do you notice?** These processes are the reverse of each other. In photosynthesis, the energy from sunlight is used to combine carbon dioxide and water to produce sugar and oxygen. When plants use the energy in food, sugar combines with oxygen to give off energy, carbon dioxide, and water.

Critical Thinking **A terrarium is a closed container with plants growing in it. How can the plants get enough water and oxygen?** Although the plants do use water and oxygen, they also produce them.

BACKGROUND

Webliography
Keyword plant adaptations
www.harcourtschool.com

Germinating Seeds Inside a seed is an embryo that contains a root and stem. After a seed absorbs water, the embryo cells can divide and grow. The seed contains enough food to keep the embryo growing until the seedling can make its own food.

Photosynthesis involves a series of chemical reactions. Light energy is converted to chemical energy in the form of glucose, which may be converted to starch for storage. Carbon, oxygen, hydrogen, and various nutrients are building blocks of the chemicals involved in these processes.

Nonvascular Plants In areas of heavy rainfall, nonvascular plants such as mosses grow abundantly. In the rain forests of Washington State and British Columbia, mosses blanket fallen logs and tree trunks. Because many mosses do not need soil, they may grow on rocks in streams or in pavement cracks.

Cultural Connection

Sprouted seeds are a staple of Asian cooking. Because sprouts can easily be grown in a few days at any time of year, they are sometimes called nature's most convenient vegetable. Sprouted mung beans add flavor and nutrition to many Chinese foods. Because of their crunchiness, the Chinese call them "teeth vegetables." In the United States today, other kinds of sprouted seeds, including pea, bean, radish, and alfalfa, are also used as foods.

Guide the Learning continued

SCIENCE IDEAS

Help students focus on the supporting details for this **main idea:** **Vascular plants have tissues that carry water and dissolved materials throughout the plant.**

- **What are the two kinds of tissue that make up the vessels of vascular plants?** xylem and phloem
- **What purpose do xylem and phloem serve?** to carry water, nutrients, and food to all plant parts

VISUAL LEARNING

Have students **compare and contrast** the photographs and captions on pages B8 and B9. Ask questions about what they can observe and infer from the illustrations.

- **Compare the three enlargements of the tree. What is similar about them?** They all are cross-sections of some part of the tree. All show plant tissues that carry water. The cells are packed tightly together.
- **What do you observe about the structure and function of xylem?** It is made up of thick-walled cells that form continuous tubes that extend the length of the tree. These tubes carry water and nutrients from the roots to the leaves.

Ongoing Assessment

Use the ✓ questions in the pupil edition.

Vascular Plants

The water and nutrients absorbed by a plant's roots travel throughout the plant. The food made in the leaves also travels through the plant to the places where energy is needed. Many kinds of plants have tissues, or specialized groups of cells, that carry water and dissolved materials. Plants that have these tissues are called **vascular plants**. The oak tree shown on these pages is a vascular plant.

The tissues that carry water form a system of vessels that reach all parts of the plant. The vessels are made of two kinds of tissue: xylem and phloem. **Xylem** (ZY•luhm) carries water and

THE INSIDE STORY

Inside an Oak Tree

Some plants, like this oak tree, have woody stems. Their stems get thicker each year. Bark forms around stems, branches, and parts of the roots. As the tree grows thicker, new xylem and phloem form. ►

The xylem tissue forms continuous, hollow tubes. These tubes reach from the roots all the way up the trunk and through the branches to the leaves. Xylem cells have very thick walls, which provide support for vascular plants.

Water and nutrients are absorbed through the root hairs. They move into the xylem. Phloem carries food to the roots, where much of it is stored, and to all other parts of the plant, where it is used in life processes. ►

THE QUESTIONS KIDS ASK!

How can water get all the way to the top of a tree through those tiny tubes?

Address Misconceptions The xylem cells form many continuous tubes that reach all the way from the roots to the leaves at the top of even the tallest trees. Water is pulled upward through these tubes. Explain that the continuous loss of water through transpiration in the leaves pulls the water up from the roots through the stems.

Reading Mini-Lesson

Distinguish Fact and Opinion

Remind students that phrases like *I believe, I think,* and *I feel* are phrases that are used to express opinions. Display the transparency. Have students use these phrases to provide opinions based on this fact: Plants need water, sunlight, oxygen, and nutrients to grow well.

Reading Mini-Lesson Transparency B1-1

Distinguish Fact and Opinion

I believe . . .
I believe that plants should always be grown in a south-facing window.

I think . . .
I think that plants should be watered every two days.

I feel . . .
I feel that oxygen is the most important thing plants need to grow.

READING SKILLS PRACTICE

To practice and apply this reading skill, see **Workbook p. WB36.**

nutrients from the roots to the leaves. **Phloem** (FLOH•uhm) carries food from the leaves to the other parts of the plant for use and storage.

Some of the food is carried to the roots, where it is stored for the winter. In the spring this food moves through the phloem to the buds, where new stems and leaves are about to grow.

One reason vascular plants can grow very large is that they have xylem and phloem to carry water, nutrients, and food to all their parts.

✓ **Why does a vascular plant need both xylem and phloem?** Answers should include the different materials the tissues carry; students may also suggest that the xylem and phloem often carry materials in opposite directions.

USE PROCESS SKILLS

Observe Allow students a few moments to observe the illustration of the tree and read the captions. Challenge them to use the terms *xylem* and *phloem* when they answer the question.

- ▶ **What are the three main parts of the tree? What are the functions of each part?** The roots absorb water and nutrients which are transported upward through the xylem. Materials are carried up and down the tree through the trunk. The leaves absorb carbon dioxide and oxygen through stomata and are the site of photosynthesis. Xylem carries water and nutrients from the roots to the leaves and phloem carries the food produced in the leaves to all other parts of the plant.

DEVELOP SCIENCE VOCABULARY

vascular plant, p. B8—Have students look up the root of *vascular* in the dictionary. Point out the connection between vessels in general and blood vessels (as in cardiovascular).

Also, make sure students understand that a vascular plant contains xylem and phloem, which form an internal transportation system.

xylem, p. B8 and **phloem**, p. B9—Remind students that xylem carries water and nutrients and phloem carries food. Point out that the words *phloem* and *food* both start with an "f" sound.

Have students tell in what direction or directions each type of tissue carries materials.

Social Studies Link

Cereal Grains Many of the world's staple foods are cereal grains, or seeds. The major cereal crops are wheat, rice, maize (corn), barley, sorghum, oats, millet, and rye. Wheat, the most important food crop in the world, is the main food of 35 percent of the population. It provides 20 percent of the calories consumed in the world, as well as a large portion of the nutrients. In tropical regions, rice is the major crop. Maize is a staple in tropical and subtropical regions. Ask students to bring in empty packages from a variety of foods. Have them read the ingredients and **use the tables or other graphic sources of information** to determine which cereal grains they contain. Survey the class to determine the three most prevalent grains found in the packaged foods.

ESL ACTIVITY

Give students fast-sprouting seeds such as alfalfa or radish seeds. Have them sprinkle 6 to 12 seeds on a moist paper towel and place it in a warm, dim place. Tell students to keep their towel damp but not wet. Ask students to observe their seeds each day and to sketch them. Have them label their sketches, using words such as *germinates, seedling, root,* and *shoot.*

Guide the Learning continued

USE PROCESS SKILLS

Compare Invite students to **compare** vascular and nonvascular plants. Record their answers in a chart on the board or have students make their own charts.

- **In what ways do nonvascular plants obtain water and nutrients?** Each cell of the plant absorbs water and nutrients directly from the soil.
- **In what ways are nonvascular plants different from vascular plants?** Nonvascular plants have no xylem or phloem; do not have true roots, stems, or leaves; are small; and often grow in moist places.

DEVELOP SCIENCE VOCABULARY

nonvascular plant, p. B10—Have students think of nonvascular plants that they have seen. Ask students to specify in what kinds of environments they may have seen nonvascular plants. moist environments, such as marshy areas, damp forests, and ditches

- Also ask students to draw logical conclusions about how the size of a nonvascular plant might be related to its structure. It is small because it has no xylem or phloem to transport water and nutrients. It must take them in directly from the environment.

Nonvascular Plants

Some plants, such as mosses and liverworts, do not have xylem and phloem. They do not have tissues through which water, nutrients, and food are transported. These plants are called **nonvascular plants**.

Nonvascular plants are different from vascular plants in several ways. Because this type of plant does not have tissues that transport water and nutrients, it does not grow very large. In addition, each cell of the plant absorbs these materials directly from the soil or air. Unlike vascular plants, nonvascular plants do not have true roots. Instead they have long, hairlike cells called *rhizoids* (RY•zoydz), which hold them in the soil.

Many nonvascular plants need a lot of water. These species thrive in moist, shady places such as woods and swamps. In some cases, they grow so close together that they actually help support one another, much as xylem helps support vascular plants.

The ground in woodlands and near streams and waterfalls is moist and shady, and is often covered with mosses.

A bed of moss is actually made up of hundreds of tiny individual moss plants. The plants grow very close together to form a thick, spongy, bright-green mat.

B10

Reaching All Learners

Visual Learners Provide visual learners with a white carnation and a container of water colored with red or blue food coloring. Have them place the flower stem in the water and observe the changes in color due to the rise of colored water through the xylem. Have students draw a picture to show what is happening inside the carnation.

INVESTIGATION CHALLENGE

Hands-On Activity: More about Seed Germination

Use this idea for a hands-on activity to extend students' understanding of key lesson concepts.

Provide students with tape, water, bean seeds, and a small plastic container with a lid. Have them fill the container with beans, add water, and tape the container shut. Then have them observe what happens over the next few days. Ask students to comment on the force of the swelling seeds. Challenge them to devise other ways to test and observe the force of swelling seeds. Students might also compare the ways different kinds of seeds swell in water.

Students may use a hand lens to observe one of the seeds after it has become swollen with water. If they open the seed at the outside curve, they will not damage the embryo.

For student self-assessment, see Assessment Guide p. AGxvii.

Nonvascular plants have parts that resemble leaves, stems, and roots, though they are less complex structures than those of vascular plants. The leaflike parts of mosses and liverworts contain chlorophyll. Although these parts have only one or two layers of cells, they are able to produce food for the plant.

✓ **Why are nonvascular plants small?** because they do not have vascular tissue through which food, water, and nutrients can be transported

Summary

Seeds germinate when they have the right amounts of oxygen, water, and warmth. Growth occurs at the tips of the roots and shoots and at the lateral buds. Plants need water, sun, and carbon dioxide to carry on photosynthesis. They need oxygen and sugar to produce energy for life processes. They also need nutrients, which they absorb from the soil through root hairs. Vascular plants have xylem, which transports water and nutrients from the roots throughout the plant. They also have phloem, which transports food from the leaves to all other parts of the plant. Nonvascular plants do not have tissues to transport materials.

Review

1. Under what conditions do seeds germinate?
2. How can you make sure that a houseplant stays healthy?
3. Why can most vascular plants grow much taller than nonvascular plants?
4. **Critical Thinking** Suppose you pound a nail into a tree trunk. Will the nail stay at the same height each year? Explain.
5. **Test Prep** Nutrients are carried through a vascular plant by —
   A seedlings
   B bark
   C xylem
   D phloem

LINKS

MATH LINK

Compute Percents A germination rate is given on the back of every package of seeds. The seed companies have tested their seeds to determine the rate. If, for example, 75 out of 100 seeds sprout, then the germination rate is 75 percent. Calculate the following germination rates as percents: 10 out of 25 seeds, 28 out of 50 seeds, 36 out of 50 seeds. 40%, 56%, 72%

WRITING LINK

Informative Writing—Compare and Contrast Write a paragraph for your classmates in which you compare xylem and phloem tissues with the delivery systems at a factory.

ART LINK

Seed Crafts Seeds show a great variety of sizes, shapes, and colors. They can be used for decoration in countless ways. People use them for jewelry, collages, sculptures, mobiles, and more. Make your own art project using seeds.

TECHNOLOGY LINK

To learn more about plant adaptations and growth, watch *Giant Plant Gene* on the **Harcourt Science Newsroom Video.**

LINKS TIPS

Links Objective Extend understanding of how plants meet their needs.

Math Link Encourage students to use calculators. Remind them that the smaller number is divided by the total number of seeds in each example. 10 ÷ 25 = 40%, 28 ÷ 50 = 56%, 36 ÷ 50 = 72%

Writing Link For prewriting, encourage students to think about and discuss how things are transported throughout a factory. Then ask students to think about the ways materials are transported through a vascular plant. Writing models are provided in the **Teaching Resources** book.

Art Link Provide students with sample pieces of art or photographs of artwork that were made with or without seeds. Have small groups brainstorm ways to use seeds in art. Encourage students to complete one of these projects.

3 Wrap Up and Assess

SUMMARY

Graphic Organizer for Chapter Concepts After reading the summary, have students complete the Lesson 1 section of the graphic organizer (shown on page B6).

REVIEW

1. Seeds germinate when they have the right amounts of warmth, air, and water.
2. You can keep a houseplant healthy by keeping its soil moist and making sure it receives nutrients and enough sunlight.
3. Vascular plants have xylem and phloem, through which water, nutrients, and food are transported. Nonvascular plants do not have these tissues or the support they provide.
4. yes, because growth in height occurs only at the tips of the branches
5. C, xylem

TEST PREP Test-Taking Tips

Item type multiple choice with a sentence fragment

To help students answer Question 5, share the following strategy with them. **Read the sentence carefully. Use facts you already know to help find the correct answer. Try to visualize how food and nutrients are transported in vascular plants. Eliminate any choices that do not fit. If you are not sure of the correct answer, make an educated guess.**

LESSON CONCEPT REVIEW/ASSESSMENT

Students can use **Workbook p. WB37** to review the lesson concepts.

LESSON 2

How Do Plants Respond to Their Environments?

Objectives

- ▶ Describe how plants respond to light, gravity, and touch.
- ▶ Explain how tropisms help plants survive.
- ▶ Compare long-day and short-day plants.

Motivate

More Fast Facts

Spark interest with these facts and questions.

- Giant kelp of California grow up to 46 cm (about 18 in.) per day. Fronds of kelp form underwater forests and can reach 100 m (about 328 ft) in length.

  What's the most a giant kelp could grow in 5 days? 230 cm (90 in.)

- The Venus' flytrap is a plant that "eats" insects. It has pairs of leaves that can close together to form traps. The leaves have three special hairs. If an insect touches two or three hairs, the trap snaps shut in less than a quarter of a second.

  How does the Venus' flytrap's response to touch help it survive? It allows it to capture insects to get the nutrients it needs.

Access Prior Knowledge

- ▶ What factors make up a plant's environment?
- ▶ How can a plant, which is rooted in one place, respond to something in its environment?

When Minutes Count . . .

Whole-Class Investigate Option Set up the investigation as a demonstration. Take out the cup each day and encourage students to make observations.

Using the Activity Video The video can be used to preview and model the investigate process with students. After viewing the video, students can conduct the investigation in small groups.

LESSON 2

How Do Plants Respond to Their Environments?

In this lesson, you can . . .

- INVESTIGATE how roots respond to the environment.
- LEARN ABOUT how plants respond to light, gravity, touch, and length of the day.
- LINK to math, writing, social studies, and technology.

INVESTIGATE

Do All Roots Grow Down?

Activity Purpose When a seed is planted, it may be upside down in the ground. If it is, will the root still grow downward and the shoot still grow upward? In this investigation you'll **experiment** to find out.

Materials

- clear plastic cup
- paper towels
- water
- sprouted bean seed (from previous investigation)

Activity Procedure

1. Line the inside of a plastic cup with several layers of paper towels. Moisten the paper towels.
2. Carefully place the sprouted bean seed between the paper towels and the side of the cup. Position it so that its root is pointing downward. (Picture A)
3. Place the cup in a cabinet or a room that is dark all the time.
4. After a day or two, turn the cup on its side. **Hypothesize** how this change might affect the growth of the sprout. Write your hypothesis. Keep the paper towels moist, and **observe** the results after another day or two. (Picture B)

◀ Desert plants can live well in places with little rainfall. They lose little moisture, and they have large shallow root systems to take in as much water as possible.

B12

BACKGROUND

Webliography
Keyword plant adaptations
www.harcourtschool.com

GO ONLINE

Tenacious Tendrils When a tendril from a vine makes contact with a solid object, it reacts by curling around it. After several twirls around a branch or stem, the tendril is securely attached. The middle section of the tendril then begins to coil up. This shortens the tendril and pulls the plant toward the object, thus helping support the plant. If the wind blows, the coil acts like a spring, and the tendril stretches rather than breaks.

Plant Hormones A plant bends toward light and toward or away from the pull of gravity because of differences in the rate of growth on opposite sides of its stem or root. When a light is placed on one side of a plant, the cells on the side of the stem away from the light continue to elongate while the cells closer to the light stop growing. Scientists think that plant hormones called **auxins,** among others, may cause these differences in cell growth.

For student self-assessment, see Assessment Guide p. AGxvi.

Picture A

Picture B

Draw Conclusions

1. What happened to the seed's root after you turned the cup on its side?
2. How did your observation support or disprove your hypothesis?
3. **Scientists at Work** Scientists have **experimented** to **observe** the different effect of gravity in outer space on germinating seeds. When wheat seeds germinated in an orbiting space shuttle, the roots and shoots grew in every direction. What can you **conclude** from this experiment?

Investigate Further Try an **experiment** like the one above to test this **hypothesis**: when you do not move the cup, the direction of the growth of shoots and roots depends on how the seed is positioned. Place four sprouted lima beans with roots pointing in different directions in a plastic cup lined with moist paper towels. Put the cup in a dark place. Keep the paper towels moist, but don't turn the cup. If your hypothesis is correct, what do you **predict** will happen to each of the roots?

Process Skill Tip

To **draw conclusions**, you need to carefully think about all of the information you have. You need to think about what you already know and what you found out in your experiment.

B13

INVESTIGATE

Time 5–20 min for setup; 10 min for observation 2–3 days later

Preparation Tips See page B2c.

Grouping pairs

Process Skill Tip draw conclusions

Other Process Skills observe, hypothesize, experiment, predict

Expected Results Students should conclude that roots grow downward and shoots grow upward.

Teaching Tips

Be sure students understand what a *hypothesis* is. Encourage students to state their hypotheses as *If . . . then* statements.

Draw Conclusions

1. The root changed direction so that it again pointed downward.
2. Observation should support this hypothesis: If a plant is placed on its side, the root will grow downward and the shoot will grow upward.
3. On Earth, roots grow downward and shoots grow upward because of gravity. This conclusion may be justified by scientists' observations that in an orbiting space shuttle, roots and shoots do not grow in any predetermined direction.

Investigate Further Students should suggest that each root will grow downward. Make sure they can support their hypotheses with details they observed from their investigation.

Informal Assessment

Classroom Observation As students work in pairs to complete the activity, observe how they interact. Are students sharing tasks equally, or does one seem to take over? Encourage students to be patient with each other.

INVESTIGATE LOG

Students can use **Workbook pp. WB38–39** to record the results of the investigation.

PROCESS SKILLS PRACTICE

To practice and apply process skills, see **Workbook p. WB40.**

Process Skill Tip

Draw Conclusions Students may not realize how many other process skills go into drawing a conclusion. Display the Process Skill Tip Transparency on an overhead projector or draw a cup on the board. Discuss what other skills are used in order to answer the question on the transparency.

Point out to students that a conclusion is the final step in the investigative process. Before drawing a conclusion, they must process data, observe, and infer. Make sure they understand that a conclusion is based on much more data than an inference.

Process Skill Tip Transparency B1-2

Draw Conclusions

- Based on the information in the drawing, what can you conclude about how this cup was oriented when the seed germinated? The pot was upside down.

1 Before Reading

SET A PURPOSE

FIND OUT Remind students that these statements will guide them in what to look for as they read.

VOCABULARY If students are doing the activity described in the Vocabulary Preview on page B3, make sure they add the definitions for these words to the chart.

Graphic Organizer for Chapter Concepts Display the transparency (shown below) or have students use Workbook page WB32. Point out that by the end of the lesson, students should be able to fill in all plant adaptations shown on the transparency.

2 Guide the Learning

SCIENCE IDEAS

Help students focus on the supporting details for this **main idea: A tropism is the turning of a plant away from or toward something in its environment.**

- **How does a plant respond to light? What is this response called?** by turning toward it; phototropism
- **To what other things in their environment do plants respond? How do they respond to these things?** Roots respond to gravity by growing toward the pull of gravity, and shoots respond by growing away from the direction of gravity. Some plants have tendrils that respond to touch by growing toward the object they touch.

Critical Thinking Lead students to hypothesize how they could investigate plant tropisms.

- **How could you find out how much light is needed for a plant to turn toward it?** Students should suggest a setup that involves exposing plants to various amounts of light.

Ongoing Assessment

Use the ✔ questions in the pupil edition.

LEARN ABOUT

Shoots and leaves grow toward light. Roots grow in the direction of gravity, and shoots grow away from gravity. Some plants have tendrils that grow toward the side that is touched.

How Plants Respond to Their Environments

FIND OUT
- how plants respond to their environments
- why different plants flower at different times of the year

VOCABULARY
tropism
phototropism
gravitropism
long-day plant
short-day plant

Tropisms

Suppose you turn a potted flowering plant on your windowsill so that the flowers face the room. Before long the plant stems will bend so that the flowers again face the light from the window.

The response of a plant to turn toward or away from anything in its environment is called a **tropism** (TROH•piz•uhm). The word *tropism* comes from a Greek word that means "to turn." A response to light, like the one shown by flowers on the windowsill, is called **phototropism**. Plants are said to show *positive phototropism* because they turn toward light.

In the investigation, you saw another kind of tropism. The root of a plant grows downward, in the direction of gravity. This is **gravitropism**. Roots show *positive gravitropism*—they bend in the same direction as the pull of gravity. The shoot of a seed shows *negative gravitropism*—it bends away from the pull of gravity.

Some plants respond to touch. For example, squash, pumpkins, and cucumbers have threadlike tendrils that are sensitive to touch. The tendrils grow toward the side that is touched. In this way, they can grow around objects they touch to gain support from them.

✔ **In what ways do some plants respond to light, gravity, and touch?**

Examples of Tropisms

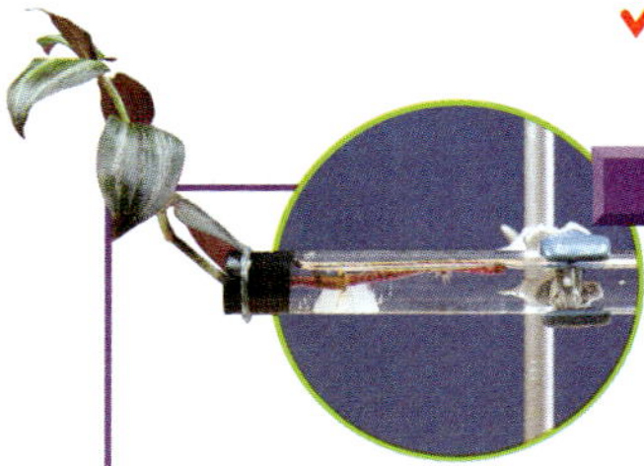

◀ The shoots of a plant grow upward. This helps plant seedlings break through soil to reach sunlight.

Plants need light, and they are able to respond to get the most available. ▶

Mimosa pudica is a tree or shrub with delicate green leaves. ▶

If you touch its leaves, they immediately close up. People call the mimosa the "sensitive plant." ▶

B14

Graphic Organizer for Chapter Concepts

Transparency B1 • Workbook, p. WB32

Plant Growth and Responses

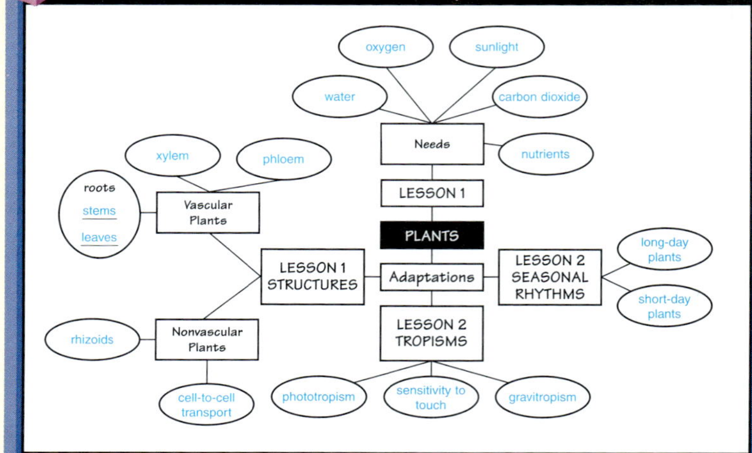

How Tropisms Affect Plants

Plants can't move around as most animals do, so plants can't respond to their environments in the same way. However, tropisms help plants get the things they need to survive.

Plant stems as well as leaves show positive phototropism—they grow toward light. The leaves of plants turn so that they get the largest amount of sunlight possible.

Because of positive gravitropism, the roots of a plant grow down into the soil, where they can take in water and minerals. The shoots show negative gravitropism—they grow upward, against the direction of the force of gravity. When they break through the surface of the ground, they get the light they need.

The tropism for touch can also help plants get light. Plants such as cucumbers have tendrils that respond to touch, allowing them to climb up supports and get more sunlight.

In about 1880 the British scientist Charles Darwin and his son Francis did experiments on tropisms in plants. They grew grass seedlings and noticed that the shoots bent toward the light just below the tips of the seedlings. This is the part of the plant where the seedlings are growing fastest. The Darwins did experiments in which they changed the seedling tips.

They cut the tips off the shoots of some seedlings and left the shoots of other seedlings whole. Then they put all the seedlings in the light. Only the shoots that were whole bent toward the light. The shoots without tips no longer bent toward the light. This happened even though the part of the shoot that usually bent toward the light was still there.

They also covered the tips of some shoots so that light could not reach them. These shoots failed to bend toward the light. Shoots that were covered with transparent caps, on the other hand, did bend toward the light.

Finally, they covered the part of the seedling just below the tip. This was the part that usually bent toward the light. However, even when this part of the seedling received no light, it still bent toward the light. The Darwins concluded that the tip of the shoot is the part that senses light and that it passes this information to the growth zone, just below the tip. Later experiments confirmed what the Darwins discovered.

The Grass Seedling Experiment

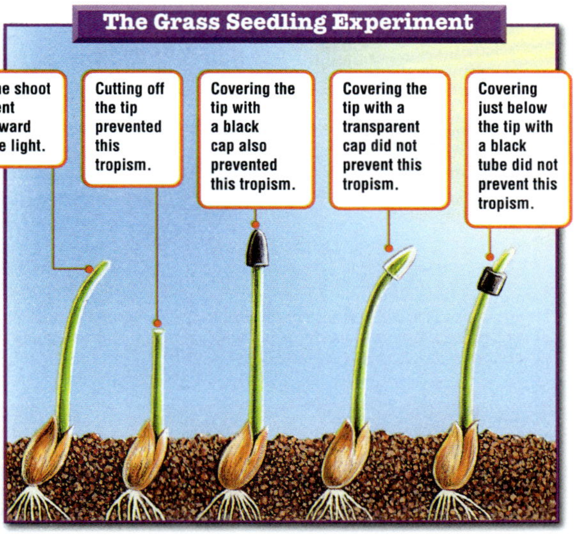

▲ The Darwins' experiments showed these results.

✓ **How do tropisms help plants survive?** Phototropism ensures that plants get as much light as possible to carry on photosynthesis. Gravitropism ensures that the roots can absorb water and minerals from the soil.

B15

USE PROCESS SKILLS

Compare, Identify Variables, and Draw Conclusions Allow a few moments for students to observe the drawings and captions of the Darwins' experiments on phototropism. Challenge them to **compare** the five setups. Ask volunteers to identify the differences among them. Encourage them to identify the variable in each setup and to **draw a logical conclusion**.

- ▶ **Which plant represents the control? How can you tell?** The first plant is the control. It is not changed in any way.
- ▶ **What conclusion can be drawn from these experiments?** The tip of the shoot is the part of the plant that senses light. It passes this information to the growth zone, just below the tip.

DEVELOP SCIENCE VOCABULARY

tropism, **phototropism**, and **gravitropism**, p. B14—Write all three words on the board. Ask students to identify what is similar about them. All three words refer to the response of a plant to something in its environment. Then ask them to identify the prefix in *phototropism* and *gravitropism*. Challenge students to **use these prefixes to determine the meanings of the words.** Have them check their answers by looking up the words in a dictionary. Phototropism is a response to light and gravitropism is a response to gravity. Ask students how the prefixes and roots of these words can help them to remember their meanings.

Reading Mini-Lesson

Make Generalizations Help students recognize the following as a generalization: All plants respond to their environments. Identify the last sentence of the first paragraph on page B15 as a generalization. Display the transparency and invite students to locate other generalizations in the text.

Reading Mini-Lesson Transparency B1-2

Make Generalizations

Page	Generalization
B14	• Some plants respond to touch.
B14	• Plants are said to show positive phototropism because they turn toward the light.
B15	• The tropism for touch can also help plants get light.

READING SKILLS PRACTICE

To practice and apply this reading skill, see **Workbook p. WB41.**

Reaching All Learners

Kinesthetic Learners Help students understand the concepts of positive and negative tropism by providing a light source and having students pretend to be plants. Have them bend toward the light. Then tell students to think of their left arm as a root and their right arm as a shoot. Have them move their arms accordingly in response to gravity.

Guide the Learning continued

SCIENCE IDEAS

Help students focus on the supporting details for this **main idea: Different plants respond to the changing seasons in different ways.**

- In most parts of the world, what changes occur as the seasons change? temperature, number of hours of daylight
- What are two ways in which plants respond to changes in the number of hours of daylight? What are the plants that respond in these ways called? Some plants, known as long-day plants, flower when the nights are short and the days are long. Other plants, known as short-day plants, need long periods of darkness. They bloom when the days are short.
- At what time of year do long-day plants bloom? What is the advantage of this? They bloom in late spring or early summer. This gives them time to make seeds before the first frost.

DEVELOP SCIENCE VOCABULARY

long-day plant and **short-day plant**, p. B16— Explain that plants are named short-day or long-day depending on the amount of darkness they need in order to bloom. Have students identify which type of plant needs long nights and which needs short nights.

Darken the room. After the students have adjusted to the darkness, quickly flick the lights on and off. Ask students how such a flash of light at night, during the time of year when nights are long, might affect the flowering of a short-day plant.

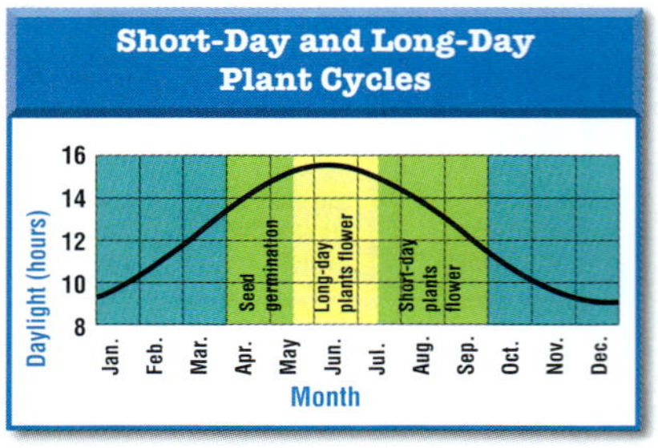

This graph shows the yearly change in day length for 43° north latitude. When do long-day plants flower at this location? When do short-day plants flower?

Strawberries are short-day plants that produce fruit in the late spring and early summer.

Some plants show daily rhythms. These Livingstone daisies open their petals in the morning and close them at night.

Plant Rhythms

As the seasons change, the temperature changes. The number of hours of daylight, or the length of the day, also changes. When it is colder, days are shorter. When it is warmer, days are longer. In many parts of the world, there are extreme changes in the number of hours of daylight.

To survive, plants must be able to change with the seasons. They have to adapt to the temperature, and they must grow when it's warm and rest when it's cold. Many plants change with the seasons by responding to changes in the length of the day.

Different plants respond in different ways. **Long-day plants** flower when there is more than a certain number of daylight hours—generally between 14 and 16 hours. The nights are short. Even though they are called long-day plants, it is the amount of darkness that determines when the plants bloom. Spinach, potatoes, clover, and lettuce are long-day plants—they bloom when nights are short. They begin to flower in late spring or early summer. This allows them enough time to make seeds before the first frost.

Short-day plants flower when there is less than a certain number of daylight hours—generally between 11 and 15 hours. They need long periods of darkness before they can bloom. Poinsettias, strawberries, and ragweed are short-day plants. There are also plants that flower

B16

INVESTIGATION CHALLENGE

Hands-On Activity: Tropisms

Use this idea for a hands-on activity to extend students' understanding of key lesson concepts.

Have students line a strainer with peat moss and spread cress or other fast-growing seeds on it. Wait for the roots to grow down through the strainer and hang it at an angle above a pan of water. Have students observe the direction of the roots' growth. Challenge them to name another environmental factor to which the roots might be responding.

Math Link

Make a Graph Have students **use reference sources** to research the seasonal changes in the number of daylight hours in their region of the country. Then have them make a graph of this data similar to the graph on page B16. On the basis of their graphs, have students determine during what months long-day plants and short-day plants probably flower in their region.

For student self-assessment, see Assessment Guide p. AGxvii.

regardless of the length of the day. They begin to flower from midsummer to autumn. These plants take less time to make seeds than long-day plants. So, like long-day plants, they produce seeds before the first frost.

✓ **What is the difference between short-day plants and long-day plants?**

Summary

A tropism is a plant's turning toward or away from something in the environment. Roots show gravitropism by growing downward. Plants show phototropism by growing toward light. Some plants are sensitive to touch. Plants also adapt to the seasons. Long-day plants flower when the number of daylight hours is more than a certain number. Short-day plants flower when the number of daylight hours is less than a certain number. These adaptations help plants survive.

Review

1. How does gravitropism help plants survive?
2. Which plants' flowers produce seeds faster, short-day plants or long-day plants?
3. Why do plants need to produce seeds before the first frost?
4. **Critical Thinking** If a potted plant is turned on its side, how will its growth change?
5. **Test Prep** If a plant's roots grow toward leaking underground pipes, what might the plant be showing a tropism toward?
   A touch
   B the length of the day
   C gravity
   D water

Short-day plants flower when day length is less than a certain number of hours. Long-day plants flower when the day length is more than a certain number of hours.

LINKS

MATH LINK

Compute Ratios A cocklebur flowers when there are at least 16 hours of daylight. What is the ratio of day to night, expressed in simplest form, that the cocklebur needs to bloom?

WRITING LINK

Informative Writing—Description Choose a single plant that you see each day. Write journal entries for your classmates describing how the plant changes during two weeks.

SOCIAL STUDIES LINK

Calendars Some ancient peoples, such as the Celts, the Egyptians, and the Mayas, developed calendars for the year. These helped them know when to plant their crops. Find out about these or other ancient ways of measuring time and how they helped agriculture.

TECHNOLOGY LINK

GO ONLINE

Learn more about plant responses to the environment by visiting this Internet Site.
www.scilinks.org/harcourt

SciLINKS

B17

LINKS TIPS

Links Objective Extend understanding of how plants respond to their environment.

Math Link For students who have difficulty with math, provide 24 small cubes to represent hours. Have them divide the cubes into two groups—one representing daylight and the other representing night. 2:1

Writing Link Students may choose a houseplant or a plant they see daily. Provide one or more plants in the classroom as well. Encourage students to carefully observe the plant before writing their first entry. Challenge them to search for small, day-to-day changes. Writing models are provided in the **Teaching Resources** book.

Social Studies Link Before students begin their research, discuss how a calendar can help people know when to plant crops. Help students make the connection between short- and long-day plants, the changing seasons, and a calendar.

3 Wrap Up and Assess

SUMMARY

Graphic Organizer for Chapter Concepts After reading the summary, have students complete the Lesson 2 section of the graphic organizer (shown on page B14).

REVIEW

1. It allows their roots to reach water deep in the ground.
2. short-day plants
3. Frost kills many plants and prevents them from producing seeds.
4. The roots will turn so that they grow downward; the shoots will turn so that they grow upward.
5. D, water

TEST PREP Test-Taking Tips

Item type multiple choice with question stem

To help students answer Question 5, share the following strategy with them. **Read the question carefully and identify key words such as *tropism*. Then use the facts you already know to find the correct answer. For example, what do underground pipes carry? Eliminate any answers that don't fit.**

LESSON CONCEPT REVIEW/ASSESSMENT

Students can use **Workbook p. WB42** to review the lesson concepts.

VOCABULARY REVIEW

Students can use **Workbook p. WB43** to review the chapter vocabulary.

SCIENCE AND TECHNOLOGY

Objectives

- Evaluate the impact of research and technology on scientific thought, society, and the environment.
- Identify careers related to science.

1 Before Reading

SET A PURPOSE

Ask students to examine the title and pictures for the article and to predict what the article is about. Then ask students to list questions they would like answered based on their predictions. Write these questions on the board and check them off as they are answered. Encourage students to use reference sources to find the answers to any unanswered questions.

2 Guide the Learning

SCIENCE IDEAS

Help students gather information about this nonfiction science article by asking these questions:

- How is genetic engineering involved in the production of insulin? Researchers insert into bacteria the human gene that allows human bodies to make insulin. The bacteria then make the insulin.
- What is a danger in using genetic engineering for changing crop plants? Possible answers: When a genetically engineered crop grows near a weed, it can transfer the altered trait to the weed. This could result in species of superweeds that cannot be killed by herbicides. The new genes could spread from plants into microorganisms in the wild. Pests and diseases could become resistant to pesticides and herbicides.
- Why do some scientists support controls on the release of genetically altered organisms into the environment? They do not want to have another situation in which scientists must find a solution after damage has already been done.

SCIENCE AND TECHNOLOGY

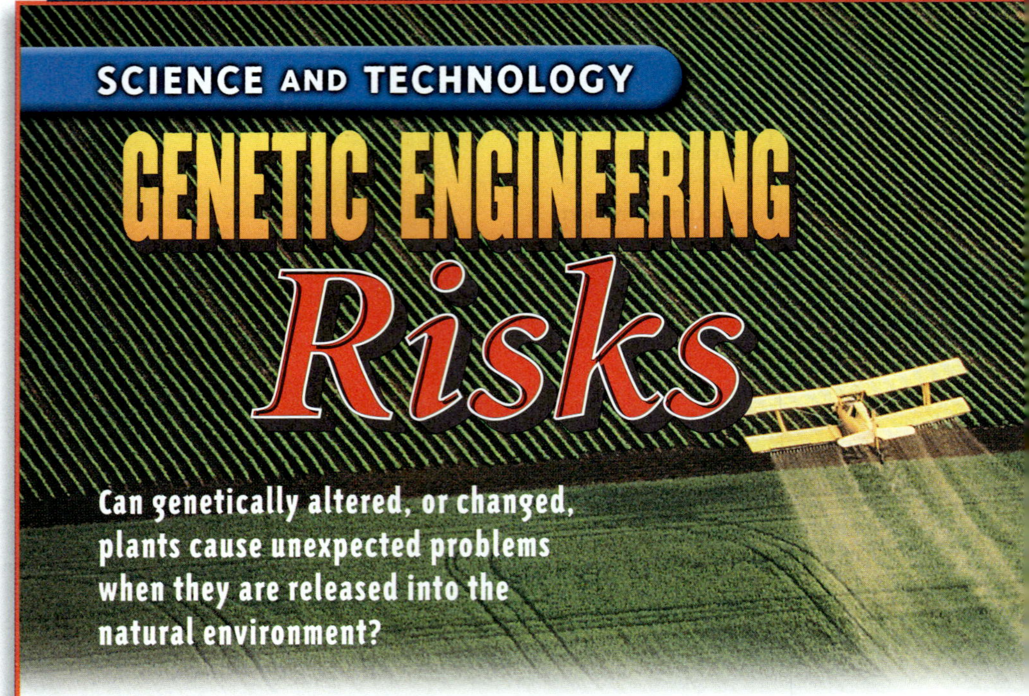

GENETIC ENGINEERING Risks

Can genetically altered, or changed, plants cause unexpected problems when they are released into the natural environment?

What is genetic engineering?

Genes are the "blueprints" that determine the characteristics of living things. Using the technology known as genetic engineering, scientists can make copies of genes from one organism and place them in another. This process produces combinations of genes that don't occur in nature. It also gives organisms new characteristics or abilities they wouldn't normally have. Genes can be designed in the lab to give a plant or animal a desired trait.

Genetic engineering has been helpful. For example, it has enabled manufacturers to increase the amounts of important medicines they make. One such medicine is insulin, which some people who have diabetes must use regularly. Researchers insert into bacteria the human gene that enables human bodies to make insulin. The bacteria then make the insulin.

How can genetic engineering help farmers?

Genetic engineering is an important tool for farmers. Scientists can alter plants grown as crops to increase the plants' resistance to chemicals called herbicides that are used to kill weeds. The herbicides can be used to kill the weeds without killing the crop. Scientists have developed genetically engineered soybean and cotton plants that are resistant to herbicides.

How might the transfer of genes cause problems?

Scientists have found that when a genetically engineered crop grows near a weed, it can transfer the altered trait to the weed. The genetically altered weeds pass on the trait to their offspring. Some scientists fear that this could result in species of superweeds that even the most powerful herbicides could not kill. In that case,

B18

Reading Mini-Lesson

Compare and Contrast Have students compare and contrast the potential benefits and risks of genetic engineering by asking them to complete a chart like this one. Encourage discussion of benefits and risks. Point out that while benefits are fairly predictable, risks are not.

Genetic Engineering of Plants	
Benefits	Risks

Genetic engineering can make plants resistant to herbicides.

stronger herbicides would be needed to kill the weeds. These could cause greater harm to the environment than the herbicides now do.

Some scientists also think it is possible for the genes to spread, or migrate, from plants into microorganisms in the wild. Unlike higher animals, bacteria and viruses can accept genes from different species and pass along genes. In this way, the pests and diseases could also become resistant to pesticides and herbicides.

What should be done about stopping the migration of genes?

Although many environmentalists are concerned, some companies that develop genetically engineered plants are not. The companies say that government controls on the release of genetically altered plants into the environment are strict enough to prevent any problems. Other scientists support greater controls on the release of genetically altered plants, animals, and microorganisms. They point to other scientific advances that were thought at first to be safe but were later found to harm the environment. For example, chemicals called chlorofluorocarbons (CFCs) were once widely used in such products as aerosol spray cans and food packaging. But scientists discovered that CFCs were destroying the ozone layer in the atmosphere, so now it is against the law to use them in aerosols. People who have concerns about migrating genes don't want to have another situation in which scientists must find a solution after damage has already been done.

THINK ABOUT IT

1. How can genetic engineering be helpful to farmers?
2. How might genetically engineered crops cause problems?

CAREERS
TREE SURGEON

What They Do
Tree surgeons plant, feed, prune, and provide pest control for trees. They must know about various tree species and diseases that can affect them. Tree surgeons then must know what to do to save or get rid of diseased trees.

Education and Training
Tree surgeons must know a lot about trees and gardening and should also enjoy working outdoors. Tree surgeons should have at least a high school education.

GO ONLINE
WEB LINK
For Science and Technology updates, visit the Harcourt Internet site.
www.harcourtschool.com

DEVELOP SCIENCE VOCABULARY

genetic engineering, p. B18—Ask students to break apart this phrase and explain the meaning of each word. For example, *genetic* relates to genes and heredity, and *engineering* has to do with building and designing things. Have students put these meanings together to define *genetic engineering*.

herbicide, p. B18—Have students break the word into its root and suffix. Have them suggest other words with the same roots or suffixes—for example, *herbaceous* or *homicide*. Have them explain the similarities and differences in the meanings.

CHAPTER CONCEPTS

This article focuses on the technology that scientists are using to alter the genetic makeup of plants and microorganisms.

- **How does genetic engineering affect the results of an organism's reproduction?** The offspring will have the new gene that has been implanted into the organism.
- **How might genetic engineering potentially affect organisms other than those it is intended to affect?** Through gene migration, genes from a genetically engineered species might move to another species, where they will be passed on from parent to offspring during reproduction.

3 Wrap Up and Assess

THINK ABOUT IT

1. Genetic engineering can create crop plants that are resistant to herbicides that kill weeds. This would allow farmers to spray weed-killing chemicals on their fields.
2. Genetically engineered crop plants can transfer the gene that gives them resistance to herbicides to weeds growing near them. The weeds would then become resistant. Farms would have to use more herbicides, or perhaps more harmful herbicides, to kill the weeds.

BACKGROUND

Webliography
GO ONLINE
Keyword careers
www.harcourtschool.com

Tree Surgeon Tree surgeons are often employed by the national park service, by arboretums, and by large cities to care for the communities' trees. The job involves having an appreciation and knowledge of trees and their needs and requirements. Tree surgeons are often self-employed and set up their own business. There are opportunities to work on a consultant basis for local authorities, landscape gardening organizations, or arboretums.

PEOPLE IN SCIENCE

Objectives

- Connect chapter concepts with the contributions of scientists.

1 Before Reading

SET A PURPOSE

Biologists study different aspects of living organisms. Have students read to find examples of Dr. Gonzalez's research and how she became a professor and researcher in botanical science.

2 Guide the Learning

THE WORK OF SCIENTISTS

Help students focus on the methods and processes scientists use and the impact of their work.

- What important skills must Dr. Gonzalez possess for her job? laboratory techniques, specialized knowledge of the subject matter, and electron micrograph imaging
- If you were a biologist like Dr. Gonzalez, would you rather be a professor or a researcher? Why? Answers will vary. Have students explain their answers.

3 Wrap Up and Assess

THINK ABOUT IT

1. biological science
2. Dr. Gonzalez's research could help scientists one day better understand how plant cells store energy and how seeds germinate. This knowledge could lead scientists to an understanding of the global carbon cycle.

PEOPLE IN SCIENCE

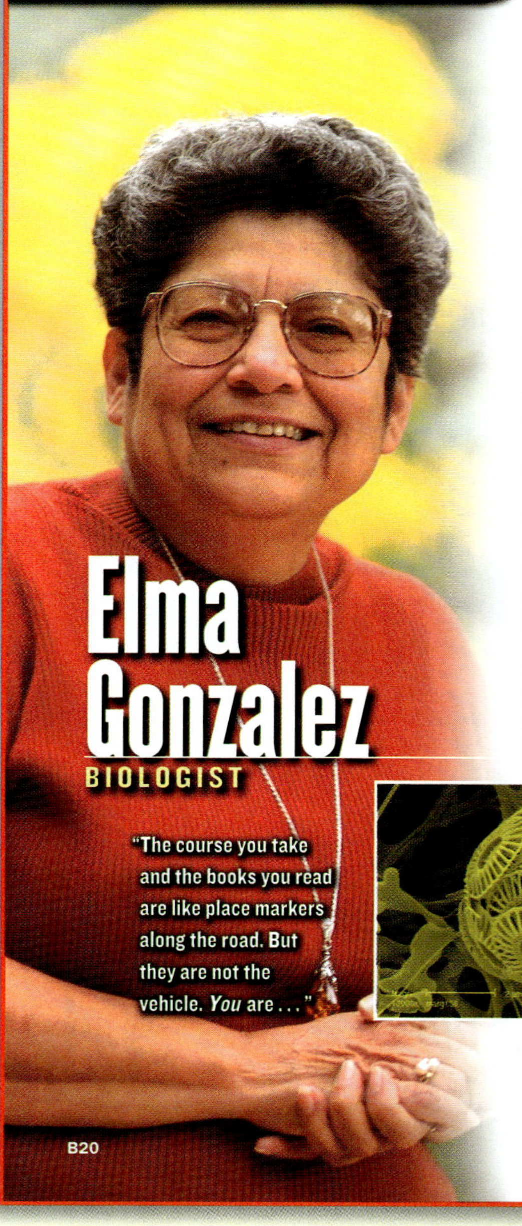

Elma Gonzalez

BIOLOGIST

"The course you take and the books you read are like place markers along the road. But they are not the vehicle. *You* are . . ."

Dr. Elma Gonzalez is a professor and researcher in the field of biological science. Her research could help scientists one day better understand how plant cells store energy and how seeds germinate. This research involves advanced theoretical knowledge of how a cell takes on and uses energy. It also requires great skill in specialized laboratory techniques.

Dr. Gonzalez grew up in southern Texas. As a youngster she and her family journeyed every summer to states such as Michigan and Wisconsin. There they harvested fruits and vegetables in order to earn money for clothing and school supplies. Although she did not attend school or speak English until she was 9 years old, Dr. Gonzalez was an excellent student who, while in high school, played basketball and sang in the choir. After graduating from high school, Dr. Gonzalez expanded her interest in science while at college. Using college loans to finance her education, Dr. Gonzalez graduated with a degree in biology and chemistry. She later earned a Ph.D. in cellular biology.

For years Dr. Gonzalez has taught courses at UCLA, where she specializes in the biochemistry of plants and plant-like organisms. Recently she studied microscopic marine phytoplankton and the beautiful crystal structures produced by them. She hopes her work will help scientists understand the global carbon cycle.

Still, teaching science to young people remains very important to Dr. Gonzalez. She is a believer in the importance of diversity in science and that the profession is made better by the many diverse people working in it.

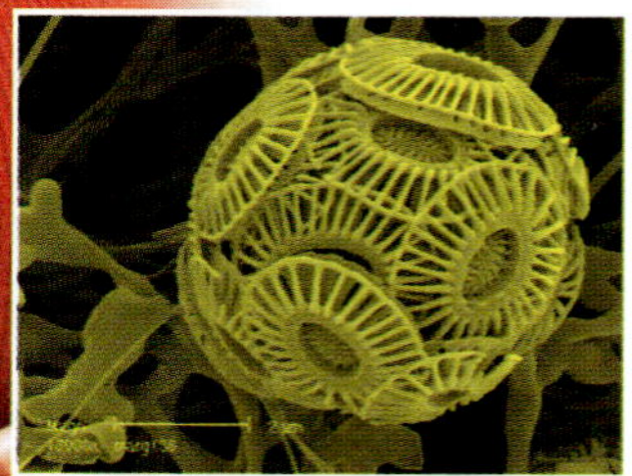

▲ Dr. Gonzalez studies tiny plantlike organisms such as this one. The circular plates you see here surround a living cell. Ordinary chalk is made up of plates from organisms like this one.

THINK ABOUT IT

1. In what branch of science does Dr. Gonzalez teach and research?
2. What might Dr. Gonzalez's research on plant cells one day teach scientists?

B20

BACKGROUND

Carbon Cycle Green plants contribute to the ongoing carbon cycle by releasing oxygen during photosynthesis. Animals and humans breathe in oxygen, use it to release energy from food, and then exhale carbon dioxide as waste. Decomposers use waste products and dead organisms as food and then release carbon dioxide back to the atmosphere, where it is used by plants in photosynthesis. Most of the oxygen in the atmosphere is produced by small plant-like protists called phytoplankton. Some of these phytoplankton produce crystals that can accumulate and form land features. Scientists such as Dr. Gonzalez have studied how phytoplankton produce and control the shape of these crystals. Dr. Gonzalez hopes this work will enable scientists to better understand the carbon cycle.

Webliography GO ONLINE

Keyword biographies

www.harcourtschool.com

ACTIVITIES FOR HOME OR SCHOOL

TRAVELING WATER

How does water move through moss and other nonvascular plants?

Materials

- small plate of dry sphagnum moss
- shallow container of water
- green food coloring
- hand lens

Procedure

1. Color the water by adding green food coloring.
2. Gently pull out a strand of moss. Make sure one end remains attached to the rest of the moss.
3. Put the end of the strand in the container of water, as shown.
4. Using the hand lens, observe what happens.

Draw Conclusions

How do you know the water has entered the moss? How did the water enter and spread through the moss? How is this movement different from water movement in vascular plants? Why does the moss feel damp?

GROWTH AND TOUCH

How do vines wrap around supports?

Materials

- clear plastic cup lined with a paper towel
- 4 morning glory or moonflower seeds
- water
- masking tape
- 4 pencils

Procedure

1. Wet the paper towel, and slip the seeds between the towel and the cup.
2. Put the cup in a warm, lighted place, and keep the towel moist.
3. Wait two to three weeks, until the seeds sprout and the stems reach the top edge of the cup.
4. Tape the pencils to the outside of the cup, as shown.

Draw Conclusions

When a stem reached a pencil, which side grew faster—the inside part touching the pencil or the outside part not touching the pencil? How does this behavior help the seedlings survive? In what direction are the seed roots growing? How does this help the seedling survive?

B21

School-Home Connection

These activities provide an excellent opportunity to assign hands-on activities for the students and their families. The materials are often easy to obtain and are safe to work with. For those activities that you do not wish to do in the classroom, encourage students to complete them at home. Remind them that communicating their results is always important, but especially so for home activities since you will not be there to observe their activities. Ask students to use diagrams, sketches, graphs, or even photographs to show their results.

ACTIVITIES FOR HOME OR SCHOOL

Traveling Water

Objectives

- **Observe how water moves through sphagnum moss.**

Suggested Time 15–20 minutes

Hints

- Have students practice using the hand lens before doing the activity.
- You may wish to give the students toothpicks to pull apart the strands of moss.

Draw Conclusions Students should see that the mound of moss gradually turns bright green as the water absorbed by the tendril spreads from cell to cell within the moss.

Investigate Further Challenge students to design and carry out an investigation to determine how much water the moss can absorb.

Growth and Touch

Objectives

- **Observe how a vine wraps around a support.**

Suggested Time 10–15 minutes on Day 1; 15–20 minutes two to three weeks later

Hints

- Make sure students regularly moisten the towels so the seeds do not dry out.
- Once the seeds have sprouted, be sure they receive adequate sunlight.
- Have students tape the pencils so that they are right beside the shoots.

Draw Conclusions Students should conclude that the outside of the stem grows faster, allowing the vine to curl around a support. The roots grow downward toward sources of water.

Investigate Further Students may want to use a computer to chart the growth of the vine over time.

ACTIVITIES FOR HOME OR SCHOOL

Reproducible copies of these activities are provided in the **Teaching Resources** book.

CHAPTER 1

Review and Test Preparation

VOCABULARY REVIEW (2 points each)

1. long-day plants; short-day plants
2. xylem
3. Gravitropism
4. vascular plant
5. phototropism
6. nonvascular plant
7. tropism
8. phloem

CONNECT CONCEPTS (4 points each)

The order of answers for Question pairs 9-10, 11-12, and 13-14 may vary.

9–10. oxygen, water
11–12. minerals, carbon dioxide
13–14. lateral buds, roots
15. gravity
16. sunlight

CHECK UNDERSTANDING (4 points each)

17. B, carbon dioxide
18. H, ferns
19. B, leaves
20. J, transparent caps were placed over the tips
21. B, short-day plants

CHAPTER 1 **Review and Test Preparation**

Vocabulary Review

Use the terms below to complete the sentences. The page numbers in ( ) tell you where to look in the chapter if you need help.

xylem (B8)
vascular plant (B8)
phloem (B9)
nonvascular plant (B10)
tropism (B14)
phototropism (B14)
gravitropism (B14)
long-day plants (B16)
short-day plants (B16)

1. When the length of daylight is greater than a certain amount, ___ flower. When day-light length is less than a certain amount, ___ flower.
2. The tissue that carries water from the roots of a plant to its leaves is called ___.
3. ___ causes the roots of a plant to grow downward.
4. An oak tree is an example of a ___, which has xylem and phloem.
5. A response that causes a plant to turn toward light, or ___, can cause houseplant stems and leaves to turn toward windows.
6. A plant that does not have xylem or phloem is called a ___.
7. A plant shows a ___ any time it turns toward or away from something in the environment.
8. The tissue that carries food to all parts of the plant is ___.

Connect Concepts

Use the words in the word bank to complete the chart.

lateral buds **gravity** **water** **carbon dioxide** **oxygen** **roots** **minerals** **sunlight**

What Seeds Need to Sprout	What Plants Need to Grow	The Parts of Plants That Grow	How Plants Respond
the right temperature	sunlight, oxygen, water	the tips of the shoots	Roots grow in the direction of the force of
9. ___	11. ___	13. ___	15. ___
10. ___	12. ___	14. ___	shoots grow toward
			16. ___

B22

CHAPTER TEST

See **Assessment Guide pp. AG19–23** for the Chapter Test. Assessment Options are provided on page B2e.

TAKE-HOME BOOK

Use the **Take-Home Book** (described on page B2d) to provide more chapter content and activities.

Check Understanding

Write the letter of the best choice.

17. When plant cells use food, which substance do they produce?
    A oxygen
    B carbon dioxide
    C sugar
    D seeds
18. Examples of vascular plants are —
    F mosses
    G liverworts
    H ferns
    J diatoms
19. What parts of a tree make food for the whole plant?
    A roots C xylem
    B leaves D phloem
20. The Darwins found that grass seedlings bent toward light if —
    F the tips of the seedlings were cut off
    G black caps were placed over the tips
    H the roots were cut
    J transparent caps were placed over the tips
21. Plants that flower when day length is less than a certain amount are —
    A long-day plants
    B short-day plants
    C tropical herbs
    D tropisms

Critical Thinking

22. When a white carnation is placed in water with red food coloring in it, its petals soon turn red. Explain why this happens.
23. If you had a bag of a plant's seeds, how could you find out whether the plant is a long-day or a short-day plant?

Process Skills Review

24. If you wanted to do an experiment on phototropism, what **variables** would you need to **control**? What variables would you need to test?
25. Suppose you tried to germinate some seeds in the refrigerator and some at room temperature. What **conclusion can you draw** if the seeds fail to germinate?

Performance Assessment

Responding to Color

Write a plan for an experiment to find out what effects the color of light has on the growth of seedlings. List the variables for the seedlings in the experiment. Indicate which of the variables you need to test by writing a *T* next to them in your list. Decide what observations you would make to conclude what effect the color of light has on plant growth. Present your plan to the class.

B23

CRITICAL THINKING (7 points each)

22. Carnations are vascular plants. The colored water travels through xylem tissue up the stem and into the flower petals.

TEST PREP Test-Taking Tips

Item type open ended format in which the student is asked to write a short answer

To help students answer Question 23, share the following strategy with them. **Turn the question into a statement with which to begin your answer, such as "You could find out if a plant is a long-day or a short-day plant by . . . " Then add the facts you know to complete the sentence.**

23. You could observe the plants to see when they flower. If they flower as the days are longer (between May and September), then they are long-day plants. If they flower as the days are shorter (between October and April), then they are short-day plants.

PROCESS SKILLS REVIEW (8 points each)

24. Variables to control include water, temperature, and nutrients. The variable to test is the direction of the light.
25. They have too much or too little water; the temperature is not correct.

Performance Assessment

Preparation Review with students the idea that variables are factors that will affect the outcome of an experiment. Point out that the variables with which they have already worked are just a few of those that can be tested.

Scoring Rubric
Performance Indicators

___ Plan includes use of different-colored lights or different-colored cellophane placed over seedlings.

___ Plan includes that all conditions such as soil, water, and temperature remain the same for each seedling.

___ Plan states that all seedlings are the same kind of plant and the same age.

Observations and Rubric Score

3 2 1 0

WRITING ABOUT SCIENCE

Have students write a composition that involves writing for a specific purpose. You may wish to use the prompt that is provided in the **Workbook.** The prompt is accompanied by a graphic organizer that will help students organize their ideas before writing. Models for writing are provided in the **Teaching Resources** book. You may wish to reproduce those for students or display them on a transparency.

WRITING PRACTICE

A chapter writing prompt and a prewriting activity are provided on **Workbook p. WB44.**

CHAPTER 2 Planning Guide Types of Plants

LESSON	PACING	OBJECTIVES	MATERIALS
1 What Are Mosses? B26–31 	2 days	▶ Compare the structures of vascular and nonvascular plants. ▶ Describe the life cycle of mosses.	**Investigate** photographs of mosses
2 What Are Ferns and Gymnosperms? B32–41 	3 days	▶ Describe the life cycles of ferns and gymnosperms. ▶ Explain adaptations for survival in ferns and gymnosperms.	**Investigate** fern frond, conifer needles, hand lens, conifer cone, safety goggles
3 How Are Angiosperms Different from Other Plants? B42–49 	3 days	▶ Recognize the features of angiosperms. ▶ Describe the structures of flowers and their functions in reproduction. ▶ List the ways in which pollination occurs.	**Investigate** flower, paper towel, hand lens
4 How Do Angiosperms Reproduce? B50–57 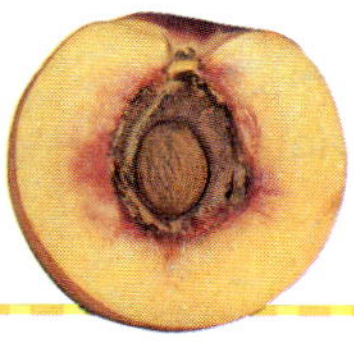	3 days	▶ List the differences between monocots and dicots. ▶ Describe the parts of fruits and seeds. ▶ Explain how seeds are distributed. ▶ Explain how plants reproduce asexually.	**Investigate** paper towels, lima bean, corn kernel, hand lens, plastic knife
End of Chapter B58–63 **Science Through Time** **People in Science** **Activities for Home or School** **Chapter Review and Test Preparation**		▶ Connect chapter concepts with the history of science. ▶ Connect chapter concepts with the contributions of scientists. ▶ Extend chapter concepts with hands-on investigations. ▶ Review chapter concepts.	

PROCESS SKILLS	VOCABULARY	RESOURCES AND TECHNOLOGY	REACHING ALL LEARNERS
Process Skill Tip observe, hypothesize **Other Process Skills** record data, infer, predict, plan a simple investigation	**moss** **asexual reproduction** **spore**	**Workbook,** pp. WB45–50 **Vocabulary Cards,** pp. TR101–108 **Transparencies** B2, B2-1 **Harcourt Science Activity Video** **Internet Site**	**ESL Activity**, p. B29 **Investigation Challenge**, p. B30 **Advanced Learners**, p. B1g
Process Skill Tip compare, infer **Other Process Skills** observe, record data, hypothesize, plan a simple investigation	**fern** **gymnosperm** **conifer** **pollen**	**Workbook,** pp. WB51–55 **Transparencies** B2, B2-2 **Harcourt Science Activity Video** **Internet Site**	**ESL Activity**, pp. B37, B38 **Investigation Challenge**, p. B40
Process Skill Tip observe, communicate **Other Process Skills** infer, compare, hypothesize, plan and conduct simple investigations	**angiosperm** **stamen** **anther** **pistil** **stigma** **petal** **style** **ovary** **pollination**	**Workbook,** pp. WB56–60 **Transparencies** B2, B2-3 **Harcourt Science Activity Video** **Harcourt Science Newsroom Video**	**Visual Learners**, p. B48 **ESL Activity**, pp. B45, B47 **Investigation Challenge**, p. B48
Process Skill Tip observe, compare **Other Process Skills** record data, infer, classify, hypothesize, plan and conduct simple investigations	**dicotyledon** **monocotyledon** **fertilization** **embryo** **runner** **grafting**	**Workbook,** pp. WB61–67 **Transparencies** B2, B2-4 **Harcourt Science Activity Video** **Internet Site**	**Kinesthetic Learners**, p. B54 **ESL Activity**, p. B53 **Investigation Challenge**, p. B56
		Take-Home Book, pp. TH21–24 *Amazing Trees* **Activities for Home or School,** p. TR68 **Chapter Test**, pp. AG26–30	**Performance Assessment**, p. AG31 **Portfolio Evaluation Sheet**, p. AGxxiv

Prepare for Activities

Use this page to help you organize and prepare materials for the Investigations, which begin each lesson of this chapter.

LESSON INVESTIGATION | **MATERIALS**

1 Where Mosses Grow
B26–27

PURPOSE Discover where mosses grow and infer the conditions they need in order to live.

PREPARATION TIPS You may want to provide students with copies of a chart similar to the one on page B27 to help them organize their observations. Be sure to leave enough room for students to make drawings of what they observe.

TIME 30 minutes

EXPECTED RESULTS Students should conclude that mosses live in shady, moist environments.

MATERIALS
- photographs of mosses

2 Comparing Ferns and Conifers
B32–33

PURPOSE Observe various parts of ferns and conifers and conclude how they are alike and how they are different.

PREPARATION TIPS You may want to provide students with a chart on which to record their observations. If conifers grow in your area, encourage students to bring in cones for the class to examine.

TIME 30 minutes

EXPECTED RESULTS Students should conclude that ferns and conifers are alike in that they are both vascular plants but different in their adaptations for survival.

MATERIALS
- fern frond
- conifer needles
- <u>hand lens</u>
- <u>conifer cone</u>
- <u>safety goggles</u>

3 The Parts of a Flower
B42–43

PURPOSE Observe some parts of a flower.

PREPARATION TIPS Purchase large flowers such as lilies or gladioluses from a flower shop. Use flowers that have large and easily observable petals, sepals, pistils, and stamens. You may wish to provide a variety of flowers and have students observe them.

TIME 20–25 minutes

EXPECTED RESULTS Students should conclude that a flower has clearly distinguishable reproductive parts.

MATERIALS
- flower
- paper towel
- <u>hand lens</u>

4 Inside Seeds
B50–51

PURPOSE Observe the insides of some seeds and draw conclusions about the beginning of life for two different kinds of angiosperms.

PREPARATION TIPS Obtain lima bean and corn seeds from a seed or a biological supply house. **Important note:** Soak the seeds overnight in water to soften them.

TIME 20–30 minutes

EXPECTED RESULTS Students should conclude that seeds have an outer covering, a tiny plant, and stored foods. Corn seeds have two different food-storage parts, and bean seeds have two similar ones.

MATERIALS
- paper towels
- <u>lima bean</u>
- <u>corn kernel</u>
- <u>hand lens</u>
- <u>plastic knife</u>

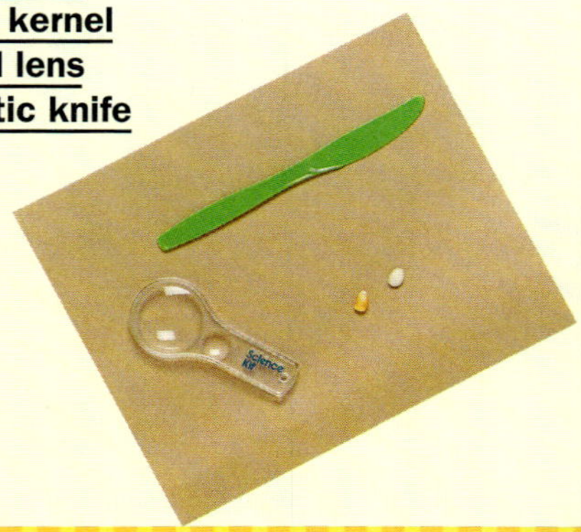

Harcourt Science Activity Videos

Preview the Investigations The Activity Video for this unit shows students doing the Investigations. You may wish to view the video for classroom management ideas. The video can also be used to model the Investigation process for students.

Equipment Kits

<u>Underlined items</u> above are provided in the equipment kits (available separately).

. . . with Trade Books

These books provide in-depth information about different types of plants.

Down to Earth by Michael J. Rosen (Harcourt, 1998) features recipes and activities related to plants and gardening from several authors and illustrators, including Lynne Cherry and Lois Ehlert.

I Didn't Know That Some Plants Grow in Midair by Claire Llewellyn (Copper Beach Books, 1998) discusses amazing facts about rain forests and contains hands-on fun projects.

***Plants* (The Kingdoms of Life)** by Alvin, Virginia, and Robert Silverstein (Twenty-First Century Books, 1996) discusses specific kinds of plants after describing the plant kingdom and its classification.

***The Visual Dictionary of Plants* (Eyewitness Visual Dictionaries)** by Deni Brown (Dorling Kindersley, 1992) not only presents the spectrum of plants from mosses to angiosperms, but discusses plant parts and processes as well, concluding with classification information.

Trade book titles are current at time of publication but may go out of print without notice.

. . . with the Take-Home Book

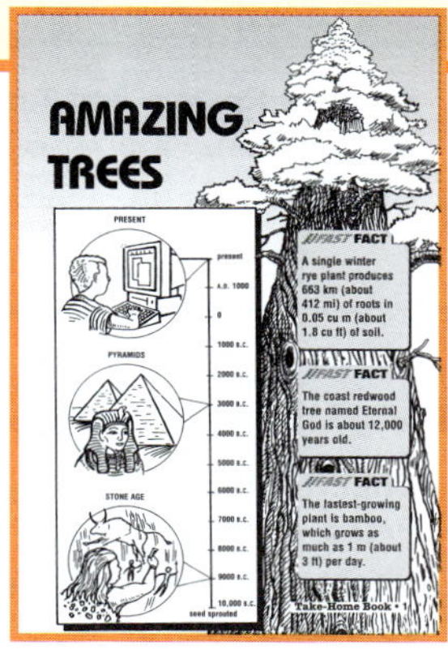

Amazing Trees

CONTENTS

. . . with More Hands-On Activities

Investigation Challenge

Activities for Home or School

The cross-curricular Links that appear at the end of each lesson in the student edition and throughout the teacher's edition also provide activities.

. . . with Technology

Students can learn more about types of plants through these components and special Internet sites. Look for Technology Links at the end of each lesson.

CNN Turner Learning

Harcourt Science Newsroom Video
Rainforest to Shuttle

Smithsonian Institution®

Harcourt Science brings the Smithsonian Institution to your classroom.
www.si.edu/harcourt/science

SCILINKS™ THE WORLD'S A CLICK AWAY

These Internet Links are provided by the National Science Teachers Association.
www.scilinks.org/harcourt

Visit The Learning Site for related links, activities, and resources: www.harcourtschool.com

WELCOME TO THE LEARNING SITE

Assessment Options ✓ Chapter 2

Formal Assessment

- **Chapter Review and Test Preparation,** PE pp. B62–63
- **Chapter Test,** pp. AG26–30

Ongoing Assessment

- ✓ **Questions,** PE pp. B28–31, B34–41, B44–49, B52–57
- **Lesson Review,** PE pp. B31, B41, B49, B57
- **Informal Assessment**
  - **Classroom Observation,** TE p. B43
  - **Portfolio,** TE p. B27, B51
- **Observation Checklist,** AGxiv

Student Self-Assessment

- **Self-Assessment–Investigate,** p. AGxvi
- **Self-Assessment–Learn About,** p. AGxvii
- **Project Summary Sheet,** p. AGxix

Performance Assessment

- **Chapter Review and Test Preparation,** PE p. B63
- **Chapter Performance Task,** pp. AG31–32
- **Project Evaluation Checklist,** p. AGxviii

My Work Portfolio Assessment

- **Science Experiences Record,** p. AGxxii
- **Guide to My Science Portfolio,** p. AGxxiii
- **Portfolio Evaluation,** p. AGxxiv

Chapter Test AG26

Types of Plants

Part I Vocabulary 1 point each

Use the letters of the terms in the Word Banks to complete the sentences.

A fern	C pollen	E gymnosperms	G spores
B conifer	D moss	F asexual reproduction	

A **1.** __D__ is a very short green plant that does not have true stems, roots, or leaves. It has a life cycle with two different stages. One generation looks plantlike. The other generation looks like a stalk that grows out of the leaflike part of the plant. This generation carries out **2.** __F__. In this process, a new plant is produced without the joining of a sperm with an egg. Instead, the stalk produces **3.** __G__, which contain cells that can grow into a new plant. The life cycle of a **4.** __A__, a plant with large, feathery fronds, also includes two different generations. These plants use spores inside brownish structures called sori to reproduce. Other kinds of plants use seeds to reproduce. These plants are divided into two groups. Plants with seeds that are not surrounded by a protective container are **5.** __E__. One type of these plants is a **6.** __B__, which means "cone bearer." These plants have male cones and female cones. The male cone produces **7.** __C__, while the female cone produces egg cells.

H angiosperms	J runners	L dicotyledons	N grafting
I pollination	K embryo	M fertilization	O monocotyledons

The other group of plants reproduces with seeds that are protected within a fruit. Plants in this group are **8.** __H__. These plants produce fruit from their flowers, which are the reproductive structures of the plants. Flowering plants are divided into two groups. Those that have two seed leaves are **9.** __L__. Those with only one seed leaf are **10.** __O__. Movement of pollen from the male reproductive organs to the female reproductive organ is called **11.** __I__. The sperm and egg join in the process of **12.** __M__ and produce an **13.** __K__, the tiny plant that forms in the seed.

Some plants reproduce without seeds or spores. Plants such as strawberries and spider plants can reproduce with long, slender stems called **14.** __J__ that put out roots and shoots along the ground. Farmers sometimes use **15.** __N__, in which they join parts of two plants to make a single plant.

Chapter Test AG27

Use the terms in the Word Bank to label the following diagram.

Anther	Stamen	Stigma	Style	Petal	Pistil	Ovary

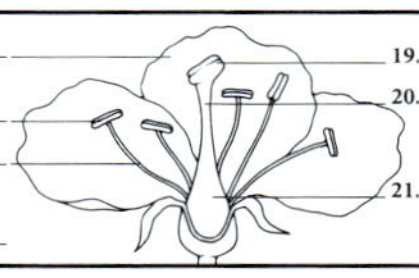

16. Petal
17. Anther
Filament
18. Stamen
19. Stigma
20. Style
21. Ovary
22. Pistil

Part II Science Concepts and Understanding 3 points each

Use the following diagram to answer Questions 23–24.

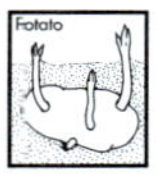

I

II

III

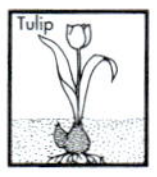

IV

__I__ **23.** Which drawing represents sexual reproduction?

__D__ **24.** Which drawings represent angiosperms?
A I and II B II and III C III and IV D I, II, III, and IV

Write the letter of the best choice.

__F__ **25.** Which of the following statements about mosses is **NOT** correct?
- F They are vascular plants.
- G They make food through photosynthesis.
- H They have rootlike structures that hold the plant in the soil.
- J They carry out both sexual and asexual reproduction.

__C__ **26.** Mosses can live only in places that are —
A sunny B dry C moist D cold

Chapter Test AG28

__G__ **27.** Which of the following plants have thin, wax-covered needles?
- F ferns
- G conifers
- H mosses
- J liverworts

__C__ **28.** Which of the following statements about ferns is **NOT** correct?
- A They reproduce with spores.
- B They have lived on Earth for about 400 million years.
- C They are nonvascular plants.
- D Their fronds are large and collect a lot of light.

__F__ **29.** When egg cells in the female cone of a conifer are mature, the cone —
- F forms a sticky liquid to trap pollen
- G splits open and deposits the egg cells on the ground
- H splits open, and the egg cells are carried off by the wind
- J forms long shoots that connect to the stamen

__D__ **30.** A structure that encloses seeds is called a —
- A cotyledon
- B ovule
- C nut
- D fruit

__H__ **31.** Plants that provide most of the food products humans use are —
- F liverworts
- G gymnosperms
- H angiosperms
- J conifers

__B__ **32.** All angiosperms have —
- A flowers and cones
- B flowers and fruits
- C spores and cones
- D spores and fruits

__G__ **33.** Angiosperms that go through their life cycle in less than a year are called —
- F perennials
- G annuals
- H hornworts
- J bulbs

__B__ **34.** For pollination to take place, pollen grains must —
- A stick to an insect's body
- B get to the stigma of a flower
- C be carried by the wind
- D become nectar

__J__ **35.** Flowers pollinated by wind are usually —
- F quite large
- G quite fragrant
- H brightly colored
- J small

Chapter Test AG29

__B__ **36.** Most of the lumber used in the United States comes from —
- A monocotyledons
- B gymnosperms and angiosperms
- C ferns and horsetails
- D annuals and perennials

__J__ **37.** Angiosperms are used for all of the following **EXCEPT** —
- F fabrics
- G medicines
- H perfumes
- J plastics

Part III Critical Thinking 4 points each

38. Why do most gymnosperms grow in the northern United States and Europe but not in tropical rain forests?

Possible answer: Most gymnosperms are adapted to harsh climates that are somewhat dry and where winters are cold. They have needles that help prevent water loss and a chemical that prevents internal water from freezing. In a rain forest, these adaptations are not only unnecessary but also undesirable. They would cause the trees to retain too much water, which could cause them to rot.

39. What adaptations of ferns make them useful for decorating inside building lobbies and offices?

Possible answer: Many ferns can live in dim light or without direct sunlight.

40. Why do flowers usually produce many more pollen grains than egg cells?

Possible answer: Pollen grains are carried away by the wind and by animals. There must be many pollen grains to ensure that some will get to another flower, fall on the stigma, and grow a pollen tube to carry the sperm to the eggs.

Chapter Test AG30

41. Why do the seeds of fruits that animals eat have tough, leathery coverings?

Possible answer: These seeds need a tough covering so they can pass through an animal's body without being digested. The seeds may germinate where the animals drop their waste, far from the original plant.

Part IV Process Skills Application

8 **42.** **Compare** monocots with dicots by completing the following chart.

	Monocots	Dicots
Flower parts	Multiples of three	Multiples of four or five
Cotyledons	One	Two
Leaf veins	Parallel	Netlike
Roots	Spread out	Taproots
Vascular tissue	Scattered	In a ring

3 **43.** **Compare** seeds with spores.

Possible answers: Both seeds and spores are reproductive structures; both are plant parts; when conditions are right, both can become new plants; seeds have their own food to start with; spores must land in the right conditions.

3 **44.** If a plant has tiny white flowers with no scent, what can you **infer** about how it is pollinated? Explain your answer.

Possible answer: It is probably wind pollinated. Plants that depend on insects or other animals for pollination usually have showy, colorful flowers and strong scents to attract the pollinators.

3 **45.** What might be the best way to **communicate** information about the reproductive structures of an angiosperm to someone who has never seen a flower?

Possible answer: You could use a drawing or diagram along with a detailed description and explanation of the parts and processes.

Performance Assessment AG31

Classify Seeds

PERFORMANCE TASK

Materials
- reference books
- several varieties of seeds
- pencil or pen

1. Obtain a seed collection. Your teacher may provide you with seeds. If time, location, and weather conditions allow, go outside and collect seeds.
2. Carefully observe the seeds. Make a chart in which to record your observations.
3. Classify the seeds as gymnosperms or angiosperms.
4. Classify the seeds as monocots or dicots.
5. Describe the environmental requirements of each type of seed.
6. Make a dichotomous key to classify the seeds.

Assessment Continued

Performance Assessment AG32

PERFORMANCE TASK

Classify Seeds

Materials Performance Task sheet, reference books, several varieties of seeds, pencil or pen

Time 30 minutes

Suggested Grouping individuals or pairs

Science Processes observe, compare, classify, communicate

Preparation Hints Assemble reference books on the typical vegetation in your area of the country. If time is a consideration, ask students to collect seeds prior to starting this task at school. You may wish to provide different groups of students with different varieties of seeds so that not all students are working with the same kinds of seeds.

Introduce the Task Review the reproductive processes of gymnosperms and angiosperms. Tell students they are going to identify and classify seeds common to your area. They will need to classify the seeds as gymnosperms or angiosperms and as monocots or dicots.

Promote Discussion When students finish, ask volunteers to share their collections with the class. Have them talk about the reproduction and seed-spreading ability of each plant. You might also want to ask students whether the adaptations that enable their plant to live in a particular environment would prevent it from living in another environment. For example, a plant with broad leaves would not survive in the desert.

Scoring Rubric

Performance Indicators

____ Demonstrates an understanding of the difference between gymnosperms and angiosperms.
____ Explains different methods of plant reproduction.
____ Distinguishes between monocotyledons and dicotyledons and communicates that gymnosperms are neither monocots nor dicots.
____ Describes the environmental requirements of the seeds.
____ Develops a functional dichotomous key.

Observations and Rubric Score

3 2 1 0

Workbook Support

Investigate Log WB46

Where Mosses Grow

Materials

photographs of mosses

Activity Procedure

1. **Observe** the mosses in Pictures A–F on page B26.
2. **Record** your **observations** in the chart below. Draw a picture of each moss, and describe it. Pay close attention to the surroundings of each moss. Describe the environment in which the moss is growing. Remember to include details such as the place where the moss grows, the surrounding living things, and the amounts of sunlight and moisture in the environment. If you can **infer** the approximate temperature from the picture, include that information in your description, too.

Observations of Mosses

Picture	Description	Environment

Investigate Log WB47

Draw Conclusions

1. By **observing**, what did you **infer** about the surroundings in which mosses can live? Students should infer that mosses can survive best in shady, damp places.
2. What can you hypothesize about what mosses need in order to grow? Students should hypothesize that mosses require a lot of moisture and little sun.
3. Where might you expect to find mosses growing? List at least three places not shown in the photographs on page B26. Answers should include places where there is moisture and shade—for example, at a woodpile near a house, in the shade of a boulder, near a pond or stream.
4. **Scientists at Work** Scientists can test a **hypothesis** by **observing** further. How could making observations help you test your hypothesis about what mosses need in order to grow? ____

Investigate Further Use what you have learned in this activity to **formulate a hypothesis** about the best growth conditions for mosses. Use your hypothesis to **predict** where in your local area you might find mosses and **plan an investigation** to check your prediction. You could look for mosses in very dry places in full sunlight to disprove the hypothesis. To support your hypothesis, you could carefully measure and monitor conditions where mosses grow.

Process Skills Practice WB48

Observe and Hypothesize

Observation is the most basic of all the skills that scientists use. It makes use of the senses to learn about an object or event. Sometimes instruments such as microscopes, metersticks, or balances are used to increase the range and precision of observations. Without careful and complete observations, valid hypotheses would be impossible.

Think About Observing and Hypothesizing

Kelly filled a bottle with water to within an inch of the top. She took a freshly cut leaf, wrapped clay around its stem, and placed the stem into the bottle. She made sure the clay fit snugly around the mouth of the bottle and the stem was in the water. Then she poked a hole through the clay and inserted a straw through the hole, making certain the straw did not touch the water. She pressed the clay around the straw and bottle opening so that no air could escape. Next, Kelly stood in front of a mirror and sucked the air out of the bottle with the straw. Watching in the mirror, she observed air bubbles appearing from the bottom of the leaf's stem. In the space, sketch Kelly's leaf in the bottle and show with arrows how you think the air was moving through the system.

1. Why do you think she needed to observe by watching in the mirror instead of observing directly? She couldn't simultaneously draw the air out and watch the bottom of the stem any other way.
2. Kelly observed air bubbles at the end of the stem. Hypothesize about where they came from. There are holes in the leaf called stomata and tiny tubes called xylem that run down the stem. The leaf and stem act as a straw for the plant. As Kelly drew air out of the bottle, more air was drawn into the bottle through the stomata and xylem. This is the same system that water moves through in a plant.
3. Design an experiment that supports your hypothesis. Designs depend on answers to Question 2. Possible design includes completely covering the leaf with a plastic bag and observing whether bubbles appear.

Reading Skills Practice WB49

Arrange Events in Sequence

Read the selection about the life cycle of liverworts. Then arrange the events in sequence by naming each stage, starting at the beginning.

The life cycle of the liverwort begins when a spore is blown to a moist location. There it develops into a *gametophyte*. In this form it most resembles the mature liverwort. From here, most gametophytes form either an *antheridium* or an *archegonium*. The spherical antheridium releases sperm that swim to the archegonium, which has the eggs. From this union develops the *sporophyte*.

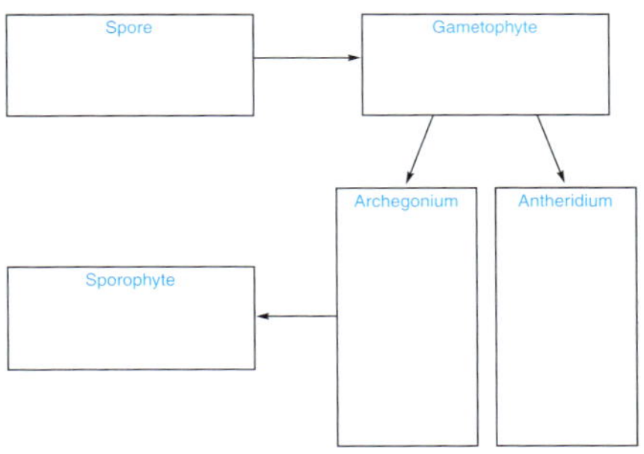

Lesson Concept Review/Assessment WB50

What Are Mosses?

Lesson Concept

Mosses, liverworts, and hornworts are nonvascular plants. They do not have tissues for transporting water and nutrients. They do not absorb much through their rootlike structures. They usually live in areas that are shady and moist. All have a life cycle with two generations, one reproducing sexually when egg and sperm unite, the other reproducing asexually by means of spores.

Vocabulary

moss (B28) **asexual reproduction** (B29) **spore** (B29)

Answer each question with one or more complete sentences.

1. The haircap moss commonly grows on sandy, open soil. When it is dry, the fronds of the moss are compressed tightly against the stalks. When it rains, special hinge cells quickly spread out the fronds. What is the purpose of this adaptation? The adaptation is to conserve moisture when it is dry and to expose as much plant surface as possible when it rains.
2. How do mosses transport water and food? They have specialized cells that move materials along slowly.
3. How is the growth of vascular and nonvascular plants different? Why? Nonvascular plants do not grow as large as vascular plants because they have no systems to get nutrients from roots to leaves.
4. How do mosses and liverworts differ from vascular plants? They have no phloem and xylem and no true leaves, stems, or roots.
5. Number the following statements about the reproductive cycle of mosses in the order of occurrence.

__1__ Spores are deposited on the ground, and buds form.
__5__ The fertilized egg produces a stalk with a spore capsule on top.
__6__ The capsule opens and releases spores.
__3__ Shoots develop into structures that produce eggs or sperm.
__2__ Buds send up shoots and send down rhizoids.
__4__ When moisture is present, the sperm swims to the egg and fertilizes it.

Workbook Support Continued

Investigate Log WB51

Comparing Ferns and Conifers

Materials

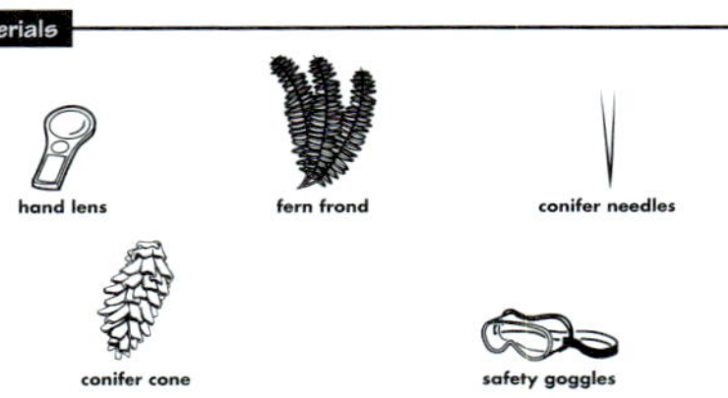

CAUTION

Activity Procedure

1. **Observe** the fern frond and a conifer needle. Both are leaves with adaptations that help the plants survive.
2. **Compare** the leaves, and **record** your observations by drawing the leaves and by describing what you see. Use a hand lens to **observe** the underside of the fern frond and the structure of the conifer needle.
3. Now **observe** and **compare** the conifer cone and the photograph of a conifer cone on page B33. They both came from the same kind of tree. **Record** your observations.
4. CAUTION **Put on your safety goggles.** Carefully break open the cone, and **observe** it. **Record** your observations.

Investigate Log WB52

Draw Conclusions

1. Are ferns vascular plants? Are conifers? How do you know? Both have veins in the leaves and thus are vascular plants.
2. Ferns can grow in shady places, and conifers can survive periods when there is little water. **Infer** the leaf adaptations that help the plants survive these conditions. Ferns' leaves are broad and spread out, which allows them to collect more light; conifer needles have a small surface area and have a waxy covering to keep water inside.
3. What did you **observe** on the underside of the fern frond? What might you **infer** about the structures you saw there? Students should point out that they observed small structures. They may infer that these have something to do with the plant's reproduction.
4. **Scientists at Work** Scientists **compare** objects to find out how they are alike and how they are different. Compare the cone you **observed** to the photograph of the conifer cone. What differences do you observe between the two cones? What can **you infer** from your observations? The larger cone contains seeds, and the smaller cone contains pollen. Students may infer that these are reproductive adaptations.

Investigate Further The seeds of conifers are adapted so that they can travel. Examine a conifer seed. **Formulate a hypothesis** about how the seed is adapted to travel. **Design an experiment** to test your hypothesis.
Conifer seeds have little "wings" on them. These wings would carry the seeds farther.

Process Skills Practice WB53

Compare and Infer

Comparing and inferring both start with careful observations. When you compare things, you make note of both their similarities and differences. Sometimes the data is arranged in a table or a graph to make the comparisons more obvious. Inferring is using logical reasoning to arrive at explanations for observations. Inferences are based on judgment and are not always correct.

Think About Comparing and Inferring

Conifers are vascular plants. Therefore, they have leaves. Although they don't look like the leaves you are familiar with, pine needles are true leaves. However, they differ in some ways from the broad leaves of the oak or maple tree. The chart below compares a pine needle to an oak or maple leaf.

Characteristic	Broad Leaf	Pine Needle
Size and Shape	broad and flat, thin	long and narrow, thick
Surface Exposed	much surface exposed	little surface exposed
Veins	many veins	no veins evident
Flexibility	leaf is soft and flexible	needle surface is hard
Season	leaves drop in winter	leaves stay on all year round

1. Because pine needles are true leaves, what four things can you infer must be present in both the broad leaf and the pine needle? phloem, xylem, chlorophyll, and stomata
2. Conifers flourish in difficult northern environments that are cold and dry, places where broad-leaved trees cannot grow. How are needles adapted to this climate? The needles have little surface exposed, so water is conserved. Their skin is thick to resist harsh conditions.
3. What could you infer by comparing the pine needle with the broad leaf? Answers should include that, although pine needles have the structures of broad leaves, they have leaf adaptations that make their leaves suitable for their environment.

Reading Skills Practice WB54

Use Prefixes and Suffixes to Determine Word Meanings

Use a dictionary to look up the meanings of the words below. Then write the meaning of the suffix.

Word	Definition
Vascular	of or related to a channel for conveying body fluids
Molecular	of, relating to, or produced by molecules
Spectacular	of, related to, or being a spectacle

–ar means...

of or relating to

Lesson Concept Review/Assessment WB55

What Are Ferns and Gymnosperms?

Lesson Concept

Ferns are simple vascular plants that grow well in low light. Ferns have a two-generation life cycle. They produce spores on the undersides of some of their leaves. Gymnosperms are nonflowering plants that produce seeds in cones. Conifers have needlelike leaves adapted for dry or freezing weather.

Vocabulary

fern (B34) **gymnosperm** (B36) **conifer** (B36) **pollen** (B37)

Answer each question with one or more complete sentences.

1. Why do you think ferns are in this lesson with gymnosperms rather than in the previous lesson with mosses? Ferns contain phloem and xylem and true leaves, stems, and roots.
2. What is the prothallium? What is its purpose? It is the tiny heart-shaped sexual generation of the fern. It produces the egg and sperm cells that unite to produce the large spore-forming generation of the fern.
3. How are ferns and gymnosperms different? Ferns reproduce first by egg and sperm cells, then by spores; gymnosperms reproduce by seeds.

Circle the letter of the best answer.

4. The structures that are not part of a large fern are ____.
   - (A) rhizoids
   - B leaves
   - C stems
   - D xylem and phloem
5. Seeds have an advantage over spores because they contain ____.
   - A sori
   - (B) a food supply
   - C phloem
   - D roots
6. Conifers can survive below-freezing temperatures because ____.
   - A the branches are flexible
   - B the trees have true roots
   - (C) the trees produce an antifreeze chemical
   - D the needles are not true leaves

Workbook Support Continued

Investigate Log WB56

The Parts of a Flower

Materials

flower — paper towel — hand lens

Activity Procedure

1. Lay the flower on the paper towel.
2. Handle the flower gently while you **observe** its parts. Be sure to look at the center between the petals.
3. Now use the hand lens to examine the parts of the flower. Draw what you **observe**.
4. The flower includes at least three different structures. Are there at least three different structures in your sketch? If not, **observe** the flower again and revise your sketch.

Investigate Log WB57

Draw Conclusions

1. Describe the structures you **observed** in the flower. Answers will vary, but students will probably describe petals, pistils, and stamens.

2. What do you **infer** are the functions of these structures? Students should infer that the structures help the plant reproduce.

3. **Scientists at Work** Scientists **communicate** information in many ways. They use drawings, diagrams, words, and measurements. **Compare** these methods. Choose one of the flower structures you drew. Which method or combination of methods do you think would be the most helpful for communicating information about the structure? Explain. Accept any valid, supported answer, such as diagrams with measurements, or actual-size, detailed drawings.

Investigate Further Use your observations of the flower in this activity to **hypothesize** what structures you would find in any flower. **Design an experiment** to test your hypothesis. ______

Process Skills Practice WB58

Observe and Communicate

Observing is the most basic of all scientific process skills. Careful observation also means recording what you observe. Good record-keeping makes communication among scientists much easier and more effective. Communicating in science means informing others of the results of your experiments in an organized fashion.

Think About Observing and Communicating

Alicia obtained two stalks of fresh celery, one with the leaves removed and one with the leaves still on. She placed the stalks together in a glass of water to which some red food coloring had been added. After four hours she checked the celery stalks. Red color was rising in both of them, but much faster in the one that still had leaves. She left the celery stalks in the colored water overnight. When she checked them the next morning, she saw little red marks in the leaves of the leafy stalk. Then she removed the stalks from the water and cut across each one about 5 centimeters from its bottom. The "strings" in each stalk were red, and there were little red dots on the outer edge of each stalk.

1. What are some ways that Alicia could record and communicate her observations? Possible answers: she could make a graph for each stalk and actually measure how far the coloring had risen at different times, she could make a chart with the same information, or she could write a narrative of the events. Accept all reasonable answers.

2. Alicia knew that plants get rid of excess moisture through the stomata in their leaves. With this knowledge, why do you think the red color rose faster in the stalk with the leaves? Colored water was pulled up faster through the veins in the leafy stalk because it was losing water through the stomata in its leaves.

3. How could Alicia effectively communicate this result to someone else? She could write a careful description, but adding drawings would make it better.

Reading Skills Practice WB59

Summarize and Paraphrase a Selection

Read the selection below, and write a one-sentence summary of each paragraph. Then summarize the entire selection in one or two sentences.

Many types of plants have adapted to the harsh environment of the desert. They have done so by altering their behavioral and physical mechanisms.

The cactus has special ways to store water and, to reduce transpiration, usually has very few leaves. The organ pipe cactus, one of the few types of cactus to have its own national monument, blooms only at night, when the temperatures are cooler. From May through July, lavender-white flowers 6 centimeters (2.4 in.) wide open after sunset. Native Americans have eaten the fruit from these flowers, which also can be made into jelly or a beverage.

Phreatophytes, such as the mesquite tree, are a type of desert plant that adapts to the dry desert environment by growing unusually long roots that reach close to the water table. The roots can grow to just over 24 meters (79 ft.) long!

There are desert perennials that become dormant during the driest periods when water is not available. One such plant is the ocotillo. Its waxy leaves maintain moisture during dormancy, and it may bloom up to five times per year.

There are cactus plants that survive desert conditions by growing few leaves, and one type blooms at night in order to avoid the heat.
Phreatophytes grow long roots that reach down to the water table.
Some desert plants become dormant to avoid the driest periods, but they may bloom up to five times per year.

↓

Numerous types of desert plants have adapted their behavioral and physical mechanisms in order to survive the desert's harsh environment.

Lesson Concept Review/Assessment WB60

How Are Angiosperms Different from Other Plants?

Lesson Concept

Angiosperms, the most common plants on land today, produce flowers and have efficient vascular systems and broad leaves. All parts of the flower contribute to the plant's sexual reproduction. Some flowers are pollinated by insects or other animals; others depend on the wind to carry pollen from one flower to another.

Vocabulary

angiosperm (B44) **petal** (B46) **stamen** (B46)
anther (B46) **pistil** (B46) **stigma** (B46)
style (B47) **ovary** (B47) **pollination** (B47)

1. Complete the following table describing animal-pollinated and wind-pollinated flowers.

Structure	Animal-Pollinated	Wind-Pollinated
Flower	large, bright, scented, produces nectar	small, unscented, not brightly colored
Pollen	heavy and sticky	fine, light, dry

2. How are the seeds of angiosperms different from those of gymnosperms? Angiosperm seeds are contained in fruits.

3. What is the purpose of the sepals and the petals of a flower? Both protect the inner parts of the flower. The petals often attract animals to help pollination.

4. Identify which statements below describe annual, *A*, plants and which describe perennial, *P*, plants.

   P Many have stiff, woody stems.
   P Plants stay alive below ground as bulbs or roots.
   A Plants usually have soft stems.
   A Plants go through their whole life cycle in a few months.

Workbook Support Continued

Investigate Log WB61

Inside Seeds

Materials

Activity Procedure

1. Cover your desk with paper towels. Lima beans and corn kernels are seeds. Use the hand lens to **observe** the two kinds of seeds. Make a sketch of each.
2. Carefully peel the outside covering from each seed. Then open the seeds. You may need to cut open the corn seed.
3. Use the hand lens to **observe** each seed's structures. **Record** your observations by drawing what you see.
4. **Compare** the structures inside the two seeds. Look for similarities and differences. From your **observations, infer** what the function of each part of the seeds might be.

Investigate Log WB62

Draw Conclusions

1. What can you **infer** about the outside covering of the seeds? Is the function the same for both seeds? Students should infer that the outside covering protects each seed.
2. Were you able to identify a tiny plant in each seed? Which part of the seed was it? On their sketches students should have identified the plant embryo in each seed.
3. What did you **infer** about the function of the other parts of the two seeds? Is the function the same for both seeds? Students should have inferred that the other parts of the seeds provide food for the embryo.
4. **Scientists at Work** To **classify** organisms, scientists **observe** and **compare** their characteristics. The seeds you observed belong to two different groups of angiosperms. What differences in the seeds did you observe? Students should have noticed that the bean seed has two similar food-storage parts (cotyledons), while the corn seed has two dissimilar food-storage parts (a cotyledon and an endosperm).

Investigate Further Look again at the two seeds. Find the tiny plant in each, and **hypothesize** which part of it will become the root and which part will become the shoot. **Plan and conduct an experiment** to test your hypothesis.

Process Skills Practice WB63

Observe and Compare

You use your senses of sight, hearing, smell, and touch to study the world around you. Sometimes you use instruments to extend your senses. When you compare observations, you look for what they have in common and how they are different.

Think About Observing and Comparing

Flowers come in a great variety of colors, sizes, shapes, and scents. Some flowers, such as those of the wheat plant, look very different from the flowers you usually find in a typical flower garden. The drawing on the left below shows the parts of what is called a complete flower. The other drawings are also of flowers, but nature has left out some parts. Those flowers are called incomplete flowers.

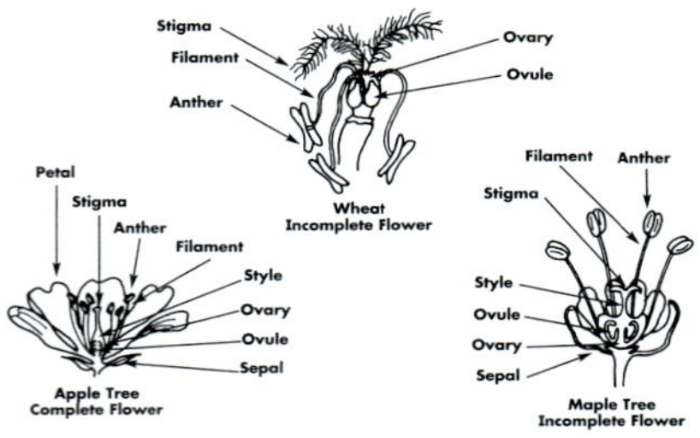

1. Observe and compare the three drawings. What parts do all the flowers have? anthers, filaments, stigma, ovary, ovules
2. Which of the flowers shown are so incomplete that the plant cannot reproduce? None; all the plants can reproduce with these flowers.
3. If insects and birds are attracted by the colors of a flower's petals, what can you infer about how the wheat flower is pollinated? Because the wheat flower has no petals for color, it is probably pollinated by some other means, such as the wind.

Reading Skills Practice WB64

Use Context to Determine/ Confirm Word Meaning

Read the paragraph below, and determine the meaning of the boldface words from the context.

Roses have been the most valued of all flowers. The rose is written about in ancient history and legend. In China, only the nobles and **dignitaries** of the court were allowed to use the oil that was pressed from roses. Any commoner who had even a small amount of the oil was condemned to death. During the Middle Ages, a rose was **suspended** from the council chamber ceiling. This reminded all the lawmakers sitting under it that they were pledged to secrecy, since the rose had become a symbol of silence.

Word: dignitaries

Clues: linked with nobles; contrasted with commoners; seems to have the root word *dignity.*

Definition: ones who possess exalted rank or hold positions of dignity or honor

Word: suspended

Clues: from the ceiling; lawmakers sat under it

Definition: hung by a support from above

Workbook Support Continued

Lesson Concept Review/Assessment WB65

How Do Angiosperms Reproduce?

Lesson Concept

Angiosperms are divided into two groups—monocots and dicots. Monocots have one cotyledon, parallel veins in their leaves, spreading root systems, and scattered vascular bundles in their stems. Dicots have two cotyledons, netlike veins, one main taproot, and vascular bundles arranged in a ring in their stems. The ovules in a pollinated flower are fertilized and become seeds encased in fruits, which can be classed as drupes, pomes, berries, aggregate fruits, or multiple fruits. Seeds can be spread by wind, animals, or water. Many plants can reproduce asexually from cuttings, roots, runners, and grafting.

Vocabulary

dicotyledon (B52)	**monocotyledon** (B52)	**fertilization** (B53)
embryo (B53)	**runners** (B55)	**grafting** (B55)

1. Identify which statements below describe dicots, *D*, and which describe monocots, *M*.

M	one cotyledon	D	flower parts in multiples of four or five
M	spread out roots	M	parallel leaf veins
D	netlike leaf veins	D	vascular tissue in a ring
M	scattered vascular tissue	M	taproot
D	two cotyledons	M	flower parts in multiples of three

2. What are cotyledons? They are the seed leaves of the plant that supply the young plant with food.

3. In a grain of pollen, what is the function of the sperm cell that doesn't join with the egg cell? This sperm cell joins with part of the ovule to form a food source.

4. What are two adaptations of seeds that enable them to be spread by animals? edible fruits and burrs

5. What are two adaptations of seeds that enable them to be spread by the wind? Any two: they have winglike structures; they are like parachutes; or they are very lightweight.

Vocabulary Review WB66

Recognize Vocabulary

Place the letter of the term next to the phrase that best describes it.

F 1. cone-producing plant
O 2. part that produces the ovules
H 3. plant that produces flowers
T 4. tiny plant that forms in the seed
L 5. female reproductive organ of the plant
C 6. structure containing cells for asexual reproduction
V 7. adding a cutting to a plant
I 8. especially colorful flower part that protects inner parts of the plant
Q 9. plant with netlike leaf veins
U 10. means by which strawberries and spider plants reproduce
D 11. primitive vascular plant
S 12. joining of the sperm and egg
G 13. spores that turn into sperm
J 14. plant part made up of the anther and filament
M 15. top part of the pistil
R 16. plant with flower parts that come in multiples of three
B 17. reproduction without the joining of sperm and egg
K 18. plant part that produces pollen grains
E 19. plant with bare seeds
P 20. event that occurs at the moment pollen reaches the stigma
A 21. small green plant without true stems, leaves, or roots
N 22. plant part that connects the stigma to the ovary

A moss
B asexual reproduction
C spore
D fern
E gymnosperm
F conifer
G pollen
H angiosperm
I petal
J stamen
K anther
L pistil
M stigma
N style
O ovary
P pollination
Q dicotyledon
R monocotyledon
S fertilization
T embryo
U runners
V grafting

Writing Practice WB67

Write About a World Without Conifers

Informative Writing–Description

Write a magazine article describing a future time when conifers have become extinct. Describe how the conifers died out. Tell what products would no longer be available. Describe how life would be different without these trees. Add details to the outline below to help you plan your article.

I. Introduction
II. How conifers became extinct
III. What products are no longer available
IV. How life is different without conifers
V. Conclusion

CHAPTER 2

Types of Plants

Generate Science Questions

Read aloud and discuss the Fast Facts. Afterward, ask students if they have any questions about types of plants which they would like to answer. Lead the class in generating a list of questions, and record the questions on chart paper or on the board.

Ask students what kinds of investigations they might design to find answers to their questions. Explain that although the chapter will have investigations and reading material to help answer these questions, they may wish to conduct investigations on their own.

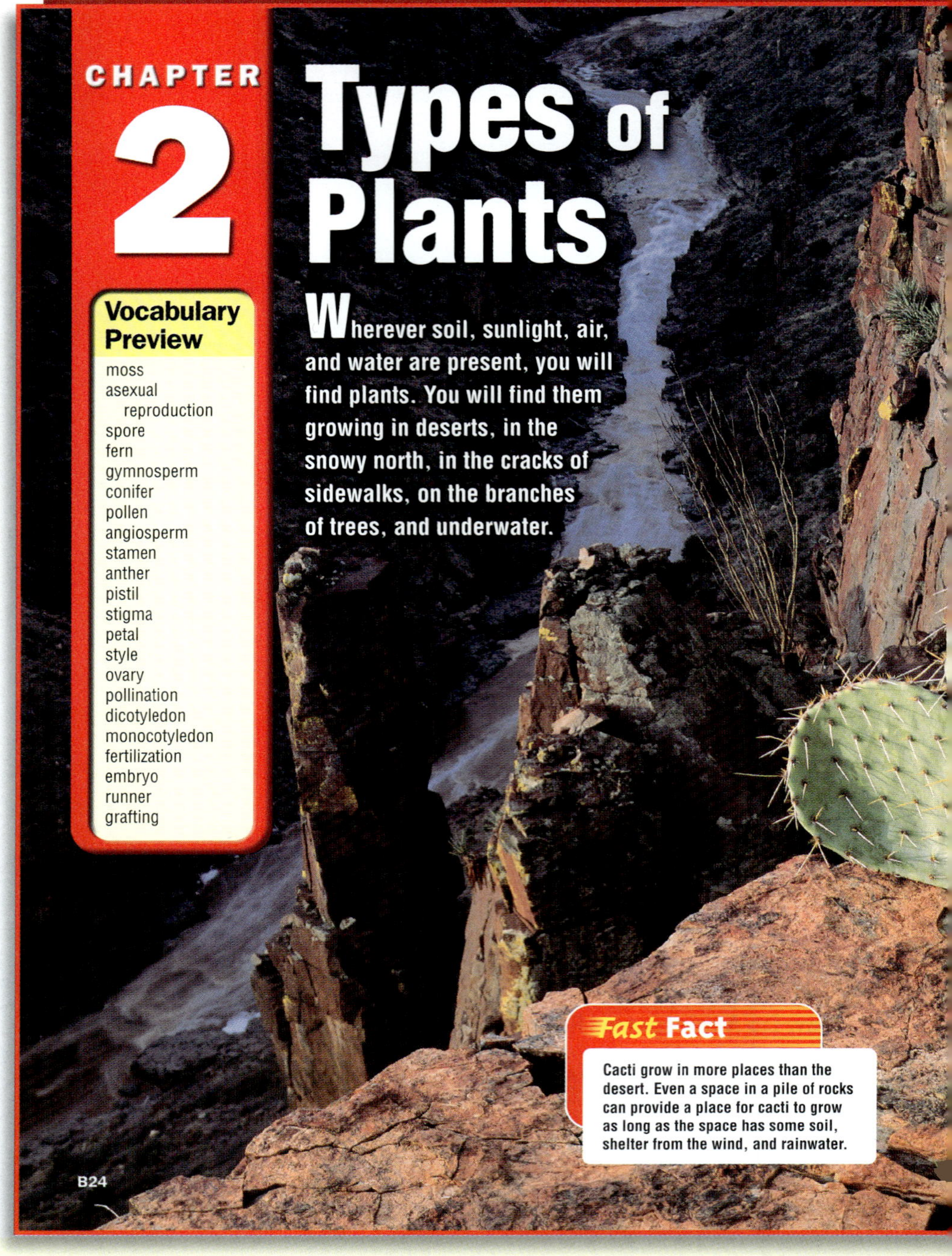

CHAPTER 2

Types of Plants

Vocabulary Preview

moss
asexual reproduction
spore
fern
gymnosperm
conifer
pollen
angiosperm
stamen
anther
pistil
stigma
petal
style
ovary
pollination
dicotyledon
monocotyledon
fertilization
embryo
runner
grafting

Wherever soil, sunlight, air, and water are present, you will find plants. You will find them growing in deserts, in the snowy north, in the cracks of sidewalks, on the branches of trees, and underwater.

Fast Fact

Cacti grow in more places than the desert. Even a space in a pile of rocks can provide a place for cacti to grow as long as the space has some soil, shelter from the wind, and rainwater.

B24

Reading Skills Checklist

Strategies for developing the following reading skills are provided in this chapter.

- ☑ use prefixes and suffixes to determine word meanings *pp. B35, B53*
- ☑ use context to determine/confirm word meaning *p. B54*
- ☑ identify supporting facts and details *pp. B28, B34, B36, B38, B44, B46, B48, B52, B54*
- ☑ arrange events in sequence *p. B30*
- ☑ identify the main idea *pp. B28, B34, B36, B38, B44, B46, B48, B52, B54*
- ☑ summarize and paraphrase a selection *pp. B45, B47*
- ☐ identify cause and effect
- ☐ predict probable future actions and outcomes
- ☑ use graphic sources for information *p. B40*
- ☐ draw logical conclusions
- ☑ compare and contrast *pp. B48, B55*
- ☐ distinguish fact and opinion
- ☐ make generalizations
- ☑ use reference sources *pp. B39, B40, B56, B58*

School-Home Connection

Distribute copies of the School-Home Connection, p. TR7. Have students take the page home to share with their families as you begin the chapter.

Follow Up Have volunteers compare the results of the activity they did at home. Additional School-Home Connections are provided by the **Activities for Home or School** (p. B61) and the **Take-Home Book** (p. B24d).

Teaching Resources, p. TR7

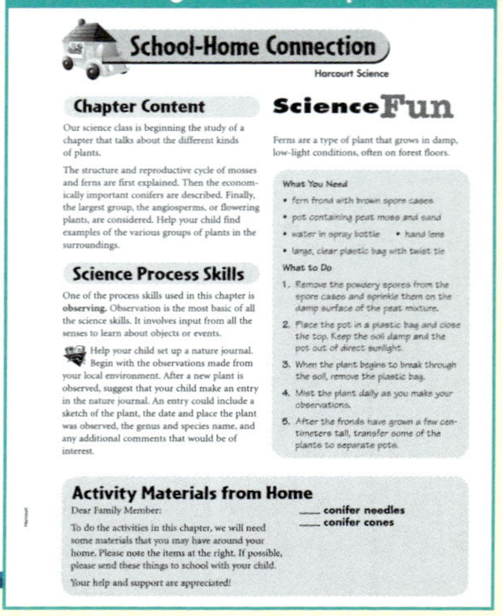

Fast Fact

The navel orange is an unusual fruit because it has no seeds. The first navel orange tree was a genetic accident. Ever since, navel oranges have grown on branches that have been grafted onto normal orange trees.

Fast Fact

The oldest seed to be grown successfully came from a 10,000-year-old Arctic lupine plant. Most seeds, however, remain able to produce a plant for only a few years.

Seed Life

Plant	Years
Cucumber	5
Turnip	5
Lettuce	4
Pumpkin	4
Tomato	3
Beet	3
Pepper	2
Sweet corn	2
Carrot	1
Pea	1

B25

Prereading Strategies

K-W-L Chart

Have students make a four-column chart headed What I Know, What I Want to Know, What I Learned, and "+" for What I'll Learn More About. Ask them to look at the chapter title and fill in the first two columns. Then have them predict what the chapter will be about. As students study the chapter, have them fill in the third column. At the end of the chapter, encourage students to fill in the last column with additional things they would like to learn.

K-W-L Chart

What I Know	What I Want to Know	What I Learned	+

Preview the Chapter

Suggest that students scan the chapter to identify the lesson titles and major headings. Have them use the headings and titles to make outlines of the chapter. Encourage students to write questions as they make their outlines.

Vocabulary Preview

Have students make a three-column chart for the vocabulary words. In the columns, students write the word, its definition, and how they remember what the word means.

Science Journal

Word	Definition	How I Remember

BACKGROUND

Webliography GO ONLINE

Keyword plants

www.harcourtschool.com

Life-Cycle Patterns Although plants display an amazing variety of individual life styles, their life cycles conform to a number of well-known patterns. These patterns help us identify and classify plants.

The simplest plants, such as the mosses and ferns, have a life cycle with two distinct generations. One generation is plantlike in appearance and reproduces sexually through the union of an egg and sperm. The other reproduces asexually by spores.

More complex plants reproduce sexually by seed. **Gymnosperms**, such as the conifers, have no flowers and produce seeds that have no covering. All flowering plants, called **anglosperms**, produce seeds enclosed in a covering, such as a fruit. The fruit is actually the mature ovary of the flower.

Fast Facts/Math Links

Choose an Appropriate Graph Have students use the data in the Seed Life table to make a graph. Ask them to decide if a circle graph, line graph, or bar graph would best display the data. Be sure students explain their reasoning. A bar graph would best display the data because bar graphs are used to compare data by category, unlike circle graphs that compare parts to the whole.

Find the Average What is the average seed life for the plants listed in the table? 3 years

LESSON

1 What Are Mosses?

Objectives

- ▶ **Compare the structures of vascular and nonvascular plants.**
- ▶ **Describe the life cycle of mosses.**

Motivate

More Fast Facts

Spark interest with these facts and questions.

- Mosses have a long fossil record. Some fossils of mosses have been found that are about 280 million years old.

  What do you think would happen if you planted mosses in a dry, sunny place? They would dry up and die.

- Sphagnum moss grows in acidic water containing few bacteria. It is sterile enough to be used as an emergency bandage.

  Are all mosses green? No. They can also be red, brown, or grayish-green.

Access Prior Knowledge

- ▶ Are mosses plants?
- ▶ Where have you seen them growing?

When Minutes Count . . .

Whole-Class Investigate Option Divide the class into three teams, and have each team record its observations on one set of pictures. Ask each team to share its observations with the class.

Using the Activity Video The video can be used to preview and model the investigate process with students. After viewing the video, students can conduct the investigation in small groups.

LESSON 1

What Are Mosses?

In this lesson, you can . . .

- INVESTIGATE where mosses grow.
- LEARN ABOUT the structures of mosses and how they reproduce.
- LINK to math, writing, social studies, and technology.

Mosses play an important part in the lives of many small animals. This beetle makes its home in mosses. ▼

B26

INVESTIGATE

Where Mosses Grow

Activity Purpose When you walk through a park or a forest, you may first notice the large living things, such as trees. Then you might look down and see small wildflowers and ferns. If you look down even lower, you can see mosses, which often grow almost flat on the ground. In this investigation you'll find out where mosses grow and the conditions they need in order to live.

Materials

- photographs of mosses

Picture A

Picture B

Picture C

Picture D

Picture E

Picture F

BACKGROUND

Webliography

Keyword plants

www.harcourtschool.com

GO ONLINE

Alternation of Generations One reason that mosses, liverworts, and hornworts must live in damp environments is that they lack specialized structures for transporting water, minerals, and food. Another reason is that they depend on a film of water to accomplish sexual reproduction. Mosses, liverworts, and hornworts reproduce sexually in one generation and asexually in the next; that is, they alternate their reproductive strategies. The asexual generation produces spores, which do not require water to complete reproduction. Spores can also remain dormant until conditions are favorable for growth. But in the sexual generation, the male sperm must swim in a continuous film of water to join the female egg. The plants that have alternation of generations have a distinct advantage because they are not dependent on a single set of conditions to ensure the continuation of the species.

For student self-assessment, see Assessment Guide p. AGxvi.

Activity Procedure

1. **Observe** the mosses in Pictures A–F.
2. **Record** your **observations** in a chart like the one shown here. Draw a picture of each moss, and describe it. Pay close attention to the surroundings of each moss. Describe the environment in which the moss is growing. Remember to include details such as the place where the moss grows, the surrounding living things, and the amounts of sunlight and moisture in the environment. If you can **infer** the approximate temperature from the picture, include that information in your description, too.

Observations of Mosses

Picture	Description	Environment

Draw Conclusions

1. By **observing**, what did you **infer** about the surroundings in which mosses can live?
2. What can you **hypothesize** about what mosses need in order to grow?
3. Where might you expect to find mosses growing? List at least three places not shown in the photographs on page B26.
4. **Scientists at Work** Scientists can test a **hypothesis** by **observing** further. How could making observations help you test your hypothesis about what mosses need in order to grow?

Investigate Further Use what you have learned in this activity to **formulate a hypothesis** about the best growth conditions for mosses. Use your hypothesis to **predict** where in your local area you might find mosses, and **plan an investigation** to check your prediction.

Process Skill Tip

If you **observe** carefully and often, you're more likely to be correct when you **hypothesize**. Observations also provide evidence to support your hypothesis.

INVESTIGATE

Time 30 min **Preparation Tips** See p. B24c.

Grouping individuals

Process Skill Tip observe, hypothesize

Other Process Skills record data, infer, predict, plan a simple investigation

Expected Results Students should conclude that mosses live in shady, moist environments.

Teaching Tips

Be sure students understand that a hypothesis is based on observations and that it should lead to more observations and further research.

Draw Conclusions

1. Students should infer that mosses survive best in shady, damp places.
2. Students should hypothesize that mosses require a lot of moisture but little sunlight.
3. Answers should include places where there is a lot of moisture and shade, for example, in a woodpile, in the shade of a boulder, or near a pond or stream.
4. You could look for mosses in very dry places in full sunlight to disprove the hypothesis. To support your hypothesis, you could carefully measure and monitor conditions where mosses grow.

Investigate Further Guide students to formulate hypotheses before they plan their investigation. Encourage them to use their hypotheses to help formulate their predictions.

Process Skill Tip

Observe and Hypothesize Display the Process Skill Tip Transparency on an overhead projector. Tell students to assume that the plants were placed next to each other. Have students identify the variables that remained constant for all three plants. Students should conclude that the amount of heat, light, and moisture for all three plants were the variables that remained constant. Then have them observe changes in the plants after two weeks.

Process Skill Tip Transparency B2-1

Observe and Hypothesize

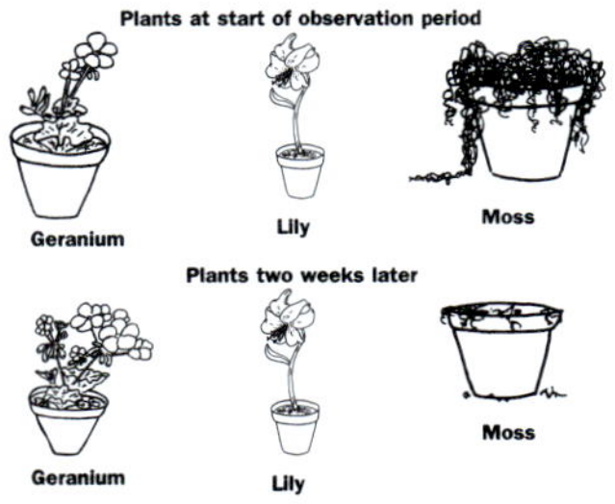

- **Observe these three houseplants.**
- **What hypothesis can you formulate about their different environmental needs?** The moss probably needs less light, more water, and lower temperatures than the other two plants.
- **How could you test your hypothesis?** with a simple experiment

Informal Assessment

Portfolio Students may add the charts to their science portfolios.

INVESTIGATE LOG

Students can use **Workbook pp. WB46–47** to record the results of the investigation.

PROCESS SKILLS PRACTICE

To practice and apply process skills, see **Workbook p. WB48.**

LEARN ABOUT

1 Before Reading

SET A PURPOSE

FIND OUT Remind students that these statements will guide them in what to look for as they read.

VOCABULARY If students are completing the activity described in the Vocabulary Preview on page B25, remind them to add the definitions for these words to the chart.

Graphic Organizer for Chapter Concepts Display the transparency or have students use Workbook p. WB45. By the end of the lesson, students should be able to fill in information about mosses, liverworts, and hornworts.

2 Guide the Learning

SCIENCE IDEAS

Help students focus on supporting details for this **main idea:** **Mosses are nonvascular plants.**

- ▶ **Why are structures on mosses described as "stemlike," "leaflike," and "rootlike"?** Mosses do not have true stems, roots, and leaves, but they do have structures that look like them and function in similar ways.
- ▶ **Where does photosynthesis occur in mosses?** in the green, leaflike structures
- ▶ **Why do mosses need to live in wet environments?** They do not have a well-developed system for absorbing water and must take in water and nutrients from the air or from nearby cells; sperm must swim through water to reach and fertilize the eggs.

Critical Thinking Lead students to draw a conclusion about mosses and pollutants.

- ▶ **Why do you think mosses are known as "indicator species"?** They are sensitive to pollutants and die if the air in their environment becomes polluted. Because of these characteristics, they serve as indicators of air quality.

VOCABULARY CARDS AND ACTIVITIES

Students can use the Vocabulary Cards to make their own graphic organizers or to add to an ongoing file of science terms. The Vocabulary Cards and a variety of strategies and activities are provided beginning on p. TR85 in the **Teaching Resources** book.

LEARN ABOUT

FIND OUT

- about the structures of mosses and how mosses meet their needs
- about how mosses reproduce
- about other nonvascular plants

VOCABULARY

moss
asexual reproduction
spore

Nonvascular Plants

Mosses

A **moss** is a very short green plant that does not have true leaves, stems, and roots. It's a nonvascular plant, so it does not have xylem or phloem to carry water, nutrients, and food.

Mosses have structures that look like leaves, stems, and roots. The leaflike structures of mosses are green. The moss makes food within these structures. Unlike true leaves, however, these structures do not have veins, and they are only one or two cells thick. The stemlike structures of mosses have cells for carrying water and food. The water and food move through them slowly. They are not like the complex tissue in vascular plants. The rootlike structures of mosses

THE INSIDE STORY

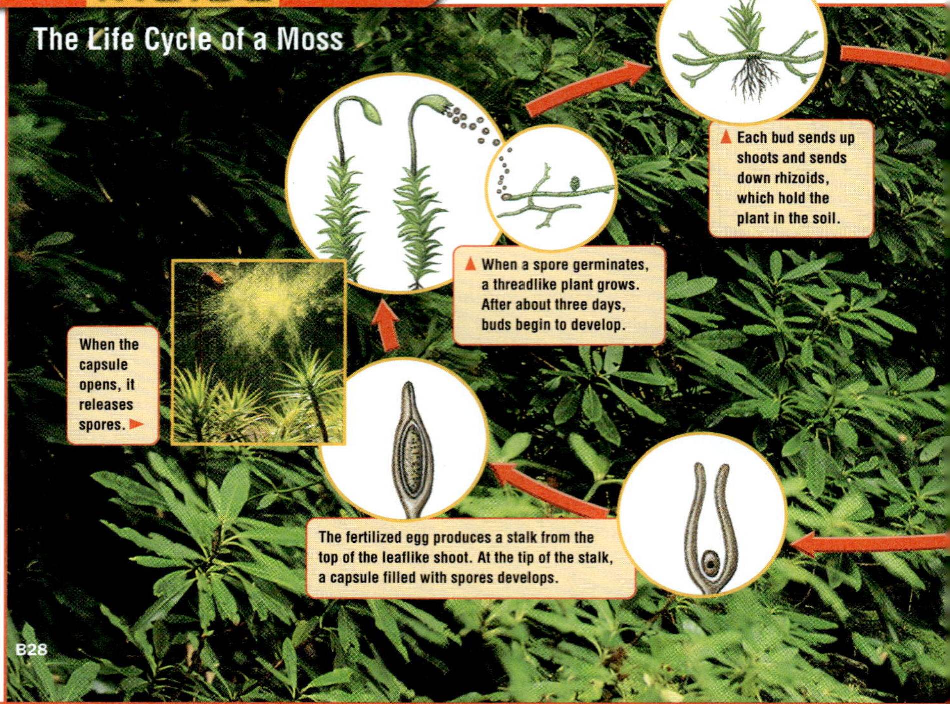

Graphic Organizer for Chapter Concepts

Transparency B2 • Workbook, p. WB45

Types of Plants

	LESSON 1	LESSON 2		LESSONS 3 & 4
	Mosses, liverworts, & hornworts	Ferns	Gymnosperms	Angiosperms
Water transport	nonvascular			
Plant size	small			
Type of reproduction	sexual and asexual			
Means of sexual reproduction	water			
New plant form	spore			
How seeds or spores are dispersed	gravity, wind			

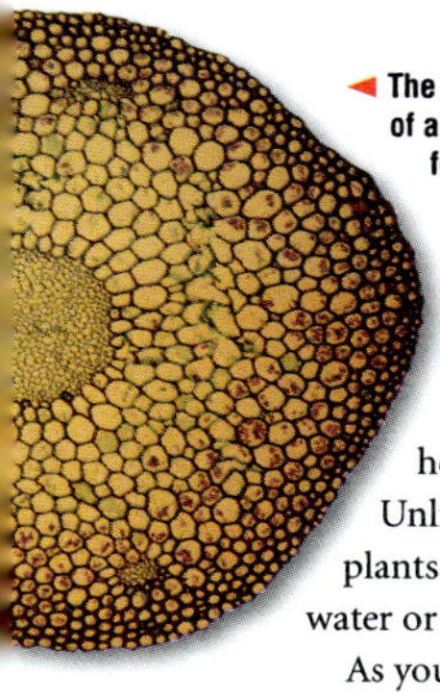

The heavy-walled cells in the center of a moss stem are specialized for moving water. The cells just outside those cells are specialized for food transport. Both are much simpler than the xylem and phloem tissues in vascular plants.

hold the plant in the soil. Unlike the roots of vascular plants, they do not absorb much water or many nutrients.

As you observed in the investigation, mosses grow best in moist, shady places. Mosses often grow on a shady forest floor or beside pools or streams. Some mosses, such as peat moss, grow in bogs or swamps and can hold quite a lot of water.

Mosses have a life cycle with two very different stages. One generation looks plantlike. The plantlike generation carries out sexual reproduction. It makes egg and sperm cells that join to produce a new moss plant.

Then the other generation begins. This generation looks like a stalk that grows out of the leaflike part of the moss. This generation carries out **asexual reproduction**, in which a new plant is produced without the joining of a sperm and an egg cell. The stalk produces spores. A **spore** is a structure that contains cells that can grow into a new plant without joining with other cells.

✔ **How do mosses differ from vascular plants?** Mosses do not have xylem, phloem, or true roots, stems, and leaves.

Some shoots grow a structure that produces eggs, and some grow a structure that produces sperm.

After rain or a heavy dew, the sperm swim through the standing water and fertilize the eggs.

Mosses Help the Environment

Mosses often grow in dense layers and cover very large areas. They are attached to the soil, so they help keep it from being eroded by the wind or rain. Mosses also store nutrients. When they die, the mosses decompose, and their matter returns to the environment. The nutrients and other matter from the mosses greatly improve the soil for other plants.

Many species of mosses are sensitive to polluted air. When the air where they live becomes polluted, many mosses die. Then they are no longer seen growing in their usual surroundings. In this way mosses can show where there is pollution.

✔ **How do mosses help the environment?** They keep soil from being washed away. After they die, they enrich the soil with nutrients and organic matter.

In Japan, where there is little space for gardens, moss gardens are popular. Mosses in different shades of green are planted around carefully arranged rocks.

B29

VISUAL LEARNING

Ask students to study the picture and captions about the life cycle of the moss on pages B28–29. Have them explain its two different stages.

- **Describe the sexual stage of the life cycle of the moss.** The plantlike generation makes eggs and sperm. The sperm swims through water to join with and fertilize the egg. A new moss plant grows.
- **Describe the asexual stage of the life cycle of the moss.** The moss plant grows a stalk out of the leaflike part. The stalk produces spores that grow into a new plant.

DEVELOP SCIENCE VOCABULARY

moss, p. B28—Have students check the dictionary for the derivation of the word. They will find it comes from a Scottish word meaning "bog" or "swamp." This may help students remember that mosses need adequate moisture.

asexual reproduction and **spore**, p. B29—Ask students to give a meaning for the one-letter prefix *a-*. "not" or "non" Therefore, an asexual form of reproduction is one that does not involve egg and sperm cells. Be sure students understand that spores are one-celled and do not represent the union of a male and a female cell.

Ongoing Assessment

Use the ✔ questions in the pupil edition.

ESL ACTIVITY

Help students understand the concept of **asexual reproduction** by having them draw and label the stalk produced at the top of the leaflike shoot and the capsule filled with spores.

THE QUESTIONS KIDS ASK!

Do mosses always grow on the north side of a tree?

Address Misconceptions Not necessarily, but quite often they do. Ask students what conditions on the north side of a tree might be favorable for mosses. shade, because there would be less evaporation and more moisture Ask what conditions on the south side might prevent mosses from growing. direct sunlight

BACKGROUND

Webliography GO ONLINE
Keyword plants
www.harcourtschool.com

Mosses as Indicators of Pollution One reason that mosses are so sensitive to pollution is that they lack roots. Soil acts as a filter that keeps some dissolved pollutants from reaching plants with roots. Instead of absorbing moisture beneath the soil, mosses absorb water from its surface. Air pollutants such as sulfur dioxide dissolve in rainwater and are absorbed directly by mosses. Mosses act like sponges; they absorb everything in the rainwater, including the pollutants. Sulfur dioxide does the most damage. When it reaches certain levels, it begins to break down the chlorophyll inside the plant. Without chlorophyll, the plant faces certain death.

Guide the Learning continued

USE PROCESS SKILLS

Compare Discuss the similarities and differences among mosses, hornworts, and liverworts.

- **How are mosses, hornworts, and liverworts similar?** They are all nonvascular, live in moist shady environments, and have two different generations.
- **How are liverworts and hornworts different from mosses?** Liverworts and hornworts are smaller, have different structures, and look different from mosses. Encourage students to examine the pictures on pages B30–31 and describe the two plants in their own words.

Other Nonvascular Plants

Mosses are not the only nonvascular plants. Others are liverworts and hornworts. They grow in the same kinds of surroundings as mosses, sometimes among them. But liverworts and hornworts are even smaller than mosses, and they usually don't grow in such large clumps. Only the most careful observers notice them. You need a hand lens to see the structures of these tiny plants.

Although nonvascular plants have many things in common, they can have different structures. Mosses have leaflike structures all the way around their stemlike structures. But liverworts have leaflike structures that are flat and scaly. Their stemlike structures lie along the ground. The word *wort* means "herb." Liverworts got their name because some of them are shaped like the human liver.

Hornworts, as you might guess, have structures that look like an animal's horns. The "horns" are reproductive structures that produce spores. The green leaflike part of a hornwort is very small and very simple. It has a flat, rounded shape and no clear stemlike structure.

Like mosses, both liverworts and hornworts have two different generations. At times, you may see the horns of a hornwort. These are on the spore-producing generation. On liverworts, you may see small green structures that look like umbrellas. These carry out sexual reproduction.

✔ **How are liverworts and hornworts similar to mosses?**

All grow in the same kinds of surroundings; all have two different generations, one of which reproduces by spores; and all are small and have no vascular systems.

This liverwort, *Marchantia*, can be found on the banks of cool streams, where it sometimes grows in large mats. Bright yellow spore capsules form on the undersides of the umbrella-like female plants. ►

The hornlike structures of this hornwort, *Anthoceros*, can continue producing spores over a long period of time. ▼

B30

INVESTIGATION CHALLENGE

Hands-On Activity: How much water can sphagnum moss hold?

Use this idea for a hands-on activity to extend students' understanding of a key lesson concept.

Tell students that sphagnum moss has amazing water-holding ability. It can hold about 20 times its dry weight in water. Challenge students to design an experiment to find out how much water a sample of dry sphagnum moss can hold. Their experiment should include measuring the dry weight of the moss, measuring the amount of water it takes to saturate the moss, and performing calculations that compare wet weight relative to dry weight.

Reading Mini-Lesson

Arrange Events in Sequence Have students review the life cycle of mosses by filling in a flowchart. Display the transparency and have students write the characteristics of each stage in the appropriate box.

READING SKILLS PRACTICE

To practice and apply this reading skill, see **Workbook p. WB49.**

Reading Mini-Lesson Transparency B2-1

Arrange Events in Sequence

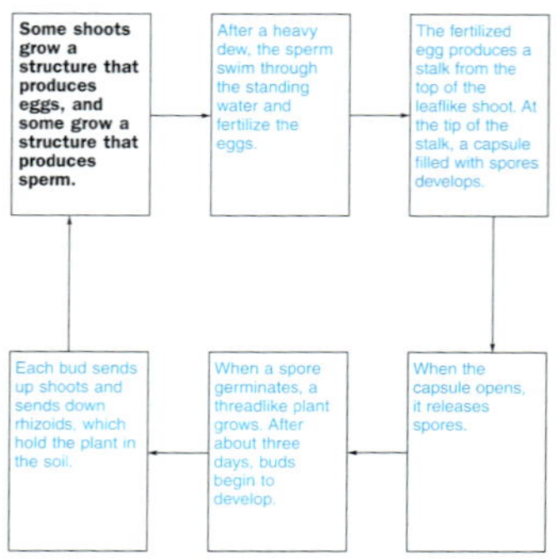

For student self-assessment, see Assessment Guide p. AGxvii.

Summary

Mosses, liverworts, and hornworts are nonvascular plants. They do not have vascular tissue to transport water or many nutrients. They do not absorb much water and nutrients through their rootlike structures. They usually live in areas that are shady and moist, where they will not dry out. All have a life cycle with two different generations. One generation reproduces sexually when egg and sperm unite. The other generation reproduces asexually by means of spores.

Review

1. In what kinds of places do nonvascular plants grow?
2. What is the difference between the two generations of a moss plant?
3. What are two other kinds of nonvascular plants?
4. **Critical Thinking** Mosses can live only in moist places because they do not have a good transport system for water. What is another reason for their living in moist places?
5. **Test Prep** Which of the following are nonvascular plants?
   - A bacteria
   - B fungi
   - C liverworts
   - D elm trees

LINKS

MATH LINK

Write a Ratio A liverwort may reach a height of 2 cm. A redwood tree may reach a height of 100 m. What is the ratio of the height of the liverwort to that of the redwood tree?

WRITING LINK

Informative Writing—Explanation Suppose you live near a volcano that has not erupted in a while. One of the signs that an eruption might occur is that the volcano releases polluting gases. One day you observe that many mosses are dying. You are not aware of any increase in local air pollution because of humans. Write a short letter to a newspaper to explain why you think the nearby volcano may erupt. Use the death of mosses as evidence.

SOCIAL STUDIES LINK

Mosses in History During World War I, soldiers used sphagnum (SFAG•nuhm) moss to cover wounds. Some Native Americans used it for babies' diapers. Today it's used for growing plants. Find out how other mosses have been used in the past and how they are used today.

TECHNOLOGY LINK

GO ONLINE

Learn more about mosses and other plants by visiting this Internet Site.
www.scilinks.org/harcourt

SCiLINKS

B31

LINKS TIPS

Links Objective Extend understanding of mosses.

Math Link Point out that the question uses two different measurements: centimeters and meters. Students should convert the meters to centimeters before they calculate the ratio. 100 m × 100 cm = 10,000 cm; 2:10,000

Writing Link As a prewriting activity, review with students the appropriate form for a letter to the newspaper. Writing models are provided in the **Teaching Resources** book.

3 Wrap Up and Assess

SUMMARY

Graphic Organizer for Chapter Concepts After reading the summary, have students complete the Lesson 1 section of the graphic organizer (shown on page B28).

REVIEW

1. moist, shady locations
2. One generation reproduces sexually when egg and sperm unite. The other generation reproduces asexually by spores.
3. liverworts and hornworts
4. The sperm needs water to swim to the egg.
5. C, liverworts

TEST PREP Test-Taking Tips

Item type multiple choice with a question stem

To help students answer Question 5, share the following strategy with them. **Read the question carefully and find key words, such as *nonvascular plants,* that are clues to the correct answer. Try to recall some examples of nonvascular plants to narrow the choices. Then select the best answer.**

LESSON CONCEPT REVIEW/ASSESSMENT

Students can use **Workbook p. WB50** to review the lesson concepts.

Language Arts Link

Naming Plants Hundreds of years ago, people who observed plants closely gave them highly descriptive names. We still use many of these names today. The names of many wildflowers end in *wort,* which means "plant." Birthwort was believed to help women during childbirth, ragwort has raggedy leaves, and soapwort produces a cleansing foam. The names of many supposedly poisonous plants end in *bane,* meaning "murder." Dogbane, cowbane, and fleabane were thought to be poisonous to those animals. Scientists have not always agreed with the claims made for these plants, but their names are a charming window into the past.

LESSON 2

What Are Ferns and Gymnosperms?

Objectives

- **Describe the life cycles of ferns and gymnosperms.**
- **Explain adaptations for survival in ferns and gymnosperms.**

Motivate

More Fast Facts

Spark interest with these facts and questions.

- The African n'tumbo tree, whose African name means "onion of the desert," gets almost all its moisture from sea-fog dew. Some specimens that have been found are more than 500 years old.

  Why do you think this plant is able to survive so long in such a harsh climate? It is well adapted to desert conditions. (See its picture on page B36, scientific name *Welwitschia mirabilis.*)

- The Carboniferous Period, which took place between 345 and 280 million years ago, is called the Age of Coal because that is when many of the world's largest coal deposits were formed.

  From what kinds of materials did coal form? from the remains of plants, mostly ferns

Access Prior Knowledge

- In what kind of environment would you expect to find ferns? pine trees?
- How do you think these two kinds of plants reproduce?

When Minutes Count . . .

Whole-Class Investigate Option Set up the investigation in the science center. Ask for small groups of students to report on their observations.

Using the Activity Video The video can be used to preview and model the investigate process with students. After viewing the video, students can conduct the investigation in small groups.

LESSON 2

What Are Ferns and Gymnosperms?

In this lesson, you can . . .

- INVESTIGATE how ferns differ from conifers.
- LEARN ABOUT the structures of ferns and gymnosperms and how they reproduce.
- LINK to math, writing, social studies, and technology.

◀ Topiary (TOHP•ee•air•ee) is the art of forming living plants into ornamental shapes. Gymnosperms are often used in topiary.

B32

INVESTIGATE

Comparing Ferns and Conifers

Activity Purpose You may have seen ferns before. You can tell them from most other plants by their large, feathery leaves, or *fronds.* Ferns often grow on the forest floor and in shady parts of gardens. You may also know about conifers—trees and bushes that have cones and needlelike leaves. Pine trees and fir trees are examples of conifers. In this investigation you'll **observe** various parts of ferns and conifers. You'll find out how they are alike and how they are different.

Materials

- fern frond
- conifer needles
- hand lens
- conifer cone
- safety goggles

Activity Procedure

CAUTION

1. **Observe** the fern frond and a conifer needle. Both are leaves with adaptations that help the plants survive.
2. **Compare** the leaves, and **record** your observations by drawing the leaves and by describing what you see. Use a hand lens to **observe** the underside of the fern frond and the structure of the conifer needle. (Picture A)
3. Now **observe** and **compare** the conifer cone and the photograph of a conifer cone on page B33. They both came from the same kind of tree. **Record** your observations.

BACKGROUND

Ferns and Gymnosperms are both vascular plants with well-developed true leaves and roots. Both have systems for carrying fluids to all parts of the plant. Both are nonflowering plants.

Like mosses, ferns alternate between sexual and asexual reproductive generations. A plant at the sexual stage of reproduction looks completely different from that same plant at the asexual stage.

Webliography GO ONLINE

Keyword plants

www.harcourtschool.com

For student self-assessment, see Assessment Guide p. AGxvi.

Picture A

Picture B

4 **CAUTION** **Put on your safety goggles.** Carefully break open the cone, and **observe** it. **Record** your observations. (Picture B)

▲ Conifer cone

Draw Conclusions

1. Are ferns vascular plants? Are conifers? How do you know?
2. Ferns can grow in shady places, and conifers can survive periods when there is little water. **Infer** the leaf adaptations that help the plants survive these conditions.
3. What did you **observe** on the underside of the fern frond? What might you **infer** about the structures you saw there?
4. **Scientists at Work** Scientists **compare** objects to find out how they are alike and how they are different. Compare the cone you **observed** to the photograph of the conifer cone. What differences do you observe between the two cones? What can you **infer** from your observations?

Investigate Further The seeds of conifers are adapted so that they can travel. Examine a conifer seed. **Formulate a hypothesis** about how a seed is adapted to travel. **Design an experiment** to test your hypothesis.

Process Skill Tip

If you **compare** the structures of different objects, you can often **infer** how they differ in what they do.

B33

INVESTIGATE

Time 30 min **Preparation Tips** See p. B24c.

Grouping individuals

Process Skill Tip compare, infer

Other Process Skills observe, record data, hypothesize, plan a simple investigation

Expected Results Students should conclude that ferns and conifers are alike in that they are both vascular plants but different in their adaptations for survival.

Teaching Tips

Discuss the idea that students can make logical inferences only after they have carefully compared and observed the fronds, needles, and cones.

Draw Conclusions

1. Yes; yes; both have veins in their leaves.
2. The leaves of ferns are broad and spread out, which allows them to collect more light; conifer needles have a small surface area and a waxy covering that helps keep water inside.
3. Students should point out that they observed small brownish spots. They may infer that these have something to do with the plant's reproduction.
4. The larger cones contain seeds, and the smaller cones contain pollen. Students may infer that these are reproductive adaptations.

Investigate Further Guide students to formulate hypotheses about how adaptations enable seeds to travel before they plan their investigation. Encourage them to use their hypothesis to help design a simple investigation to prove their hypothesis.

Process Skill Tip

Compare and Infer Show the Process Skill Tip Transparency on an overhead projector. Have students compare the maple and the spruce seeds. Ask students to infer how each of these seeds travels.

Process Skill Tip Transparency B2-2

Compare and Infer

- **Compare the two seeds. How are they alike? How are they different?** Students should notice that the maple seeds are paired and the spruce seed is single. Both have winglike structures.
- **From their structure, what can you infer about how these seeds are dispersed?** They are dispersed by wind.
- **How might their movements be different? Why do you think so?** The maple seed twists or rotates through the air like a propeller, and the spruce seed floats on the wind.

INVESTIGATE LOG

Students can use **Workbook pp. WB51–52** to record the results of the investigation.

PROCESS SKILLS PRACTICE

To practice and apply process skills, see **Workbook p. WB53.**

LEARN ABOUT

1 Before Reading

SET A PURPOSE

FIND OUT Remind students to use these topics to help them focus on what to look for as they read.

VOCABULARY If students are completing the Vocabulary Preview activity described on page B25, remind them to add the definitions for these words to their chart.

Graphic Organizer for Chapter Concepts Display the transparency (shown below) or direct students to use Workbook page WB45. Point out that by the end of the lesson, students should be able to fill in information about ferns and gymnosperms on the transparency.

2 Guide the Learning

SCIENCE IDEAS

Help students focus on supporting details for this **main idea: Ferns are well adapted for survival.**

- **What adaptations have allowed ferns to thrive for more than 400 million years?** Large surface area lets them catch a lot of light; alternation of sexual and asexual generations ensures the continuation of the species; spores can remain dormant until conditions are favorable for growth; ferns can grow in many different climates; in some, tropisms for touch allow them to cling to trees and to collect water and nutrients.
- **What are sori? How do they help identify a fern?** Sori are clusters of spores found on the undersides of fern fronds. The sori on each type of fern are arranged in a distinctive pattern.
- **What is the prothallium? What is its function?** The prothallium grows from the fern spore. It is the sexual generation of the fern. It produces egg and sperm cells.

Ongoing Assessment

Use the ✓ questions in the pupil edition.

LEARN ABOUT

Ferns and Gymnosperms

FIND OUT

- the structures and life cycle of ferns and gymnosperms
- how conifers are adapted to their habitats
- the uses of conifers

VOCABULARY

fern
gymnosperm
conifer
pollen

Ferns

Ferns are vascular plants that first appeared on Earth about 400 million years ago. Like other very ancient groups of vascular plants, ferns reproduce without seeds. Today most fern species live in moist, shady locations. Their feathery fronds spread out and have a large surface area. Because of their shape, fronds can collect a lot of light, so ferns are able to live on the dimly lit forest floor.

As with mosses, the life cycle of a fern includes two different generations. The asexual generation reproduces by spores that grow on the undersides of some of its fronds. The spores are inside brownish structures called *sori*. On different ferns the sori have different shapes and are arranged in different patterns. The sori may be round or shaped like rods or crescents. They are arranged either on the edges of the fronds or along the middle. The spores in the sori are released when the spores are mature.

Fern spores do not dry out easily. They can survive for a long time before they germinate. The spores are often blown to new locations by the wind.

When a fern spore falls to the ground, it doesn't begin to grow fronds right away. Instead, the spore grows into a tiny heart-shaped plant called a *prothallium*. The prothallium does not look like a fern frond at all. If you observe closely, you may find prothallia growing in

Spores are produced in the brown sori on the undersides of some fern fronds. ▼

◀ The heart-shaped prothallium (proh•THAL•ee•uhm) is the sexual generation in the fern's life cycle. It produces sperm and egg cells. Prothallia don't have vascular tissues.

◀ When fern fronds break through the ground, they are tightly coiled. They are called fiddleheads because they look like the curl at the top of a violin. As they grow, the fiddleheads unroll and spread out.

B34

Graphic Organizer for Chapter Concepts

Transparency B2 • Workbook, p. WB45

Types of Plants

	LESSON 1	LESSON 2		LESSONS 3 & 4
	Mosses, liverworts, & hornworts	Ferns	Gymnosperms	Angiosperms
Water transport	nonvascular	vascular	vascular	
Plant size	small	varies	varies	
Type of reproduction	sexual and asexual	sexual and asexual	mainly sexual	
Means of sexual reproduction	water	water	wind	
New plant form	spore	spore	seed	
How seeds or spores are dispersed	gravity, wind	gravity, wind	gravity, wind	

▲ Staghorn ferns can do well indoors in conditions like those of a rain forest. They need warm, moist air, lots of water, and some light, but not direct sunlight.

▲ Tree ferns are most common in damp rain forests in mountainous regions in the tropics. *Cyathea australis* is found on Norfolk Island in the South Pacific. It can reach a height of 28 m (about 92 ft).

damp, shady places where there are ferns. Prothallia are often less than 1 cm (about 0.4 in.) across.

Underneath the prothallium little structures grow that produce sperm and eggs. This is the sexual generation of the fern plant. The sperm must swim through water to the egg cell. When a sperm cell and an egg cell join, a new fern plant with fronds begins to grow. This is the beginning of the other generation of the fern's life cycle.

There are more than 12,000 species of ferns in the world today. Most of the species grow in tropical rain forests, but many live in cooler areas. They are quite different in size and shape. Some are only 2.5 cm (about 1 in.) in length. Others look like trees and can grow to be about 28 m (92 ft) tall!

On ferns that grow in temperate climates, the structures that look like stems are actually parts of the fronds. The stems lie on or under the surface of the soil. On a few ferns, such as the large tree ferns of the tropics, the stem is upright and forms a trunk. Tree ferns like these are the tallest kinds of ferns.

Some ferns have tropisms for touch. For example, staghorn ferns, which grow in tropical rain forests, cling to trees. A lower frond of the staghorn fern is like a suction cup that holds the plant to a tree's bark. This frond clings tightly to the tree except at the top of the frond, where it opens out. Nutrients that fall into the opened part of this frond supply the fern with materials it needs to make food. The plant's other fronds look like a male deer's (a stag's) horns, from which the plant gets its name.

✔ **What structures are involved in the asexual generation of a fern?**
sori and spores

B35

VISUAL LEARNING

Have students study the pictures of the life cycle of the fern on page B34.

- ▶ **Describe the asexual stage of the fern's life cycle.** The fern produces spores inside the sori, which are located on the underside of the fronds.
- ▶ **Describe the sexual generation of the life cycle.** The spore sprouts into a heart-shaped prothallium that produces eggs and sperm. The sperm cell joins with an egg cell, and the fertilized egg grows into a fern plant.

DEVELOP SCIENCE VOCABULARY

fern, p. B34—Have a volunteer draw a large feather on the blackboard. Tell students that the word *fern* comes from the Anglo-Saxon word *fearn,* meaning "feather" or "plume." Ask students to compare the shape of a feather with that of a fern frond. How are the two similar? Both have a midrib and small structures branching out on either side.

Reading Mini-Lesson

Use Prefixes and Suffixes to Determine Word Meanings Point out the prefix *pro-* in the word *prothallium.* Display the transparency and have students define these words that contain this prefix: *proceed, produce,* and *progress.* Ask students to think about what the prefix *pro-* means and how it contributes to the meaning of *prothallium.*

Reading Mini-Lesson Transparency B2-2

Use Prefixes and Suffixes to Determine Word Meanings

Word	Definition
Proceed	to come before or continue
Produce	to make or present
Progress	to move forward or toward a goal

Pro- means...
before, forward

READING SKILLS PRACTICE

To practice and apply this reading skill, see **Workbook p. WB54.**

Cultural Connection

The Victorian Fern Fad During the reign of Queen Victoria of England (1837–1901), ferns were in vogue. Their graceful fronds were used as designs on carpets, fabric, and silverware, and the passion for live specimens was so intense that collectors stripped the countryside of plants. But ferns could not tolerate the polluted, sooty air in London gardens, and they quickly died.

Nathaniel Ward, a London physician, discovered that ferns flourished when placed in a sealed glass case and protected from polluted air. The fern case, later called a Wardian case in the doctor's honor, became the new craze. Eventually, Londoners added mosses, rocks, and even small reptiles to the Wardian cases. We now use simpler versions, called **terrariums**.

Guide the Learning continued

SCIENCE IDEAS

Help students to focus on supporting details for this **main idea: Gymnosperms are vascular plants that produce seeds in cones.**

- **Why is it an advantage for a plant to be able to produce seeds?** Seeds contain the two things that a plant needs to begin life: an embryo and a food supply.
- **Describe the seed of a gymnosperm.** It has no outer container. Plants that produce flowers, by contrast, surround their seeds with an outer container called an ovary.

VISUAL LEARNING

Have students examine the pictures of the three examples of gymnosperms on page B36 and the picture of the blue spruce tree on page B37.

- **What are the four groups of gymnosperms?** conifers (the largest group), ginkgoes, cycads, and welwitschia

Critical Thinking Lead students to draw a conclusion about the diversity of gymnosperms.

- **Observe the pictures of the four examples. What do you notice about sizes, shapes, and leaf structures of gymnosperms?** They are very different from one another.
- **What do you notice about their environments?** All are different—desert, tropics, city street.
- **What can you conclude about the diversity of gymnosperms?** They are a diverse group that differ in appearance and in their adaptations to different environmental conditions.

Gymnosperms

Nonvascular plants reproduce by spores. Ferns, which are vascular plants, also reproduce by spores. However, the most common plants on Earth are those that reproduce by seeds.

Seeds might be called survival packages for new plants. Inside a seed is a tiny plant, called an embryo, as well as food for the embryo. An outer coating protects the seed. Making seeds is a great advantage for plants that live on land. The seeds can stay as they are for a long time, until conditions are good for growth.

There are two groups of plants that have seeds. A **gymnosperm** (JIM•noh•sperm) is one kind of plant with seeds. This word comes from ancient Greek words meaning "naked" and "seed." Gymnosperm seeds are said to be naked because they are not surrounded by a "container." The other group of plants with seeds are flowering plants. These plants have seeds that are surrounded by a container called an *ovary*. Gymnosperms do not produce flowers and ovaries, so they are called nonflowering plants.

One kind of gymnosperm is a **conifer**. *Conifer* means "cone bearer." These plants grow cones like the ones you observed in the investigation in this lesson. The cones are the reproductive structures that produce seeds. Most trees with needlelike leaves that stay on the tree all year are conifers.

Conifers are the largest of the four groups of gymnosperms. The other groups have far fewer

◄ ***Ginkgo biloba*** **is the only remaining species of the ginkgoes. Ginkgoes are sometimes called living fossils because they are very similar to plants that lived millions of years ago. A ginkgo's leaves are unusual—they grow on the ends of shoots, are fan-shaped, and have parallel veins. Ginkgoes are often planted along city streets because they can survive disease, insects, and pollution.**

◄ **Cycads grow in warm areas. They look like short palm trees and are sometimes called sago palms. But they are not really palms at all. The seeds are in cones at the top of the stem.**

Welwitschia mirabilis **grows in the deserts of southern Africa. Many of these plants live for more than 100 years. This gymnosperm produces only two leaves, although there seem to be more because the leaves split as they grow. Every year, stiff, stemlike growths develop where the leaves join the trunk. The growths bear small male cones and bright-red female cones.** ▼

B36

BACKGROUND

Webliography

Keyword plants

www.harcourtschool.com

GO ONLINE

Cones Conifers have distinctly different male and female cones. Male cones are generally smaller and less woody than female cones. After male cones release their pollen, they shrivel up and die.

In many conifers, both male and female cones are present on the same plant. They are called **monoecious**. In other conifers, the individual plants are either male or female, and fertilization requires one plant of each sex. These conifers are called **dioecious**.

Not all conifers have cones. Junipers and yews, for example, bear their seeds in small, berrylike structures.

Life Cycle of a Gymnosperm

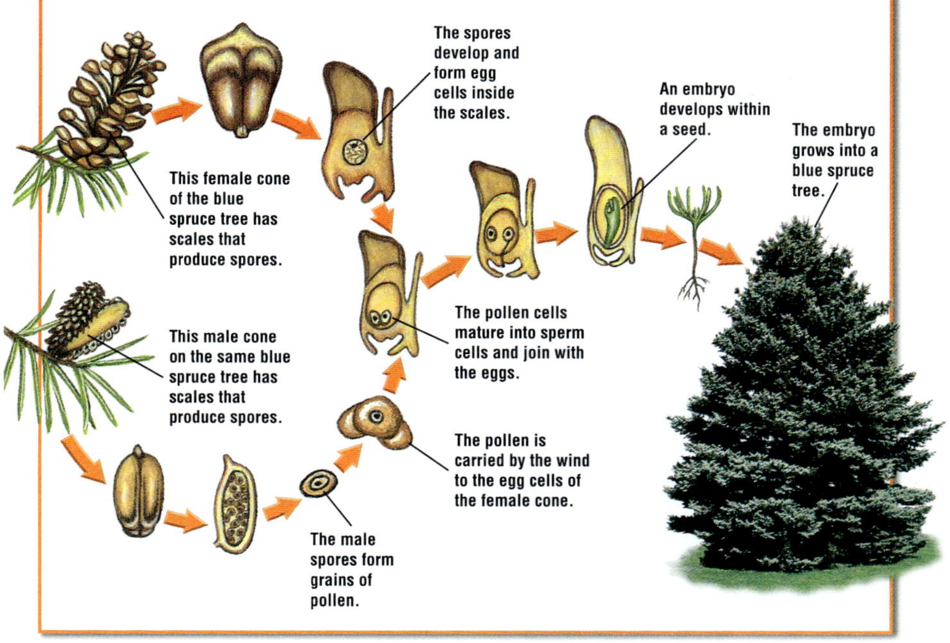

species than the conifers. Those groups are ginkgoes (GING•kohz), cycads (SY•kadz), and welwitschia (wel•WICH•ee•uh). They are shown in the photographs on page B36.

✔ **How do gymnosperms differ from ferns?** Gymnosperms reproduce by seeds; ferns reproduce by spores as well as by the union of egg and sperm.

Life Cycle of Conifers

Conifers have two kinds of cones—small male cones and large female cones. Both kinds form spores. Spores from the male cone develop into **pollen**. Each grain of pollen has two male sex cells. Spores from the female cone develop into egg cells. The egg cells stay inside the cone.

When the egg cells in a female cone are mature, the cone forms a sticky liquid. Grains of pollen, carried on the wind, are trapped in this liquid. The pollen grains mature into sperm cells and fertilize the female egg cells. Unlike the sperm cells of nonvascular plants or ferns, the sperm cells of conifers do not need water to travel to the egg cells.

The fertilized egg in the female cone slowly forms an embryo within a seed. When the seed is mature, the female cone opens, releasing the seeds. Conifer seeds often have "wings" that help them travel on the wind. When conditions are right, the seeds germinate and a new plant starts to grow.

✔ **How do the two kinds of cones on a conifer differ?** Male cones produce pollen grains, which are blown to the female cones. The female cones produce egg cells that become the seeds.

B37

USE PROCESS SKILLS

Compare Have students study the life cycle of the blue spruce tree on page B37 and compare it with that of the fern on page B34.

- **What major differences are there between the two life cycles?** Ferns produce spores and have two different generations (sexual and asexual). Conifers produce seeds and have only one generation (sexual).
- **In the two types of plants, how do the sperm reach the egg?** In ferns, the sperm must swim to the egg. In conifers, the sperm develops in the pollen grains in the male cone and is carried by the wind to the egg in a female cone.

DEVELOP SCIENCE VOCABULARY

gymnosperm, p. B36—Explain to students that the ancient Greek word *gymnos* means "naked." Ask students to describe a "naked" seed as opposed to one that is covered. They may compare the conifer seed with a fruit seed enclosed in a protective container, such as an apple.

conifer, p. B36—Have students research the derivation of the word. They will find it comes from two Latin words: *conus* meaning "cone," and *ferre* meaning "to bear" or "carry."

pollen, p. B37—Although microscopic in size, particles of pollen are released in great clouds, and for those who are sensitive, can cause discomfort. Ask students to describe any allergies they know about that are triggered by pollen.

ESL ACTIVITY

Suggest that students work in pairs to decide what kinds of **gymnosperms** — **conifers**, ginkgoes, cycads, or welwitchia—are found near where they live. Encourage them to identify particular species found in their region. Students with cultural connections to another country or region might also find out and tell others what gymnosperms are typical of that region.

Guide the Learning continued

SCIENCE IDEAS

Help students focus on supporting details for this **main idea: Conifers are well adapted for survival.**

- **What adaptations make conifers well suited to live in cold regions?** Their seeds can survive low temperatures and germinate only when conditions are right for growth; chemicals in the wood act like antifreeze to prevent water in the tree from freezing; sloping, slippery branches let heavy snow slide off; branches are flexible and bend under the weight of snow.
- **Why are needles better than broad leaves in a dry climate?** Needles have a smaller surface area, so less water evaporates from them. Needles are also covered with a thick, waxy layer that retains water.

VISUAL LEARNING

Have students examine the pictures of the different conifers on page B38 and pay special attention to the needles and seed-bearing structures.

- **If you saw one of these plants outdoors, what features would help you identify it? How would you know what kind of conifer it was?** The presence of needles and cones would identify two of the plants as conifers. (The exception is the yew, which has unusual, waxy, berrylike cones.) To identify the species, you would need to observe the shape and size of the whole plant, of the cones, and of the needles. The needles are an especially good visual clue because each species bears needles of a particular length and shape, arranged in distinctive patterns.

▲ The leaves of most conifers are very thin and pointed. They are coated with a thick, waxy covering that helps the plant keep its water.

White spruce trees have short leaves and long cones. ►

Pitch pines have needles in bunches of twos and threes and long, rounded cones. The seeds have "wings" that can be carried by the wind. ►

Yews can be low shrubs or small trees. Their seeds, unlike those of most other conifers, grow in small cups that are open on one end. These cups look like berries, and they are highly poisonous to humans. ►

Adaptations

Because of their adaptations, conifers can live in dry environments and in places that are sometimes freezing cold. Their seeds can survive long dry or cold periods and then can germinate and grow when conditions are better.

Many people call conifers evergreens. This is because most conifers have narrow, hard, green leaves, most of which stay on the tree all year. Pines and firs have needlelike leaves. Cedars have scalelike leaves. The small surface area of conifer leaves and their thick, waxy coating help prevent loss of water. These adaptations help the tree survive when water is not available.

Conifers that grow in cold climates also produce a chemical that prevents the water in the tree from freezing. This is like adding antifreeze to the water so that it can flow even when the temperatures are below freezing.

Many conifers are also adapted to heavy snow. The snow falls off their sloping branches and smooth, flexible leaves. As a result, the branches don't break under the weight of the snow.

✓ **How are conifers adapted to dry conditions?** The seeds are protected so they can survive dry conditions; the leaves have a small surface area and a waxy covering, which prevent water loss.

B38

ESL ACTIVITY

Have small groups of students work together to review the concept of adaptation. Have them point to different parts of the plants in the pictures on page B38 and explain how each part functions to help the conifer live in dry environments or freezing cold places.

Sequoia trees include redwoods, the tallest trees in the world. Unlike some other conifers, sequoias require frequent rain. They grow in a narrow band along the west coast of the United States. Giant sequoias grow in small groups scattered throughout the western slopes of the Sierra Nevada of central California. Some reach heights of more than 110 m (about 360 ft).

Pitch pines are the main species in the Pine Barrens of New Jersey. They have an unusual ability to produce new sprouts along their trunk following fires, which can be common in these dry areas.

Bristlecone pines are bushy evergreens that are among the oldest living trees on Earth. Some bristlecone pines in the mountains of California and Nevada are more than 4500 years old. These may be the world's oldest living organisms.

Where Conifers Grow

Conifers grow all around the world, but especially in cold regions. They encircle the far north in a solid band of forest that stretches across North America, Scandinavia, and Siberia. These northern forests of conifers, called the taiga (TY•guh), have long, very cold winters with lots of snow. Conifers can also be found on the steepest slopes of high mountains in more southerly areas.

Conditions in the taiga make it ideal for conifers. However, these forests are not good for many other kinds of plants. The soil is not very rich, because the cold temperatures slow down the decay of materials. As a result, fewer nutrients enter the soil. Also, the shade from the conifers prevents many other vascular plants from growing on the forest floor.

Not all conifers grow in cold regions. Many can be found in warmer areas that have dry, sandy soil in which other trees can't grow. The pitch pines shown on this page, for example, grow in the Pine Barrens of New Jersey.

✓ **Where are most conifers found, and why?** in northern regions, because conifers are well adapted to areas that have freezing winters

B39

USE PROCESS SKILLS

Observe Ask students to study the map on page B39. Have them locate the area in which they live. Then ask the following questions.

- **Do sequoias grow where you live? Why or why not?** Answers will depend on where students live. Sequoias grow in a narrow band along the west coast of the United States. They also grow in small groups scattered throughout the western slopes of the Sierra Nevada of central California.
- **Do pitch pines grow where you live? Why or why not?** Answers will depend on where students live. Pitch pines grow in the Pine Barrens of New Jersey.
- **Do bristlecone pines grow where you live? Why or why not?** Answers will depend on where students live. Bristlecone pines grow in the mountains of California and Nevada.

BACKGROUND

Webliography GO ONLINE

Keyword plants

www.harcourtschool.com

The taiga is a subarctic coniferous forest region that begins where the tundra ends. Centuries ago, glaciers covered the taiga. They left behind depressions that filled with water, and the area is now a maze of lakes, swamps, and bogs. The glaciers also scraped away much of the topsoil, so the plant life in the taiga must survive in some of the world's poorest soil.

It is a harsh environment. Winters are very cold and summers are short. The average temperature remains below freezing for six months of the year. Although the upper levels of soil are not frozen all year, much of the taiga region rests on **permafrost**, a permanently frozen layer beneath the surface of Earth. The permafrost chills the active thawed layer of soil above it. Because water cannot penetrate the permafrost, the upper layer of soil becomes waterlogged.

Social Studies Link

Trees Have students **use reference sources** to check out the trees of the Pacific Northwest. Develop a dichotomous key to identify a mystery tree of your choice.

If students have not used a dichotomous key, they will need an introduction on how to use it. Explain that the key is really a series of paired descriptions. At each stage, they must decide which description better fits the mystery tree and then follow directions to move on to the next pair of descriptions.

Guide the Learning continued

ANALYZE AND INTERPRET DATA

Ask students to read and interpret the **graphic source** showing the uses of conifer wood on page B40.

- What accounts for the greatest use of conifer wood? lumber, 57 percent
- What is the second most common use? pulp products, including paper, 27 percent
- How much of the conifer wood is burned for fuel? 4 percent
- On the basis of your reading, what might be included in the "Other industrial products" category on the graph? alcohols, medicines, explosives

How People Use Conifers

The wood from conifers is usually much softer than wood from trees such as oak, maple, and hickory. It is also not as strong. Nevertheless, softwood, as the wood from conifers is called, is the most common wood used to build houses and other buildings in North America. In fact, so much conifer wood is used that the trees are farmed in many places. Two of the conifers most often used in building are the Douglas fir and the loblolly pine. These trees grow tall and straight. They grow fast, so they are often planted for lumber.

Conifer wood is also used for plywood, which is made by gluing together several thin layers of wood with their grains in opposite directions. Plywood is strong, stiff, and inexpensive to make. It is often used for cabinets and as parts of floors and walls.

Most of the paper made in North America is also made from conifer softwood. The wood is ground into a fine pulp, mixed with water, and then spread evenly over a screen. After the water drains off, the pulp is pressed into sheets of paper. Spruces are especially good for producing paper pulp because they have very strong, long fibers.

Chemists have developed many useful items from the chemicals in conifers. Some products, such as alcohols, are used in manufacturing, medicines, and explosives. Acetylene is used as a fuel. Another product made from conifers is the fabric called rayon. Rayon is used to make clothing. It's also used in automobile tires and even bowling balls.

Conifers are often planted in rows on the edges of fields to block the wind. Around homes they give year-round shade. Conifers such as hemlocks, yews, pines, junipers, and cedars are often planted because they are beautiful and stay green in cold climates all year.

This circle graph shows how conifer wood is used in the United States.

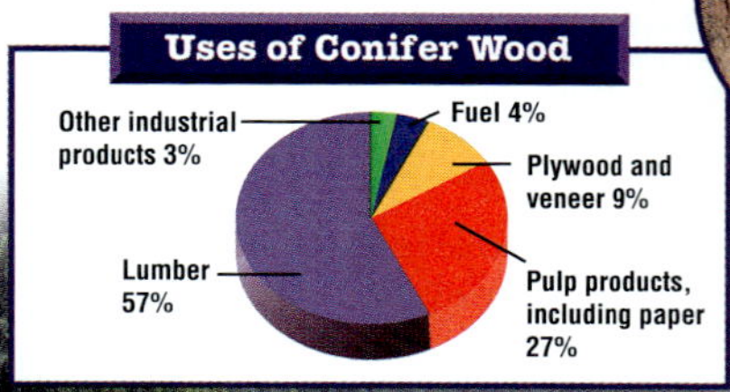

Piñon (PEEN•yahn) pines are small pine trees that grow in dry areas. Their small cones contain seeds called pine nuts, or pignoli (peen•YOHL•ee). These are good to eat and are used in foods such as pesto.

Spruce trees provide much of the material used to make the fabric known as rayon. Many different types of clothes are made of rayon or rayon blended with other materials.

B40

Math Link

Conifers Have students **use reference sources** to find information on conifers. Have them find the average heights of five types of conifers and then make a bar graph to compare the heights of the trees. Tell students that they may use a computer graphing program such as ***Data ToolKit*** to make their bar graphs.

INVESTIGATION CHALLENGE

Hands-On Activity: Use Pine Cones to Determine Humidity

Use this idea for a hands-on activity to extend students' understanding of key lesson concepts.

Pine cones are one of nature's weather indicators. In dry weather, the scales on a pine cone open; in damp weather, the scales absorb moisture, become soft and pliable, and close up.

Challenge students to demonstrate this natural occurrence. They might place a pine cone on an outdoor window ledge and observe it during wet and dry weather. Or they might dampen a pine cone, observe the results, and then place the cone in a warm, dry place and observe what happens as it dries.

For student self-assessment, see Assessment Guide p. AGxvii.

Conifer forests are peaceful, beautiful places to visit. Because these forests have little growth under the trees, they often have good hiking trails. Thousands of visitors enjoy America's majestic conifer forests every year.

✔ **What are some uses of conifers?**

Summary

Ferns are some of the simplest vascular plants. The large surface area of their fronds allows them to grow well in low light. Ferns produce spores on the undersides of some of their leaves. The sexual generation of a fern is very small. The plants usually identified as ferns are the much larger spore-producing generation.

Gymnosperms are nonflowering plants that produce seeds in cones. They have male cones that produce pollen and female cones that produce egg cells and then seeds. Conifers, the most common gymnosperms, have needlelike leaves and are well adapted for dry or freezing weather. Most are found in northern regions.

Review

1. Why do ferns need a damp environment to reproduce?
2. What adaptation enables ferns to grow well in environments with little light?
3. Why are conifers found mostly in northern regions?
4. **Critical Thinking** Conifers produce large amounts of pollen—many more grains than are used to fertilize the conifer egg cells. Why is this necessary?
5. **Test Prep** Which is NOT a major use of conifers?
   A lumber
   B food
   C chemicals
   D fibers

They are used as lumber for building; for making paper, chemicals, and rayon; and to shade homes.

LINKS

MATH LINK

Use Formulas The diameter of General Sherman, a giant sequoia in California, is about 10 m. What length of rope would you need to completely encircle the trunk of the tree? Remember that the perimeter of a circle is found by multiplying the diameter by pi. Use 3.14 for pi.

WRITING LINK

Informative Writing—How-to Papermaking is an important industry in the United States. Find out how paper is manufactured. Then write a short report for your classmates telling how paper is made from trees.

SOCIAL STUDIES LINK

How Native Americans Used Conifers Native Americans once used lodgepole pines for lodge and tepee supports. Find out about these and other uses of conifers among different Native American tribes. Write a paragraph to describe your findings.

TECHNOLOGY LINK

GO ONLINE

Visit the Harcourt Learning Site for related links, activities, and resources.
www.harcourtschool.com

Welcome to The Learning Site

B41

LINKS TIPS

Links Objective Extend understanding of ferns and gymnosperms.

Math Link Remind students that *pi*, or π, $\approx$ 3.14. 31.4m

Writing Link Challenge students to find how many tons of paper are used daily in the United States and what is being done to replace the trees required to meet this demand. Writing models are provided in the **Teaching Resources** book.

3 Wrap Up and Assess

SUMMARY

Graphic Organizer for Chapter Concepts After they read the summary, have students complete the Lesson 2 section of the graphic organizer (shown on page B34).

REVIEW

1. The sperm cells of a fern need water to reach the egg cells.
2. Fern fronds have a large surface area that enables them to take in much of the available light.
3. Their adaptations enable them to survive cold and dry weather.
4. Wind carries the pollen away, and much of it never reaches a female cone.
5. **B,** food

TEST PREP Test-Taking Tips

Item type multiple choice with a question stem using a negative such as *not*

To help students answer Question 5, share the following strategy with them. **Read the question carefully and identify the words that help to qualify the correct answer choice, such as "NOT a major use of conifers." Eliminate any choices that do not fit the question. Try to narrow possible answers to two. Then select the better one.**

LESSON CONCEPT REVIEW/ASSESSMENT

Students can use **Workbook p. WB55** to review the lesson concepts.

LESSON 3

How Are Angiosperms Different from Other Plants?

Objectives

- Recognize the features of angiosperms.
- Describe the structures of flowers and their functions in reproduction.
- List the ways in which pollination occurs.

Motivate

More Fast Facts

Spark interest with these facts and questions.

- The rafflesia found in Indonesia is the largest flower in the world. It grows up to 91 cm (about 36 in.) across. The petals are 3 cm (1 in.) thick and 46 cm (18 in.) long. The flowers weigh 7 kg (about 15 lb) and give off an acrid smell that attracts flies.

  What is the function of this flower? reproduction

- The bright colors of Himalayan balsam flowers attract bees. The bees land on a platform of petals and then crawl deep inside the flower, guided by markings on the sepals, to reach the nectar that is their food. They also pick up pollen that they carry to the next flower.

  How are bees important to the Himalayan balsam? The plants need bees to reproduce.

Access Prior Knowledge

- What is a flowering plant?
- How do flowering plants differ from other plants?

When Minutes Count . . .

Whole-Class Investigate Option Set up a class science center with the materials for the activity. Allow pairs of students time to make observations.

Using the Activity Video The video can be used to preview and model the investigate process with students. After viewing the video, students can conduct the investigation in small groups.

LESSON 3

How Are Angiosperms Different from Other Plants?

In this lesson, you can . . .

- INVESTIGATE the parts of a flower.
- LEARN ABOUT the functions of the angiosperms.
- LINK to math, writing, technology, and other areas.

Gladioluses (glad•ee•OH•luhs•uhz) have tall, slender stems with large, silky flowers and long, sword-shaped leaves. They reproduce in two ways. The flowers make seeds that develop into new plants. Underground stems also produce new shoots.

B42

INVESTIGATE

The Parts of a Flower

Activity Purpose When you think about flowers, you probably think about the shape and color of their petals and maybe their scent. The most noticeable part of most flowers is the petals, but there are other flower parts that are very important. In this investigation you will **observe** some parts of a flower.

Materials

- flower
- paper towel
- hand lens

Activity Procedure

1. Lay the flower on the paper towel.
2. Handle the flower gently while you **observe** its parts. Be sure to look at the center between the petals. (Picture A)
3. Now use the hand lens to examine the parts of the flower. Draw what you **observe**. (Picture B)
4. The flower includes at least three different structures. Are there at least three different structures in your sketch? If not, **observe** the flower again and revise your sketch.

BACKGROUND

Webliography
Keyword plants
www.harcourtschool.com
GO ONLINE

Flowers are often arranged in groups. The blossoms may be attached along a stalk, arranged in broad clusters, or packed together so tightly that they look like a single flower. The **puya**, called a living fossil because it is so ancient, is a tall, treelike angiosperm that grows high in the Andes. Its flowers form spikes of up to 8000 bright-green blossoms. It can grow to be about 9 m (30 ft) tall.

Unusual Pollinators The tiny Australian honey possum feeds on the pollen and nectar from desert flowers such as the banksia and, in the process, pollinates the plants. Some kinds of bats and rodents also pollinate flowers. Flies pollinate many flowers, most of which give off a bad odor. For example, the South American creeper produces a smell of rotting fish. When flies enter, they are trapped overnight. In the morning the flower fades, and the pollen-covered flies escape.

For student self-assessment, see Assessment Guide p. AGxvi.

Picture A

Picture B

Draw Conclusions

1. Describe the structures you **observed** in the flower.
2. What do you **infer** are the functions of these structures?
3. **Scientists at Work** Scientists **communicate** information in many ways. They use drawings, diagrams, words, and measurements. **Compare** these methods. Choose one of the flower structures you drew. Which method or combination of methods do you think would be the most helpful for communicating information about the structure? Explain.

Investigate Further Use your observations of the flowers in this activity to **hypothesize** what structures you would find in any flower. **Design an experiment** to test your hypothesis.

Process Skill Tip

Sometimes it is easier to **communicate** what you **observe** in drawings or diagrams rather than in words. Scientific drawings should be as detailed and as accurate as possible in order to clearly present what you observed.

B43

Process Skill Tip

Observe and Communicate
Display the Process Skill Tip Transparency on an overhead projector or draw a similar flower on the board. Have students think about the answer to the first question. Then brainstorm with the class a list of ways to communicate what they observed. Have students choose one method from the list and use it. Have them test how well they have communicated their observations by sharing their finished products with the class.

Process Skill Tip Transparency B2-3

Observe and Communicate

- What parts can you see on this flower?
- In what ways could you communicate what you observed? written, visual, or oral
- Choose one method from the list and communicate to your class what you observed.

INVESTIGATE

Time 20–25 min **Preparation Tips** See p. B24c.

Grouping pairs

Process Skill Tip observe, communicate

Other Process Skills infer, compare, hypothesize, plan and conduct simple investigations

Expected Results Students should conclude that a flower has clearly distinguishable reproductive parts.

Teaching Tips

Caution students to treat the flowers gently so that the parts do not get broken.

Draw Conclusions

1. Answers will vary, but students will probably describe petals, sepals, pistil, and stamens.
2. Students should infer that these structures help the plant reproduce.
3. Accept any valid, supported answer, such as diagrams with measurements or detailed life-sized drawings.

Investigate Further Guide students to formulate hypotheses about the structures of flowers before they plan their investigation. They should design and conduct a simple investigation to prove their hypothesis.

Informal Assessment

Classroom Observation As students work in pairs to complete the activity, observe how they interact. Note whether students are sharing tasks equally or whether one seems to take over. Encourage students to exercise patience with those who are slower at making observations.

INVESTIGATE LOG

Students can use **Workbook pp. WB56–57** to record the results of the investigation.

PROCESS SKILLS PRACTICE

To practice and apply process skills, see **Workbook p. WB58.**

LEARN ABOUT

1 Before Reading

SET THE PURPOSE

FIND OUT Remind students that these statements will guide them in what to look for as they read.

VOCABULARY If students are doing the activity described in the Vocabulary Preview on page B25, remind them to add the definitions for these words to the chart.

Graphic Organizer for Chapter Concepts Display the transparency (shown below) or have students use Workbook page WB45. Point out that by the end of the lesson, students should be able to fill in the information on the transparency about water transport and plant size in angiosperms.

2 Guide the Learning

SCIENCE IDEAS

Help students focus on the supporting details for this **main idea: Angiosperms are vascular plants that have flowers and produce fruit.**

- **How do angiosperms differ from gymnosperms?** The seeds of the angiosperm are contained in a structure known as a fruit.
- **What is the function of a flower?** It is the reproductive part of a plant and produces fruit.
- **How do the flowers of angiosperms vary?** Some are small and hard to see; others are large and brightly colored.
- **How do annuals and perennials differ?** An annual goes through its entire life cycle in less than a year, after which the entire plant dies. The underground parts of a perennial remain alive in winter, even though the part of the plant that is above ground dies.
- **What are two functions of fruits?** to protect and spread their seeds

Ongoing Assessment

Use the ✓ questions in the pupil edition.

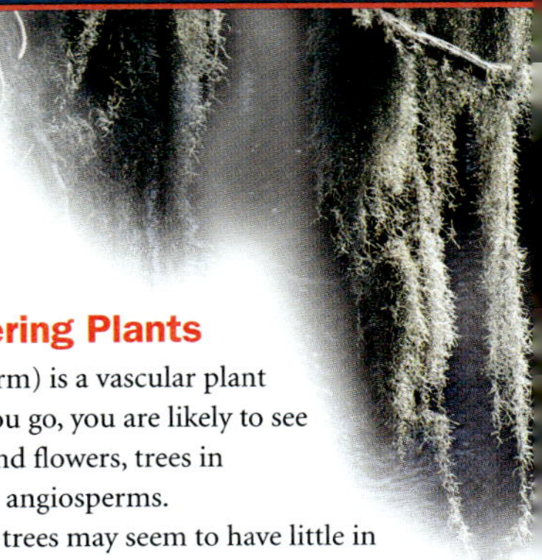

LEARN ABOUT

Angiosperms

FIND OUT

- the features of angiosperms
- the structures of flowers and their function in reproduction
- the ways in which pollination occurs

VOCABULARY

angiosperm
stamen
anther
pistil
stigma
petal
style
ovary
pollination

Angiosperms Are Flowering Plants

An **angiosperm** (AN•jee•oh•sperm) is a vascular plant that produces flowers. Wherever you go, you are likely to see angiosperms. Grasses, vegetables and flowers, trees in orchards, and crops in fields are all angiosperms.

Grasses, vegetables, flowers, and trees may seem to have little in common. But all angiosperms have basic similarities. Their name comes from Greek words that together mean "seeds in a vessel." Unlike gymnosperm seeds, which are not contained in a structure, the seeds of angiosperms are contained in fruit. Tomatoes, acorns, and bean pods are all fruit. So are the foods that we usually call fruit, such as apples, pears, and oranges. In biology, any structure that encloses seeds is a fruit.

Angiosperms produce fruit from their flowers, which are the reproductive structures of the plants. After the egg cells in a flower have been fertilized, the organ that surrounds the egg cells develops into a fruit that contains the plant's seeds. Many angiosperms produce flowers that are so small you might not notice them. The grass in a lawn, for example, produces tiny flowers whose petals are not colorful. The flowers on some trees, such as cherry trees, dogwoods, and magnolias, are brilliant and obvious. The flowers on many other trees, however, are hard to see. On angiosperms that we call "flowers"—roses, daffodils, and pansies, for example—the flower is usually the most noticeable part.

Angiosperms and gymnosperms also have different kinds of leaves. Most angiosperms have flat, wide leaves. Most gymnosperms have needlelike leaves.

The variety of sizes and shapes of angiosperm leaves is huge. Some leaves measure 1 m (about 3 ft) from stalk to tip. Others, like the leaves of the duckweed that floats on ponds, are tiny. Whatever their shape, most angiosperm leaves have a network of veins through which food and water are transported.

◄ Many of the foods that people eat are produced by angiosperms. We eat angiosperm roots (carrots and parsnips), stems (celery), leaves (lettuce and cabbage), flower buds (broccoli), seeds (peas), and fruit (tomatoes and apples).

B44

Graphic Organizer for Chapter Concepts

Transparency B2 • Workbook, p. WB45

Types of Plants

	LESSON 1	LESSON 2		LESSONS 3 & 4
	Mosses, liverworts, & hornworts	Ferns	Gymnosperms	Angiosperms
Water transport	nonvascular	vascular	vascular	vascular
Plant size	small	varies	varies	varies
Type of reproduction	sexual and asexual	sexual and asexual	mainly sexual	
Means of sexual reproduction	water	water	wind	
New plant form	spore	spore	seed	
How seeds or spores are dispersed	gravity, wind	gravity, wind	gravity, wind	

Spanish moss isn't a moss at all. It's an angiosperm that produces tiny flowers.

In angiosperms the xylem and phloem that transport materials are made up of highly specialized cells. The cells function like tubes. They transport materials better than the xylem and phloem in ferns and gymnosperms.

The adaptations that allow angiosperms to survive from year to year are different from those of gymnosperms. For example, *annual* plants are angiosperms that go through their entire life cycle, from seed to flower to seed, in less than a year. Annuals usually have soft stems. Although annuals die when winter comes, they have produced seeds that will sprout the following year.

Other soft-stemmed angiosperms, including many spring flowers, do not die in winter, although they appear to. The parts of the plant above the ground wither and die. Underground, the plants are alive as bulbs or fleshy roots. When the warm weather returns, they send up shoots and they flower again. These angiosperms are called *perennials* (puhr•EN•ee•uhlz).

Many angiosperms have woody stems. These plants include broad-leaved trees such as oaks, maples, and beeches, as well as shrubs such as roses. They are perennials, but they do not die back to the ground. Their tough stems protect both their above-ground growth and their roots. As colder weather approaches, most of these plants drop their leaves. The plants become dormant during the winter. They grow new shoots and leaves when warm weather returns.

With about 235,000 known species, angiosperms are the most successful plants on Earth. They have both the greatest number of species and the greatest number of individual plants. The adaptations of their xylem and phloem make their transport systems the most efficient of any plants. They are well adapted for many kinds of environments. The fruit they produce not only protect the seeds but also help spread them. The fruits attract animals, which eat them. The seeds pass through the animals' digestive systems. The animals deposit the seeds far from the parent plants. All of these adaptations have resulted in the success of angiosperms.

✔ **What are the two main characteristics of all angiosperms?** They produce flowers and make fruits.

The genus *Magnolia* includes trees and shrubs that have large, sweet-smelling flowers.

This barrel cactus makes food and stores water in its stem. Its leaves have adapted by becoming spines, which don't perform photosynthesis. The cactus flowers for a short time after a brief rainy season.

B45

USE PROCESS SKILLS

Communicate Challenge students to communicate either visually or orally what they have learned from reading pages B44–45. Remind students that it is important that they organize their information in order to communicate effectively. Use the following questions to help students **summarize and paraphrase the selection.** Encourage students to describe supporting details for each subtopic. Then challenge them to communicate the information they have outlined.

- What is the main topic covered on these pages? the characteristics of angiosperms
- What subtopics are covered? angiosperm fruits, angiosperm flowers, angiosperm leaves, transport systems, adaptations to survive from year to year

DEVELOP SCIENCE VOCABULARY

angiosperm, p. B44—Ask students to divide this word into its prefix and root. Have them look up the prefix *angio-* in the dictionary and find its definition. the ripened and variously modified walls of a plant ovary Then have students give the dictionary definition of *angiosperm.* any of a class of vascular plants having the seeds in a closed vessel

BACKGROUND

Webliography GO ONLINE

Keyword plants

www.harcourtschool.com

Hibiscus is a genus of plants of the warm temperate, subtropical and tropical areas of the Northern Hemisphere. They can be small trees, shrubs, or herbaceous plants. Some hibiscus plants grow in the wild; others are cultivated. The hibiscus is the state flower of Hawai'i. Three species are native to the Islands and 33 species have been imported. Horticulturists have crossed these species to produce more than 5000 varieties. The most common species is a red-flowered hibiscus from India that produces a purple dye. Hummingbirds pollinate most hibiscuses.

Waterlily is the common name for a family of about 70 species of aquatic plants. Many have floating leaves. Waterlilies generally grow in quiet waters. Because the soil in which they are rooted is oxygen-poor, they obtain oxygen through air passages from the leaves. Amazon waterlily leaves can be 2 m (over 6 ft) across.

ESL ACTIVITY

Take students on a nature walk and encourage them to spot as many **angiosperms** as they can. Remind students that not all angiosperms produce decorative flowers. Point out that grasses, many trees, vegetables, and numerous other plants are also angiosperms.

Guide the Learning continued

SCIENCE IDEAS

Help students focus on the supporting details for this **main idea: A flower contains specialized structures that aid in reproduction.**

- **What is the flower's male reproductive structure called?** the stamen
- **What are the two parts of the stamen, and what do they do?** Anther and filament; the anther produces pollen. The filament holds the anther to the plant.
- **What is the flower's female reproductive structure called?** the pistil
- **What are the parts of the pistil, and what do they do?** Stigma, style, and ovary; the stigma is the tip of the pistil. It is sticky and traps pollen grains. The style connects the stigma to the ovary, in which the egg cells develop.

VISUAL LEARNING

Have students observe the illustrations in The Inside Story on pages B46–47 and read the captions.

- **What structures do you observe on the large flower? How many of each do you see?** five petals; five stamens; one pistil
- **What is the function of each of these structures? How does their shape, color, or size aid in their function?** Petals are bright and showy in order to attract the insects that pollinate the flowers. The stamens have large anthers, where pollen is produced. The pistil has a long style that raises the stigma so that it can easily collect pollen needed to fertilize the eggs in the ovary.

Flower Parts

You observed some of the parts of a flower in the investigation. You may recognize them in the diagram of a lily that has been cut to show its parts.

Clustered around the center of a flower are several stemlike structures called stamens. Each **stamen** (STAY•muhn) is one of the flower's male reproductive organs. A stamen has two parts. The **anther** (AN•thuhr) produces pollen grains. The *filament* (FIL•uh•muhnt) connects the anther to the plant.

At the center of a flower is a structure called the **pistil** (PIS•tuhl). This is the female reproductive organ. It consists of three main parts. At its tip is the **stigma** (STIG•muh). The stigma

THE INSIDE STORY

The Anatomy of a Flower

The largest and most visible parts of a flower are usually its petals. Each **petal** helps protect the other parts of the flower. Petals are often colorful. The colors help attract the insects and other animals that are important to the plant's reproduction. Below the petals are *sepals* (SEE•puhlz). They are usually smaller than the petals and resemble green leaves. Before a flower blooms, the sepals completely enclose the bud. They protect the inner parts of the flower. After the flower blooms, the sepals spread apart and may drop off.

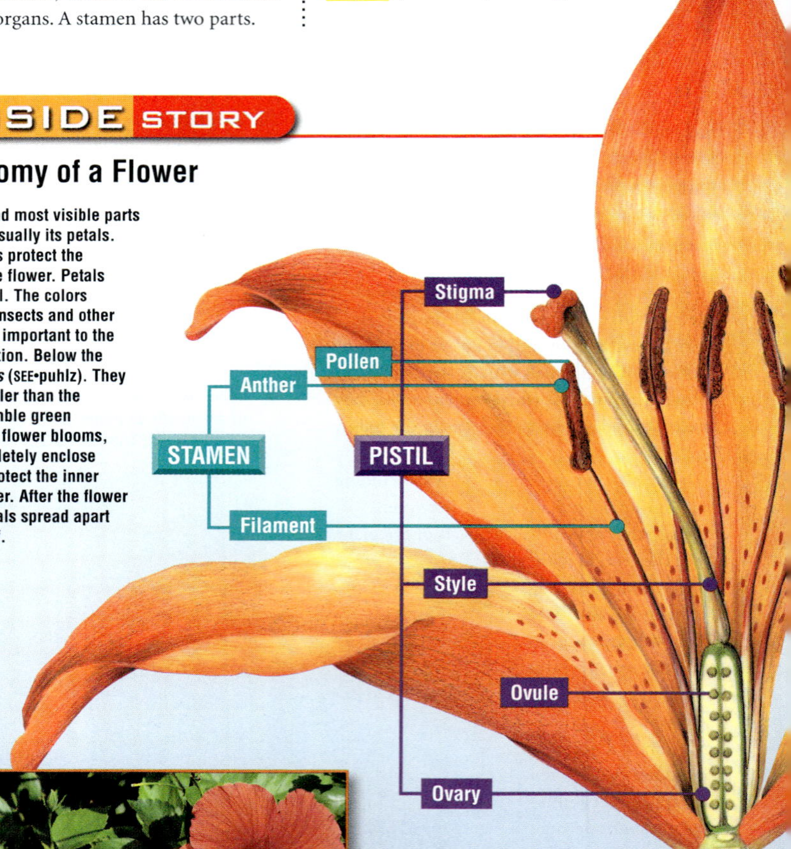

The large, bell-shaped flowers of hibiscus have five sepals and five large petals. The stamen and the pistil are fused together.

B46

Art Link

The Botanical Museum at Harvard University is home to a unique collection of more than 3000 colored-glass models of some 800 plant species. After heating and softening the glass, artists Leopold and Rudolph Blaschuka shaped the flowers by hand. Details were added using lacquer mixed with ground colored glass. The larger pieces were reinforced with wire. The men used sketches and live and preserved plants to ensure that their glass flowers were scientifically accurate. The Blaschuka collection is used in teaching.

Give students materials such as colored clay, colored paper, and pipe cleaners. Challenge them to make accurate model flowers.

is sticky, so when pollen grains land there, they stick.

Below the stigma is the stemlike part of the pistil called the **style**. It connects the stigma to the **ovary** below. Inside the ovary are *ovules* (AHV•yoolz), in which the flower's egg cells develop.

Pollination occurs when pollen from the anthers lands on the stigma. The pollen grains grow tubes that stretch through the style to the ovary. When the pollen tubes reach the ovary, sperm cells from the pollen grains enter the ovules. The fertilized egg cells develop into seeds.

✓ **What are the functions of a flower's pistil and stamens?** The pistil is the female reproductive organ, and the stamens are the male reproductive organs.

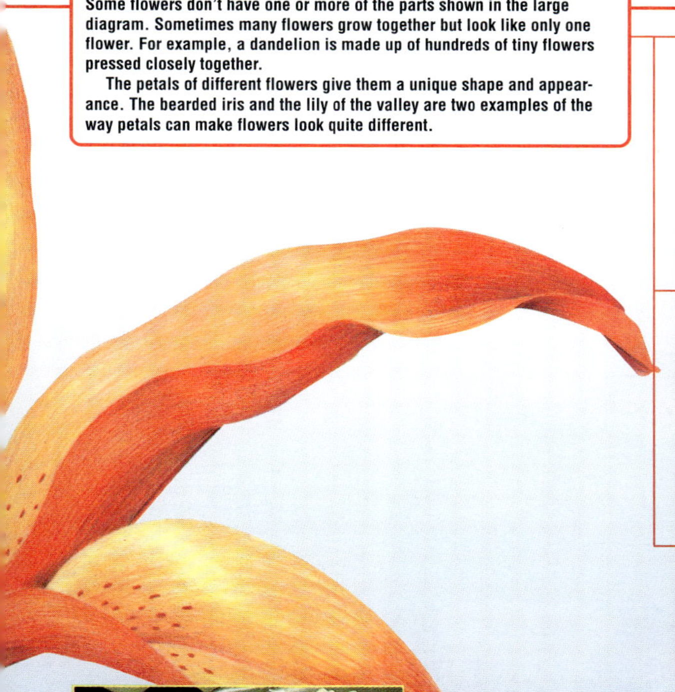

Angiosperms vary in the number, shape, and color of their flower parts. Some flowers don't have one or more of the parts shown in the large diagram. Sometimes many flowers grow together but look like only one flower. For example, a dandelion is made up of hundreds of tiny flowers pressed closely together.

The petals of different flowers give them a unique shape and appearance. The bearded iris and the lily of the valley are two examples of the way petals can make flowers look quite different.

Dandelion

Bearded iris

Lily of the valley

◀ The beautiful flowers of waterlilies have four sepals and many petals. In the center of a waterlily, a group of pistils is fused together in a saucerlike shape, surrounded by many stamens.

Critical Thinking Lead students to draw a conclusion about variations in flower structure.

▶ **What kinds of differences can be seen among flowers?** Number, size, shape, and color of sepals and petals vary; flowers may lack one or more flower parts; flowers may grow singly or in small or large groups.

Encourage students to describe variations they have seen among flowers in the classroom or in their neighborhoods.

DEVELOP SCIENCE VOCABULARY

stamen, **anther**, **pistil**, **stigma**, **petal**, **style**, and **ovary**, pp. B46–47—Point out that the anther is the functional part of the flower's male reproductive organ (the stamen), which is the site of the formation of pollen grains. The ovary is the functional part of the flower's female reproductive organ (the pistil), which is the site of the formation of the egg cells. The style is the female equivalent of the filament of the stamen. The stigma is the site where pollen grains land. Each petal provides protection for these parts and often color for attraction purposes in reproduction. Invite students to make side-by-side drawings of a stamen and a pistil and label the parts.

pollination, p. B47—Write the words *pollen*, *pollinate*, and *pollination* on the board and challenge students to explain the differences among them. Remind students that *pollen* is a noun that describes the substance produced by the stamen. *Pollination* is the transfer of pollen from the stamen to the pistil. *Pollinate* is a verb that describes the act of pollination.

Reading Mini-Lesson

Summarize and Paraphrase a Selection Ask students to read the section on page B48 called *Flowers and Pollination*. Display the transparency and have students write a one-sentence summary of each paragraph in the selection. Then ask them to write one or two sentences that summarize all five of those sentences.

Reading Mini-Lesson Transparency B2-3

Summarize and Paraphrase a Selection

Because pollination only occurs when pollen grains reach the stigma of a flower, certain flowers have adapted to make sure that pollination takes place.

Some flowers attract insect pollinators with bright colors and scents.

Flowers that are pollinated by insects have pollen that sticks to the bodies of the insects so that it is carried to the stigmas of other flowers.

Other animals, such as bats and birds, also pollinate flowers.

Pollination can occur in many different ways. Depending on the type of flower, pollen can travel to other flowers by insect, animal, or wind.

READING SKILLS PRACTICE

To practice and apply this reading skill, see **Workbook p. WB59.**

ESL ACTIVITY

Help students clarify the main idea in this lesson by illustrating the anatomy of a flower, including the **stamen**, **anther**, **pistil**, **stigma**, **petals**, **style**, and **ovary**. Have students draw and label these parts of a flower. Then have students discuss how the parts are related.

Guide the Learning continued

SCIENCE IDEAS

Help students focus on the supporting details for this **main idea: Flowers have various adaptations that help pollination occur.**

- **Name two main ways by which flowers are pollinated.** insects or wind
- **What are the main adaptations of flowers that are pollinated by insects? What is the function of each?** brightly colored petals and strong scents to attract insects; sweet nectar that provides food for insects; heavy, sticky pollen that gets caught on an insect's body
- **Describe the usual adaptations of flowers that are pollinated by the wind.** small, unscented, barely noticeable
- **What are characteristic adaptations of the pollen of wind-pollinated plants?** large amounts of light, dry pollen that the wind can easily carry

Critical Thinking Why is it important for wind-pollinated plants to produce large amounts of pollen? Much of the pollen will probably be lost because it will not land on another plant of the same species.

VISUAL LEARNING

Have students observe the illustration of the catkin on page B48 and read the caption. Point out that willows also produce catkins. Ask students to **compare and contrast** the terms *catkin* and *flower.* Make sure they understand that a catkin contains a large number of tiny flowers.

▲ Many trees, including oaks, birches, and maples, have very small flowers that grow in tight bunches called *catkins*. The male and female parts are in separate catkins. The wind carries pollen from the male catkins to the female catkins.

Flowers and Pollination

For pollination to take place, pollen grains must reach the stigma of a flower. Flowers have various adaptations to help pollination occur.

Flowers that are pollinated by insects often have brightly colored petals and strong scents. They produce a sweet liquid called *nectar* that some animals use as food. The bright colors and scents of the flowers are signs to the insects that the flowers have nectar. Some flowers pollinated by bees have markings that aren't visible to most other animals.

Flowers pollinated by insects have heavy, sticky pollen that gets caught on an insect's body. The pollen rubs off on the stigma of the next flower the insect visits.

Insects aren't the only animals that pollinate flowers. Birds as well as bats and other small mammals also carry pollen from one flower to another. Many of the flowers pollinated by birds are red, a color to which birds are attracted. In many cases one species of animal pollinates only one species of flower. The plant and the animal depend on each other.

Flowers not pollinated by animals usually rely on the wind to carry pollen from their anthers to their stigmas. These include the flowers of birches, oaks, corn, and grasses. The flowers are usually small, unscented, and barely noticeable. They don't need to attract animals. A wind-pollinated flower often produces a lot of light, dry pollen that the wind can carry to the stigmas of other flowers.

✔ **How do wind-pollinated flowers and insect-pollinated flowers differ?** Flowers pollinated by wind are not bright or large and produce no scent or nectar; flowers pollinated by insects are brightly colored with strong scents.

◄ Insects are not the only animals that pollinate flowers. This bat eats the sweet nectar of the cactus flower. As it pushes its face into the flower to get food, grains of pollen from a previous flower rub off and pollinate the cactus flower.

B48

Reaching All Learners

Visual Learners Allow these learners to answer the review questions by drawing pictures or making models. Be sure students respond to all five questions.

INVESTIGATION CHALLENGE

Hands-On Activity: More About Pollen

Use this idea for a hands-on activity to extend students' understanding of key lesson concepts.

Have students prepare microscope slides of different kinds of pollen. Provide several different flowers. Instruct students to spread a thin layer of clear glue in the center of several slides and shake a different flower over each slide. The pollen should stick to the glue. Have students place a cover slip on each slide and observe the pollen under a microscope.

Challenge students to determine whether the shapes and sizes of the various kinds of pollen are the same or different. Encourage them to make labeled drawings of their observations.

For student self-assessment, see Assessment Guide p. AGxvii.

▲The honeycreeper lives in tropical America and Hawai'i. It pokes its bill into flowers to get nectar. As it does, it also pollinates the flower.

Summary

Angiosperms produce flowers. They have efficient vascular systems and broad leaves. Their seeds are enclosed in fruit. These adaptations have resulted in their being the most successful plants on land today. All parts of a flower contribute to the plant's sexual reproduction. Some flowers are pollinated by insects or other animals. Others depend on the wind to carry pollen from one flower to another.

Review

1. What are three adaptations that help angiosperms live on land?
2. What parts of a flower produce pollen?
3. How do insects pollinate flowers?
4. **Critical Thinking** Some hummingbirds that eat nectar have very long, thin beaks. What can you tell about the flowers that provide them with nectar?
5. **Test Prep** The parts of a flower that enclose a bud before it blooms are the —
   - A filaments
   - B stamens
   - C pollen grains
   - D sepals

LINKS

MATH LINK

Write a Fraction Suppose a flowering plant produces 120 seeds, but only 16 of the seeds sprout and grow. What fraction of the plant's seeds become new plants? Express your answer in simplest form.

WRITING LINK

Expressive Writing—Poem Many poets have written famous poems about flowers. Write a poem for your classmates that tells what flowers in general, or a favorite flower, means to you.

SOCIAL STUDIES LINK

Flowers as Emblems Since earliest times, nations and states have used flowers as their emblems, or symbols. The official flower of the United States is the rose. Find out what your state flower is and why it was chosen.

ART LINK

Flowers in Art Most cultures have used flowers as subjects of paintings or designs. Study paintings of flowers. Draw or paint a flower in the style of your favorite flower painting.

TECHNOLOGY LINK

Learn more about plants and their uses by watching *Rain Forest to Shuttle* on the **Harcourt Science Newsroom Video.**

CNN Turner Learning

B49

3 Wrap Up and Assess

SUMMARY

Graphic Organizer for Chapter Concepts After reading the summary, have students complete the Lesson 3 section of the graphic organizer (shown on page B44).

REVIEW

1. They produce flowers and fruit, and they have very efficient vascular tissues.
2. anthers (or stamens)
3. Insects are attracted to flowers that produce nectar. Pollen rubs off on the insects' bodies and is deposited on the stigmas of the other flowers they visit.
4. The flowers must be long and narrow.
5. **D,** sepals

TEST PREP Test-Taking Tips

Item type multiple choice with a sentence fragment

To help students answer Question 5, share the following strategy with them. **Read the question and use facts you already know to help find the correct answer. Make a list of the parts of a flower and the functions of each part. Read each answer choice and select the one that fits best.**

LESSON CONCEPT REVIEW/ASSESSMENT

Students can use **Workbook p. WB60** to review the lesson concepts.

LINKS TIPS

Links Objective Extend understanding of angiosperms.

Math Link Remind students that they must check to see if the numerator and denominator have a common factor in order to determine whether or not their answer is in simplest form. Students should find that $\frac{16}{120}$ written in simplest form is $\frac{2}{15}$.

Art Link Display a variety of styles of artwork that have flowers as subjects. Images may be found on the Internet as well as in art books and encyclopedias. You may wish to provide reproductions of the works of Claude Monet, in particular his waterlilies; *Flowers* by William H. Johnson; and *Bowl with Sunflowers, Roses and Other Flowers* by Vincent Van Gogh.

LESSON 4

How Do Angiosperms Reproduce?

Objectives

- ▶ List the differences between monocots and dicots.
- ▶ Describe the parts of fruits and seeds.
- ▶ Explain how seeds are distributed.
- ▶ Explain how plants reproduce asexually.

Motivate

More Fast Facts

Spark interest with these facts and questions.

- Only about 1000 of the more than 230,000 species of angiosperms are of major economic importance. Fifteen of these provide most of the world's food.

  What are some other uses of angiosperms? Answers may include wood for furniture and flowers for decorations.

- Flowering plants have been found in the Himalayan mountains at altitudes greater than 6000 m (about 20,000 ft). These hardy plants find shelter between rocks.

  What difficulties might these plants have in reproducing? Possible answer: Pollination could be difficult if the plants are spaced too far apart.

Access Prior Knowledge

- ▶ What are the functions of seeds and fruits?
- ▶ How can angiosperms reproduce without seeds and fruits?

When Minutes Count . . .

Whole-Class Investigate Option Set up a class science center with the materials for the activity. Allow pairs of students time to make observations. Gather as a class to discuss students' findings.

Using the Activity Video The video can be used to preview and model the investigate process with students. After viewing the video, students can conduct the investigation in small groups.

LESSON 4

How Do Angiosperms Reproduce?

In this lesson, you can . . .

INVESTIGATE the structures of seeds.

LEARN ABOUT how angiosperms reproduce.

LINK to math, writing, art, and technology.

The mouse-sized honey possum of Australia eats nectar, thereby playing an important part in the way this flowering plant reproduces.

INVESTIGATE

Inside Seeds

Activity Purpose The pollination of an angiosperm flower results in a fruit that contains seeds. Under the right conditions, a tiny seed can grow into a much larger plant. An acorn is a seed that is about the size of one of your fingertips. Yet it may grow into a towering oak tree. In this investigation you'll **observe** the insides of some seeds. You'll find out about the beginning of life for two different kinds of angiosperms.

Materials

- paper towels
- lima beans (soaked overnight)
- corn kernels (soaked overnight)
- hand lens
- plastic knife

Activity Procedure

1. Cover your desk with paper towels. Lima beans and corn kernels are seeds. Use the hand lens to **observe** the two kinds of seeds. Make a sketch of each. (Picture A)
2. Carefully peel the outside covering from each seed. Then open the seeds. You may need to cut open the corn seed.
3. Use the hand lens to **observe** each seed's structures. **Record** your observations by drawing what you see. (Picture B)
4. **Compare** the structures inside the two seeds. Look for similarities and differences. From your **observations**, **infer** what the function of each part of the seeds might be.

B50

BACKGROUND

Plants, Poisons, and Medicines Flowering plants have been used for medicinal purposes since ancient times. A drug used during eye surgery and in treatment of stomach problems comes from the poisonous, deadly nightshade plant. The drug digitalis comes from the colorful foxglove plant. In small doses, it improves the pace and strength of the heartbeat. In large doses, digitalis produces palpitations and dizziness that can be fatal. In fact, plant chemicals that are used for medicines are often poisonous if consumed in large quantities.

Webliography GO ONLINE

Keyword plants

www.harcourtschool.com

For student self-assessment, see Assessment Guide p. AGxvi.

Picture A

Picture B

Draw Conclusions

1. What can you **infer** about the outside covering of the seeds? Is the function the same for both seeds?
2. Were you able to identify a tiny plant in each seed? Which part of the seed was it?
3. What did you **infer** about the function of the other parts of the two seeds? Is the function the same for both seeds?
4. **Scientists at Work** To **classify** organisms, scientists **observe** and **compare** their characteristics. The seeds you observed belong to two different groups of angiosperms. What differences in the seeds did you observe?

Investigate Further Look again at the two seeds. Find the tiny plant in each, and **hypothesize** which part of it will become the root and which part will become the shoot. **Plan and conduct an experiment** to test your hypothesis.

Process Skill Tip

The more carefully you **observe** two different organisms, the better able you will be to **compare** them and to identify their similarities and differences.

B51

INVESTIGATE

Time 20–30 min **Preparation Tips** See p. B24c.

Grouping pairs

Process Skill Tip observe, compare

Other Process Skills record data, infer, classify, hypothesize, plan and conduct simple investigations

Expected Results Students should conclude that seeds have an outer covering, a tiny plant, and stored foods. Corn seeds have two different food-storage parts, and bean seeds have two similar ones.

Teaching Tips

Caution students to remove the seed coat carefully in order not to damage the seed itself. To facilitate their comparisons, have students make a chart to show the similarities and differences in structure of the two seeds.

Draw Conclusions

1. Students should infer that the outside covering protects both seeds.
2. Students should have identified the plant embryo in each seed.
3. Students should have inferred that the other parts of both seeds provide food for the embryo.
4. Students should have noticed that the bean seed has two similar food-storage parts (cotyledons), and that the corn seed has two dissimilar food-storage parts (a cotyledon and an endosperm).

Investigate Further Students should base hypotheses on what they have learned in this investigation.

Process Skill Tip

Observe and Compare Point out to students that in order to be able to compare objects, it is first necessary to make careful observations. Display the Process Skill Tip Transparency, draw similar seeds on the board, or show students peanut and coconut seeds. Discuss the questions on the transparency. Remind students that it is also possible to make observations with other senses, such as touch or smell.

Process Skill Tip Transparency B2-4

Observe and Compare

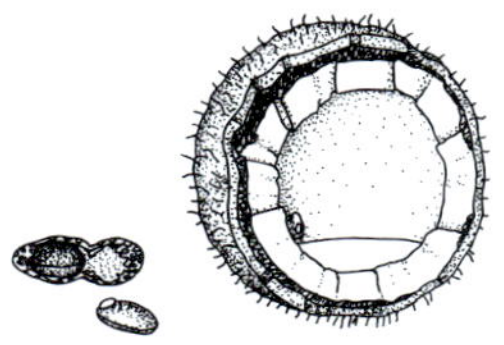

Make Observations

- Carefully observe the seeds. Write at least two things you observe.

Make a Comparison

- How do these seeds compare with those you observed in the activity? The coconut seed is similar to the corn seed and the peanut seed is similar to the bean seed.

Informal Assessment

Portfolio Students may wish to keep their completed drawings in their science portfolios.

INVESTIGATE LOG

Students can use **Workbook pp. WB61–62** to record the results of the investigation.

PROCESS SKILLS PRACTICE

To practice and apply process skills, see **Workbook p. WB63.**

LEARN ABOUT

1 Before Reading

SET A PURPOSE

FIND OUT Remind students that these statements will guide them in what to look for as they read.

VOCABULARY If students are doing the activity described in the Vocabulary Preview on page B25, make sure they add the definitions for these words to the chart.

Graphic Organizer for Chapter Concepts Display the transparency (shown below) or have students use Workbook page WB45. Tell students that by the end of the lesson, they should be able to fill in the remaining information on the transparency on angiosperm reproduction.

2 Guide the Learning

SCIENCE IDEAS

Help students focus on supporting details for this main idea: **Flowering plants are classified as monocotyledons and dicotyledons.**

- **What is a cotyledon?** a seed leaf that can store food for the young plant within a seed
- **How many cotyledons are found in a seed?** one or two
- **Describe five characteristics of monocots.** one cotyledon, parallel leaf veins, spread-out roots, flower parts in threes, and scattered vascular tissue
- **What are some examples of monocots?** grasses, corn, and palm trees
- **Name the five characteristics of dicots.** two cotyledons, netlike leaf veins, taproot, flower parts in fours or fives, vascular tissue in a ring
- **What are some examples of dicots?** oak trees, roses, beans, and dandelions

Ongoing Assessment

Use the ✓ questions in the pupil edition.

LEARN ABOUT

cotyledons, number of flower parts, leaf veins, vascular arrangement, and roots

Seeds and Fruit

FIND OUT

- the differences between monocots and dicots
- the parts of fruit and seeds and how they are classified
- how seeds are distributed
- how plants reproduce asexually

VOCABULARY

dicotyledon
monocotyledon
fertilization
embryo
runner
grafting

Two Kinds of Flowering Plants

In the investigation, you may have observed two similar parts under the outer coating of the lima bean. These parts are called *cotyledons* (kaht•uh•LEED•uhnz). They are seed leaves that store food for the young plant within the seed. The lima bean is a **dicotyledon** (dy•kaht•uh•LEED•uhn), because it has two seed leaves. Dicotyledons are often called dicots. Oak trees, roses, and beans are dicots.

A corn kernel, like the one you observed, does not have two similar parts. It has only one cotyledon, a leaflike structure wrapped around the young plant. Corn is a **monocotyledon** (mahn•oh•kaht•uh•LEED•uhn), because it has only one seed leaf. Monocotyledons are often called monocots. Grasses, corn, and palm trees are monocots.

The flower parts of monocots occur in multiples of three. For example, a monocot flower may have six petals, three sepals, and nine stamens. The flower parts of dicots occur in multiples of four or five.

The leaves of the two groups also show differences. Monocot leaves have veins that run side by side without crossing each other. Dicot leaves have netlike veins that branch out in many directions and cross each other. The vascular tissues in a monocot are scattered throughout the stem. In a dicot they are arranged in a ring going up the stem. Monocots have slender roots that spread out in all directions. Some dicots, such as carrots, have one main root, called a *taproot*.

✓ **What characteristics of monocots and dicots differ?**

Differences Between Monocots and Dicots

B52

Graphic Organizer for Chapter Concepts

Transparency B2 • Workbook, p. WB45

Types of Plants

	LESSON 1	LESSON 2		LESSONS 3 & 4
	Mosses, liverworts, & hornworts	Ferns	Gymnosperms	Angiosperms
Water transport	nonvascular	vascular	vascular	vascular
Plant size	small	varies	varies	varies
Type of reproduction	sexual and asexual	sexual and asexual	mainly sexual	mainly sexual
Means of sexual reproduction	water	water	wind	wind, insects, other animals
New plant form	spore	spore	seed	seeds
How seeds or spores are dispersed	gravity, wind	gravity, wind	gravity, wind	gravity, wind, animals

From Flower to Fruit

Pollination is the first step in angiosperm reproduction. The reproductive process is complete when a seed sprouts into a new plant.

Pollination occurs when pollen grains from a flower are deposited on the stigma of a flower of the same species. Then the pollen grains grow tubes. These go through the style and reach the ovules in the flower's ovary. There are usually many pollen grains on a stigma, and each one grows its own tube.

Each pollen grain contains two sperm cells. When the tube from a pollen grain reaches an ovule, the sperm cells travel down the tube. One sperm cell joins with an egg cell in the ovule. This process is called **fertilization**. The fertilized egg develops one or two cotyledons and an **embryo**, the tiny plant that forms in the seed. The other sperm cell joins with a different cell in the ovule to form a food source. This nourishes the embryo as it germinates. Angiosperms are the only living things in which sperm cells function in these different ways.

The flower's fertilized egg cells develop into seeds. As this happens, the ovary swells and grows into a fruit. A fruit is a structure that surrounds and protects the seeds. Some fruit serve as food for animals that eat the fruit and spread the seeds. Other fruit are adapted to travel on the wind. When winter is over, seeds that have reached areas with the right conditions germinate and grow into new plants.

Once a flower has been pollinated, tubes begin to grow down the style to the ovary.

When the cells inside the ovules have been fertilized, the flower begins the process of becoming a fruit. The petals are no longer needed, and they fall off the plant.

As the seeds develop, the ovary develops into a fruit.

The fertilized ovules develop into seeds, each with an embryo, a source of food, and a seed coat.

Each seed contains a tiny embryo, which will grow into a new plant when conditions are right.

✓ **What is the function of the tubes that grow from pollen grains?** They are the means by which sperm cells from pollen travel to the ovules of a flower.

B53

VISUAL LEARNING

Allow students time to observe the illustration on page B53 and read the captions. Then, write the heading *Plant Reproduction* on the board, and under it write the subheadings *Pollination, Fertilization, Fruits, Seeds,* and *New Plants.* Invite five volunteers to describe what happens at each of these stages in the reproduction of a plant. Encourage students to use the correct terminology to describe the parts of a plant. As they use these terms, list them under the appropriate heading on the board. Invite students to add any prior knowledge they might have as they describe each stage in the process.

DEVELOP SCIENCE VOCABULARY

monocotyledon and **dicotyledon**, p. B52—Write these words on the board and ask students to identify the prefix and root of each word. Encourage students to **use the prefixes to determine the word meanings**. Invite them to identify other words with these prefixes.

fertilization, p. B53—Ask students to identify the root *fertil* and then brainstorm other words with the same root. Challenge students to explain how the meaning of each word is related to the root.

embryo, p. B53—Have students look at the drawings of seeds in their science portfolios and identify the embryo in each drawing. Remind them that the embryo will grow into an adult plant.

BACKGROUND

Webliography GO ONLINE

Keyword plants

www.harcourtschool.com

Seeds Big and Small The largest and heaviest seed in the world is the Seychelles coconut, which weighs 30 kg (about 66 lb). The smallest seeds in the world are from orchids. One million of these tiny seeds weigh only 0.38 g (0.013 oz).

Seed Dispersal Milkweed, dandelions, and thistles produce seeds with fine hairs that extend in all directions to form a lightweight ball. These seeds can be carried high into the air by an updraft and can float in the breeze far from the parent plants. Seed pods of plants such as the touch-me-not, meadow cranes bill, and cow vetch act like catapults and fling their seeds in all directions. Seeds of raspberries, blueberries, and apples are dispersed by the animals that eat these fruits. Nuts are also seeds that are dispersed in this way. The shell of a walnut or pecan is the fruit of these trees.

ESL ACTIVITY

Have students work in pairs. Provide art supplies and index cards. Ask students to draw simple pictures of at least four different stages in the reproduction of angiosperms. Then ask students to exchange cards and try to put the pictures in the correct order.

Guide the Learning continued

SCIENCE IDEAS

Help students focus on the supporting details for this **main idea: A fruit is the part of an angiosperm that contains seeds.**

- **What is a dry fruit? Give an example.** those with seeds in pods and nuts; possible answers: peas, beans, walnuts
- **What is a fleshy fruit? Give an example.** those with pulp; possible answers: apples, cucumbers
- **What are the five different kinds of fleshy fruits? Describe and give an example of each.** drupe (plum, cherry), single seed inside a hard covering; pome (apple, pear), several seeds in a core; berry (blueberry, tomato), many seeds scattered throughout; aggregate fruit (strawberry, raspberry), groups of small fruitlets; multiple fruit (pineapple, fig), groups of fruitlets that form a single fruit

Critical Thinking Help students better understand the structure of aggregate fruits.

- **Where are the seeds located in an aggregate fruit? Explain.** The seeds are scattered throughout. Each fruitlet forms from a separate pistil of a single flower and therefore has one or more seeds.

VISUAL LEARNING

Instruct students to observe the illustrations on page B54 and read the captions. Write the words *tomato, pineapple, peach, blackberry,* and *apple* on the board. For each, have students tell what kind of fruit the food is and why it is classified in this way.

◄ A berry has seeds throughout its flesh. That's why a tomato is classified as a berry.

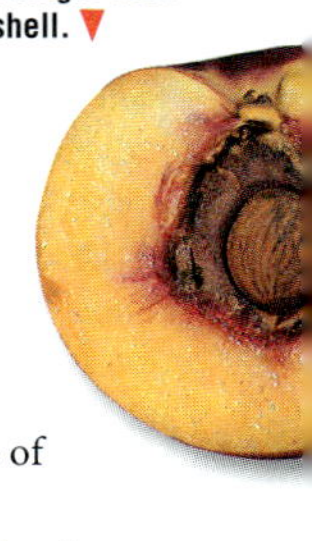

A drupe, such as this peach, has a single seed with a hard shell. ▼

Fruit

If you asked for some fruit, you might be surprised if you got pea pods, cucumbers, or walnuts. But each of these items really is a fruit. So are apples and oranges and most of the other sweet, juicy plant foods that people usually call fruit.

In science the word *fruit* is used for the part of an angiosperm that contains seeds. Tomatoes are fruit, and so are peanuts and raspberries. They all contain seeds. *Dry fruit* include peas, beans, and other seeds in pods, as well as nuts. *Fleshy fruit* include apples, cucumbers, and other fruit with pulp.

Scientists classify fleshy fruit by their structure. Different kinds of fruit have different structures because they develop from different flower parts. Some even develop from several flowers.

A fleshy fruit that has a single seed in a hard covering is classified as a drupe. The seed developed from a single pistil, and the fleshy pulp came from a single ovary. Plums, cherries, and peaches are drupes.

Some fleshy fruit have cores that contain several seeds with leathery seed cases. These fruit are classified as pomes. They include apples and pears. A pome's core developed from a flower's ovary. The fleshy part around the core developed from parts of the flower that surrounded the ovary.

A berry also contains many seeds. But they aren't grouped together in a core. The seeds are scattered throughout the flesh of a single enlarged ovary. Blueberries, tomatoes, and watermelons are berries. But strawberries, blackberries, and raspberries are not berries. They are aggregate fruit. Each fruit is made up of groups of smaller fruitlets. Each fruitlet developed from a pistil of one flower that has many pistils.

Sometimes groups of fruitlets that form a single fruit develop from the pistils of several flowers of the same species. The results are classified as multiple fruit. These include pineapples and figs.

✔ **Why isn't a raspberry a true berry?**

▲ Blackberries are groups of tiny fruitlets that formed from a single flower. They are classified as an aggregate fruit.

The seeds in an apple are all in its core, so it is a pome. ▼

◄ Many different pineapple flowers combined their fruit to form this pineapple. It is classified as a multiple fruit.

A berry contains seeds scattered throughout the flesh of the fruit, but a raspberry is an aggregate fruit, made up of many fruitlets.

B54

Reaching All Learners

Kinesthetic Learners Provide these learners with a variety of fruits. Challenge them to locate the seeds and to identify what kind of fruit each one is. Encourage students to infer how the seeds are spread.

Reading Mini-Lesson

Use Context to Determine/Confirm Word Meaning Students can get clues to a word's meaning from the words that come before or after it. In the third paragraph on page B53, have students locate the words *ovule* and *germinate* and read each word in context. Display the transparency and have students define the words based on the context using surrounding words.

Reading Mini-Lesson Transparency B2-4

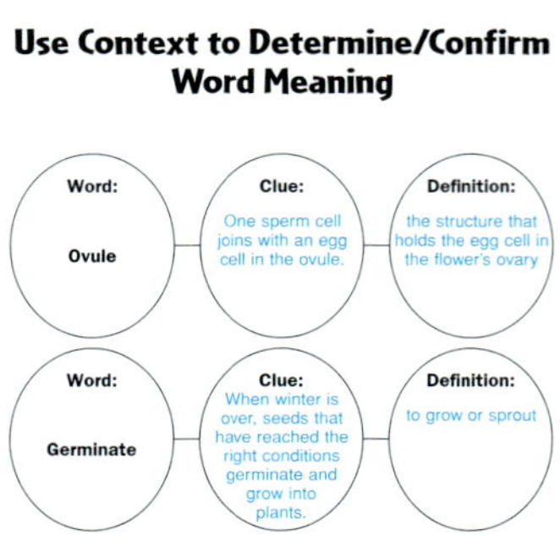

READING SKILLS PRACTICE

To practice and apply this reading skill, see **Workbook p. WB64.**

Spreading of Seeds

Suppose seeds just fell to the ground under their parent plants. If they did, they might not get enough light, water, and nutrients to survive. But plants have adaptations for reaching other places.

Fleshy fruit are often brightly colored. These colors attract animals that eat the fruit. The soft part of the fruit is digested. The seeds in the fruit pass through the animals' digestive systems unharmed. The seeds are carried to wherever the animals' droppings fall.

Animals also help spread burs. Burs are seeds with hooks that catch on the animals' fur. Later the burs are rubbed or scratched off.

Plants such as grasses have seeds that are light enough to be carried by the wind. Other plants scatter their seeds themselves. As their seed cases grow, tension is built up that finally causes the seed cases to burst and to throw the seeds through the air. Sometimes the bursting is set off by the wind or by an animal touching the seed cases.

✔ **How does the spreading of seeds help them germinate?** Seeds may reach locations where conditions are better for growth.

Squirrels don't always eat the nuts and other seeds they buried to store for the winter. In the spring these seeds may germinate and grow. ▼

A thick, hard shell protects the seed of a coconut. Coconuts often fall into the ocean. Then they wash ashore at other locations, where they can germinate and grow. ►

Strawberry plants produce long runners that spread out over the ground. New plants grow from buds on the runners and send roots into the soil. ►

Asexual Reproduction

Some plants reproduce without using seeds, a process called asexual reproduction. Asexual reproduction produces new plants that have exactly the same genes as the parent plant.

Many plants can grow from leaf cuttings. Leaves from the parent plant that are placed in water or in rich soil can develop into new plants.

Plants like strawberries and spider plants reproduce by **runners**. These are long, slender stems that grow close to the ground and put out roots and shoots. The new plants can grow to live without the parent plant. Some garden plants, such as irises, produce underground stems that divide. Each stem grows into a new plant.

Tubers are plants with swollen underground stems that store food. Potatoes are tubers. Their sprouting "eyes" grow into new plants.

Farmers grow certain plants by a form of asexual reproduction called **grafting**. They join parts of two plants together to make a single plant. The new plant has some characteristics of both of the parent plants. By grafting, farmers can grow plants faster and change the time of year that they bear fruit. Grafting can also produce plants with stems and roots that grow in poor soil but with fruit that would normally require rich soil.

✔ **What are three different kinds of asexual reproduction?** Possible answers: reproduction by leaf cuttings, runners, dividing underground stems, tubers, and grafting

B55

USE PROCESS SKILLS

Compare Write the terms *leaf cuttings, runners, tubers,* and *grafting* on the board. Challenge students to describe each method of asexual reproduction. Record their answers under each term. Then invite students to **compare** these methods of asexual reproduction.

DEVELOP SCIENCE VOCABULARY

runner, p. B55—Encourage students to relate the word *runner,* meaning someone who runs, to the biological term *runner,* meaning a plant stem that runs along the ground and puts out roots and shoots. Challenge students to describe how this can help them remember the meaning of the biological term *runners.*

grafting, p. B55—Make sure students understand that grafting is the process by which parts of two plants are joined to make a single plant with characteristics of both parent plants. Write the term on the board and invite students to list the advantages of grafting.

Health Link

Vegetables are the edible parts of soft-stemmed plants. They can be grouped by plant part. Lettuce is a leaf vegetable; celery is a stalk; carrots and beets are roots; potatoes are tubers; and onions are bulbs. Broccoli is actually flowers, green beans and cucumbers are fruit, and peas are seeds. Vegetables are important sources of vitamins and minerals. They provide fiber, which helps keep the digestive system healthy. The United States Department of Agriculture (USDA) recommends eating three to five servings of vegetables a day. Display the USDA Food Guide Pyramid and various fresh vegetables. Challenge students to identify the foods as fruits, flowers, or seeds. Have them determine how much of each food constitutes one serving.

Social Studies Link

Origins of Fruit Explain to students that peaches and cherries were brought to this country by immigrants. Have students find out the origins of these or any other fruits that are not native to the United States.

Guide the Learning continued

VISUAL LEARNING

Divide the class into six groups. Have each group read a different caption on page B56 and observe the associated picture. Then have them decide together how to communicate to the rest of the class the information in their section. Have students work cooperatively to prepare a presentation for the class. Encourage them to organize their materials carefully and to add information from their previous knowledge. Students may wish to **use additional reference sources** for information. As the groups make their presentations, allow class members to ask questions. Encourage careful and respectful listening.

Uses of Angiosperms

WOOD

Broad-leaved trees such as oak, maple, and walnut are sometimes called hardwoods. Hardwoods are used to make furniture and flooring that needs to be strong and long-lasting.

FOOD

People depend on flowering plants for food, including almost all grains, fruits, and vegetables. Many of the animals that we use for food also live on flowering plants.

CLOTHING

Cotton, a fabric often used for clothing, is made from the fruit of cotton plants. Linen is a fabric made from the fibers of flax plants.

MEDICINES

Since ancient times, people have known that some flowering plants are useful for medicines. For example, the juice of aloe plants has been used to treat burns. The plant in the picture is a rosy periwinkle, from which two cancer-fighting medicines are made.

DECORATION

Many decorative trees and shrubs are angiosperms, and most gardens are planted with flowers. People use cut flowers in their homes, for celebrations, and as gifts.

SCENTS

Oils from flowers such as roses and lilies are used in perfumes. Dried flowers are sometimes used to freshen rooms, drawers, and closets.

THE QUESTIONS KIDS ASK!

What is the difference between hardwood and softwood?

Address Misconceptions The wood from angiosperm trees is hardwood. The wood from gymnosperm trees is softwood.

Have students compare and contrast how these two different types of wood are used. Hardwood is used to make furniture and flooring. Softwood is used for lumber, plywood, and paper.

INVESTIGATION CHALLENGE

Hands-On Activities: More About Plant Reproduction

Use these ideas for hands-on activities to extend students' understanding of key lesson concepts.

Challenge students to grow plants from fruits or vegetables. Have them cut off the top of a pineapple or several sprouted eyes from a potato and plant them. Or, they may want to cut off the top of a carrot or parsnip and see whether they can get it to grow.

Take students on a short walk through a field or empty lot where the soil is moist. When they return to class, have them scrape the mud from their shoes, put it in a pot with sterile potting soil, and keep it moist. When small plants begin to grow, challenge students to explain why this happened.

For student self-assessment, see Assessment Guide p. AGxvii.

Summary

Angiosperms are divided into two groups—monocots and dicots. Monocots have one cotyledon, flower parts in threes, parallel veins in their leaves, scattered vascular bundles in their stems, and spreading root systems. Dicots have two seed leaves, flower parts in fours or fives, netlike veins, vascular bundles arranged in a ring in their stems, and often one main taproot. The ovules in a pollinated flower are fertilized and become seeds encased in fruit. Fleshy fruit include drupes, pomes, berries, aggregate fruit, and multiple fruit. Seeds can be spread by wind, animals, or water. Many plants can reproduce asexually from cuttings, underground stems, runners, and grafting.

Review

1. How do monocot and dicot seeds differ?
2. Where are the egg cells in a flower located?
3. How does the bright color of some fruit help spread their seeds?
4. **Critical Thinking** Suppose some farmers wanted to produce new apple trees just like the ones already in their orchard. What method of reproduction would be best?
5. **Test Prep** Which of these is NOT a fruit?
   - A a raspberry
   - B a tomato
   - C a potato
   - D an apple

LINKS

MATH LINK

Find Multiples Flowers may be monocots or dicots. The number of petals on a monocot is a multiple of 3. The number of petals on a dicot is a multiple of 4 or 5. If you counted 12 petals on a flower, could you tell whether it was a monocot or a dicot? Explain your answer.

WRITING LINK

Informative Writing—Description Write a paragraph for a friend, describing a plant you would like to invent by grafting two other plants together.

ART LINK

Dried Flowers Many cut flowers stay beautiful indefinitely if you hang them upside down until they are completely dry. Then you can use them in arrangements. You can also press flowers between sheets of paper that absorb the moisture. Put some flowers and absorbent paper between two heavy books for a couple of weeks. Then use them to make an arrangement or picture.

GO ONLINE TECHNOLOGY LINK

Learn more about plant reproduction by visiting the Smithsonian Institution Internet Site. **www.si.edu/harcourt/science**

Smithsonian Institution®

LINKS TIPS

Links Objective Extend understanding of how angiosperms reproduce.

Math Link If students do not know how to approach the problem, suggest that they begin by finding all the factors of 12. Students should conclude that since both 3 and 4 are factors of 12, you could not tell whether it was a monocot or dicot.

Writing Link For prewriting, have students determine the function that their plant would serve. Then have them list all the characteristics that they would give it. Encourage students to write in a lively style. Writing models are provided in the **Teaching Resources** book.

3 Wrap Up and Assess

SUMMARY

Graphic Organizer for Chapter Concepts After reading the summary, have students complete the Lesson 4 section of the graphic organizer (shown on page B52).

REVIEW

1. Monocot seeds have one cotyledon, or seed leaf, and dicot seeds have two cotyledons.
2. in ovules within the ovary at the base of the pistil
3. The color identifies the fruit and attracts animals that will eat it and then spread the seeds to other locations.
4. grafting (asexual reproduction)
5. C, a potato

TEST PREP Test-Taking Tips

Item type multiple choice with question stem using the negative *not*

To help students answer Question 5, share the following strategy with them. **Read the question carefully and isolate key information such as "is NOT a fruit." Eliminate any unreasonable answer choices. If necessary, choose the most reasonable answer.**

LESSON CONCEPT REVIEW/ASSESSMENT

Students can use **Workbook p. WB65** to review the lesson concepts.

VOCABULARY REVIEW

Students can use **Workbook p. WB66** to review the chapter vocabulary.

HEALTH HANDBOOK

To reinforce or extend the science concepts in this lesson, you may want to use the content and activities in the Health Handbook, pp. R8–R11 (T47–T48).

Objectives

- ▶ **Connect chapter concepts with the history of science.**

1 Before Reading

SET A PURPOSE

Hold a brainstorming session to find out what students already know about the history of agriculture and what they would like to find out. Focus the discussion on the question "How has agriculture changed from prehistoric times to the present day?" Record students' ideas and questions. Encourage students to look for answers to their questions in the reading selection and, if necessary, to use additional reference materials.

2 Guide the Learning

SCIENCE IDEAS

Help students gather information about the key advances in agriculture. Ask:

- ▶ **What was agriculture like in prehistoric times?** First, people weeded out non-edible wild plants. Then, in about 9000 B.C., they began planting seeds and using fertilizer. They used tools such as hand-pulled plows, sickles, and grinding stones.
- ▶ **Why was the Roman conquest important to the advancement of agriculture?** The Romans brought new farming methods, such as crop rotation, to all the lands they conquered.
- ▶ **How did the agricultural revolution in the 1700s result in a move to the cities?** New inventions and improved farming techniques resulted in larger crops, so fewer farmers could make a living. Many moved to the cities to find work.

USE PROCESS SKILLS

Use Time/Space Relationships Have students list the advances in technology related to the plow and note their approximate dates. Then ask students to insert the items on their lists into the time line.

Students should list the following advances in technology related to the plow: hand-pulled plow (exact date unknown, but before 5000 B.C.), ox-pulled wooden plow (between 5000 and 3000 B.C.), horse-drawn plow (900s), gasoline-powered tractors (1920s).

SCIENCE THROUGH TIME

Advances in AGRICULTURE

Prehistoric Farming

It's hard to think of a time when people did not practice agriculture—the planting, tending, and harvesting of plants for food. In fact, archaeologists agree that agriculture may have had its beginnings as early as 10,000 B.C., when prehistoric people began destroying plants they could not eat that competed with plants they could eat. A thousand years later (about 9000 B.C.), people realized that some of the wild plants they ate could be grown from seeds. Grains such as wheat and barley were most likely the first crops. The earliest farming tools included hand-pulled plows, curved knives called sickles, and grinding stones. These inventions, as simple as they may seem, were important for the development of agriculture.

Farming in Early Civilizations

The next two great agricultural advances were made between 5,000 and 3,000 B.C. in Egypt and Mesopotamia. The first was the invention of a wooden plow that could be pulled by an ox. With an ox instead of the farmer pulling the plow, a much greater area could be plowed in less time. The second was the development of irrigation, or watering, systems that could be used on large areas of land. Irrigation made it possible for people to grow crops in regions with little rainfall.

The Spread of Advanced Agriculture

During the time of the Roman Empire, the Romans took new methods of farming to every land they conquered. They may have been the first to realize the importance of rotating crops and of leaving a field unplanted, or fallow, for a period of time. Different crops use different amounts of nutrients, and some crops add nutrients to the soil, so crop rotation keeps nutrients in the soil from being used up.

Agriculture Through the Ages

8000 B.C. Prehistoric people in the Middle East grow grains. They use wood ash and manure as fertilizer.

275 B.C.–A.D. 476 Roman Empire spreads new farming methods, including crop rotation, through Europe.

1400–1700 European explorers bring back food plants from around the world. Spanish explorers introduce potatoes from Peru to Europe.

1620s Native Americans show the Pilgrims how to plant crops with fish as a fertilizer to help the plants grow.

1701 Jethro Tull invents the seed drill.

1700 Agricultural Revolution: Farmers begin using a four-field crop rotation method.

8000 B.C. | A.D. 1000 | A.D. 1600 | A.D. 1700 | A.D. 1800

B58

Reading Mini-Lesson

Use Reference Sources Give students a chance to investigate the history of domesticated grains by having them work in groups to research corn, wheat, rice, and other cereal crops. Have each group research one grain. Students in each group should develop questions about the food they are researching. Ask each group member to use reference sources to answer two or three questions. Students can then prepare a poster or other visual display presenting the information collected in their group.

Advances in agriculture continued through the Middle Ages, each leading to the production of larger crops with less work. For example, in the 900s, a harness was invented that allowed horses to pull plows. The faster horse soon replaced the slow-moving ox in the field.

The Agricultural Revolution

The Agricultural Revolution began in Great Britain in the 1700s, thanks to an Englishman named Jethro Tull. Tull invented the horse-drawn hoe and the seed drill. The seed drill was the first machine that could drill even rows of holes in a field and drop seeds into them. Farmers also improved their methods of crop rotation. Since larger crops meant fewer farms were needed, many farmers left the countryside and moved to cities to find work.

The Machinery of Modern Agriculture

In 1892 American farmers began to use steam-powered tractors. By the 1920s, however, gasoline-powered tractors had begun to replace steam-powered tractors and animals for pulling equipment. Today, farmers use a variety of powerful machines such as combines (KAHM•bynz) that do several jobs and seed drills that can drill up to 12 rows at a time.

High-Tech Plants

Over the centuries farmers have done many things to change crop plants. People have grafted trees to get better fruit. They have also crossbred different varieties of the same plant. The hybrid plants that result have better taste, produce larger crops, and resist insects and diseases. Today many crops, including most types of corn, are hybrids.

For centuries farmers have tried putting different materials on the soil to improve their crops and on the plants to kill pests. Today farmers have a great variety of products, both natural and human-made, that they can use. There are pesticides to kill insects, herbicides to kill weeds, and fertilizers to help plants grow.

Since the 1970s scientists have worked to change the genes of crop plants. With the new genes they are given, plants are better able to resist insect pests and plant diseases. In the future, this genetic engineering may allow farmers to raise crops without the use of chemicals, making it cheaper to farm.

THINK ABOUT IT

1. Why would pulling nonedible weeds be considered the beginning of agriculture?
2. Why was the seed drill such an important part of the Agricultural Revolution?

A combine does several jobs—it cuts a plant stalk, separates the seed from the straw and other waste materials, collects the seeds in a bin, and returns the waste to the ground.

1945 The first chemical herbicide is patented in the United States.

1900 — 2000

Today Scientists study the genes of plants to find ways to make them better able to resist pests, diseases, and drought. They also look for ways to increase crop yield.

B59

DEVELOP SCIENCE VOCABULARY

graft, p. B59—Explain that grafting involves inserting a stem of one woody plant into a slot cut in the stem of another woody plant. If the procedure is successful, the grafted stem continues to grow. Ask students to think about why grafts are successful only in closely related plant species. Why might a plant "reject" a graft?

crossbreed and **hybrid**, p. B59—Ask students to explain what they understand about the term *crossbreed* as applied to plants. It means to interbreed two plants of the same species. The offspring resulting from the crossbreeding is called a *hybrid*. Ask students to give reasons why plants are crossbred.

CHAPTER CONCEPTS

This feature focuses on how humans have improved agricultural tools, farming methods, and even the plants themselves.

► **How have people changed plants in order to produce larger crops, better flavors, or resistance to pests and diseases?** People have grafted trees, crossbred plants to produce new hybrids, and are now working on genetic engineering.

3 Wrap Up and Assess

THINK ABOUT IT

1. People started to care for a crop and help it grow rather than just waiting for it to grow and harvesting it. By weeding, they eliminated competing plants.
2. It gave people the ability to plant a great many more seeds much more carefully. People could cultivate much larger fields and bring in much larger crops.

PEOPLE IN SCIENCE

Objectives

- **Connect chapter concepts with the contributions of scientists.**

1 Before Reading

SET A PURPOSE

Plant physiologists study the functions and chemistry of the parts of plant cells. Have students read to find out the focus of Dr. Camellia Okpodu's research and the contributions she has made both to plant science and to student interest in science.

2 Guide the Learning

THE WORK OF SCIENTISTS

Help students focus on the methods and processes scientists use and the impact of their work.

- **How might Dr. Okpodu's research on the effects of acid rain and air pollution on plant cells be useful?** It could lead to developing plant varieties that are more resistant to the effects of air pollution and acid rain. It also could help scientists further understand the dangers of pollution and encourage the need for cleaner air standards, laws, and programs.
- **If you were to study plants, on what aspect of plant life would you concentrate?** Answers will vary. Encourage students to discuss the types of plants they find most interesting.

3 Wrap Up and Assess

THINK ABOUT IT

1. Dr. Okpodu's efforts at discovering the ideal conditions in which to develop disease-free oak trees could help save endangered tree populations.
2. Dr. Okpodu teaches plant molecular biology at a university, establishes research projects for students, and has contributed her science experiences to a book that encourages women's endeavors in many careers.

PEOPLE IN SCIENCE

"Use your own yardstick to measure your success, not your neighbor's. You never know when your neighbor's yardstick is in inches when yours is in feet."

Camellia Moses Okpodu

PLANT PHYSIOLOGIST

Dr. Camellia Okpodu's interest in plants dates to her experiences growing up on the family farm. In school Dr. Okpodu developed an interest in science. Her teachers encouraged her to pursue further this interest in college. While studying at North Carolina State University, Dr. Okpodu discovered her field of specialty, plant biochemistry. There she became one of the first African-American women to attain a Ph.D. in plant physiology and biochemistry. Today she is an assistant professor of plant molecular biology at Hampton University in Virginia.

Part of Dr. Okpodu's research focuses on the effects of air pollution and acid rain on plants and on the best conditions for cloning plants in a laboratory. Like the teachers who helped her develop an interest in science, Dr. Okpodu has been active in helping students expand their interest in the field. In fact, one recent project has enabled Dr. Okpodu to blend her research with her desire to raise student interest in science.

She and some of her students have aided in the spread of the seeds of the Emancipation Oak. Virginia oak trees such as the historic Emancipation Oak, under which the Emancipation Proclamation first was read, recently have suffered from the effects of pollution. In Project O.A.K. students from across the country have researched ways to develop disease-free oak seedlings. For their part in the project, Dr. Okpodu and her students have been growing seedlings from plant tissue cultures under varying conditions.

One focus of Dr. Okpodu's research has been discovering ideal conditions for growing copies of the Emancipation Oak, a historic tree of the *Quercus virginiana* species.

Dr. Okpodu's efforts at developing student interest in science include writing a chapter for the recent book *How Jane Won*. The book examines factors that contribute to women's success in their chosen careers, including the field of science.

THINK ABOUT IT

1. How might Dr. Okpodu's work make a difference in the survival of oaks and other trees?
2. In what ways has Dr. Okpodu helped develop student interest in science?

B60

BACKGROUND

Webliography
Keyword biographies
www.harcourtschool.com
GO ONLINE

Micropropagation Micropropagation is the process of mass plant multiplication by growing tissue cultures *in vitro,* or in test tubes under controlled laboratory conditions. Since their development more than 25 years ago, micropropagation techniques have allowed large numbers of plants to be produced in much shorter times than have conventional growing methods. Micropropagation has allowed scientists to establish populations of plants that are endangered, rare, difficult to grow, or in high demand by farmers or gardeners.

ACTIVITIES FOR HOME OR SCHOOL

GROWING PLANTS FROM SPORES

How does a fern reproduce from spores?

Materials

- fern with spores that have turned dry and brown
- pot of potting soil with peat moss
- plastic wrap
- saucer of water

Procedure

1. Sprinkle the spores over the soil, and cover the pot with plastic wrap.

2. Move the pot to a light but not sunny location. Place the pot in the saucer.
3. Keep water in the saucer.
4. When the ferns are 2.5–4 cm tall, move clumps of them to their own pots.

Draw Conclusions

How does fern reproduction differ from reproduction in other plants with which you may be familiar? In what ways do you think that spores have helped ferns survive?

GROWING PLANTS FROM SEEDS

Which parts of seeds are essential for growth?

Materials

- 10 lima bean seeds, soaked overnight
- knife and cutting board
- 3 plastic pots filled with potting soil
- saucers for the pots
- pot labels and a marker
- water

Procedure

1. Have an adult help you split all but two of the seeds in half. Keep the two best cotyledons (halves), with embryos (tiny plants). Throw the rest away.

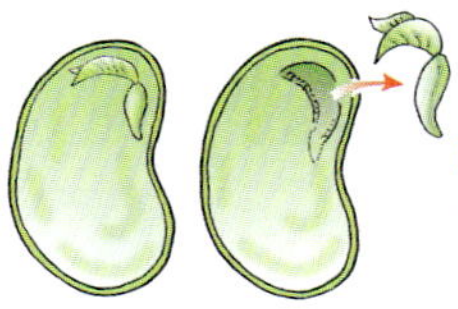

2. Use your fingernail to remove the embryos from two cotyledons.
3. Label the pots *Complete seeds*, *No embryos*, and *No cotyledons*.
4. Plant the two complete seeds in one pot, the two cotyledons with no embryos in another, and the two embryos in the third pot. Water the pots and place them in a warm, sunny spot. Observe their growth.

Draw Conclusions

Which pot of seeds grew best? Why did the cotyledons without embryos fail to grow? Why did the embryos without cotyledons grow less well than the complete seeds?

B61

School-Home Connection

These activities provide an excellent opportunity to assign hands-on activities for the students and their families. The materials are often easy to obtain and safe to work with. For those activities that you do not wish to do in the classroom, encourage students to complete them at home. Remind them that communicating their results is always important, but especially so for home activities since you will not be there to observe their activities. Ask students to use diagrams, sketches, graphs, or even photographs to show their results.

ACTIVITIES FOR HOME OR SCHOOL

Growing Plants from Spores

Objectives

- **Observe fern reproduction.**

Suggested Time 45 minutes to set up the experiment; 10 minutes each week to observe and record results

Hints

- To collect spores, select a frond that has well-developed, yellow to dark brown spore cases. Put the frond in a brown paper bag for a week. Empty the bag onto a sheet of white paper. Discard the leaves. The spores look like dust.
- Be patient. It may take months for the spores to germinate. Keep the soil moist.

Draw Conclusions Students should note that fern reproduction takes longer and that ferns go through two stages. They may conclude that spores can lay dormant for a long time until conditions are right for germination.

Growing Plants from Seeds

Objectives

- **Experiment to discover which parts of seeds are essential for growth.**

Suggested Time 45–60 minutes to set up the experiment; 10 minutes daily thereafter to observe and record results

Hints

- Remind students to keep all conditions the same for all three pots throughout the experiment. Each pot should get the same amount of water, light, and heat.

Safety Tip Soaked seeds may be soft enough for students to pry open with their fingernails. If not, an adult should use the knife to split them open.

Draw Conclusions Students should find that only the complete seeds grew because the other two groups lacked an essential part of the seed necessary for growth.

Investigate Further Students could repeat the experiment with other types of seeds. They could try other beans, pea seeds, or peanuts.

ACTIVITIES FOR HOME OR SCHOOL

Reproducible copies of these activities are provided in the **Teaching Resources** book.

CHAPTER 2

Review and Test Preparation

VOCABULARY REVIEW (1 point each)

1. Angiosperms
2. monocotyledon, dicotyledon
3. Conifers, gymnosperms
4. fern, spores
5. anther, stamen
6. Pollination, stigma
7. Asexual reproduction
8. runners
9. pistil, style
10. Fertilization
11. moss
12. grafting
13. embryo
14. ovary
15. pollen

CONNECT CONCEPTS (4 points each)

16. vascular
17. liverworts
18. gymnosperms
19. spores
20. seeds

CHECK UNDERSTANDING (4 points each)

21. D, xylem
22. G, in the female cone
23. D, both A and B
24. J, spread-out roots
25. C, asexual reproduction
26. F, stigmas

CRITICAL THINKING (5 points each)

27. Remove the stamens from the first flower and then dust pollen from the other flower onto the first flower's stigmas.

TEST PREP Test-Taking Tips

Item type open-ended format in which the student is asked to write a short answer

To help students answer Question 28, share the following strategy with them. **Turn the question into a statement to begin your answer, such as "Fruit farmers often keep beehives in their orchards because..." Then complete the sentence.**

28. The bees help ensure that the flowers on the fruit trees are pollinated.
29. Ferns or mosses; they need moisture in order to reproduce and survive.

CHAPTER 2 Review and Test Preparation

Vocabulary Review

Use the terms below to answer questions 1 through 15. The page numbers in ( ) tell you where to look in the chapter if you need help.

moss (B28)
asexual reproduction (B29)
spores (B29)
fern (B34)
gymnosperms (B36)
conifers (B36)
pollen (B37)
angiosperms (B44)
stamen (B46–B47)
anther (B46–B47)
pistil (B46–B47)
stigma (B46–B47)
style (B46–B47)
ovary (B46–B47)
pollination (B47)
fertilization (B53)
embryo (B53)
monocotyledon (B52)
dicotyledon (B52)
grafting (B55)
runners (B55)

1. ___ have seeds enclosed in fruits.
2. An angiosperm can be either a ___ or a ___.
3. ___ are cone-bearing trees that are ___, producing naked seeds.
4. A ___ is a vascular plant that produces ___ rather than seeds.
5. Pollen forms on the ___ at the top of the ___ of a flower.
6. ___ occurs when pollen lands on a sticky ___.
7. ___ is a kind of plant reproduction without sex cells.
8. Trailing stems that form new plants are called ___.
9. A flower's ___ consists of the stigma, ___, and ovary.
10. ___ occurs when sperm cells and egg cells unite.
11. A ___ is a plant without true stems, leaves or roots.
12. When two different plants are joined, it is called ___.
13. The tiny plant inside a seed is an ___.
14. Ovules are located within the ___.
15. Animals or the wind can carry ___ to another location.

Connect Concepts

Use the terms from the word bank to complete the concept map below.

seeds **spores** **liverworts**
vascular **gymnosperms**

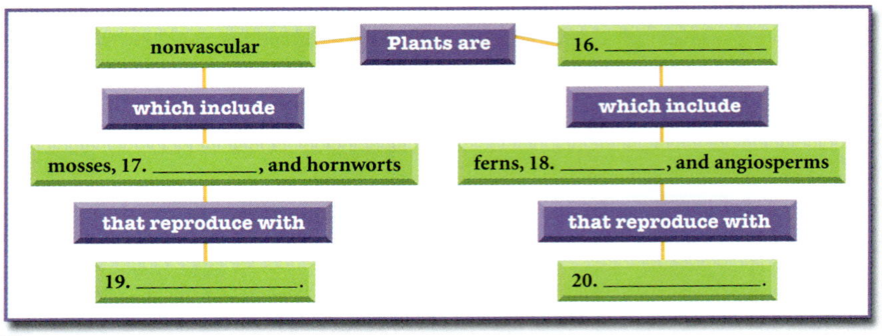

B62

CHAPTER TEST

See **Assessment Guide pp. AG26–30** for the Chapter Test. Assessment Options are provided on p. B24e.

TAKE-HOME BOOK

Use the **Take-Home Book** (described on p. B24d) to provide more chapter content and activities.

Check Understanding

Write the letter of the best choice.

21. What is the name of the tissue that plants have to transport water from roots to leaves?
    - A spore
    - B phloem
    - C pollen
    - D xylem
22. Where do the seeds of a conifer develop?
    - F in the male cone
    - G in the female cone
    - H in the pollen
    - J in the xylem
23. Examples of gymnosperms include —
    - A conifers
    - B cycads
    - C ferns
    - D both A and B
24. Monocots have —
    - F netlike veins
    - G two cotyledons
    - H vascular tissue in rings
    - J spread-out roots
25. Grafting is a method of —
    - A using runners
    - B leaf cutting
    - C asexual reproduction
    - D sexual reproduction
26. The female parts of the flower include —
    - F stigmas
    - G anthers
    - H petals
    - J sepals

Critical Thinking

27. How would you try to cross a hibiscus that has purple flowers with a hibiscus that has red flowers?
28. Why do fruit farmers often keep beehives in their orchards?
29. Suppose you had a damp, shady spot in your yard. What kinds of plants would do best there? Explain.

Process Skills Review

30. If a tree has needlelike leaves, what kind of plant would you **hypothesize** it is? What other characteristics would you look for to support your hypothesis?
31. **Compare** monocots and dicots in regard to cotyledons, leaf veins, roots, and flower parts.
32. If a plant has bright flowers, what can you **infer** about its pollination?
33. Suppose you have discovered a new plant near your home. How would you **communicate** information about your discovery?

Performance Assessment

Designing a Plant

If you had a fern, a gymnosperm, a monocot, and a dicot that each measured 0.5 meters (1.5 ft) tall, what procedures would you follow to identify which plant was which? List the methods you would use, and the characteristics of each kind of plant you would compare.

B63

PROCESS SKILLS REVIEW
(5 points each)

30. Gymnosperm; you could look for cones and seeds without coverings.

31. cotyledons: one in monocots and two in dicots; leaf veins: parallel in monocots and netlike in dicots; roots: spread-out roots in monocots and taproot in dicots; flower parts: in threes in monocots and in fours or fives in dicots

32. The flower is probably pollinated by an insect, bird, or other animal.

33. Students' answers should include communicating through drawing a picture or diagram, as well as communicating a written description of the characteristic and location.

Performance Assessment

Preparation Encourage students to begin answering the question by making lists of the characteristics of each type of plant.

Scoring Rubric

Performance Indicators

___ Includes making observations of three or more plant parts such as leaves, roots, vascular tissue, petals, and seeds.

___ Includes making observations of one or two plant parts.

___ Gives an appropriate method.

Observations and Rubric Score

3   2   1   0

WRITING ABOUT SCIENCE

Have students write a composition that involves writing for a specific purpose. You may wish to use the prompt that is provided in the **Workbook.** The prompt is accompanied by a graphic organizer that will help students organize their ideas before writing. Models for writing are provided in the **Teaching Resources** book. You may wish to reproduce those for students or display them on a transparency.

WRITING PRACTICE

A chapter writing prompt and a prewriting activity are provided on **Workbook p. WB67.**

CHAPTER 3 Planning Guide Invertebrates

LESSON	PACING	OBJECTIVES	MATERIALS
1 What Are Vertebrates and Invertebrates? B66–73 	3 days	▶ Recognize that all animals have the same basic needs. ▶ Conclude that animals meet their basic needs in different ways. ▶ Distinguish between vertebrate and invertebrate animals.	**Investigate** cardboard box, art supplies and natural materials for a diorama
2 What Are Sponges, Cnidarians, and Worms? B74–83 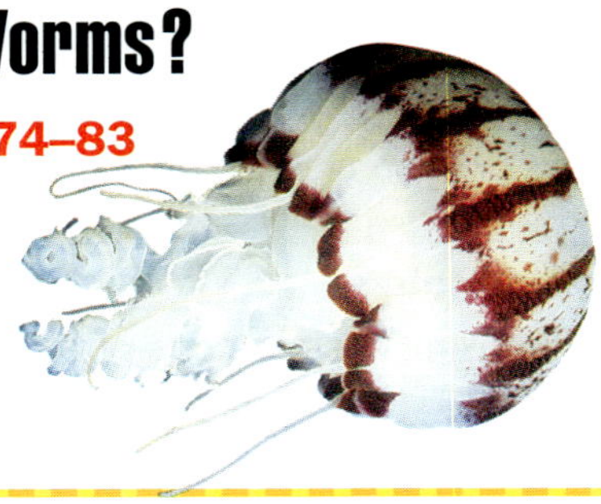	3 days	▶ Describe the important characteristics of sponges, cnidarians, and worms. ▶ Compare and contrast the structures in these three types of invertebrates.	**Investigate** hand lens, sponge, living planarian, Microslide Viewer and slides of hydras
3 What Are Mollusks, Arthropods, and Echinoderms? B84–93 	3 days	▶ Describe the distinguishing characteristics of mollusks, arthropods, and echinoderms. ▶ Identify examples of each of these three types of invertebrates.	**Investigate** clam shell, spirula shell, snail shell, grasshopper study sheet, sea urchin shell, starfish
End of Chapter B94–99 **Science and Technology** **People in Science** **Activities for Home or School** **Chapter Review and Test Preparation**		▶ Evaluate the impact of research and technology on scientific thought, society, and the environment. ▶ Identify careers related to science. ▶ Connect chapter concepts with the contributions of scientists. ▶ Extend chapter concepts with hands-on investigations. ▶ Review chapter concepts.	

PROCESS SKILLS	VOCABULARY	RESOURCES AND TECHNOLOGY	REACHING ALL LEARNERS
Process Skill Tip plan and conduct simple investigations **Other Process Skills** infer, compare	**vertebrate** **invertebrate** **endoskeleton** **exoskeleton**	**Workbook,** pp. WB68–73 **Vocabulary Cards**, pp. TR107–112 **Transparencies** B3, B3-1 **Harcourt Science Activity Video** **Internet Site**	**ESL Activity**, p. B71 **Investigation Challenge**, p. B72 **Advanced Learners**, p. B1g
Process Skill Tip communicate **Other Process Skills** compare, observe, record data	**sponge** **cnidarian** **flatworm** **roundworm** **parasite** **segmented worm**	**Workbook,** pp. WB74–78 **Transparencies** B3, B3-2 **Harcourt Science Activity Video** **Internet Site**	**Visual Learners**, p. B75 **ESL Activity**, pp. B79, B81 **Investigation Challenge**, p. B82
Process Skill Tip classify **Other Process Skills** observe, compare, record data, infer	**mollusk** **arthropod** **arachnid** **insect** **echinoderm**	**Workbook**, pp. WB79–85 **Transparencies** B3, B3-3 **Harcourt Science Activity Video** **Harcourt Science Newsroom Video**	**Kinesthetic Learners**, p. B89 **ESL Activity**, p. B89 **Investigation Challenge**, p. B92
		Take-Home Book, pp. TH27–30 *The Ant Colony* **Activities for Home or School,** p. TR69 **Chapter Test**, pp. AG33–37	**Performance Assessment**, p. AG38 **Portfolio Evaluation Sheet**, p. AGxxiv

Prepare for Activities

Use this page to help you organize and prepare materials for the Investigations, which begin each lesson of this chapter.

LESSON INVESTIGATION	MATERIALS

1 Designing Habitats
B66–67

PURPOSE Infer how one animal adapts to meet its basic needs in its environment.

PREPARATION TIPS Ask students or parent volunteers to bring in sturdy cardboard boxes for the dioramas. Shoe boxes and computer-paper boxes would be appropriate. You may want to take students outdoors to gather natural materials such as rocks, leaves, and twigs to include in their dioramas.

TIME 120 minutes

EXPECTED RESULTS Students should design an appropriate environment for one of the three animals.

Materials
- cardboard box
- art supplies and natural materials for a diorama

2 Three Groups of Invertebrates
B74–75

PURPOSE Compare three invertebrates to discover their similarities and differences.

PREPARATION TIPS Be sure to follow the recommendations for maintaining the live specimens.

TIME 60 minutes

EXPECTED RESULTS Students should record their observations of the similarities and differences among the three organisms.

Materials
- hand lens
- sponge
- living planarian
- Microslide Viewer
- slides of hydras

3 Compare Three Groups of Animals
B84–85

PURPOSE Compare the characteristics of animals from these groups—mollusks, arthropods, and echinoderms.

PREPARATION TIPS To help students develop a classification system, provide field guides or dichotomous keys for mollusks, arthropods, and echinoderms.

TIME 60 minutes

EXPECTED RESULTS Students should classify the organisms into three groups.

Materials
- clam shell
- grasshopper study sheet
- spirula shell
- sea urchin shell
- snail shell
- starfish

Harcourt Science Activity Videos

Preview the Investigations The Activity Video for this unit shows students doing the Investigations. You may wish to view the video for classroom management ideas. The video can also be used to model the Investigation process for students.

Equipment Kits

Underlined items above are provided in the equipment kits (available separately).

. . . with Trade Books

These books provide in-depth information about invertebrates.

An Extraordinary Life: The Story of a Monarch Butterfly by Laurence Pringle (Orchard, 1997) chronicles the life cycle, feeding habits, migration, predators, and mating of a monarch butterfly named Danaus. *NSTA Trade Book*

Invertebrate Zoology by Ellen Doris (Thames and Hudson, 1993) examines seven phyla of invertebrates and includes ideas for projects, field trips, and other fun suggestions for investigating these "spineless" creatures.

Wings, Stings, and Wriggly Things by Martin Jenkins (Candlewick, 1996) introduces readers to the world of such invertebrates as bees, snails, and butterflies.

Trade book titles are current at time of publication but may go out of print without notice.

. . . with More Hands-On Activities

Investigation Challenge

Activities for Home or School

The cross-curricular Links that appear at the end of each lesson in the student edition and throughout the teacher's edition also provide activities.

. . . with the Take-Home Book

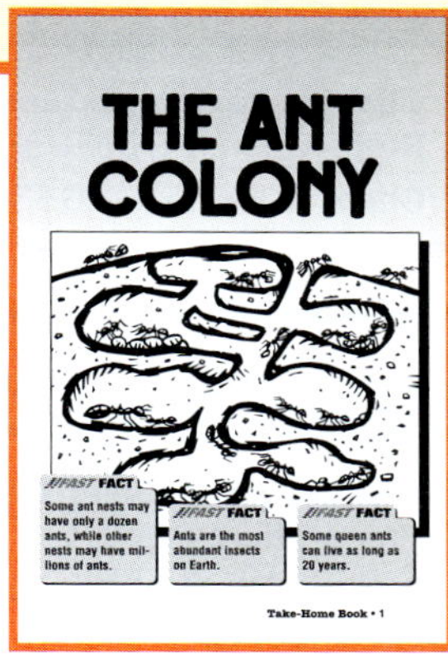

The Ant Colony

CONTENTS

. . . with Technology

Students can learn more about invertebrates through these components and special Internet sites. Look for Technology Links at the end of each lesson.

Harcourt Science Newsroom Video
Predatory Snails

Smithsonian Institution®
Harcourt Science brings the Smithsonian Institution to your classroom.
www.si.edu/harcourt/science

Visit The Learning Site for related links, activities, and resources: www.harcourtschool.com

WELCOME TO THE LEARNING SITE

Assessment Options ✓ Chapter 3

Formal Assessment

- **Chapter Review and Test Preparation,** PE pp. B98–99
- **Chapter Test,** pp. AG33–37

Ongoing Assessment

- ✓ **Questions**, PE pp. B68–73, B76–83, B86–93
- **Lesson Review**, PE pp. B73, B83, B93
- **Informal Assessment**
  - **Performance,** TE p. B67
  - **Portfolio,** TE p. B75
- **Observation Checklist,** AGxiv

Student Self-Assessment

- **Self-Assessment–Investigate**, p. AGxvi
- **Self-Assessment–Learn About**, p. AGxvii
- **Project Summary Sheet**, p. AGxix

Performance Assessment

- **Chapter Review and Test Preparation**, PE p. B99
- **Chapter Performance Task,** pp. AG38–39
- **Project Evaluation Checklist**, p. AGxviii

Portfolio Assessment

- **Science Experiences Record**, p. AGxxii
- **Guide to My Science Portfolio**, p. AGxxiii
- **Portfolio Evaluation,** p. AGxxiv

Chapter Test AG33

Invertebrates

Part I Vocabulary 1 point each

Match each term in Column B with its meaning in Column A.

	Column A		Column B
D	1. Animals with a backbone	A	sponge
K	2. Animals without a backbone	B	echinoderms
F	3. Skeleton within an animal's body	C	cnidarians
E	4. Skeleton outside an animal's body	D	vertebrates
C	5. Animals that have tentacles with the ability to sting	E	exoskeleton
A	6. Animal with a skeleton made of protein rods	F	endoskeleton
J	7. Worm with a body made from connected sections	G	mollusk
O	8. Type of worm with a single opening to its digestive system	H	roundworm
M	9. Harmful animal that lives in or on another animal	I	arthropod
H	10. Worm with a tubelike body and lengthwise muscles	J	segmented worm
G	11. Animal that uses a muscular foot to move	K	invertebrates
N	12. Largest group of arthropods and largest group of animals	L	arachnids
I	13. Animal with a jointed exoskeleton	M	parasite
L	14. Arthropods with eight legs and no antennae	N	insects
B	15. Animals with an interior skeleton and spiny skin	O	flatworm

Chapter Test AG34

Part II Science Concepts and Understanding 3 points each

Write the letter of the best choice.

B 16. Most animals that live in water get oxygen from —
A food
B the water
C the air
D air tubes

J 17. The exoskeleton provides —
F food and energy
G air and oxygen
H water and life processes
J protection and body shape

C 18. Why do cnidarians sting other animals?
A to make the animals go away
B to make themselves invisible
C to stun the animals before eating them
D to protect their own fertilized eggs

G 19. Which of the following is **NOT** true about cnidarians?
F They are meat-eating animals.
G They live in water and on land.
H Their body parts are arranged like wheel spokes.
J They all have stinging cells.

A 20. The flatworm and the roundworm —
A can harm humans and other animals
B have complex digestive systems
C live only in freshwater habitats
D can be seen only with a microscope

F 21. The earthworm has —
F two openings to its digestive system
G slits for taking oxygen into its lungs
H muscles that criss-cross
J many stinging cells

Chapter Test AG35

D 22. Which of the following is a mollusk without a shell?
A sea anemone
B hydra
C sponge
D octopus

G 23. Mollusks use their muscular feet to —
F move food to their mouths
G move or to burrow in the sand
H release fertilized eggs in the water
J drill into hard surfaces

D 24. Which of the following animals is **NOT** an arthropod?
A grasshopper
B spider
C shrimp
D snail

F 25. How does an arthropod protect itself after it molts?
F It conceals itself in its habitat.
G It changes colors to match its surroundings.
H It squirts a dark, inky fluid.
J It covers itself with slime.

B 26. Sea urchins —
A eat dead coral
B eat algae
C keep sharks away
D fertilize sponges

Use the following pictures to answer Questions 27–29.

B 27. Which is a vertebrate?

C 28. Which is an echinoderm?

A 29. Which has an exoskeleton?

Chapter Test AG36

Part III Critical Thinking 6 points each

30. In what two ways do earthworms help plants grow?

Answers might include that the waste products of earthworms fertilize the ground and provide the nutrients the plants need. Their burrowing in the ground helps keep it from compacting, so water and oxygen can get to the roots of plants.

31. Why did people first classify sponges as plants? When and why did this classification change?

Possible answer: People classified sponges as plants because sponges do not move and seem rooted to the ground. After the invention of the microscope, people could see cells with structures that pulled water into the sponge's body.

32. Describe two ways in which a bivalve can protect itself when it closes its shells.

Possible answer: Closing the shells provides protective cover for the body of the bivalve. Clapping its shell together can also provide a push, moving the bivalve away from danger.

33. Some invertebrates have the ability to regenerate body parts. How does this ability help protect a group? Give an example.

Answers may include that regeneration helps the group survive and reproduce. If regeneration were not possible, the animal might die if it were cut in two or lost a body part. If enough animals died, the whole group might eventually die out. Examples might include planarian or starfish.

Chapter Test AG37

Part IV Process Skills Application

Use the following picture to answer Question 34.

7 **34.** What characteristics would you use to **classify** each of the animals in this aquarium?

Possible answer: The fish has a backbone, so it is a vertebrate. The anemone has stinging tentacles, so it is a cnidarian. The shrimp has a jointed exoskeleton, legs, and antennae, so it is an arthropod.

6 **35.** You may have observed earthworms on the sidewalk after a heavy rain. **Infer** how earthworms behave during floods. Describe how you could **plan and conduct a simple investigation** to test your inferences.

Possible answer: Fill a plastic tub with soil and place healthy earthworms on top. Observe the behavior of the worms on the first day. Over the next few days, add water to the tub until the soil is fully saturated. Observe the behavior of the earthworms each day.

6 **36.** You have been asked to give a presentation on the relationship between arthropods and humans. Give three key points that you would cover, and describe how you would **communicate** this information to your audience.

Key points might include the following: Some arthropods damage crops (insects); some may be venomous (spiders, scorpions). Some arthropods (shrimps, crabs, lobsters) are used for food. Many have economic uses for humans (such as shrimpers, crabbers, exterminators, beekeepers, honey salespeople). Ways to communicate might include graphs, charts, photographs, projection slides, and videotapes.

Performance Assessment AG38

PERFORMANCE TASK

Invertebrates Everywhere

Materials

- scrap paper
- drawing paper
- markers or colored pencils
- written description of area to search
- reference books

1. Search for invertebrates in a designated area of your school. Find at least three kinds.
2. On scrap paper, draw each invertebrate you find. Classify it by type and habitat. Use the reference books for help in classifying.
3. On drawing paper, make a chart like the one on this page, only much larger.
4. Copy your drawings onto the chart. List the type, distinguishing characteristics, and habitat of each invertebrate you draw.

Invertebrates in Our School Environment

Drawing	Type	Characteristics	Habitat

Assessment Continued

Performance Assessment AG39

Invertebrates Everywhere

PERFORMANCE TASK

Materials Performance Task sheet, scrap paper, drawing paper, markers or colored pencils, written descriptions of areas to search, reference books

Time 30–40 minutes

Suggested Grouping groups of three or four

Science Processes observe, classify, compare, communicate

Preparation Hints Identify enough areas within the school or schoolyard to allow groups of students to be in different locations. Select areas that are likely to contain different kinds of invertebrates. Prepare a description of each group's location. Obtain permission for students to search the areas selected.

Introduce the Task Explain to students that they will search for invertebrates in their school environment. Each group will search a different location. They will need to classify at least three different kinds and describe the characteristics and habitat of each one. They will prepare a chart on which to compare the organisms they find. Suggest that group members make a plan for carrying out the task and that they assign responsibilities so that they finish within the time allowed.

Promote Discussion When students finish, have each group present its findings. Ask students to identify the type of invertebrate found most frequently. Have students compare findings for the different locations to determine the common characteristics and habitats of the invertebrates.

Scoring Rubric

Performance Indicators

_____ Locates and sketches three or more kinds of invertebrates.
_____ Lists characteristics of each invertebrate.
_____ Classifies each invertebrate by phylum, class, or both.
_____ Identifies the habitat of each invertebrate.

Observations and Rubric Score

3 2 1 0

Workbook Support

Investigate Log WB69

Designing Habitats

Materials

Activity Procedure

1. Choose one of the following animals: the Chilean rose tarantula, the giant green iguana, or the green tree frog.
2. Read in the appropriate chart about the animal's natural environment and about its habits. **Infer** what the animal needs to survive and how its environment meets these needs.
3. Draw a terrarium that would provide the animal with what it needs to survive. Make sure that you consider all the animal's needs.
4. Build a diorama that shows your animal in the terrarium.

Chilean Rose Tarantula
- lives in semidry environments
- temperature of environment is 21°–24°C
- lives in burrow during day
- eats at night
- eats insects
- eats only once a week

Giant Green Iguana
- lives in humid environments
- can survive temperatures as low as 27°C
- often basks in direct sunlight when the temperature is above 32°C
- lives in trees
- eats in the daytime
- eats leaves and flowers
- eats daily

Green Tree Frog
- lives in rain-forest environments
- temperature of environment about 29°C all the time
- lives in trees
- eats during the day
- eats insects
- needs one or two insects each day to survive

Investigate Log WB70

Draw Conclusions

1. **Compare** your diorama to those of other students. How are they alike and how are they different? How do these similarities and differences reflect the needs of the animals? Each should include food, water, and shelter. Differences will include the kinds of food, the temperature and humidity, and the kinds of shelter the animals need.
2. From what you have learned in this activity, what can you **infer** about the needs of all animals? They all need food, water, shelter, and the correct temperature and humidity.
3. **Scientists at Work** Scientists **plan and conduct simple investigations** to test hypotheses and inferences they have made. Write a plan for a simple investigation you might be able to do to find out if your terrarium would be a suitable habitat for your animal. Make certain that your investigation would not harm the animal. Answers may include plans for placing the animal in the terrarium for a carefully supervised test period to see if it can live in the environment.

Investigate Further Find out more about the animal you chose for this investigation. Use reference books or encyclopedias. Change your diorama if you need to, based on what you find out. ______

Process Skills Practice WB71

Plan an Investigation

Designing experiments is often an important part of a scientist's job. The outcome of an experiment can be doomed before the first observation is made if the procedure is poorly thought out. You must state a clear hypothesis and then design the procedure to specifically test it. Choose your independent and dependent variables carefully, and then control all other factors.

Think About Planning an Investigation

Plan an investigation to determine what kind of pet food an animal prefers. First, state your hypothesis. Then, decide what your variable will be. You should try several brands or types of pet food. Decide which factors must be controlled and for how long you need to run the experiment. Remember, your experiment must be designed so it in no way harms the pet.

1. State your hypothesis. Answers will vary, but may predict a preference for one type or brand of food. Hypotheses are always positive statements.
2. What is the variable in this investigation? the brand or type of pet food
3. What factors need to be controlled? Factors should include time of feeding, number and sequence of feedings per brand or type of food, amount of food per feeding.
4. Write a procedure. ______
5. How are the results of the investigation dependent on the plan? If the investigation doesn't really test the hypothesis, or all factors but the variable aren't controlled, the results won't be valid.

Reading Skills Practice WB72

Identify Supporting Facts and Details

Read the following selection about polar bears. Then fill in the chart by indicating the main idea and the supporting facts and details.

Polar bears are uniquely suited for cold temperatures and an aquatic habitat. Although the bears appear to have white fur, the individual hairs are actually transparent and hollow. The hair is waterproof and traps air, keeping the bears warm and dry. Under their fur, polar bears have black skin, which absorbs more heat than any other color does. A thick layer of fat under the skin helps insulate the bears. These bears also have small ears, which prevent a major heat loss.

Polar bears are excellent swimmers. They have been tracked swimming continuously for 100 kilometers (62 mi.) and have been clocked at speeds of 10 kilometers per hour (6.2 mph). The front paws of polar bears act as paddles, while the rear paws serve as rudders. Polar bears tend to get overheated, so they swim to maintain a body temperature of 37°C (98.6° F).

Main Idea: Polar bears are well suited for cold temperatures and an aquatic environment.

A. cold temperatures		
1	2	3
waterproof, transparent, hollow fur	black skin, fat layer	small ears

B. aquatic environment		
1	2	3
swim far and fast	front paws as paddles, rear paws as rudders	swim to maintain even body temperature

Lesson Concept Review/Assessment WB73

What Are Vertebrates and Invertebrates?

Lesson Concept

All animals have the same basic needs: food, water, oxygen, and protection. An animal can be either a vertebrate, which means it has a backbone, or an invertebrate, which means it has no backbone. Ninety-six percent of all animal species are invertebrates.

Vocabulary

vertebrates (B70) **invertebrates** (B70) **endoskeleton** (B70) **exoskeleton** (B71)

Answer each question by using one or more complete sentences.

1. How are animals different from plants? Plants can make their own food. Animals must find and eat food to survive.
2. Why do all living things need oxygen? They need it to change food into useful energy.
3. Do animals that live in the desert need water? Explain why or why not. Yes; all animals need water because most life processes take place in watery solutions.
4. Vertebrates are chordates. What distinguishes a chordate? It has a notochord, a nerve cord, gill slits, and a tail.
5. What is a notochord? Why is it important to vertebrates? It is a rod of specialized cells alongside the nerve cord. In vertebrates it develops into a backbone that surrounds the nerve cord.
6. How do the skeletons of vertebrates differ from the skeletons of most invertebrates? Vertebrates have endoskeletons, which are connecting bones inside their bodies. Most invertebrates have exoskeletons on the outside of their bodies.
7. Invertebrates have no internal means of regulating their body temperature. Where would you expect to find the fewest invertebrates? The most? The fewest would be in the polar regions and the hottest deserts. The most would be in the tropics or in the ocean.

Workbook Support Continued

Investigate Log WB74

Three Groups of Invertebrates

Materials

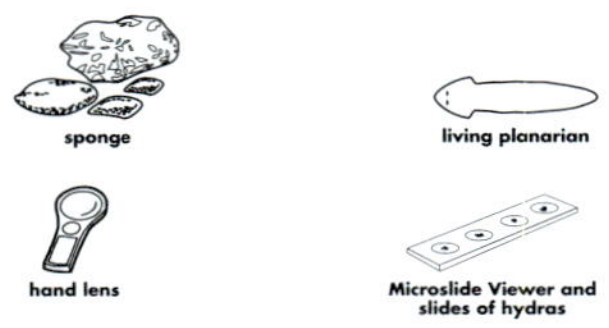

Activity Procedure

1. Use your hand lens to **observe** the sponge and the planarian. Draw a picture of each. Include all of the structures you see.
2. **Observe** the hydra by using the Microslide Viewer and slides.
3. **Compare** the three animals, and **record** your **observations** by writing a description of each one. Describe the structures you observe.

Investigate Log WB75

Draw Conclusions

1. What similarities and differences among the animals did you **record** in your descriptions? Similarities: They are invertebrates without hard outer coverings; the hydra and planarian have tissues and may have been observed moving. Differences: the hydra's mouth and tentacles; the planarian's eyespots, mouth, and feeding tube; and the sponge's inner and outer layer of cells.
2. As you **compared** the animals, which animal seemed to have the simplest body structure? Explain. Most students will suggest the sponge because it has no organs.
3. **Scientists at Work** When scientists discover information through their investigations, they **communicate** their findings to others who study the same thing. One way to communicate **observations** is to write a description. How else might scientists communicate their observations? with drawings, diagrams, or photographs; by oral descriptions

Investigate Further If you have access to a microscope, use it to **observe** animals you can find in a drop of pond water. Students might observe *Daphnia*, hydras, and many types of worms, including planarians, for example.

Process Skills Practice WB76

Communicate

Communicating is passing information on to others. You tell other people about your data and your conclusions. You can tell about your work with words, with pictures, and with tables and graphs. When you do an experiment, you must present your results to other scientists in a clear, organized manner so they can know what you have done and can repeat your work.

Think About Communicating

Fruit flies are tiny insects scientists study in laboratories to learn more about inherited traits. They are used because they have several easily distinguished inherited characteristics. One such trait is eye color. Some fruit flies have white eyes and some have red eyes. Imagine an experiment in which two red-eyed fruit flies are bred. You observe that all of their offspring have red eyes. When two flies from this first generation are bred, you observe that one-fourth of their offspring have white eyes. Next, a red-eyed fly and a white-eyed fly from this second generation are bred. The offspring (the third generation) are 50 percent red-eyed and 50 percent white-eyed. Communicate the results of this experiment by filling in the three Punnett squares below.

	R	r
R	RR	Rr
R	RR	Rr

First Generation: 4 Red, 0 White

	R	r
R	RR	Rr
r	Rr	rr

Second Generation: 3 Red, 1 White

	r	r
R	Rr	Rr
r	rr	rr

Third Generation: 2 Red, 2 White

1. If you were asked to write an account of this experiment, what conclusions would you report? Answers will vary but could include that red eyes were dominant and that at least one original parent had a white-eye gene.
2. Why is it important to clearly communicate the results of an experiment? Other scientists must be able to repeat your work. No matter how great your work, it means very little if no one knows about it.

Reading Skills Practice WB77

Identify Cause and Effect

Read each statement below about coral reefs. Indicate which part of the sentence indicates cause and which part indicates effect.

	Cause	Effect
Statement 1 When tropical forests are cleared, topsoil washes into coastal ecosystems, blocking the light.	when tropical forests are cleared	topsoil washes into coastal ecosystems, blocking the light
Statement 2 Coral skeletons are popular in home saltwater aquariums because they are natural biological filters.	because they are natural biological filters	coral skeletons are popular in home saltwater aquariums
Statement 3 Fishers have damaged coral reefs by carelessly handling lobster traps, lines, and nets.	by carelessly handling lobster traps, lines, and nets	fishers have damaged coral reefs

Lesson Concept Review/Assessment WB78

What Are Sponges, Cnidarians, and Worms?

Lesson Concept

Sponges are animals that filter water through many canals for feeding. Cnidarians have simple tissues and bodies that are hollow containers with poisonous tentacles circling a mouth. Flatworms have tissues and sense organs, while roundworms also have a digestive system with two body openings. Many roundworms and flatworms are parasites. Segmented worms have more complex body systems than other worms, including a circulatory system with five hearts.

Vocabulary

sponge (B76) **cnidarian** (B78) **flatworm** (B80) **roundworm** (B81) **parasite** (B81) **segmented worms** (B82)

Answer each question by using one or more complete sentences.

1. Your book tells you that the name *cnidarian* comes from a Greek word that means "sting." Why is this a good name for this group? All cnidarians can sting.
2. If you looked carefully at the water around a colony of sponges, it might appear somewhat turbulent. Why? Certain sponge cells have long, whiplike structures that beat to force water through the sponge. This stirs up sediment and debris in the surrounding water.
3. Why would reproducing sexually to form larvae that can swim away from the parent be an advantage to sponges and cnidarians? The offspring can move to other locations so they won't compete with the parents for food.

Place *S* for sponge, *C* for cnidarian, *R* for roundworm, or *E* for earthworm in the blank in front of the following descriptions.

C bodies have radial symmetry
E body has two sets of muscles for motion
S body functions carried out by individual cells
C body has a polyp form
C body functions carried on by tissues
C digestive system has just one opening
S body contains spicules
S body has no symmetry
E body has a heart and circulatory system
C all are carnivores

Investigate Log WB79

Compare Three Groups of Animals

Materials

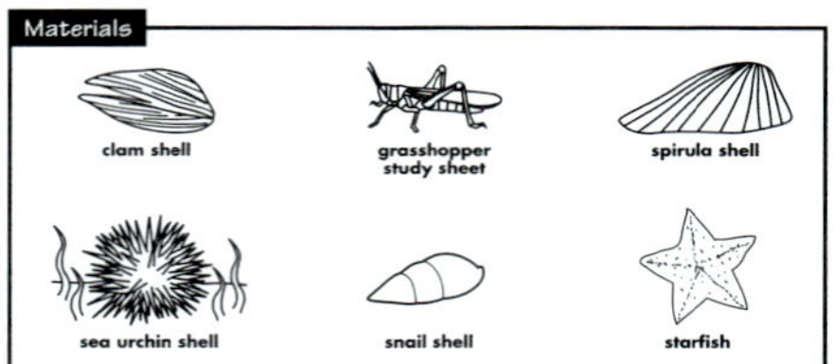

Activity Procedure

1. **Observe** the animal samples. Use the pictures in your textbook on page B85 to add to your observations. The cuttlefish is similar to the animal that contains the coiled spirula shell.
2. **Compare** the characteristics of the animals, and **record** your findings. Note any similarities and differences that you see among the animals.
3. Use your observations to **classify** the animals into three groups. (Hint: one group will have three animals, one group will have two, and one group will have one.)
4. Describe the characteristics that the animals within each group share. For the group that has only one animal, describe or name another animal that belongs in the group.

Workbook Support Continued

Investigate Log WB80

Draw Conclusions

1. What shared characteristics did you **observe** among the clam, the grasshopper, and the starfish? The clam has a soft body and is covered by a hard shell; the grasshopper has wings, eyes, a hard skeleton covering its body, and jointed legs; the starfish has radial symmetry and spiny skin.

   In what ways was each animal unlike the others?

2. Which of these characteristics did you use to **classify** the animals? Answers should include appropriate differentiating characteristics.

3. Name one more animal that might belong in each of your three groups. Answers may include brittle stars and sand dollars for the group including starfish and sea urchins; octopods, squid, and other snails and clams for the group including snails, limpets, and clams; and any insect, crabs, lobsters, or spiders for the group that includes grasshoppers.

4. **Scientists at Work** Scientists **classify** animals in the same group if those animals have important characteristics in common. How are the **inferences** you made to answer Questions 2 and 3 similar to a scientist's method of classification? The characteristics that students chose should be common to all the animals in each group, just as a scientist's would be.

Process Skills Practice WB81

Classify

Classifying means arranging objects together according to similarities in their characteristics.

Think About Classifying

For centuries people have been arranging and rearranging living things into categories based on common characteristics. The problem has always been choosing which characteristics to use. Below, you will find drawings of six different animals. In the questions that follow, you will be asked to classify the animals based on different characteristics.

1. Which of the animals are able to live in water? Because they all have that characteristic, should they be grouped together as closely related animals? Two and possibly three: the shark, the dolphin, and possibly the turtle. Students should be aware that the three animals are very different from one another and are not closely related just because they live in water.

2. What characteristics are shared by the turtle, the tiger, and the wolf? They all walk on four legs, and they can live on land. Students may think of other shared characteristics. Accept all reasonable answers.

3. What other characteristics of these animals can be used for classifying them? Some possible answers are the ability to fly; the ability to swim; having fur, feathers, or a shell; or having a dorsal fin. Students may suggest others. Accept all reasonable answers.

Reading Skills Practice WB82

Compare and Contrast

After reading the selection below, fill in the chart with details about how the octopus and the squid are alike and different.

Octopods and squid belong to the class Cephalopoda, which means "head-footed." Both of these unusual sea creatures have blue blood, which is caused by the oxygen-carrying molecule in their blood that contains copper. All cephalopods eat meat.

The octopus lives alone in a den at the bottom of the sea. This sea animal is known for its intelligence. In one experiment, an octopus removed a cork from a jar to get to the food inside.

The squid is known as a fast swimmer. It has been clocked at speeds up to 20 knots (23 mph 37 kph). It swims in schools and lives in the open seas. Squid have been used in medical studies of the nervous system.

Shared Characteristics	Different Characteristics
belong to the same class—Cephalopoda; have blue blood; eat meat; live in the ocean	octopus lives alone in dens; squid travels in schools and lives in open sea; octopus is known for its intelligence; squid is known for its fast swimming

Lesson Concept Review/Assessment WB83

What Are Mollusks, Arthropods, and Echinoderms?

Lesson Concept

Mollusks are soft-bodied invertebrates, such as clams, octopods, oysters, slugs, and squid. Many kinds of mollusks have external protective shells. Arthropods are invertebrates with jointed legs and an exoskeleton. They include insects, crustaceans, and arachnids. Spiny-skinned invertebrates called echinoderms have tube feet in rows on body parts that surround the animal's center. Echinoderms include starfish, brittle stars, sand dollars, sea urchins, and sea cucumbers.

Vocabulary

mollusk (B86) **insects** (B89) **echinoderms** (B91)
arthropod (B88) **arachnids** (B89)

1. What is a characteristic of all mollusks? They all have soft bodies with no bones.

2. What does the mantle do in clams? In slugs? In clams it secretes shell-forming material. In slugs it forms a tough body covering.

3. How are oysters and snails different? Oysters have two shells; snails have one shell and a radula.

4. How are the animals in this lesson different from most of the animals you studied in the last lesson? They have identifiable organ systems, and they reproduce almost exclusively sexually.

5. What are the characteristics of arthropods? They have an exoskeleton and jointed legs, a digestive system with two openings, well-developed organ systems, and a definite head.

6. Soft-shelled crab is considered quite a delicacy by many people. Crabs normally have hard shells. Suggest why at certain times of the year the crab's shell is soft. The crab is molting.

7. Arthropods include the only group of invertebrates that can fly. What group is this? insects

Vocabulary Review WB84

Recognize Vocabulary

In front of each statement, write *True* if the statement is true and *False* if the statement is false. If the statement is false, replace the underlined word to make the statement true. Write the correct word in the space provided after the statement.

False 1. The largest phylum of animals is mollusks. arthropods
False 2. Cnidarians have no special body symmetry. sponges
False 3. Invertebrates are part of the phylum Chordata. Vertebrates
False 4. Animals that harm their hosts and help themselves are arachnids. parasites
False 5. A planarian is an example of a roundworm. flatworm
False 6. Insects have endoskeletons. exoskeletons
False 7. Starfish are examples of arthropods. echinoderms
True 8. Invertebrates are animals without backbones.
True 9. Cnidarians have radial symmetry and tissues to carry out body functions.
False 10. Spiders are a type of insect. arachnid
False 11. Squid and octopods are echinoderms. mollusks
False 12. Earthworms are a kind of roundworm. segmented worm
True 13. Endoskeletons are on the interior of an animal's body.
False 14. Spiders are the largest group of animals in the world. Insects
True 15. Many roundworms are parasitic, including the *Ascaris* that harms humans.
False 16. Most arthropods have an endoskeleton. exoskeleton
True 17. Among the chordates, the vertebrates are the largest group.

Writing Practice WB85

Compare and Contrast Invertebrates

Informative Writing–Compare and Contrast

Write a handout to help young marine biologists tell the difference between sponges and cnidarians. Use the Venn diagram below to compare and contrast these two invertebrates in preparation for writing.

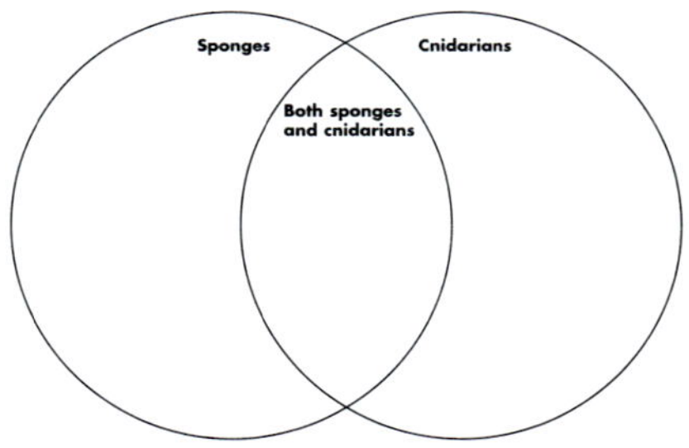

CHAPTER 3

Invertebrates

Generate Science Questions

Read aloud and discuss the Fast Facts. Afterward, ask students if they have any questions about invertebrates which they would like to have answered. Lead the class in generating a list of questions, and record the questions on chart paper or on the board.

Ask students what kinds of investigations they might design to find answers to their questions. Explain that although the chapter will have investigations and reading material to help answer these questions, they may wish to conduct investigations on their own.

CHAPTER 3

Invertebrates

Being vertebrates ourselves, we might think that a backbone is essential to all animals. The truth, however, is that more than nine out of ten species of animals do not have a backbone. Crabs, spiders, butterflies, and bees—to name only a few animals—are invertebrates.

Vocabulary Preview

- vertebrate
- invertebrate
- endoskeleton
- exoskeleton
- sponge
- cnidarian
- flatworm
- roundworm
- parasite
- segmented worm
- mollusk
- arthropod
- arachnid
- insect
- echinoderm

Fast Fact

There are more species of beetles than of any other animal. But if you could weigh all the animals on Earth, ants would account for 10 percent of the weight.

Number of Animal Species

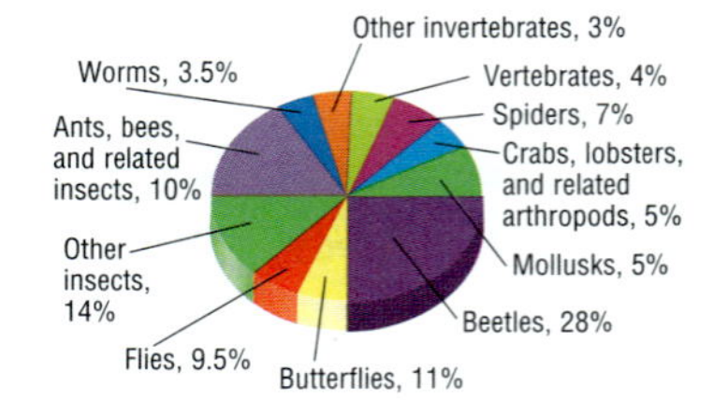

B64

Reading Skills Checklist

Strategies for developing the following reading skills are provided in this chapter.

- ☑ use prefixes and suffixes to determine word meanings *p. B71*
- ❑ use context to determine word meaning
- ☑ identify supporting facts and details *pp. B68, B70, B76, B78, B80, B81, B86, B88, B91, B92*
- ❑ arrange events in sequence
- ☑ identify the main idea *pp. B68, B70, B76, B78, B80, B81, B82, B86, B88, B91, B92*
- ❑ summarize and paraphrase a selection
- ☑ identify cause and effect *p. B78*
- ☑ predict probable future outcomes *p. B94*
- ❑ use graphic sources for information
- ☑ draw logical conclusions *p. B70*
- ☑ compare and contrast *p. B90*
- ☑ make generalizations *p. B87*
- ☑ use reference sources *p. B73*
- ☑ distinguish fact and opinion *p. B90*

School-Home Connection

Distribute copies of the School-Home Connection, p. TR9. Have students take the page home to share with their families as you begin the chapter.

Follow Up Have volunteers compare the results of the activity they did at home. Additional School-Home Connections are provided by the **Activities for Home or School** (p. B97) and the **Take-Home Book** (p. B64d).

Teaching Resources, p. TR9

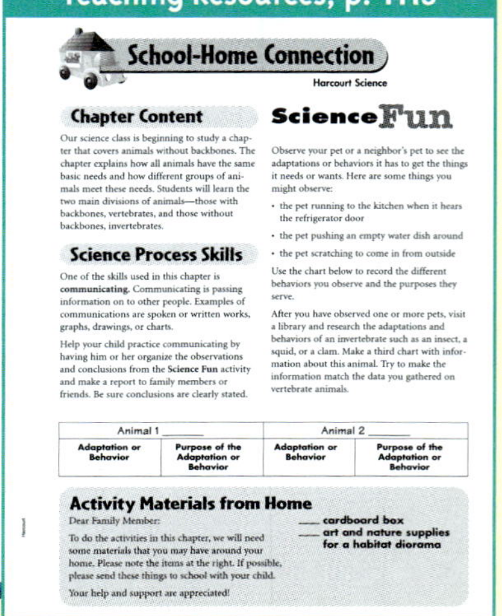
School-Home Connection

Harcourt Science

Chapter Content

Our science class is beginning to study a chapter that covers animals without backbones. The chapter explains how all animals have the same basic needs and how different groups of animals meet these needs. Students will learn the two main divisions of animals—those with backbones, vertebrates, and those without backbones, invertebrates.

Science Process Skills

One of the skills used in this chapter is **communicating.** Communicating is passing information on to other people. Examples of communications are spoken or written works, graphs, drawings, or charts.

Help your child practice communicating by having him or her organize the observations and conclusions from the **Science Fun** activity and make a report to family members or friends. Be sure conclusions are clearly stated.

Science Fun

Observe your pet or a neighbor's pet to see the adaptations or behaviors it has to get the things it needs or wants. Here are some things you might observe:

- the pet running to the kitchen when it hears the refrigerator door
- the pet pushing an empty water dish around
- the pet scratching to come in from outside

Use the chart below to record the different behaviors you observe and the purposes they serve.

After you have observed one or more pets, visit a library and research the adaptations and behaviors of an invertebrate such as an insect, a squid, or a clam. Make a third chart with information about this animal. Try to make the information match the data you gathered on vertebrate animals.

Animal 1 _______		Animal 2 _______	
Adaptation or Behavior	Purpose of the Adaptation or Behavior	Adaptation or Behavior	Purpose of the Adaptation or Behavior

Activity Materials from Home

Dear Family Member:

To do the activities in this chapter, we will need some materials that you may have around your home. Please note the items at the right. If possible, please send these things to school with your child.

Your help and support are appreciated!

___ cardboard box
___ art and nature supplies for a habitat diorama

Fast Fact

Can something without a backbone grow to be very large? The truth is that some invertebrates grow to a remarkable size.

Big Invertebrates

Animal	Size
Goliath spider	11 in.
Starfish	38 in.; 11 lb
Giant spider crab	body 12–14 in., claw span 8–9 ft
Lobster	42 lb 11 oz
Clam	49 in.; 660 lb
Jellyfish	diameter 7.5 ft, tentacles 120 ft
Sponge	diameter 3 ft, height 3.5 ft

Fast Fact

A strand of spider silk is five times as strong as a strand of steel of the same weight and thickness.

B65

Prereading Strategies

K-W-L Chart

Have students make a four-column chart headed What I Know, What I Want to Know, What I Learned, and "+" for What I'll Learn More About. Ask them to look at the chapter title and fill in the first two columns. As students study the chapter, have them fill in the third column. At the end of the chapter, encourage students to fill in the last column.

K-W-L Chart

What I Know	What I Want to Know	What I Learned	+

Preview the Chapter

Have students scan the chapter. They should identify the lesson titles and major headings and use them to make an outline. They can write questions as they make their outlines.

Vocabulary Preview

Have students make a four-column chart for the words in the vocabulary list. In the first column, students should write each word; and in the second column, the date they heard or saw the word. In the third column, students should write what the word might mean. In the fourth column, they should write the definition.

Vocabulary Chart

Word	What I Think It Means	Definition

BACKGROUND

Webliography

Keyword animals

www.harcourtschool.com

GO ONLINE

Attitudes Toward Invertebrates

In spite of the fact that invertebrates make up about 96 percent of all known species of animals on Earth, remarkably few of them are appreciated. In fact, many of them are feared by humans. Bees, ants, and jellyfish can inflict painful stings. Worms, spiders, and octopods are often viewed as ugly or threatening. Flies spread disease and tapeworms make us ill.

Invertebrates get a lot of "bad press." Yet they play vital roles in their particular ecosystems. Arthropods such as bees and flies pollinate. Others, such as spiders and ladybugs, help control populations of pests. Clams, oysters, crabs, and lobsters are important food sources. If nothing else, perhaps humans should acknowledge invertebrates for their sheer numbers and for their tenacity in all kinds of environments.

Fast Facts/Math Links

Organize Data Have students use the circle graph of Number of Animal Species to make an organized list of the number of invertebrate species, from greatest to least. The list would begin with beetles.

Analyze Data Discuss the data shown on the Big Invertebrates table. Have students give reasons why the data cannot be represented in a table or a graph. They should conclude that because the data includes several different types of measurements, such as inches, pounds, feet, and diameters, there is no logical basis for comparison. Help students understand that in order to make comparisons, they need like data.

LESSON 1

What Are Vertebrates and Invertebrates?

Objectives

- **Recognize that all animals have the same basic needs.**
- **Conclude that animals meet their basic needs in different ways.**
- **Distinguish between vertebrate and invertebrate animals.**

Motivate

More Fast Facts

Spark interest with these facts and questions.

- Humans have seven bones (vertebrae) in the neck.

  How many vertebrae do you think giraffes have in their long necks? They also have just seven neck vertebrae, but the vertebrae are greatly elongated.

- The average weight of the Blue whale is 84 tons. Its greatest length is 30 meters.

  Do whales have a backbone? yes

Access Prior Knowledge

- Review the basic needs of all animals.
- Name some animals with a backbone. Name some animals without a backbone.

When Minutes Count . . .

Whole-Class Investigate Option Set up the activity in the science center. Have students visit the center individually or in pairs. When the dioramas are done, have students discuss their similarities and differences.

Using the Activity Video The video can be used to preview and model the investigate process with students. After viewing the video, students can conduct the investigation in small groups.

LESSON 1

What Are Vertebrates and Invertebrates?

In this lesson, you can . . .

INVESTIGATE how to design an environment for an animal.

LEARN ABOUT how animals meet their needs and how animals are classified.

LINK to math, writing, physical education, and technology.

Moth caterpillars don't look like insects, but they are.

B66

INVESTIGATE

Designing Habitats

Activity Purpose Each environment on Earth has its own conditions. The animals in each environment may have different adaptations that enable them to meet their needs in the environment. However, all animals have some basic needs in common.

In this investigation you will find out how one animal is adapted to meet its basic needs in its environment. Then you'll design a terrarium (tuh•RAIR•ee•uhm) that could also enable the animal to live and grow.

Materials

- cardboard box
- art supplies and natural materials for a diorama

Activity Procedure

1. Choose one of the following animals: the Chilean rose tarantula, the giant green iguana, or the green tree frog.
2. Read in the appropriate chart about the animal's natural environment and about its habits. **Infer** what the animal needs to survive and how its environment meets these needs.
3. Draw a terrarium that would provide the animal with what it needs to survive. Make sure that you consider all the animal's needs.
4. Build a diorama that shows your animal in the terrarium. (Picture A)

Chilean Rose Tarantula

- lives in semidry environments
- temperature of environment is 21°–24°C
- lives in burrow during day
- eats at night
- eats insects
- eats only once a week

BACKGROUND

Webliography

Keyword animals

www.harcourtschool.com

GO ONLINE

Basic Needs Whether they are vertebrates or invertebrates, all animals have the same basic needs. All animals need food for energy, growth, and maintenance. They need oxygen to carry out respiration. Water is essential to transport elements into and out of cells and throughout their body systems. Climate and temperature must be appropriate. Finally, all animals need enough space to live.

The environment must provide for all the needs of animals that live in the wild. Humans must satisfy the needs of animals living in captivity.

For student self-assessment, see Assessment Guide p. AGxvi.

Giant Green Iguana

- lives in humid environments
- can survive temperatures as low as 27°C
- often basks in direct sunlight when the temperature is above 32°C
- lives in trees
- eats in the daytime
- eats leaves and flowers
- eats daily

Picture A

Green Tree Frog

- lives in rain-forest environments
- temperature of environment about 29°C all the time
- lives in trees
- eats during the day
- eats insects
- needs one or two insects each day to survive

Draw Conclusions

1. **Compare** your diorama to those of other students. How are they alike and how are they different? How do these similarities and differences reflect the needs of the animals?
2. From what you have learned in this activity, what can you **infer** about the needs of all animals?
3. **Scientists at Work** Scientists **plan and conduct simple investigations** to test hypotheses and inferences they have made. Write a plan for a simple investigation you might be able to do to find out if your terrarium would be a suitable habitat for your animal. Make certain that your investigation would not harm the animal.

Investigate Further Find out more about the animal you chose for this investigation. Use reference books or encyclopedias. Change your diorama if you need to, based on what you find out.

Process Skill Tip

Scientists **plan and conduct simple investigations** to test ideas they have about how things relate or work. A simple investigation should focus on only one question.

B67

INVESTIGATE

Time 2 hr **Preparation Tips** See p. B64c.

Grouping groups of four

Process Skill Tip plan and conduct simple investigations

Other Process Skills infer, compare

Expected Results Students should design an appropriate environment for one of the three animals.

Teaching Tips

Encourage students to draw preliminary sketches of the dioramas.

Draw Conclusions

1. Each diorama should include food, water, and shelter. Differences will include the kinds of food, the temperature and humidity, and the kinds of shelter.
2. All need food, water, shelter, and the correct temperature and humidity.
3. Answers may include plans for placing the animal in the terrarium for a carefully supervised test period to see how healthy it stays in the environment.

Investigate Further Provide research materials. Let students know that scientists often make changes when they discover new information.

Informal Assessment

Performance Check students' understanding of concepts by having them describe how the environment in their diorama provides for all the needs of the animal for which it is designed.

INVESTIGATE LOG

Students can use **Workbook pp. WB69–70** to record the results of the investigation.

PROCESS SKILLS PRACTICE

To practice and apply process skills, see **Workbook p. WB71.**

Process Skill Tip

Plan and Conduct Simple Investigations Display the Process Skill Tip Transparency on an overhead projector. Invite students to propose simple plans for proving or disproving the theory. Plans should include offering the cat one meat meal and one vegetable meal at the same time, observing the cat's preference, recording data for a period of time, and repeating the tests. Students should also ensure that the cat will not be harmed by the experiment.

Process Skill Tip Transparency B3-1

Plan and Conduct Simple Investigations

- Cats are said to be strictly carnivorous. Plan a simple experiment to prove or disprove this theory.
- Can you think of a situation where the cat might eat vegetables?

LEARN ABOUT

1 Before Reading

SET A PURPOSE

FIND OUT Remind students that these topics will help them know what to look for as they read.

VOCABULARY If students are completing the Vocabulary Preview activity described on page B65, remind them to add the definitions for these words to the chart.

Graphic Organizer for Chapter Concepts Display the transparency (shown below) or have students use Workbook page WB68. Point out that by the end of the lesson, they should be able to list the needs of vertebrates and invertebrates.

2 Guide the Learning

SCIENCE IDEAS

Help students focus on supporting details for this **main idea: All animals have the same basic needs.**

- ▶ **What basic needs do all animals have in common?** food, water, oxygen, appropriate climate, enough space
- ▶ **In addition, what do prey animals need?** They need ways to protect themselves from predators. These might include shelters that the animal constructs (beaver lodge, beehive); natural armor (turtle shells, porcupine spines); poison (wasps, scorpions, snakes); camouflage; and traveling in schools (fish) or herds (zebras).
- ▶ **Describe some different ways animals get oxygen.** from the air (mammals, birds, reptiles, insects); through their skin (worms); from the water they live in (fish)
- ▶ **How do an animal's teeth provide clues to what it eats?** Meat eaters (carnivores) such as members of the dog and cat families have sharp, pointed teeth that help them puncture and tear apart flesh. Plant eaters (herbivores) such as the hoofed animals have broad, flat teeth that are suitable for grinding.

VOCABULARY CARDS AND ACTIVITIES

Students can use the Vocabulary Cards to make their own graphic organizers or to add to an ongoing file of science terms. The Vocabulary Cards and a variety of strategies and activities are provided beginning on p. TR85 in the **Teaching Resources** book.

LEARN ABOUT

Two Types of Animals

FIND OUT

- about animal needs and how animals meet them
- how animals can be separated into groups based on the presence of a backbone

VOCABULARY

vertebrate
invertebrate
endoskeleton
exoskeleton

All Animals Have the Same Basic Needs

An adult giant green iguana may grow to be almost 1.5 m (about 5 ft) long. A young tarantula grows until it is about the size of your hand. Animals grow larger as they mature. Some change shape as they grow. Frogs start life as eggs and hatch into tadpoles that do not look much like adult frogs. As a tadpole grows, it loses its tail and develops legs. It finally becomes an adult frog. Animals can reach adulthood only if all their basic needs are met. These needs include food, oxygen, water, and protection from predators and other dangers in the environment.

Animals need food so they can have the energy to grow and to perform all life processes. Unlike plants, which make their own food, animals must find and eat food to survive. Some animals eat only plants. Others eat animals. And many animals eat both plants and animals. You can infer an animal's diet by observing its teeth. Teeth with broad surfaces can grind up plants. Long and sharp teeth can tear meat.

Sharks depend on their strength and speed as they hunt for food. Many other ocean animals strain tiny plants and animals from the water around them. Sponges pull water into their pores and strain bacteria, algae, or decaying plant or animal matter from the water. The largest animal on Earth, the blue whale, eats some of the smallest animals and plants. Blue whales strain ocean water through sievelike structures called baleen. In this way, a blue whale can consume as much as 3500 kg (about 7700 lb) of food in a single day.

Adult foxes care for their young and teach them how to protect themselves. ▼

B68

Graphic Organizer for Chapter Concepts

Transparency B3 • Workbook, p. WB68

Invertebrates

LESSON 1

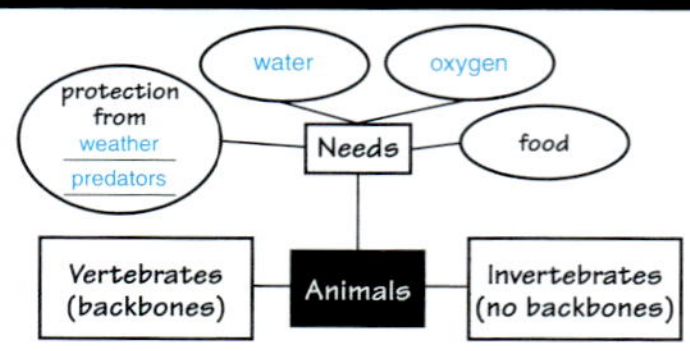

LESSON 2	Symmetry	Reproduction	Environment	Getting Food/Processing Food
Sponges		mostly asexual	shallow ocean	
Cnidarians	radial			
Planarians				
Roundworms				
Segmented Worms				
LESSON 3				
Mollusks				
Arthropods				
Echinoderms				

◀ Australian termites build mounds from bits of soil mixed with saliva. A mound can be 6 m (about 20 ft) tall and is divided into many chambers. It protects the termites from the weather and from predators. It is also a safe place in which young termites can hatch and grow.

Animals eat food for energy, but they can't use the energy unless they also have oxygen. Animals that live on land get oxygen from the air. Animals such as worms absorb the oxygen they need through the surface of their bodies. Most animals have more complex adaptations for getting oxygen. Insects have air tubes that extend to the body organs from openings on the insects' sides. Frogs and snakes have small, simple lungs. Birds and mammals, which are more active, have larger, more specialized lungs. These lungs enable them to take in the amount of oxygen they need.

Animals that live in water take in oxygen directly from the water. Many of these animals have gills, thin tissues that absorb oxygen from the water that passes over them.

Water is a basic need of all animals, not just those that live in water. All animals are made up of cells, which are mostly water. Most life processes take place in a watery solution inside the cells. Animals that live in water can take it in directly from their surroundings. Land animals must find water in order to stay alive.

Even animals that have enough food, oxygen, and water need one other thing. They need protection from the weather and from predators. Many animals build shelters or use naturally formed shelters. Some animals, such as armadillos and turtles, have armorlike shells for shelter and protection. Scorpions and wasps protect themselves by stinging their attackers. Some animals are protected because they look like their surroundings, which keeps predators from finding them. Other animals defend themselves by traveling in large groups or by running to escape danger.

✔ **What do all animals need to survive?**

food, oxygen, water, and protection from the weather and predators

The axolotl (AKS•uh•laht•uhl) is an amphibian. Unlike many other amphibians, it lives almost all of its life in the water. It gets oxygen from the water through its gills. ▶

B69

VISUAL LEARNING

Using the photos on pages B68–69, have students discuss the variety of ways in which the three animals meet their needs in the environment.

- **How do foxes learn to protect themselves?** adult foxes teach their young
- **How do Australian termites provide a safe habitat for the colony?** by building a mound
- **How is the axolotl adapted to living in water?** It can swim and has large external gills for obtaining oxygen from the water.

Ongoing Assessment

Use the ✔ questions in the pupil edition.

BACKGROUND

Webliography

Keyword animals

www.harcourtschool.com

Continuation of the Species

Not all animals actively defend themselves against predators. Very weak or vulnerable animals, such as the young or those in larval stages, depend on passive strategies for survival. For example, fish, frogs, and insects may release hundreds of eggs at a time. Most are eaten by predators, but many do survive. Some animals depend on camouflage; the flounder blends in with the floor of the sea, and the dappled fawn hides in a forest glade. Some animals, such as rabbits and gazelles, can outrun their predators. Regardless of the strategy, it is important that sufficient numbers of each of these species be able to reach maturity and reproduce, and thereby ensure the continuation of that particular form of life.

Guide the Learning continued

SCIENCE IDEAS

Help students focus on supporting details for this **main idea:** **Vertebrates are animals that have a backbone and invertebrates are animals that lack a backbone.**

- ▶ **What is the main difference between a vertebrate and an invertebrate?** A vertebrate has a backbone and an invertebrate does not.
- ▶ **Vertebrates are members of the phylum Chordata; they are chordates. What characteristics do all chordates have in common at some stage of their development?** All chordates have, at some time, a notochord, a nerve cord, gill slits, and a tail.
- ▶ **What two kinds of symmetry might you see in the invertebrates?** Two-sided (bilateral) symmetry, where each side is a mirror image of the other, is common in arthropods. Radial symmetry, in which similar parts are arranged around a central axis, is common in starfish, sea urchins, and sand dollars.
- ▶ **What is the difference between an exoskeleton and an endoskeleton?** An exoskeleton is on the outside of the animal; an endoskeleton is on the inside of the animal.

Critical Thinking Lead students to **draw a conclusion** about how animals are classified.

- ▶ **Are all vertebrates chordates? Are all chordates vertebrates? Explain your ideas.** Yes. All vertebrates belong to the phylum Chordata. But not all chordates are vertebrates. Some, like the lancelet, never develop a backbone.

Vertebrates

A **vertebrate** (VER•tuh•brit) is an animal that has a backbone. Scientists use the term **invertebrate** (in•VER•tuh•brit) to refer to any one of many different groups of animals that lack a backbone.

Vertebrates are part of the animal phylum known as chordates. At some stage of their development, all chordates have four characteristics. First, every chordate has a *notochord*—a long, flexible rod of specialized cells that runs along much of the body. Second, beside the notochord is a nerve cord that connects the animal's brain with the other parts of its body. Third, a chordate has gill slits. Finally, a chordate has a tail. Although some mature vertebrates do not have some of these characteristics, they all begin their lives as embryos that do have them. For example, cows do not have gill slits, but cow embryos do.

Among the chordates, the vertebrates are the largest group. Vertebrates include fish, amphibians, reptiles, birds, and mammals.

Vertebrates differ from other chordates because their notochord develops into a backbone. A mammal, for example, begins life as an embryo with a notochord. But soon bones develop around the notochord and replace it. The bones are *vertebrae* (VER•tuh•bray), individual parts of a backbone. They surround and protect the nerve cord.

The backbone is the core of a vertebrate's skeleton. It supports the other bones that form the animal's skeleton. A vertebrate has an **endoskeleton**—a system of connecting bones that lie within the animal's body. (The prefix *endo-* means "inside.") The skeleton supports the body and protects its most important organs. It also provides a system of levers, operated by muscles, that enables the body to move.

✓ **What is a vertebrate, and what are some examples of vertebrates?** A vertebrate is an animal with a backbone. Fish, amphibians, reptiles, birds, and mammals are all vertebrates.

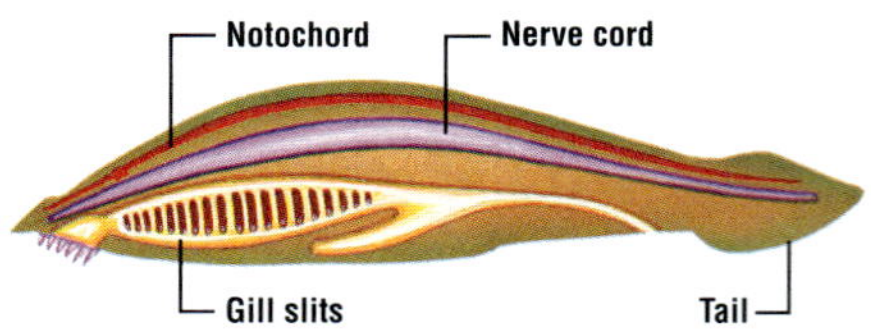

A lancelet is a chordate that lives in the water. It has a notochord, a nerve cord, gill slits, and a tail. But its notochord doesn't develop into a backbone. So a lancelet is not a vertebrate.

With its long string of vertebrae, a snake can bend, crawl, climb, and swim. Its backbone is made up of as many as 300 vertebrae. Humans have 33 vertebrae. ▶

B70

Reading Mini-Lesson

Identify Supporting Facts and Details Display the transparency and have students identify the main idea of this lesson: There are two types of animals. Then have students identify supporting facts and details for this main idea.

Reading Mini-Lesson Transparency B3-1

Identify Supporting Facts and Details

Main Idea	
There are two types of animals.	
Vertebrates have a backbone.	Invertebrates do not have a backbone.
Supporting Details	**Supporting Details**
1. The backbone is made of individual parts called vertebrae. 2. The backbone is the core of a vertebrate's skeleton called the endoskeleton. 3. The skeleton supports the body and protects its most important organs.	1. Invertebrates have many adaptations that enable them to survive without a backbone. 2. Some invertebrates have unusual body shapes called radial symmetry. 3. Many invertebrates have an exoskeleton on the outside of their bodies to protect tissues and organs.

READING SKILLS PRACTICE

To practice and apply this reading skill, see **Workbook p. WB72.**

◀ Dragonflies are insects that are often found near water. Dragonflies might be called living fossils because fossils of dragonflies look like the living animals. A dragonfly has four wings.

Invertebrates

Vertebrates make up about 4 percent of the animal kingdom. The other 96 percent of animal species are invertebrates.

Sponges, worms, centipedes, mollusks, and insects are just some of the animals that are invertebrates. Invertebrates have many adaptations that enable them to survive without a backbone.

Some invertebrates, such as sponges, have no head, mouth, or internal organs. Their body walls are only a few cells thick. However, the adaptations of sponges enable them to survive in their environments.

Some invertebrates, such as starfish, sea urchins, and sand dollars, have an unusual body arrangement. Their bodies are like the spokes of a wheel around a hub. This is called *radial symmetry*. They are not like most other animals, which have right and left sides that mirror each other. Among the animals with sides that mirror each other, the largest group is the arthropods, which include lobsters, crabs, spiders, and insects.

Most arthropods have an exoskeleton. An **exoskeleton** is a skeleton that is outside of an animal's body. (*Exo-* means "outside.") An exoskeleton protects the tissues and organs of an arthropod's body and gives it support. The skeleton is jointed, allowing arthropods a greater variety of movements than many other invertebrates.

✔ **How do vertebrates and invertebrates differ?**

The Portuguese man-of-war looks like a jellyfish, but it is not. It floats on the surface of the sea. Its mouth is at the center of its hollow body, surrounded by stinging tentacles that capture prey.

Diversity in the Animal Kingdom

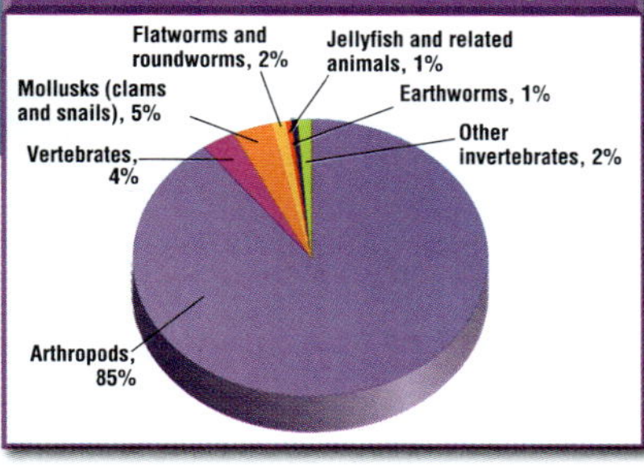

▼ The queen conch is a sea snail with a spiral shell. Conchs are mollusks, as are clams, snails, octopods, and squid. Most of these invertebrates have soft bodies that are protected by hard outer shells.

Invertebrates do not have a backbone; many invertebrates have an exoskeleton rather than an endoskeleton.

B71

ANALYZE AND INTERPRET DATA

Have students study the circle graph "Diversity in the Animal Kingdom" on page B71.

▶ **What percent of animals are invertebrates?**
96 percent

▶ **Of the invertebrates, what is the largest group?**
the arthropods, 85 percent

DEVELOP SCIENCE VOCABULARY

vertebrate and **endoskeleton**, p. B70—Suggest to students that *vertebrate* has the same word part as *vertical* ("straight up and down"), and that backbones help keep humans vertical. Then make the point that all animals with backbones have endoskeletons. Have students give examples of vertebrates to confirm this concept.

invertebrate and **exoskeleton**, pp. B70–B71—Review with students the **prefixes** *in-*, meaning "not," and *exo-*, meaning "on the outside." Have them give examples of other words that use the same prefixes, such as *inactive* or *inaccurate*, and *exhale* or *exit*. Lead them to conclude that most invertebrates have an outside skeleton. Have students give examples of invertebrates, and then have them name things that have a protective coating. They can name things found in nature or artifically made objects.

Art Link

Spider webs are amazing in their design and geometrical precision. Each species of spider builds a distinctive web, and webs can be a clue for identification. Spider webs are very delicate and are easier to see when they are coated with dew. You can also sprinkle a tiny bit of corn starch on a web to make it more visible.

Encourage students to find and observe a real spider web. Then provide them with string, glue, and a dark piece of paper to construct their own webs. CAUTION: **Some spiders can inflict painful and even dangerous bites. Do not touch them.**

ESL ACTIVITY

Students may find it confusing that one meaning of the prefix *in-* is "not," and the prefix *endo-* means "inside." Have students work in groups to review some other words with the *in-* prefix meaning "not" that may be more familiar, such as *invisible, ineligible,* or *inappropriate.* Have them use these words in sentences.

Guide the Learning continued

USE PROCESS SKILLS

Observe and Infer Help students infer that invertebrates inhabit a wide range of environments.

- **Using the reading selection and the photographs on pages B72–73, make a list of the environments that invertebrates inhabit. Add your own ideas to the list.** Lists will vary, but environments should include soil, plants, fresh water, salt water, a wide range of temperature and humidity, and interior as well as exterior locations.

Acceptable answers include almost every indoor or outdoor environment.

Invertebrate Environments

Almost any environment that supports vertebrates also supports invertebrates. In fact, there are far more invertebrate habitats than vertebrate habitats.

Invertebrates are Earth's most common animals both in numbers of species and in numbers of individuals. Look almost anywhere and you will find invertebrates such as insects, worms, and sow bugs. You may not see them at first, because many of them are well hidden. They may be under rocks or in the soil. Some live in moist, shady places, and others can be found in rotting logs.

Some invertebrates survive best in dry air and bright sunshine. They can be found above the ground. They feed on plants, other invertebrates, or small vertebrates. Many fly through the air, landing to eat or to rest. The leaves and stems of weeds, flowers, bushes, and trees are also environments in which many invertebrates live. Even when these animals are hard to find, you may see signs of them, such as spider webs, chewed leaves, or tiny mounds of dirt.

Invertebrates don't live just on land. Millions of these animals live in the oceans. At low tide, holes and trails show where they've been. Invertebrates such as marine worms and clams bury themselves in the wet sand. In deeper water, horseshoe crabs, starfish, and lobsters move across the ocean floor.

Tropical seas provide habitats for invertebrates such as corals and sponges. Freshwater environments also supply the needs of invertebrates. Lakes and ponds, and the mud at the bottom of them, provide many habitats for invertebrates.

✔ **Where would you look for invertebrates in your area?**

▲ An orb weaver spider makes complicated webs between tree branches or flower stems. Threads of dry silk extend from the center of the web, like spokes of a wheel. Coiling lines of sticky silk connect the spokes. The web is the spider's habitat and a trap for its prey.

A water spider lives most of its life under water. It breathes air from bubbles that it holds close to its body. Its underwater nest is a bell-shaped silk web that it fills with air bubbles.

B72

Math Link

Find Symmetry Explain to students that most animals have bilateral symmetry—one side of the animal mirrors the other side. Explain that some animals, like jellyfish, have radial symmetry—the body forms a circle around a point at the center.

Have students invent an animal that has bilateral symmetry and an animal that has radial symmetry, and have them make a drawing of each. As a predrawing activity, show pictures of animals that have bilateral symmetry (any mammal, bird, reptile, or fish) and animals that have radial symmetry (jellyfish, starfish).

INVESTIGATION CHALLENGE

Hands-On Activities: A Bone Collection

Use this idea for a hands-on activity to extend students' understanding of key lesson concepts.

Set up a spot in the classroom for a bone collection. Have students bring in a variety of bones for the class to examine. They might contribute bones from beef, chicken, fish, lamb, or pork. Instruct them to clean off the bones thoroughly before they bring them in. With adult supervision, they could remove the grease from the bones by immersing them in a weak bleach solution for a few days. Provide an identification chart. Challenge students to identify the bones and to compare the size, weight, and shape of the bones.

For student self-assessment, see Assessment Guide p. AGxvii.

▲ A trapdoor spider digs a burrow in the ground and covers the entrance with a lid, or trapdoor. It uses the burrow as protection and as a nest for its young. The trapdoor is made of silk and is attached to the lining of the burrow by silk hinges.

Summary

All animals have the same basic needs, which include food, water, oxygen, and protection. Animals can be divided into two groups—vertebrates and invertebrates. Vertebrates include all animals with a backbone. They have an endoskeleton. Invertebrates include all animals without a backbone. Many of these animals have an exoskeleton. Ninety-six percent of all animal species are invertebrates.

Review

1. Why is water a basic need of all animals?
2. How does an insect get the oxygen it needs?
3. What do all vertebrates have?
4. **Critical Thinking** A chameleon is a lizard that can change its color. When it's surrounded by leaves, the chameleon is green, and when it's on bark, it is brown. What basic need does this adaptation meet?
5. **Test Prep** Which animal has an exoskeleton?
   A a dragonfly
   B a snake
   C a cheetah
   D a shark

LINKS

MATH LINK

Compare Graphs Different kinds of graphs can show the same data. Draw a bar graph or a pictograph for the data in the circle graph on page B71. Which graph better helps you compare the percentages?

WRITING LINK

Informative Writing—How-to Go to a pet store and interview a salesperson about the care a particular invertebrate needs. Use the information to write a pamphlet for pet owners about taking care of that animal.

PHYSICAL EDUCATION LINK

Exercise for Your Pet One of the things people need for good health is exercise. Find out what sort of exercise your favorite pet animal needs. Make a plan that would help the pet get the right exercise.

TECHNOLOGY LINK

GO ONLINE

Visit the Harcourt Learning Site for related links, activities, and resources. **www.harcourtschool.com**

Welcome to The Learning Site

B73

3 Wrap Up and Assess

SUMMARY

Graphic Organizer for Chapter Concepts After reading the summary, have students complete the Lesson 1 section of the graphic organizer (shown on page B68).

REVIEW

1. Cells are composed mainly of water, and water is needed for most life processes.
2. Air travels through openings in the insect's body and along tubes to its organs.
3. a backbone
4. protection from enemies
5. **A,** a dragonfly

TEST PREP Test-Taking Tips

Item type multiple choice with question stem

To help students answer Question 5, share the following strategy. **Read the question carefully. Identify terms such as *exoskeleton* and try to recall the meaning. Then think of animals that have an exoskeleton, and try to find a similar animal in the list of possible answers.**

LESSON CONCEPT REVIEW/ASSESSMENT

Students can use **Workbook p. WB73** to review the lesson concepts.

LINKS TIPS

Links Objective Extend understanding of vertebrates and invertebrates.

Math Link Review the format for bar graphs with students. Make sure they understand what the horizontal and vertical axes represent.

Physical Education Link You may wish to have **reference sources** available from which students can find information about pets.

LESSON 2

What Are Sponges, Cnidarians, and Worms?

Objectives

- Describe the important characteristics of sponges, cnidarians, and worms.
- Compare and contrast the structures in these three types of invertebrates.

Motivate

More Fast Facts

Spark interest with these facts and questions.

- A large sponge can filter as much as 1500 liters (almost 400 gallons) of sea water through its body in one day.

  Why do you think sponges need to keep pumping water through their bodies? To get food; they trap and eat the plankton and bacteria in the water.

- In one acre of soil, you might find 50,000 earthworms.

  Are earthworms good for the soil? Yes. They aerate it by digging tunnels, and they fertilize it with their waste.

Access Prior Knowledge

- Have you ever seen a natural sponge? How was it different from a manufactured sponge?
- Name some kinds of worms you have seen. In what kinds of environments do they live?

When Minutes Count . . .

Whole-Class Investigate Option Divide the class into three teams and assign each team one animal to observe and record. Then bring the class together for a discussion to communicate their findings.

Using the Activity Video The video can be used to preview and model the investigate process with students. After viewing the video, students can conduct the investigation in small groups.

LESSON 2

What Are Sponges, Cnidarians, and Worms?

In this lesson, you can . . .

- INVESTIGATE sponges, planarians, and hydras.
- LEARN ABOUT the body structures and life cycles of sponges, cnidarians, and worms.
- LINK to math, writing, social studies, and technology.

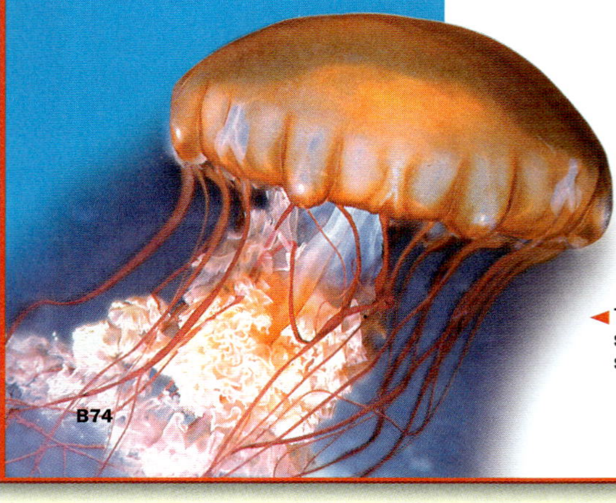

The sea nettle has tentacles with stinging cells to trap and sting small ocean organisms.

INVESTIGATE

Three Groups of Invertebrates

Activity Purpose Sponges, planarians, and hydras are three kinds of invertebrates that live in water. As invertebrates, these share some characteristics, but they also differ greatly. In this investigation you will **compare** these animals to discover their similarities and differences.

Materials

- hand lens
- sponge
- living planarian
- Microslide Viewer and slides of hydras

Activity Procedure

1. Use your hand lens to **observe** the sponge and the planarian. Draw a picture of each. Include all of the structures you see. (Picture A)
2. **Observe** the hydra by using the Microslide Viewer and slides. (Picture B)
3. **Compare** the three animals, and **record** your **observations** by writing a description of each one. Describe the structures you observe.

B74

BACKGROUND

Webliography
Keyword invertebrates
www.harcourtschool.com

Invertebrates exhibit extraordinary diversity, appearing in myriad colors, sizes, shapes, and body plans. To make sense of this diverse group of organisms, whose sole defining feature is the absence of a backbone, can be an overwhelming task.

One approach is to study invertebrates by moving from those with simpler structures and body plan to those with more complex structures and body plan. Sponges, cnidarians, and worms demonstrate increasing levels of complexity in structure and function. For example, sponges have no tissues or organs and just two layers of cells. Cnidarians have tissues, tentacles and stinging cells, and a net of nerve cells. Flatworms add a simple nervous system. Roundworms add a digestive system with two openings and a muscular system. The segmented worms have a circulatory system and more complex muscular and nervous systems.

For student self-assessment, see Assessment Guide p. AGxvi.

Picture A

Picture B

Sponge

Planarian

Hydra

Draw Conclusions

1. What similarities and differences among the animals did you **record** in your descriptions?
2. As you **compared** the animals, which animal seemed to have the simplest body structure? Explain.
3. **Scientists at Work** When scientists discover information through their investigations, they **communicate** their findings to others who study the same things. One way to communicate **observations** is to write a description. How else might scientists communicate their observations?

Investigate Further If you have access to a microscope, use it to **observe** animals you can find in a drop of pond water.

Process Skill Tip

To **communicate** your observations and conclusions so that others can understand them, include as much detail as possible in your descriptions. Use correct spelling and clear language.

B75

INVESTIGATE

Time 60 min **Preparation Tips** See p. B64c.

Grouping individuals

Process Skill Tip communicate

Other Process Skills compare, observe, record data

Expected Results Students should record the similarities and differences among the three organisms.

Teaching Tips

Remind students to use clear, descriptive language to record their observations.

Draw Conclusions

1. Similarities: They are invertebrates without hard coverings; hydra and planarian may have been moving. Differences: the hydra's mouth and tentacles; the planarian's eyespots, mouth, and feeding tube; the sponge's layers of cells
2. The sponge has a simpler body plan because it has no organs.
3. with drawings, diagrams, or photographs; by oral descriptions

Investigate Further Have students draw pictures of the animals they see in the water.

Informal Assessment

My Work

Portfolio Students may want to add their observations and descriptions of the three organisms to their science portfolios.

Process Skill Tip

Communicate Display the Process Skill Tip Transparency. Explain that scientific descriptions should be clear and detailed. They should be based on observations, not on emotional reactions or inferences. Have students write and discuss their observations. They should mention that developing embryos are visible through the transparent skin. If students use inappropriate terms like "ugly" or "gross," take the opportunity to discuss how these are reflections of opinion, not observable features.

Process Skill Tip Transparency B3-2

Communicate

- **Observe the worm**
- **Describe what you observe in detail.** smooth, slender shape, pointed at both ends
- **What are some ways to communicate your observations?** using as much detail as possible, correct spelling, and clear language

Reaching All Learners

Visual Learners These students may find it easier to use descriptive language after they have observed and drawn the organism.

INVESTIGATE LOG

Students can use **Workbook pp. WB74–75** to record the results of the investigation.

PROCESS SKILLS PRACTICE

To practice and apply process skills, see **Workbook p. WB76.**

LEARN ABOUT

1 Before Reading

SET A PURPOSE

FIND OUT Remind students to use these topics to guide them in what to look for as they read.

VOCABULARY If students are doing the activity described in the Vocabulary Preview on page B65, remind them to add the definitions for these words to their chart.

Graphic Organizer for Chapter Concepts Display the transparency (shown below) or have students use Workbook page WB68. Point out that by the end of the lesson, students should be able to list important characteristics of sponges, cnidarians, and worms.

2 Guide the Learning

SCIENCE IDEAS

Help students focus on the supporting details for this **main idea: A sponge is an invertebrate animal.**

- **In what kinds of environments do sponges live?** aquatic, both salt water and fresh water
- **Why did biologists once think that sponges were plants?** They appear inactive, stay attached to one place, and sometimes reproduce by forming buds.
- **Do sponges move?** Yes. As larvae, they are free-swimming. Once attached, they move specialized whiplike structures that push water into the sponge.
- **What stiffens the sponge's skeleton?** spicules or protein

Critical Thinking Lead students to draw a conclusion about the simple organization of the sponges.

- **Sponges have no sense organs. Why do you think they can live without them?** Students should conclude that sponges can meet their biological needs without using any senses. They remain in one place as adults, and depend on water currents to bring them food.

Ongoing Assessment

Use the ✔ questions in the pupil edition.

LEARN ABOUT

Sponges, Cnidarians, and Worms

FIND OUT

- about sponges, cnidarians, and worms
- how each of these animals has adaptations that enable it to survive

VOCABULARY

sponge
cnidarian
flatworm
roundworm
parasite
segmented worm

Sponges

A **sponge** is one type of animal that filters the water it lives in to get food. Unless you observe a sponge under very high magnification, it seems to be inactive and to have no animal-like characteristics at all. In fact, biologists 200 years ago thought that sponges were plants.

A sponge has only two layers of cells. The outer surface of a sponge's body is dotted with tiny openings. These openings lead to canals, which are made up of specialized cells with whiplike structures that beat rapidly. This motion helps pump water into the canals. These cells also have tiny threadlike structures that remove small food particles from the water. The water leaves the sponge's body through an opening called a vent.

Between the outer layer of a sponge and the inner layer is a jellylike layer that contains cells that have different functions. They carry food from the collecting cells to the other cells in the sponge. They also make materials that form a kind of skeleton for the sponge. The skeleton is made up of tiny rodlike structures of tough protein called *spicules* (SPIK•yools). It takes thousands of connected spicules, or pieces of protein, to form a sponge's skeleton.

Scientists know of about 5000 species of sponges. Most of these live in shallow ocean waters where they attach themselves to hard underwater surfaces, such as the ocean floor, rocks, other animals, and human-made structures. Different species of sponges have

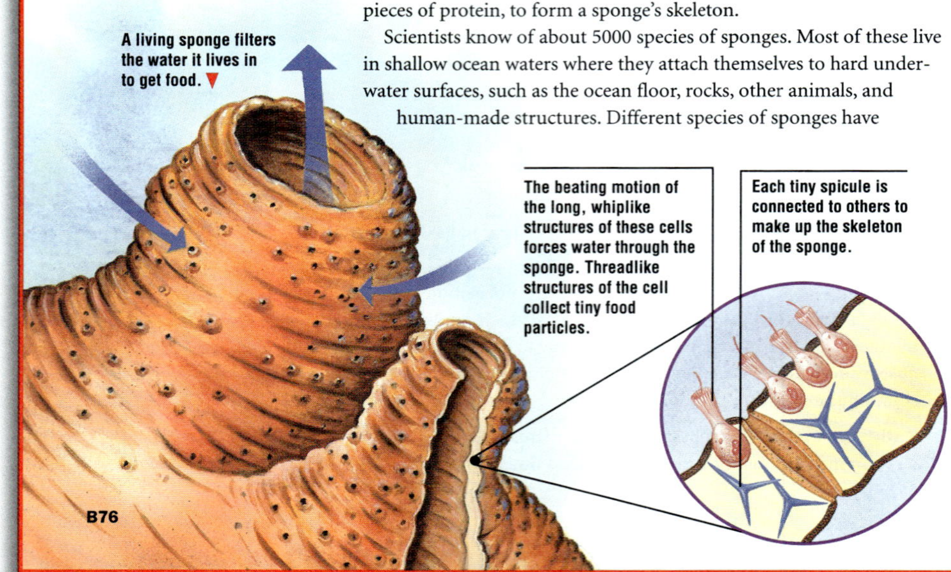

A living sponge filters the water it lives in to get food. ▼

The beating motion of the long, whiplike structures of these cells forces water through the sponge. Threadlike structures of the cell collect tiny food particles.

Each tiny spicule is connected to others to make up the skeleton of the sponge.

B76

Graphic Organizer for Chapter Concepts

Transparency B3 • Workbook, p. WB68

Invertebrates

LESSON 1

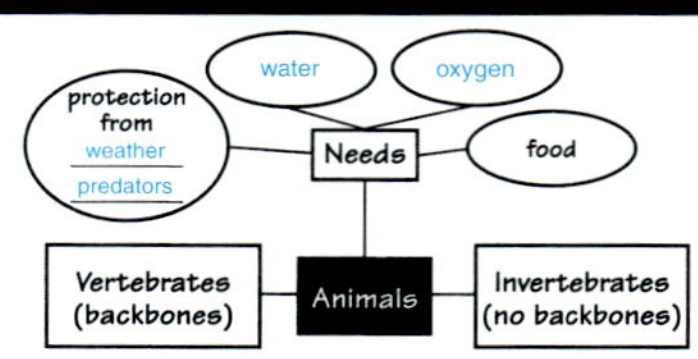

LESSON 2	Symmetry	Reproduction	Environment	Getting Food/Processing Food
Sponges	None or radial	mostly asexual	shallow ocean	filter food from water; no real digestive system
Cnidarians	radial	sexual/asexual	water	stingers and tentacles; one opening to digestive system
Planarians	bilateral	sexual	fresh water	mostly capture prey; one opening to digestive system
Roundworms	bilateral	sexual	land and water	capture prey or eat plants; two openings to digestive system
Segmented Worms	bilateral	sexual	land and water	digest food in soil; two openings to digestive system
LESSON 3				
Mollusks				
Arthropods				
Echinoderms				

different colors, shapes, and sizes. Some are black, brown, or gray, and others are bright shades of red, blue, purple, and yellow. Some sponges are shaped like fingers, some look like the branches of a tree, and others are round. The smallest sponges are about 1 cm (about $\frac{1}{2}$ in.) across, and the largest are 2 m (about 6 ft) across.

People use the word *sponge* to name the soft plastic material used for cleaning. Sponges that are sold as "natural" or "real" sponges, however, are sponge skeletons made of protein. In this country, the natural-sponge industry is centered in Tarpon Springs, Florida. Divers collect the sponges from underwater beds growing on the ocean floor. The sponges are then left to dry until only their protein skeletons remain.

Some of the smaller species of sponges live in fresh water. They can be found in lakes, ponds, and streams, growing on rocks or sticks under water. Because they are usually covered with algae, they are hard to see.

Most often sponges reproduce asexually. Buds and branches grow from a parent sponge. These buds break off and grow into new sponges. Some sponges also reproduce sexually, forming egg and sperm cells. A fertilized egg cell forms a larva that leaves the parent sponge and travels through the water to another part of the ocean floor. There it attaches itself to a hard surface, where it develops into an adult sponge.

✓ **Why is a sponge classified as an animal?** It takes in food and has moving, whiplike structures.

◄ This is the skeleton of a Venus's-flower-basket sponge. Its spicules are made of silica, or glass. These sponges grow at great depths, where ocean currents move slowly.

Sponges in tropical waters can be brightly colored. They are often found growing on or near coral reefs. ▼

USE PROCESS SKILLS

Observe Have students observe the pictures of sponges on page B77 and describe the diversity of shapes and colors.

► **What colors do sponges have? What shapes?** Sponges may be dull in color, such as brown, gray, or black; or they may be vivid shades of red, blue, orange, purple, and yellow. Shapes are very diverse: tall and thin, short and squat, and rounded or branching.

DEVELOP SCIENCE VOCABULARY

sponge, p. B76—Ask students to describe a sponge cake and to speculate on how it got its name. They should mention that a sponge cake is light and full of holes, and that it can absorb liquid sauces.

BACKGROUND

Webliography

Keyword invertebrates

www.harcourtschool.com

GO ONLINE

Porifera Sponges belong to the phylum Porifera (poh•RIF•er•uh). The name is appropriate because it means "pore bearers." A sponge's body is a network of tiny pores and canals that allow water to filter through. With the water come the small particles of food and the oxygen the animal needs. Commercially produced sponges are designed to simulate the porous structure of real sponges. Plastic, rubber, or fabric ingredients are whipped to a froth to make tiny air pockets that can absorb and hold large amounts of water.

Guide the Learning continued

SCIENCE IDEAS

Help students focus on supporting details for this **main idea: Cidarians are invertebrates that share some common characteristics.**

- ▶ **Name some cnidarians.** hydra, sea anemone, jellyfish, coral
- ▶ **In what kind of environment do all cnidarians live?** aquatic, mostly salt water
- ▶ **What adaptations help cnidarians catch their prey?** All cnidarians can sting their prey. Then they use their tentacles to move the prey into the mouth opening.
- ▶ **What two body forms can cnidarians take?** polyp or medusa

VISUAL LEARNING

Give students time to study the pictures and read the captions about cnidarians on pages B78–79.

- ▶ **Describe the reef-building cnidarians.** Descriptions will vary but should include some of the following: They are called coral, live in huge colonies in sea water, have many different shapes and colors, and show radial symmetry. Coral polyps live on the skeletons of dead coral. They continue to build the reef by forming their own skeletons.
- ▶ **Which cnidarians in the pictures show the medusa form?** The adult jellyfish and the sea nettle. They are free-floating and have umbrella-shaped bodies and long, hanging tentacles.
- ▶ **How is the comb jelly different from the cnidarians?** It does not have stinging cells.

Cnidarians

When you observed a hydra, you may have seen tentacles extending from the large part of its body. Even though it is small, a hydra can sting its prey or predators with poison found in some of its tentacle cells. This ability to sting is a characteristic of all the animals called **cnidarians** (ny•DAIR•ee•uhnz).

Cnidarians are carnivorous, or meat-eating, invertebrates. They have tentacles all around the mouth. After the prey is stung and stunned, the tentacles pull the prey toward the mouth. The mouth is connected directly to the middle of the body, where the food is digested. Once the food is digested, wastes leave through the mouth.

A cnidarian does not have organs in its body. Instead, cnidarians have cells that are organized into tissues—groups of cells that work together. Many of their body functions are carried out by tissues or cells. For example, the outer layer of tissue of a cnidarian protects the organism. The inner layer of tissue digests its food. In addition to their stinging cells, cnidarians also have other specialized sensory cells.

All cnidarians live in water. Most live in sea water, as do most species of sponges. Hydras belong to the only class of freshwater cnidarians.

Cnidarians have radial symmetry. Their body parts are arranged around a center, like spokes around the hub of a wheel. Even so, different cnidarians have different forms. Jellyfish shaped like umbrellas, sea anemones (uh•NEM•uh•neez)

The largest structure built by organisms on Earth isn't the Superdome in New Orleans or the Great Wall of China—it's Australia's Great Barrier Reef. The reef, which is about 2000 km (about 1250 mi) long, was formed by corals. ▼

Coral is a cnidarian that lives in groups called colonies. A coral produces a skeleton to protect its soft body. New coral polyps grow on top of the skeletons of dead corals. ▶

As a colony of corals grows, it forms a coral reef. Coral reefs are the habitats of many kinds of other animals, both vertebrates and invertebrates.

B78

Reading Mini-Lesson

Identify Cause and Effect Explain that a cause is an action or a condition that brings about an effect, or a result. Display the transparency and have students identify the cause and effect in each statement.

READING SKILLS PRACTICE

To practice and apply this reading skill, see **Workbook p. WB77.**

Reading Mini-Lesson Transparency B3-2

Identify Cause and Effect

	Cause	Effect
Statement 1 The development of powerful microscopes revealed that sponges were animals, not plants.	The development of powerful microscopes	revealed that sponges were animals, not plants.
Statement 2 New sponges grow when buds and branches break off from a parent sponge.	buds and branches break off from a parent sponge.	New sponges grow
Statement 3 Planarians stay out of the light, and so during daylight they can be found under rocks or in the mud of freshwater streams, lakes, and ponds.	Planarians stay out of the light	during daylight they can be found under rocks or in the mud of freshwater streams, lakes, and ponds.

that look like flowers, and corals shaped like vases are all cnidarians. They all have a hollow pouch with a single opening—the mouth—and a circle of tentacles around it.

A cnidarian that has a vase-shaped form like the hydra is called a polyp (PAHL•ip). Sea anemones and corals are also polyps. The other form of cnidarian is called a medusa. A medusa has a bowl-shaped body from which its tentacles hang down.

A medusa is the adult stage in the life cycle of some cnidarians. Adult jellyfish are medusas that develop from polyps as they mature. Not all cnidarians develop into medusas. Sea anemones and hydras, for example, remain as polyps when they are mature.

Cnidarians reproduce both sexually and asexually. In jellyfish the medusa releases sperm and egg cells into the water, where fertilization takes place. Fertilized egg cells develop into larvae that can move through the water. The larvae settle on the ocean floor and develop into polyps. In its polyp form, the jellyfish reproduces asexually. It forms medusas that break off from the parent and swim away.

✔ **What characteristics do all cnidarians have in common?**

All are carnivores, have stinging cells, live in water, and have radial symmetry.

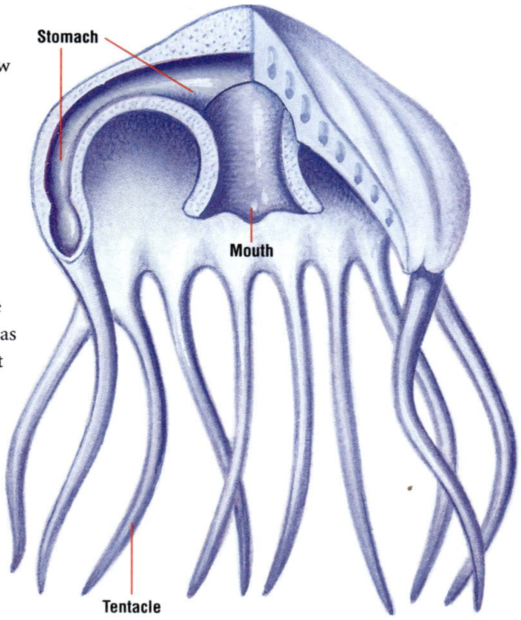

▲ An adult jellyfish, or medusa, is bowl-shaped. Some of the cells of its tentacles contain a coiled thread that has tiny hooks, or barbs, on it. The barbs can inject a poison that can paralyze a small fish.

▲ Sea nettles are a common jellyfish often seen in the water or washed up on shore. They swim by expanding and contracting their bodies. The word *cnidarian* comes from a Greek word that means "sting." Like all cnidarians, sea nettles can sting.

A comb jelly belongs to a phylum related to the cnidarians. Comb jellies have eight bands of comblike organs on the sides of their bodies. The combs are rows of tiny hairs that move the animal through the water. Like fireflies and glowworms, some comb jellies can produce light. They may or may not have tentacles, and they don't have stinging cells. ▼

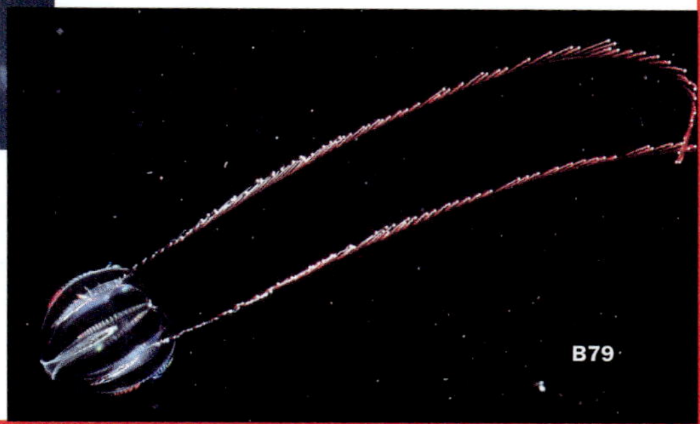

USE PROCESS SKILLS

Compare Ask students to review the information on the sponges. Then have them compare sponges with cnidarians.

▶ **How do sponges and cnidarians differ?** Cnidarians have specialized stinging cells, tentacles, a digestive cavity, cells that are organized into tissues, and sensory cells.

DEVELOP SCIENCE VOCABULARY

cnidarian, p. B78—Explain to students that the phylum name Cnidaria is derived from two roots—the Greek *knide,* meaning "nettle," plus the Latin suffix *-aria* meaning "like." Have students look up *nettle* in the dictionary. They will find that nettles are prickly or stinging plants. Ask students to explain how cnidarians are like nettles.

BACKGROUND

Webliography

Keyword invertebrates

www.harcourtschool.com

GO ONLINE

Stinging Cells There are about 9000 species of cnidarians, and all have the ability to sting using special cell structures called **nematocysts**. Nematocysts are usually found on the tentacles. The nematocyst is a tiny chamber containing a coiled thread, sometimes with barbs or spines on the end. The chamber is covered by a little lid topped with a sensitive, hairlike trigger. When a prey animal comes in contact with the trigger, the lid of the nematocyst bursts open and a barbed thread shoots out and buries itself in the skin of the animal. The tiny weapon injects poison into the prey and stuns it.

Most nematocyst stings are not very harmful to humans, but they can be painful.

ESL ACTIVITY

Have small groups of students work together to review **cnidarian**. Have students with intermediate levels of fluency draw a simple picture of a cnidarian, label the tentacles, and discuss how the drawing shows radial symmetry.

Guide the Learning continued

SCIENCE IDEAS

Help students focus on the supporting details for this **main idea: All flatworms have a number of characteristics in common.**

- **In what kind of environment do most flatworms live?** aquatic, both fresh water and salt water
- **How are flatworms, like the planarian, more complex than sponges?** Flatworms have bilateral symmetry, sense organs and nerve cords, a mouth and a digestive system, and sex organs.
- **What observable characteristics help identify a planarian?** a flattened body, a triangle-shaped head, and two eyespots that make it look cross-eyed

USE PROCESS SKILLS

Compare Ask students to compare flatworms and roundworms.

- **Observe the pictures of roundworms and flatworms. Compare their appearance.** Flatworms are flattened; roundworms are round and tubular.
- **On the basis of your reading, what other features distinguish roundworms from flatworms?** Roundworms have a digestive system with two openings. They take in food at one end and eliminate waste from the other. They have muscles that run the length of the body.
- **Compare reproduction in roundworms and flatworms.** Both reproduce sexually.

Worms

There are many phyla of worms. Some types of worms, such as segmented worms, have complex body systems. But some, such as flatworms and roundworms, have a simpler body organization. These organisms differ from sponges and cnidarians in the shape of their bodies. While sponges may have no symmetry and cnidarians have radial symmetry, flatworms and roundworms have *bilateral* (by•LAT•er•uhl) *symmetry*. Each side of the body of one of these worms is a mirror image of the other side.

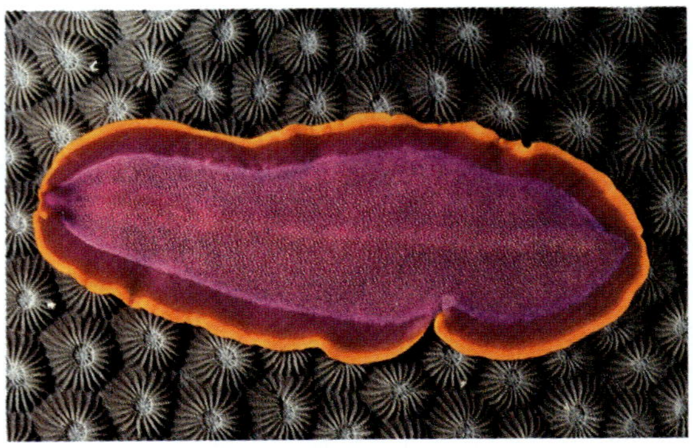

▲ This is another example of a flatworm. It lives in ocean environments. Often it is part of coral reef ecosystems.

Flatworms

A **flatworm** has a flattened body, a digestive system with only one opening, and a simple nervous system. A planarian is one kind of flatworm.

Planarians live in freshwater streams, lakes, and ponds. They are usually small. Most are less than 30 mm (about 1 in.) long. Planarians stay out of the light. During daylight they can be found under rocks or in the mud at the bottom of their freshwater environments.

Planarians are not familiar to most people. However, as you saw in the investigation, you can easily identify them by their triangle-shaped heads and their eyespots. When seen under a microscope, planarians look cross-eyed.

A planarian has body organs made up of tissues. It also has sense organs in its head. Two nerve cords run down the length of its body. A planarian has a mouth at the end of a long tube. The tube is near the middle of the underside of the planarian's body. It is used to take in food and to push out wastes.

Unlike cnidarians, planarians have reproductive systems with sex organs and can reproduce by mating and laying eggs. They also can regenerate body parts that have been cut off. If a planarian is cut in two, the part with the head will grow a new tail and the part with the tail will grow a new head.

✓ **What characteristics do all flatworms share?** flattened body, digestive system with only one opening, simple nervous system

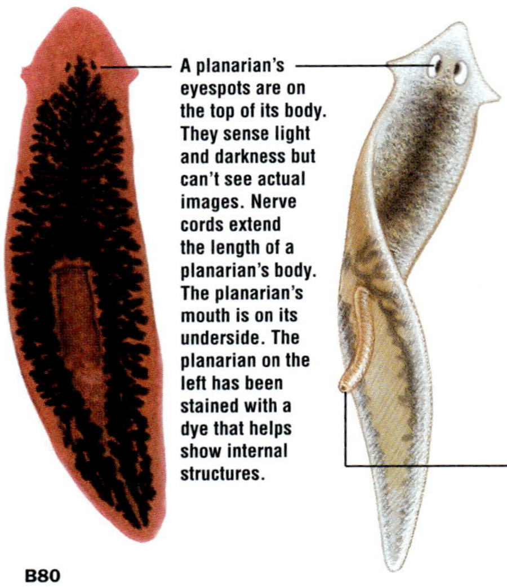

A planarian's eyespots are on the top of its body. They sense light and darkness but can't see actual images. Nerve cords extend the length of a planarian's body. The planarian's mouth is on its underside. The planarian on the left has been stained with a dye that helps show internal structures.

To eat, a planarian extends a tube from its mouth. The tube works like a straw, sucking small particles and liquids into the animal's stomach.

B80

THE QUESTIONS KIDS ASK!

Why are there dead earthworms all over the sidewalk after it rains?

Address Misconceptions A heavy rain may flood a worm's burrow, and many worms come out of their burrows. Some worms are killed by the sun's rays. Ask students how they think sunlight can kill a worm. Their answers may include that the sun may dry it out or may burn it.

Roundworms

A **roundworm** has a round, tubelike body. Most roundworms are too small to be seen without a microscope. Roundworms are very numerous and live in many types of environments. It has been estimated that a shovelful of soil can contain about 1 million roundworms!

Unlike flatworms, roundworms have a digestive system with two openings. At one end is a mouth where food is taken in. At the other end is an opening through which the worm gets rid of wastes.

The muscles in a roundworm run lengthwise through the worm's body. Roundworms are the only group of worms with this arrangement of muscles. The contraction of these muscles results in the whiplike motion of roundworms.

A roundworm has a nervous system with a brain and other simple sense organs. It also has sex organs that enable it to reproduce sexually. Many roundworms harm other organisms in which they live and feed.

✓ **How do roundworms and flatworms differ?** Roundworms have a digestive tract with two openings, muscles running lengthwise; flatworms can regenerate and have eyespots.

▲ The sheep liver fluke is a flatworm that needs two hosts—sheep and snails—to complete its life cycle. This parasite travels to the liver of the sheep, where it lays eggs.

Parasitic Worms

Some worms survive by living on or in another animal and feeding on that animal. An organism that lives this way is called a **parasite** (PAR•uh•syt). It harms the host animal, or the animal that it is feeding on, while it benefits itself.

Flatworm parasites include tapeworms and flukes, both of which harm mammals, including humans. Roundworm parasites include hookworms and *Trichinella*, which are roundworms that live in pigs. They can pass into people who eat undercooked pork.

Many parasitic worms have a life cycle that needs two species of host animals. The fluke *Schistosoma*, for example, begins life in a freshwater snail. Many of these parasites then reach human hosts, where they mature within the humans' bodies. The serious disease caused by these parasites affects more than 200 million people in tropical parts of the world.

✓ **Why do parasites need a host animal?** Parasites survive by living in other animals' bodies, where they get protection and food.

◀ The roundworm *Ascaris* is a parasite that harms people. It is estimated that about 650 million people worldwide have the disease caused by this roundworm.

B81

SCIENCE IDEAS

Help students focus on supporting details for this **main idea: Some roundworms and flatworms are parasites.**

▶ **What kind of relationship does a parasite have with its host?** The host provides the parasite with food and a safe habitat. The parasite harms its host by feeding on it and by causing weakness and disease. Only the parasite benefits from the relationship.

▶ **Name some parasitic worms.** the sheep liver fluke, the roundworm *Ascaris*, hookworms, *Trichinella*, the fluke *Schistosoma*

DEVELOP SCIENCE VOCABULARY

flatworm and **roundworm**, pp. B80–81—Each of these creatures has a distinctive outward appearance. Have students name other items that may be classified as being either round or flat. They might mention pasta shapes, for example.

parasite, p. B81—Emphasize that a parasitic relationship is one that harms the host and benefits the parasite. Ask students to think of other examples of parasitic relationships, such as a dog and fleas.

Health Link

Schistosomiasis (shis•tuh•soh•MY•uh•sis) is a parasitic disease that occurs in tropical areas of Africa, Asia, Latin America, and the Middle East. People who live in or travel to those areas are at risk of infection if their skin comes in contact with fresh water in lakes, streams, or canals where the parasite *Schistosoma* is present. The parasites penetrate the skin, and within weeks, worms are producing eggs in the bloodstream. The skin becomes intensely itchy, and fever, cough, and muscle aches develop. Although parasitic worms can cause serious illness, it is important to realize that the vast majority of worms are not parasitic, and many are beneficial. Have students research the benefits of worms.

ESL ACTIVITY

ESL students will benefit from seeing the characteristics of sponges, cnidarians, and worms summarized in simple, graphic form. Write the following chart on the board and have students help complete the graphic.

Animal Group	Body	Reproduction
Sponges	Openings and canals that move food in and out of the body	Usually asexual
Cnidarians		
Flatworms		
Roundworms		
Segmented Worms		

Guide the Learning continued

SCIENCE IDEAS

Help students focus on this **main idea: Segmented worms have more complex body structures than roundworms and flatworms.**

- **Describe the segmented worm's circulatory system.** Five enlarged tubes act like hearts to pump blood through vessels to every body part.
- **Does the segmented worm have a nervous system?** Yes. It has a simple nerve center in the head, connected to a nerve that runs the length of its body.
- **Do earthworms have lungs?** No. They carry on respiration through their skin.
- **How does an earthworm move?** It has two sets of muscles. The set that runs lengthwise contracts the body; the set that runs around the body elongates it and pushes the head forward.

VISUAL LEARNING

Ask students to use the picture of the earthworm on page B82 to trace the movement of food.

- **Describe how food moves through the worm's digestive system.** Food enters the mouth and is stored in the crop, is ground up in the gizzard, and is absorbed by the intestine.

DEVELOP SCIENCE VOCABULARY

segmented worm, p. B82—belongs to the phylum Annelida, whose name means "little rings" in Greek. Explain why this is a fitting name. Ask students to name other living things or nonliving things that are divided into segments.

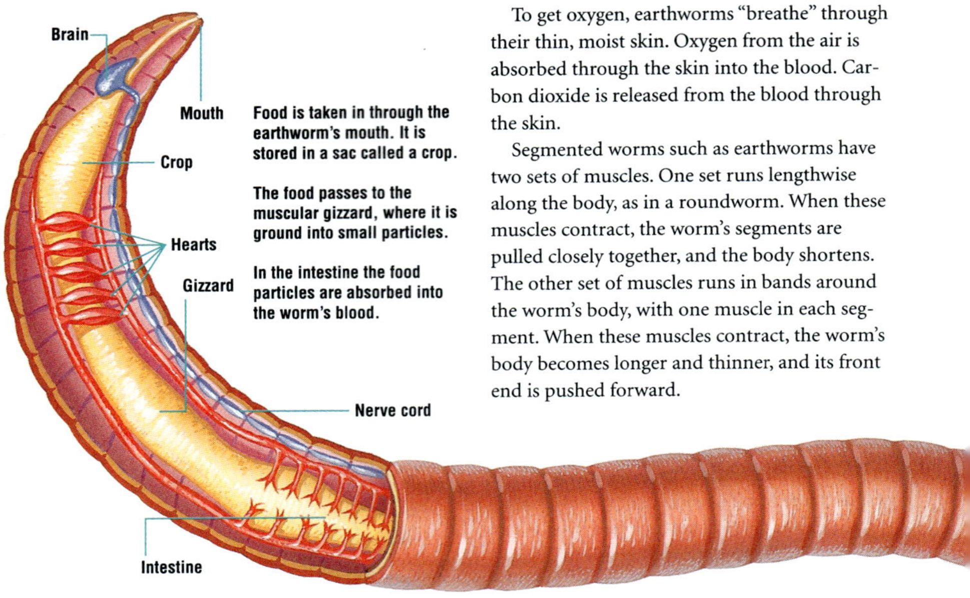

▲ Segmented worms, such as this earthworm, have bodies formed of connected segments. Some of the worm's body systems are shown in this picture.

Segmented Worms

Earthworms are very different from all the other worms you have learned about. They are part of a group of animals called **segmented worms**. These are worms whose bodies are made up of connected sections, or segments.

An earthworm's body systems are more complex than those of other worms. For example, earthworms have a circulatory system. Five enlarged tubes act as hearts. The tubes pump blood through the major vessels of the worm's body. These major vessels branch into smaller and smaller vessels, supplying blood to every part of the worm's body. An earthworm also has a nerve center, or brain, in its head. This brain is connected to a main nerve that runs the length of the worm's body.

To get oxygen, earthworms "breathe" through their thin, moist skin. Oxygen from the air is absorbed through the skin into the blood. Carbon dioxide is released from the blood through the skin.

Segmented worms such as earthworms have two sets of muscles. One set runs lengthwise along the body, as in a roundworm. When these muscles contract, the worm's segments are pulled closely together, and the body shortens. The other set of muscles runs in bands around the worm's body, with one muscle in each segment. When these muscles contract, the worm's body becomes longer and thinner, and its front end is pushed forward.

The two sets of muscles enable the earthworm to move through the soil, where it takes in food through its mouth. This burrowing not only enables the worm to get food, but it also lets air into the spaces where the worm has traveled. The wastes of the earthworm enrich the soil.

Every earthworm has both male and female sex organs. After worms mate, both partners lay eggs and produce a slimy covering that forms a cocoon protecting the eggs. Fertilized eggs develop, and soon worms hatch from them.

✓ **What is the major difference between an earthworm and other worms?**

An earthworm has body systems that are more complex than those of other worms.

B82

INVESTIGATION CHALLENGE

Hands-On Activity: Observing Earthworm Casts

Use this idea for a hands-on activity to extend students' understanding of key lesson concepts.

Earthworms play an important role in improving the soil. Their waste matter, called casts, is rich in nutrients that plants can use. Their tunnels help air reach plant roots and help drain away rainwater.

Challenge students to find worm casts and to describe them in words and pictures. They should describe the casts as small mounds of muddy soil that are made up of small particles and are found close to a worm hole.

For student self-assessment, see Assessment Guide p. AGxvii.

Summary

Sponges have just two layers of cells separated with a jellylike layer. Cnidarians, which include sea anemones, corals, and jellyfish, all have simple tissues and bodies that are hollow containers with venomous tentacles circling a mouth. Flatworms have tissues and sense organs and a digestive system with only one opening. Roundworms have a digestive system with two body openings. Many roundworms and flatworms are parasites. Segmented worms, such as the earthworm, have body systems that are more complex than those of other worms, including a circulatory system with five hearts.

Review

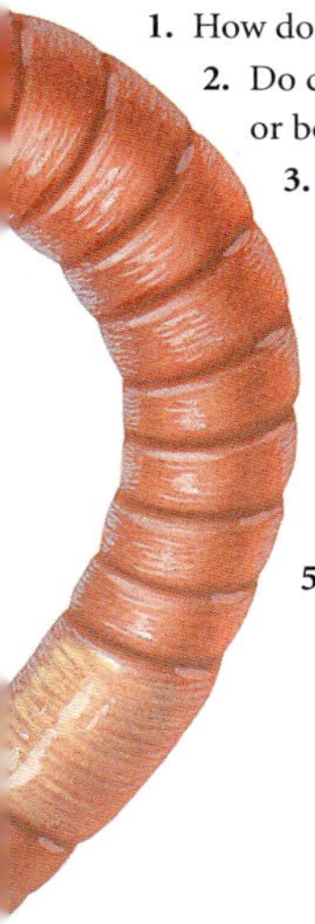

1. How do sponges get food?
2. Do cnidarians eat plants, other animals, or both?
3. How does an earthworm differ from a roundworm?
4. **Critical Thinking** A tapeworm is a kind of parasite that has no mouth. It takes in food directly through its body walls. How might these adaptations help a tapeworm survive?
5. **Test Prep** Roundworms have —
   A a digestive system with one opening
   B a circulatory system
   C a digestive system with two openings
   D tentacles

LINKS

MATH LINK

Use Mental Math A pork tapeworm can grow to be 7 m long inside its host. The tapeworm has as many as 1500 segments, each containing about 80,000 eggs. How many eggs can one tapeworm contain?

WRITING LINK

Informative Writing—Explanation Find out where the word *medusa* comes from, and write a paragraph for your teacher to explain why it is a good name for an adult jellyfish.

SOCIAL STUDIES LINK

Dams and Reservoirs To meet their needs, people have built lakes and reservoirs that hold huge amounts of fresh water. However, this progress has led to unexpected effects. Find out how the Aswan Dam in Egypt or Lake Volta in Ghana has caused an increase in dangerous diseases among the people who live nearby.

GO ONLINE — TECHNOLOGY LINK

Learn more about coral reefs by visiting the Smithsonian Institution Internet Site.
www.si.edu/harcourt/science

Smithsonian Institution®

Anno for Math Link
80,000 × 1,000 = 80,000,000
80,000 × 500 = 40,000,000
120,000,000

B83

3 Wrap Up and Assess

SUMMARY

Graphic Organizer for Chapter Concepts After they read the summary, have students complete the Lesson 2 section of the graphic organizer (shown on page B76).

REVIEW

1. They absorb food from the water that they propel through canals in their bodies.
2. other animals
3. The earthworm has a circulatory system.
4. The tapeworm lives in an environment that supplies its food. It can take in food with its entire body, without having to hunt for it or digest it.
5. C, a digestive system with two openings

TEST PREP Test-Taking Tips

Item type multiple choice with a sentence fragment

To help students answer Question 5, share the following strategy with them. **Read the sentence carefully. Try to recall some facts about roundworms that will help you narrow the possible answers to two. If you are not sure which answer is correct, make an educated guess.**

LESSON CONCEPT REVIEW/ASSESSMENT

Students can use **Workbook p. WB78** to review the lesson concepts.

LINKS TIPS

Links Objective Extend understanding of sponges, cnidarians, and worms.

Writing Link As a prewriting activity, you may want to tell students about the Greek myth of the snake-haired monster, Medusa. Once she was a beautiful maiden, proud of her charming ringlets. The goddess Athena became so jealous of Medusa's beauty that she changed her ringlets into hissing serpents. Medusa was so terrifying to behold that no living thing could look at her without being turned into stone. The warrior Perseus killed her while she slept. He took care not to look at her directly, but only at her reflection in his shield. Writing models are provided in the **Teaching Resources** book.

Social Studies Link Before students begin their research, it may be helpful to inform them that the construction of the Aswan High Dam (1970) precipitated thousands of cases of Rift Valley fever. The damming that made Lake Volta (1964) led to an outbreak of schistosomiasis.

LESSON **3**

What Are Mollusks, Arthropods, and Echinoderms?

Objectives

- ▶ **Describe the distinguishing characteristics of mollusks, arthropods, and echinoderms.**
- ▶ **Identify examples of each of these three types of invertebrates.**

Motivate

More Fast Facts

Spark interest with these facts and questions.

- An octopus can be taught to recognize colors and shapes, and can remember for a long time what it has learned.

  What physical features do you think allow an octopus to learn and remember? It has a well-developed brain and excellent eyesight.

- In the United States, more than 40 different crops, with an annual economic value of $30 billion, depend on insects in order to reproduce.

  What do insects do for the crops? They pollinate them.

Access Prior Knowledge

- ▶ Have you ever eaten any invertebrates?
- ▶ Do you have any feelings or opinions about insects, spiders, or slugs? Explain.

When Minutes Count . . .

Whole-Class Investigate Option Set up the investigation in the science center. When everyone has completed the task, discuss the classification systems as a class.

Using the Activity Video The video can be used to preview and model the investigate process with students. After viewing the video, students can conduct the investigation in small groups.

LESSON 3

What Are Mollusks, Arthropods, and Echinoderms?

In this lesson, you can . . .

INVESTIGATE the structure of some mollusks, arthropods, and echinoderms.

LEARN ABOUT the adaptations of mollusks, arthropods, and echinoderms.

LINK to math, writing, art, and technology.

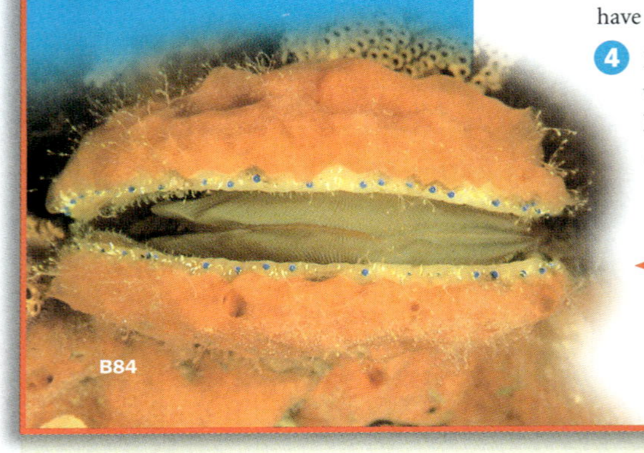

◀ This scallop is an invertebrate that has two matching hinged shells. Along the tissue that lines the edges of the shells are two rows of tiny blue eyes.

INVESTIGATE

Compare Three Groups of Animals

Activity Purpose Some of the invertebrates that you've studied so far may not have been familiar to you. But almost all of the other invertebrates belong to just three groups. You've probably seen examples of them. In this investigation you will compare the characteristics of animals from these groups—mollusks, arthropods, and echinoderms.

Materials

- clam shell
- spirula shell
- snail shell
- grasshopper study sheet
- sea urchin shell
- starfish

Activity Procedure

1. **Observe** the animal samples. Use the pictures on page B85 to add to your observations. The cuttlefish is similar to the animal that contains the coiled spirula shell. (Picture A)
2. **Compare** the characteristics of the animals, and **record** your findings. Note any similarities and differences that you see among the animals.
3. Use your observations to **classify** the animals into three groups. (Hint: One group will have three animals, one group will have two, and one group will have one.)
4. Describe the characteristics that the animals within each group share. For the group that has only one animal, describe or name another animal that belongs in the group.

B84

BACKGROUND

Webliography

Keyword invertebrates

www.harcourtschool.com

Mollusks, Arthropods, and Echinoderms are three phyla of invertebrates that have body systems and arrangements that are much more complex than those of sponges, cnidarians, and most worms. Mollusks, for example, have well-developed organ systems. Many also have a muscular foot, a filelike organ called a **radula**, and a mantle that secretes a shell. Arthropods, the largest group of animals on Earth, are even more well developed. They have jointed appendages adapted to a wide variety of uses, a head with a brain, and behavior patterns that are much more advanced than those of most other invertebrates. The echinoderms are a fascinating group. They exhibit such unique features as spiny skin and a system of tube feet that they use for locomotion and predation. Most often, they have a five-rayed arrangement of limbs. They have no head and no brain.

For student self-assessment, see Assessment Guide p. AGxvi.

Draw Conclusions

1. What shared characteristics did you **observe** among the clam, the grasshopper, and the starfish? In what ways was each animal unlike the others?
2. Which of these characteristics did you use to **classify** the animals?
3. Name one more animal that might belong in each of your three groups.
4. **Scientists at Work** Scientists **classify** animals in the same group if those animals have important characteristics in common. How are the **inferences** you made to answer Questions 2 and 3 similar to a scientist's method of classification?

Process Skill Tip

Sharing several characteristics does not always place animals in the same group. For animals to be **classified** in the same group, they must share unique similarities that they—and no other group of animals—have.

INVESTIGATE

Time 60 min **Preparation Tips** See p. B64c.

Grouping individuals

Process Skill Tip classify

Other Process Skills observe, compare, record data, infer

Expected Results Students should classify the organisms into three groups.

Teaching Tips

Tell students to base their classification systems on characteristics that they can observe, rather than on any preconceptions or opinions they have about the animals.

Draw Conclusions

1. The clam has a soft body covered by a hard shell; the grasshopper has wings, eyes, a hard outer skeleton, and jointed legs; the starfish has radial symmetry and spiny skin.
2. Answers should include appropriate differentiating characteristics.
3. Answers may include brittle stars and sand dollars for the group that includes starfish and sea urchins; octopods, squids, and other snails and clams for the group that includes snails, limpets, and clams; and any insects, crabs, lobsters, or spiders for the group that includes grasshoppers.
4. The characteristics that students choose should be common to all the animals in a group.

INVESTIGATE LOG

Students can use **Workbook pp. WB79–80** to record the results of the investigation.

PROCESS SKILLS PRACTICE

To practice and apply process skills, see **Workbook p. WB81.**

Process Skill Tip

Classify Display the Process Skill Tip Transparency on an overhead projector. Ask students to make observations of the five invertebrates and then to classify them into two groups based on their observations. Their classification system should be based on observable characteristics. For example, the two spiders have eight legs, two body segments, and no antennae. The three insects have six legs, two antennae, three body segments, and wings.

Process Skill Tip Transparency B3-3

Classify

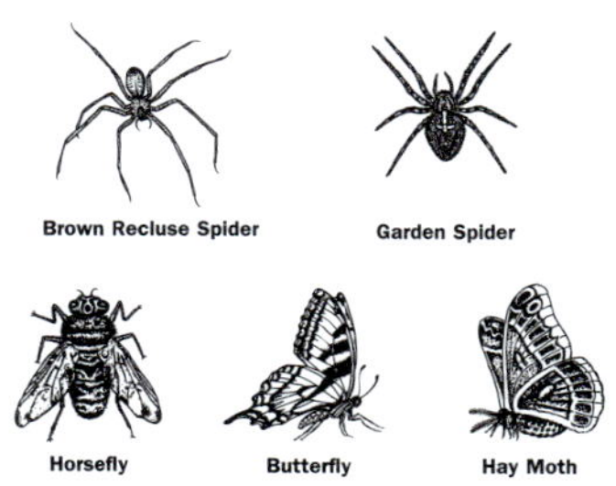

- Observe the five invertebrates.
- Classify the invertebrates based on characteristics you can observe.

LEARN ABOUT

1 Before Reading

SET A PURPOSE

FIND OUT Prompt students to focus on these topics to guide them in their reading.

VOCABULARY If students are doing the activity described in the Vocabulary Preview on page B65, remind them to add the definitions for these words to their charts.

Graphic Organizer for Chapter Concepts Display the transparency (shown below) or have students use Workbook page WB68. Point out that by the end of the lesson, students should be able to list characteristics of mollusks, arthropods, and echinoderms.

2 Guide the Learning

SCIENCE IDEAS

Help students focus on supporting details for this **main idea: Mollusks are invertebrates that have certain characteristics in common.**

- **Name some mollusks.** clams, oysters, scallops, slugs, snails, squid, octopods
- **Do all mollusks have shells? How many do they have?** No; a mollusk may have one shell, two shells (for a bivalve), or no shell. For example, slugs and octopods have no shells, and squid have only an inner shell.
- **What are characteristics of all mollusks?** a soft body and no bones
- **What are some of the ways in which mollusks protect themselves from predators?** Many can pull their bodies completely into their shells, some can burrow into sand or mud, and some can move quickly. The octopus can both move quickly and eject a dark, inky fluid to confuse its predators.

Critical Thinking Lead students to draw the conclusion that the octopus is a complex type of mollusk.

- **What physical features does the octopus have that other mollusks do not?** The octopus has a well-developed brain and nervous system, excellent vision, and eight legs that it uses to catch prey. Because these features are not present or not present to the same degree in other mollusks, students should conclude that the octopus is a complex mollusk.

LEARN ABOUT

Mollusks, Arthropods, and Echinoderms

FIND OUT

- the characteristics of mollusks, arthropods, and echinoderms
- how mollusks, arthropods, and echinoderms reproduce
- how these animals are important to people

VOCABULARY

mollusk
arthropod
arachnid
insect
echinoderm

Mollusks

The groups of invertebrates you looked at in the investigation have a larger number of complex body systems than sponges, cnidarians, and most worms. For example, like the earthworms, the three groups you looked at have circulatory systems—sponges, cnidarians, and other worms do not. The animals in each group have a number of characteristics in common. However, each group also has within it a great variety of forms and adaptations.

A **mollusk** is an animal with a soft body and no bones. The name comes from the Latin word *mollus*, meaning "soft." Mollusks include clams, oysters, slugs, snails, squid, and octopods. Most have hard shells that protect their bodies. But some, such as squid, have no outer shell. Their shells grow within their bodies, providing support. Some mollusks, such as octopods and slugs, have no shell at all, inside or out.

The shell of a mollusk is formed by an organ called the mantle, which lies between the animal's body and its shell. The mantle produces materials that harden into the shell. The shell can grow as the mollusk grows. Mollusks that have no external shells also have mantles.

Giant clams live on the coral reefs of the southern Pacific Ocean. They can weigh up to 300 kg (about 660 lb). The algae that live in their tissues often give the soft part of the clam a green or blue color. ▼

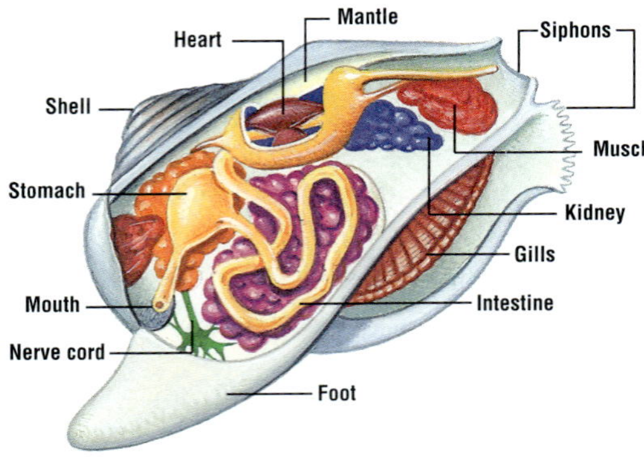

▲ The clam is a mollusk with a hard shell outside its soft body. Clams are filter feeders. They take in and push out water through organs called siphons. Food particles are caught in the gills and are passed to the clam's mouth.

B86

Graphic Organizer for Chapter Concepts

Transparency B3 • Workbook, p. WB68

Invertebrates

LESSON 1

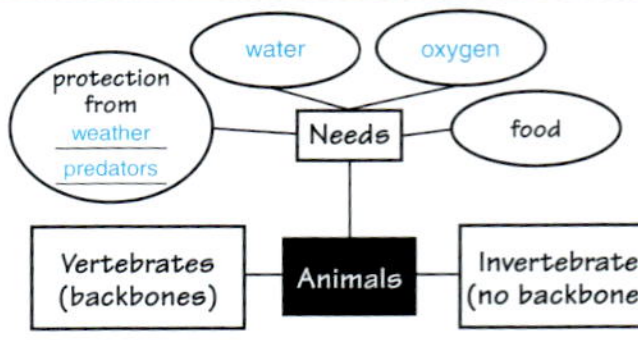

LESSON 2	Symmetry	Reproduction	Environment	Getting Food/Processing Food
Sponges	None or radial	mostly asexual	shallow ocean	filter food from water; no real digestive system
Cnidarians	radial	sexual/asexual	water	stingers and tentacles; one opening to digestive system
Planarians	bilateral	sexual	fresh water	mostly capture prey; one opening to digestive system
Roundworms	bilateral	sexual	land and water	capture prey or eat plants; two openings to digestive system
Segmented Worms	bilateral	sexual	land and water	digest food in soil; two openings to digestive system
LESSON 3				
Mollusks	varies, mostly bilateral	sexual	land and water	varies, filter food and capture prey; two openings to digestive system
Arthropods	varies, mostly bilateral	sexual	land and water	varies, capture prey and eat plants; two openings to digestive system
Echinoderms	radial	sexual	salt water only	tube feet, capture prey or filter food; two openings to digestive system

This sea snail has gills that absorb oxygen from sea water. Other marine snails have lungs rather than gills and must bob up to the surface every once in a while to collect air in cavities in their shells. ►

▼ The common garden snail, like all other snails, has a spiral shell protecting the main organs of its body.

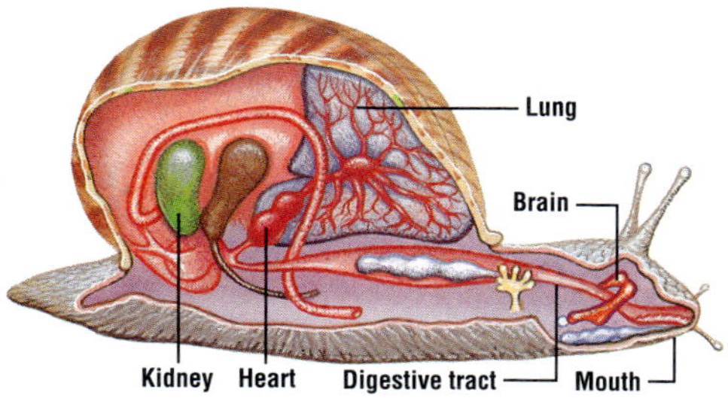

On these animals the mantle is a tough layer that forms a covering for the soft body.

A mollusk has a muscular foot. Some mollusks use it to burrow into the sand on the ocean floor. Many mollusks use the foot to move across a surface. In octopods and squid, the foot is in the form of several tentacles. Octopods often use these to crawl across the ocean floor.

Mollusks reproduce sexually. They release eggs and sperm into the water, where the eggs are fertilized. The fertilized eggs develop into larvae. The larvae are covered with hairlike structures and move like spinning tops. Each larva can develop into an adult.

Mollusks such as clams and oysters are bivalves. This means that they have two hinged shells that fit together. When a bivalve senses danger, sometimes it pulls the shells together. At other times, a bivalve may use its foot to burrow into the sand quickly. Some bivalves clap the shells together to produce a push that moves the animal away from danger.

▼ The large brain of an octopus controls the movement of each tentacle and the many suckers on the tentacles. The tentacles capture prey. The brain also controls the eyes, which have adaptations similar to those of human eyes, including eyelids.

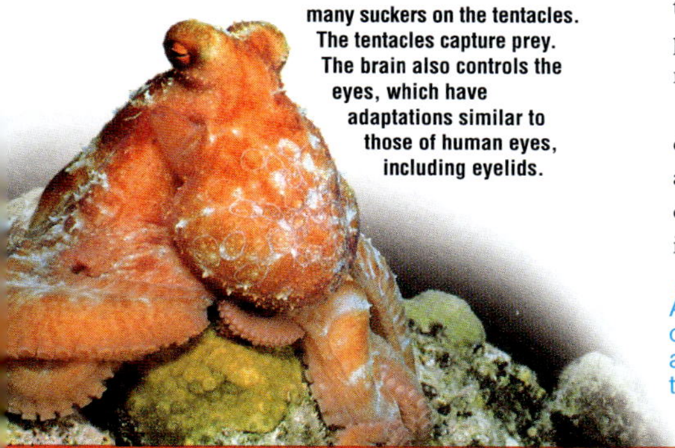

Many mollusks, such as snails, have only one shell. Some snails live on land, and some live in salt water or in fresh water. Like many other mollusks, snails have a foot they use for movement. They also have two pairs of tentacles on their heads. One pair senses the snail's surroundings, and the other supports the snail's eyes.

Snails have an organ that is called a radula (RAJ•oo•lah). The radula is shaped like a tongue and is coated with ribbons of teeth. It is used like a drill to cut through hard surfaces such as clamshells so the snail can get food.

Octopods and squid are the most active and most intelligent mollusks. They live in the ocean, where they prey on other marine animals. Their brains and well-developed nervous systems enable them to react quickly. Their well-developed sense of sight helps them locate their prey. The feet of octopods and squid are in the form of tentacles. An octopus has eight tentacles, and a squid has ten. These tentacles catch prey and bring it to the animal's mouth. The mouth has strong, beaklike jaws.

An octopus or a squid has two adaptations to defend itself. It can use its jet motion to get away quickly in any direction. It can also squirt out a dark, inky fluid. This fluid forms a cloud in the water and hides the animal as it gets away.

✔ **Name two characteristics of snails.**
Any two of the following: a soft body without bones; sexual reproduction; on some, an outer shell; a muscular foot; a radula; two pairs of tentacles on the head

VISUAL LEARNING

Ask students to use both the text and the photos to **make generalizations** about the diversity of feet in mollusks.

► **Observe the photos and read the captions and text on page B87. Discuss the variety of feet in mollusks and how they are adapted for different uses.** All are muscular. Some (clam, snail) are broad and flat. Others (octopus, squid) take the form of long tentacles. All the feet are adapted for locomotion (burrowing, walking). In the octopus and squid, the feet are also adapted for catching prey.

DEVELOP SCIENCE VOCABULARY

mollusk, p. B86—Ask students to recall the Latin root for the word, which is *mollus,* meaning "soft." Have them explain why some mollusks do not appear to be soft. because these mollusks are covered by hard protective shells Ask students to name mollusks that do appear soft. They should recall that the octopus and the slug have no hard shells.

Ongoing Assessment

Use the ✔ questions in the pupil edition.

BACKGROUND

Webliography
Keyword invertebrates
www.harcourtschool.com

Octopods is the accepted plural form for *octopus*. The word is derived from Octopoda, the name of the order to which the octopus belongs. It means "eight legs."

Gastropoda is the most numerous class of mollusks. It includes about 35,000 species as diverse as snails, limpets, and conchs—each with just one shell. The class name means "stomach foot" (*gastro* + *pod*), and these animals do indeed walk on their stomachs.

The radula is a rasping, filelike tongue found in all mollusks except the bivalves. There may be up to 250,000 teeth on the radula. The radula not only files and cuts particles of food but also acts as a conveyor belt that carries food particles to the mollusk's digestive tract.

Cultural Connection

Wampum Native Americans in some Iroquois-speaking tribes of the Northeast crafted delicate blue and white beads from clamshells. They called the beads wampum or wampumeag. The Native Americans placed these beads on strings or sewed them into patterns on belts. Wampum belts were used as a medium of exchange and to send messages between tribes. Later, they were used in trade with the European settlers.

Guide the Learning continued

SCIENCE IDEAS

Help students focus on the supporting details for this **main idea: Arthropods make up a widely diverse phylum of invertebrates, but they also display many common characteristics.**

- **What do all arthropods have in common?** an exoskeleton, jointed legs, a jointed body, a digestive system with two openings, a circulatory system, and a brain
- **Even though an arthropod is covered by an exoskeleton, it can move. Explain why.** The exoskeleton is jointed. A system of muscles attaches to the inside of the exoskeleton.
- **Why must an arthropod molt?** The exoskeleton does not grow, but the soft body inside does. When the animal sheds its old exoskeleton, a new, larger one takes its place.

VISUAL LEARNING

Have students focus on the Arthropod Adaptations on pages B88–89.

- **Compare the arthropods' legs. How are they adapted to their environments?** The lobster's legs all look very different, and are variously adapted for walking on the sea floor, for capturing and crushing prey, for providing self-defense, and for cutting up food and bringing it to the mouth. The grasshopper's powerful back legs are adapted for jumping; the four smaller legs give it balance and enable it to walk. The tarantula has eight legs that look very similar. Their even placement provides great stability for walking and climbing.

Arthropods

An **arthropod** is an animal that has a jointed exoskeleton. The hard skeleton covers an arthropod's body, protects it, and gives it support. Like a coat of armor, the jointed exoskeleton enables an arthropod to bend the parts of its body, such as its legs.

Muscles attached to the inside of the exoskeleton contract and relax, which allows for complex movements. These movements are similar in many ways to those of a vertebrate.

The scorpion uses its large claws to seize prey. Its jointed tail can reach over its head and inject its prey with a poison that stuns or kills. ▼

THE INSIDE STORY

Arthropod Adaptations

Arthropods can be found in every environment that supports life. The tremendous variety of adaptations in arthropods has allowed them to survive almost everywhere on Earth.

Mouthparts of a lobster ▲

Crustaceans

Crustaceans (kruhs•TAY•shuhnz) are one kind of arthropod. Lobsters are crustaceans that live in water. They show many of the characteristics of crustaceans. Lobsters have three body segments—a head, a thorax, and an abdomen. They have two pairs of antennae that sense their surroundings by feel. The legs around a lobster's head are adapted for capturing prey and carrying it to the animal's mouth. A lobster's claws can crush the shell of an oyster or a clam. The claws are usually not the same size. The larger one is for defense and the capture of food. The smaller claw helps to tear the prey into smaller pieces.

B88

Social Studies Link

Cultured pearls are just as genuine as natural pearls. The difference is that humans have "encouraged" an oyster to produce the pearl. Have students find out how oysters can be made to produce cultured pearls.

An arthropod's exoskeleton does not grow as the animal gets larger. Several times during its life, an arthropod grows a new, larger exoskeleton. This new exoskeleton remains soft until the animal sheds its old exoskeleton. Then the new one begins to harden. During this process, the animal has little protection against predators. It may conceal itself until its new exoskeleton has hardened.

Every arthropod has legs, a digestive system with two openings, a circulatory system, and a brain. Many arthropods have well-developed organs for sight, sound, and touch. Arthropods reproduce sexually. The female often lays fertilized eggs from which young hatch. Though they have these features in common, arthropods include such different animals as beetles, butterflies, spiders, shrimp, centipedes, and scorpions. Arthropods are by far the largest phylum of animals. Scientists know of more than a million different arthropod species.

✓ **Name two characteristics arthropods share.** jointed exoskeleton; legs; digestive system with two openings; circulatory system; brain; organs for sight, sound, and touch; sexual reproduction

Arachnids

Arthropods also include **arachnids** (uh•RAK•nidz). Spiders, scorpions, ticks, and mites are arachnids. Arachnids have eight legs and no antennae. They breathe through organs called book lungs. Book lungs look like the loosely opened pages of a book. Many arachnids have sharp body parts that inject poison. Most eat other animals, using their pincers or fangs for capturing prey. Almost all arachnids live on land.

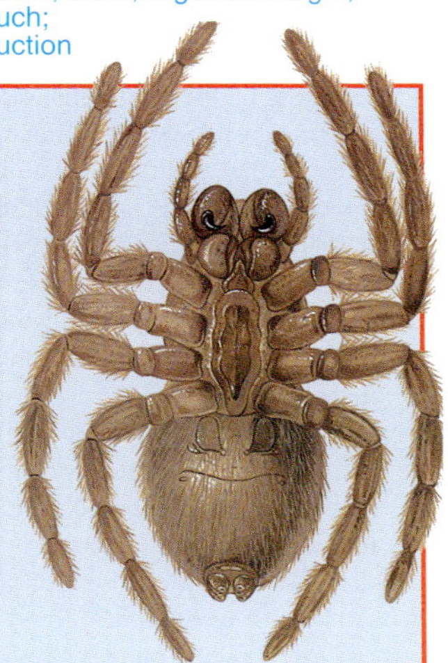

▲ The arachnid in these pictures is a tarantula. Tarantulas are sometimes thought of as dangerous animals, but they are actually not as venomous as many other spiders.

The grasshopper has many characteristics in common with all insects. However, its mouthparts are adapted for crushing food. Other insects have mouthparts that pierce, suck, or slice. ▼

Mouthparts of a grasshopper ▲

Insects

Insects, animals with six jointed legs, are the largest group of arthropods. Many insects have wings. These are the only invertebrates that can fly. Insects get oxygen through air ducts along the sides of their bodies.

B89

DEVELOP SCIENCE VOCABULARY

arthropod, p. B88—Ask students to find the meaning of the root *arthro*. They will find that it means "joint." Have them explain what part of the human body is affected by arthritis.

arachnid and **insect**, p. B89—Make sure that students understand that although arachnids and insects belong to the same phylum, Arthropoda, they represent two different types of arthropods. Ask them to recall how the tarantula is different from the grasshopper. They should mention that the tarantula has two main body segments and eight legs and that the grasshopper has three body segments and six legs. In addition, the tarantula has no antennae.

Reaching All Learners

Kinesthetic Learners To help these students better appreciate the importance of jointed limbs, have them use chenille sticks to make models of jointed and unjointed limbs.

ESL ACTIVITY

Knowing how to pronounce words can give students more confidence in using them. Help students master the pronunciation of the words **mollusk**, **arthropod**, and **arachnid** by pronouncing them slowly, syllable by syllable, and by pointing out rhymes for individual syllables. For example, the second syllable of *mollusk* rhymes with *tusk*, and the last syllable of *arthropod* rhymes with *rod* or *nod*.

Guide the Learning continued

USE PROCESS SKILLS

Draw Conclusions Guide students to draw conclusions about whether most arthropods are harmful or beneficial to humans. Encourage them to do research to back up their conclusions.

- ▶ **Are most arthropods harmful or beneficial? Use the information on page B90 as a starting point for developing your argument. Do research to strengthen your conclusions.** Students may initially be divided on the relative worth of arthropods and may list arguments on both sides of the question. Further research should provide them with data that leads to the conclusion that, on the whole, arthropods are beneficial. In fact, the process of pollination alone outweighs the damage that arthropods do.

◀ Locusts travel in huge groups called swarms. Because of the number of locusts, a swarm can destroy a field of crops in a very short time. Some locust swarms are so large that they block out sunlight.

Helpful and Harmful Arthropods

Insects are the most numerous arthropods and the most numerous animals on Earth. Often, they seem to be pests. But without insects, most plants would not be able to reproduce. Insects pollinate most of the angiosperms that other animals, including humans, use for food. Arthropods such as ladybugs help farmers and gardeners by eating many of the insects that destroy crops or flowers.

Arthropods are a food source for other animals. Much of the seafood that people enjoy comes from the arthropod phylum. Shrimp, lobster, and crab are regarded by many people as treats. Arthropods also produce goods that can be sold. Bees produce honey and the beeswax used in some candles. Bees are also among the most important pollinators of plants.

Fewer than one percent of all arthropod species are harmful, but that's a large number of species. Many insects attack and injure almost all types of plants, including food crops. Some insects destroy corn, some spoil potatoes, and some devour wheat. Other insects are household pests. Moths may ruin clothing. Termites attack the wooden beams and floors of buildings. Ants, cockroaches, and flies infect food with germs.

Some arthropods are venomous to people. Wasps, scorpions, and spiders sometimes inject people with poison that is painful or dangerous. Other insects may carry diseases and infect people.

✔ **How do arthropods help people?**

They help people by pollinating crops, providing food, and eliminating other arthropods that destroy plants.

Beekeepers provide bees with habitats and sell honey and beeswax. The bees in a hive inform each other of the location of a food source, the nectar in flowers. They do this through a "dance" that shows the direction and distance of the food source. ▶

B90

Language Arts Link

Debating Arthropods After students have researched the issue of whether arthropods are beneficial or harmful to human society, have them divide into teams and debate their findings. Remind them to **distinguish fact and opinion** when formulating their arguments. For example, have students reread page B90 to determine if the following statement is fact or opinion: Some of the most important pollinators of plants are bees. fact

Reading Mini-Lesson

Compare and Contrast Have students read through the lesson for similarities and differences in mollusks, arthropods, and echinoderms. Display the transparency and have students state their findings. After students complete the diagram, ask how many of the characteristics are shared by other invertebrates.

Reading Mini-Lesson Transparency B3-3

Compare and Contrast

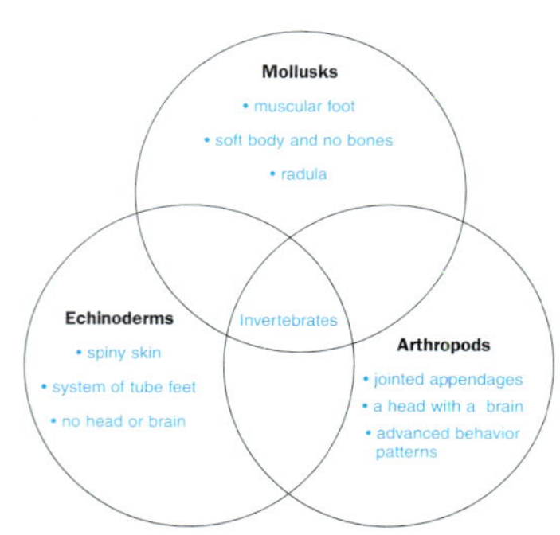

READING SKILLS PRACTICE

To practice and apply this reading skill, see **Workbook p. WB82.**

When a starfish finds a mussel, the starfish wraps its arms around the mussel, using suction cups to pull the shells apart. The starfish then extends its stomach into the space between the shells. The starfish's stomach surrounds the mussel's body and digests it. Then it absorbs the digested material. ►

Echinoderms

Not all invertebrates have exoskeletons. **Echinoderms** (ee•KY•nuh•dermz) are invertebrates that have internal skeletons and spines sticking out from their bodies. Although these spines may seem to be part of an exoskeleton, they are actually part of the skin. The word *echinoderm* comes from the Latin words meaning "spiny" and "skin."

Echinoderms live only in the ocean. They get oxygen from sea water. Examples of echinoderms are starfish, sand dollars, sea urchins, and sea cucumbers.

Although an echinoderm has no brain, it has nerves that enable it to move and feed. Most adult echinoderms appear to have radial symmetry, with five arms extending from the center of the body. Sea urchins and sand dollars do not have arms like those of starfish. However, they have five paired rows of feet extending symmetrically from the mouth.

The mouth of an echinoderm is at the center of the underside of its body. Many echinoderms have powerful jaws and, hidden among their spines, poison glands. These adaptations enable them to get enough food and to defend themselves.

Although a sea cucumber does not look like other echinoderms, it has five rows of feet. The tentacles around its mouth are used for gathering food. Some tropical sea cucumbers grow to be 3 m (about 10 ft) long.

A sea urchin's spines protect it from predators. On this echinoderm's underside are five teeth, which are used to graze on algae. Sea urchins can often be found in tide pools at the seashore. ►

SCIENCE IDEAS

Help students focus on supporting details for this **main idea: Echinoderms are invertebrates that exhibit some unique features.**

- ► **Name some echinoderms.** starfish, sand dollars, sea urchins, sea cucumbers, brittle stars, feather stars
- ► **What characteristics do echinoderms have in common?** They have spiny skin, an endoskeleton, no head or brain, and tube feet. They are adapted to living in the sea.

VISUAL LEARNING

Have students look at the photos and read the captions on page B91. Have them compare the ways these three animals get their food.

- ► **Using the photos and captions, describe how a starfish, a sea cucumber, and a sea urchin get their food.** The starfish forces open a clamshell, inserts its stomach, and digests the clam. The sea cucumber gathers food with its tentacles. The sea urchin uses its teeth to eat seaweed.
- ► **Which is more flexible—the starfish or the sea urchin?** The starfish is; it can wrap its arms around a clam.

DEVELOP SCIENCE VOCABULARY

echinoderm, p. B91—Prompt students to recall that *derm* means "skin." Ask them what a dermatologist treats and what a dermestid beetle might eat. skin ailments; leather, among other things

THE QUESTIONS KIDS ASK!

Do you mean that a starfish can actually push its stomach out of its body?

Address Misconceptions Exactly! Ask students why this adaptation is useful to an animal that eats bivalves.

Guide the Learning continued

SCIENCE IDEAS

Help students focus on supporting details for this **main idea: Echinoderms have an unusual adaptation called tube feet.**

- **What is the function of the tube feet in echinoderms?** locomotion, feeding, respiration, and sensing
- **How do the tube feet work?** Water pumps into the tiny feet and makes them expand; they eventually become firm enough to walk on. When water is pumped out of the tube foot, it causes a suction-cup effect, and the foot sticks to whatever it is touching. That is why a starfish can pry open a clam.

Critical Thinking Lead students to draw the conclusion that starfish can multiply by regeneration.

- **In the past, fishers who found starfish feeding in the clam and oyster beds where they were working would chop each starfish in half and throw it back into the sea. That did not solve the starfish problem; in fact, it made it worse. Explain.** Students should conclude that the starfish didn't die but actually multiplied because they have the ability to regenerate a whole new body from parts (in some cases, from a single arm).

Echinoderms have an unusual adaptation called tube feet. Most echinoderms have rows of tube feet along the underside of their arms or in rows extending from their mouths. They use these tube feet for moving, feeding, sensing, and respiration.

Each foot ends with a sucker. A system of canals within the animal pumps water to each of its feet. When the water is forced into a tube foot, the foot expands and becomes firm enough to walk on. When water is withdrawn from the tube foot, the sucker grasps the surface it is touching. The suckers on a starfish's feet are strong enough to force open the shells of bivalve mollusks, their main food source.

Starfish generally reproduce sexually. Each starfish is either female or male, and each of its arms has sex organs. Females release millions of egg cells, and males release millions of sperm

A brittle star looks like a starfish, but it has thinner arms and no suckers on its tube feet. It moves by waving its arms from side to side. Brittle stars are, in fact, brittle. Their arms can snap off easily.

A feather star, or crinoid, is an echinoderm whose arms, tube feet, and mouth are directed upward.

B92

INVESTIGATION CHALLENGE

Hands-On Activity: An Invertebrate Museum

Use this idea for a hands-on activity to extend students' understanding of key lesson concepts.

Have students bring in and display specimens of invertebrates, such as shells, insects, spiders, crabs, and starfish. Some students may have collections they are willing to share. Others may find dead insects on windowsills. Those who have enjoyed a seafood dinner may be able to bring in exoskeletons from crabs or lobsters.

Challenge students to try to bring in a specimen that no one can identify. Discourage students from bringing in live specimens unless they are able to care for them properly.

For student self-assessment, see Assessment Guide p. AGxvii.

cells into the water. Fertilized eggs hatch into larvae. These drift in the ocean currents for weeks. They then attach themselves to a surface as they mature into adults.

Some echinoderms can also reproduce asexually by regenerating, or forming new parts. Some starfish can regenerate if their bodies are divided in half. Each half grows into a new starfish. A few species of starfish can grow an entire animal from a single arm.

✔ **In what ways are echinoderms different from other invertebrates?** Echinoderms have an endoskeleton, tube feet, and spines that protrude from their skin.

Summary

Mollusks are soft-bodied invertebrates. They include clams, octopods, oysters, slugs, snails, and squid. Many kinds of mollusks have external protective shells. Arthropods are invertebrates with jointed exoskeletons and legs. They include insects, crustaceans, and arachnids. Spiny-skinned invertebrates, the echinoderms, have tube feet in rows on body parts that surround the animal's center. Echinoderms include starfish, brittle stars, sand dollars, sea urchins, and sea cucumbers.

Review

1. How does an octopus propel itself through the water?
2. What problem does an exoskeleton cause an arthropod?
3. How do the water canals in an echinoderm's body affect the animal's tube feet?
4. **Critical Thinking** What advantages might a winged arthropod have over an arthropod without wings?
5. **Test Prep** An example of an arthropod is —
   - A a bivalve
   - B an octopus
   - C a spider
   - D a starfish

LINKS

MATH LINK

Write Ratios The largest invertebrate is the giant Atlantic squid, which can be up to 20 m long. What is the ratio, in simplest form, of the height of a 1.5-m-tall student to the squid's length?

WRITING LINK

Narrative Writing—Story Write a short story for a friend in which a knight in armor faces an arthropod that is the same size as the knight.

ART LINK

Sepia Sepia is a brown pigment that originally came from the ink sac of a cuttlefish. Drawings and photographs in this color are sometimes called sepiatones. Find some pictures of sepiatones in reference books, and try drawing one yourself. You can use thinned brown watercolor or a pen with brown ink.

TECHNOLOGY LINK

To learn more about how invertebrates are helping people, watch *Predatory Snails* on the **Harcourt Science Newsroom Video.**

CNN Turner Learning

3 Wrap Up and Assess

SUMMARY

Graphic Organizer for Chapter Concepts After they read the summary, have students complete the Lesson 3 section of the graphic organizer (shown on page B86).

REVIEW

1. It squirts a jet of water from its internal cavity through a siphon.
2. As the animal grows, it must shed its old skin. The animal is nearly defenseless while its new exoskeleton is hardening.
3. The water canals pump water into or out of the animal's tube feet, making them rigid or causing the suckers on the feet to cling to a surface.
4. Wings allow it to travel long distances in search of food or a mate and to escape predators.
5. C, a spider

TEST PREP Test-Taking Tips

Item type multiple choice with a sentence fragment

To help students answer Question 5, share the following strategy with them. **Read the sentence carefully. Recall some facts that you know about arthropods. Try to remember what arthropods look like. Then select the best answer choice.**

LESSON CONCEPT REVIEW/ASSESSMENT

Students can use **Workbook p. WB83** to review the lesson concepts.

VOCABULARY REVIEW

Students can use **Workbook p. WB84** to review the chapter vocabulary.

LINKS TIPS

Links Objective Extend understanding of mollusks, arthropods, and echinoderms.

Math Link You may want to suggest that the class arrive at an average height for the class and then find the ratio. Discuss how to round the class's average height to a convenient measurement (about 1.5 m). Remind students how to express a ratio ($\frac{3}{40}$, 3:40, or 3 to 40).

Writing Link As a prewriting activity, ask students to discuss how the insect's exoskeleton is similar to and different from the knight's suit of armor. Writing models are provided in the **Teaching Resources** book.

Art Link Suggest that students draw sepiatones of different vertebrates and invertebrates. Be sure they label the organisms.

Objectives

- Evaluate the impact of research and technology on scientific thought, society, and the environment.
- Identify careers related to science.

1 Before Reading

SET A PURPOSE

Have students examine the title and pictures in this article. Ask them to brainstorm a list of different methods that farmers use to control insect pests. Discuss the list, and talk about which methods (such as chemical insecticides) might be harmful to the environment and why. Record any questions students might have. Encourage them to look for answers in the reading selection and, if necessary, to use additional reference materials.

2 Guide the Learning

SCIENCE IDEAS

Help students gather information about this nonfiction science article by asking these questions:

- How do insects attract mates? by color, sounds, and scents
- How do some moths harm fruit crops? The worm-like stage, the larva, eats the fruit.
- Why do these moths depend on pheromones to find mates? Because they mate at night and cannot locate a mate by sight, they use scents instead.
- Why do the male moths become confused by twist ties that smell like pheromones? There may be hundreds of ties wrapped around trees in an orchard, and they all smell like a female moth to the male moth. The male cannot tell where the real female is.
- Can farmers use pheromones to control all insect pests? no, only some species

Controlling Pests with Pheromones

Scientists are learning to use scents produced by insects to control insect populations.

What are pheromones, and how do they work?

Animals use color and sound to attract mates. Some also use smell. Pheromones (FAIR•uh•mohnz) are chemical signals or scents given off by some insects to attract mates so they can reproduce. But scientists have learned to use the insects' own signals to stop them from destroying food crops.

Why are insects a problem for farmers?

Not all insects are pests. Some, such as bees, help farmers by pollinating plants. Many types of insects,

Farmers use devices like these to send pheromones through their orchards.

Reading Mini-Lesson

Predict Probable Future Actions and Outcomes Ask students to predict probable outcomes for phenomena described in this article by having them complete the following sentences:

Codling moths and oriental fruit moths could produce worms that attack growing fruit.

When female moths send out a pheromone scent, male moths will follow the scent to find the female and mate.

If pheromone-scented ties are placed all over an orchard, males will get confused and be unable to find the females.

Pheromones that control pests could reduce pollution by diminishing use of pesticides.

however, are destructive. They burrow into fruits or stems to lay eggs, or they eat crops growing in the field. Farmers can lose billions of dollars' worth of crops each year because of destruction by insect pests.

How can pheromones be used against pests?

Apple growers in Washington state and peach growers in California use pheromones to control codling moths and oriental fruit moths. These moths produce worms that attack growing fruit.

Because moths are active at night and mate in the dark, they use pheromones to find each other. The female sends out a pheromone scent, which the male follows to her location.

Farmers confuse the insects by putting the pheromone scent on something that looks like a twist tie. They wrap the ties around parts of their apple or peach trees. Male insects then pick up the pheromone scent all over the orchard but cannot figure out where to find actual females. The insects can't mate, so females don't produce eggs that become young worms.

Why is using pheromones a good idea?

Using pheromones is a good way to reduce pests because it does not harm the environment. It doesn't work for every pest, but when it does, it helps reduce the use of pesticides that can poison soil and water or harm other organisms.

THINK ABOUT IT

1. How can pheromones help control pests?
2. Why is the use of pheromones good for the environment?

CAREERS

PEST CONTROL TECHNICIAN

What They Do

Pest control technicians use chemicals and traps to get rid of insects, rodents, and other pests that cause problems for homes and businesses.

Education and Training

Pest control technicians should have a background in chemistry, mathematics, and health. They must be good at closely following directions, because mistakes could cause harm to people. Pest control technicians should have a high school education. They must also be licensed.

GO ONLINE **WEB LINK**
For Science and Technology updates, visit the Harcourt Internet site.
www.harcourtschool.com

B95

DEVELOP SCIENCE VOCABULARY

pesticide, p. B95—Have students derive a meaning for the suffix *-cide*. killer A pesticide kills pests. Have students give meanings for other words using the same suffix, such as *herbicide* and *insecticide*.

CHAPTER CONCEPTS

The article focuses on technology that scientists have developed to control some insect pests without using chemicals that may harm the environment. The technology was developed because scientists learned about the behavior of moths.

▶ **What kinds of moth behavior did scientists have to study before they could develop pheromones to reduce their populations?** Scientists had to find out when the moths were most active, when they mated, and how males located females. Then they had to analyze the scented chemicals the females released and produce the chemicals artificially. Finally, they had to figure out how much of the artificial pheromones to use to confuse the males.

3 Wrap Up and Assess

THINK ABOUT IT

1. Farmers can use pheromones to confuse male moths so they cannot find females with which to mate. Then the females cannot produce eggs that become new moths.
2. Pheromones replace pesticides that can be harmful to the environment. Pheromones are not environmentally harmful.

BACKGROUND

Webliography GO ONLINE
Keyword careers
www.harcourtschool.com

Pest Control Technician

These technicians help eliminate pests such as rodents and insects that may cause damage to crops, businesses, and homes. Pest control technicians use traps, insecticides, and other chemicals to get rid of pests. These technicians should have knowledge in chemistry, health, and mathematics. The position requires at least a high school education and a license to practice pest control.

PEOPLE IN SCIENCE

Objectives

- Connect chapter concepts with the contributions of scientists.

1 Before Reading

SET A PURPOSE

Zoologists, such as Ernest Everett Just, study only animals and look for common characteristics among animals to help group them. Have students read to find out how Dr. Just made important discoveries about cells and cell parts.

2 Guide the Learning

THE WORK OF SCIENTISTS

Help students focus on the methods and processes scientists use and the impact of their work.

- What misconception about the cell nucleus was Dr. Just able to set straight? that the nucleus was the only important part of the cell in terms of cell function
- What are some examples of Dr. Just's great determination? What examples can you think of in your own life that show similar determination? deciding to go to school and working hard to get the money to do so

3 Wrap Up and Assess

THINK ABOUT IT

1. He helped us better understand the way cells function.
2. He studied the egg cells of certain ocean-dwelling animals.

PEOPLE IN SCIENCE

Ernest Everett Just

ZOOLOGIST/ EMBRYOLOGIST

Dr. Just studied cells in the eggs of invertebrates that lived in the sea. These are squid eggs and a young squid.

The discoveries in cell biology made by Dr. Ernest Everett Just in the 1920s and 1930s changed the way scientists think about cells and cell parts. He was one of the first biologists to realize the importance of the cytoplasm of a cell. Before his discoveries, many biologists thought that the cell nucleus controlled all cell activities. They thought that the cytoplasm did little and that the outer surface of the cytoplasm did almost nothing. However, after careful observation of the egg cells of certain invertebrates that live in the ocean, Dr. Just discovered that a cell depends on smooth cooperation between the nucleus and the cytoplasm of the cell. He also found that the outer part of the cytoplasm of a cell determines the relationship the cell has with its environment.

Ernest Everett Just was determined to succeed. As a child he worked after school in farm fields near his home in South Carolina to help earn money for his family. His dream was to attend Kimball Academy, an excellent school in New Hampshire. When he was 17, he went to New York City, earned the money he needed, and enrolled at Kimball. He finished its four-year program in just three years. Then he went on to earn a Ph.D. in embryology (em•bree•AHL•uh•jee)—the study of embryos and their development.

For much of his career, Dr. Just taught biology at Howard University in Washington, D.C. During summer breaks, he worked at the Marine Biological Laboratory at Woods Hole, Massachusetts. It was there that he made most of his important discoveries while working with the cells of certain sea animals.

THINK ABOUT IT

1. Identify Dr. Just's main scientific contribution.
2. What research did Dr. Just do that led to his main contribution?

B96

BACKGROUND

Research at Woods Hole, Massachusetts The Woods Hole Oceanographic Institution has more than 350 research projects underway at any given time. Oceanographers are studying all aspects of marine science, such as ocean physics, engineering, chemistry, biology, geology, and geophysics. Researchers also work on social and legal problems connected with the ways that humans use the world's oceans. To conduct field research, the institution maintains a fleet of vessels that includes four ships, the submersible *Alvin*, several small boats, and some remote-controlled deep-sea vehicles.

Webliography GO ONLINE

Keyword biographies

www.harcourtschool.com

ACTIVITIES FOR HOME OR SCHOOL

SEEKING SHELTER

What kind of environment do sowbugs prefer?

Materials

- wide shoebox with a lid
- sheet of construction paper
- scissors
- 8–10 sowbugs
- dry soil
- water
- tape
- flashlight

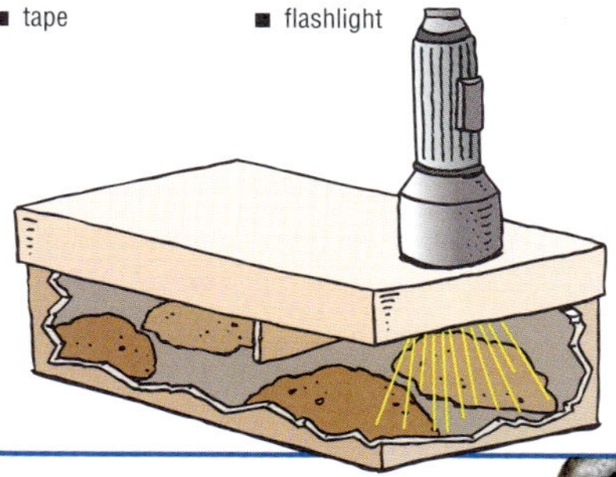

Procedure

1. Tape a paper divider to the top of the box, as shown. Cut it so the sowbugs can crawl under it.
2. Arrange four areas of soil, each covering one-fourth of the bottom of the box. Dampen one area on each side of the divider.
3. Cut a round hole in the center of one end of the lid. Balance the lighted flashlight over the hole.
4. Place the sowbugs on the lighted side of the box. Leave the box and return after 30 minutes.

Draw Conclusions

Which of the four environmental conditions do the sowbugs prefer? Where do you think you would find sowbugs in nature?

EARTHWORM MOVEMENT

How do earthworms move?

Materials

- earthworm
- square of vinyl tile
- sandpaper
- hand lens

Procedure

1. Place the vinyl tile and the sandpaper next to each other.
2. Set the earthworm on the sandpaper. Use the hand lens to watch it move. Sketch what you see.
3. Set the earthworm on the vinyl tile. Use the hand lens to watch it move. If the earthworm is moving differently this time, sketch what you see.

Draw Conclusions

Which body parts do earthworms use to move? Do earthworms move more quickly over a smooth surface or a rough surface? Why do you think that is?

B97

School-Home Connection

These activities provide an excellent opportunity to assign hands-on activities for the students and their families. The materials are often easy to obtain and safe to work with. For those activities that you do not wish to do in the classroom, encourage students to complete them at home. Remind them that communicating their results is always important, but especially so for home activities since you will not be there to observe their activities. Ask students to use diagrams, sketches, graphs, or even photographs to show their results.

ACTIVITIES FOR HOME OR SCHOOL

Seeking Shelter

Objectives

- **Experiment to find out what conditions of light and moisture sowbugs prefer.**

Suggested Time 30 minutes to set up the experiment; 30 minutes to run the experiment

Hints

- Soil in the damp area should be moist but not soaked. There should be no puddles.
- Remind students to handle the sowbugs gently.

Safety Tip Let students know that sowbugs do not bite. Have them wash their hands after handling the materials, for hygiene purposes.

Draw Conclusions Students should find that sowbugs prefer dark, moist environments. This might lead to the conclusion that sowbugs prefer dark, moist places in nature, such as on the underside of rocks or in moist, rotting logs and mulch.

Investigate Further Interested students could replicate the experiment but with other small insects, such as mealworms, crickets, or ants, to discover their environmental preferences.

Earthworm Movement

Objectives

- **Experiment to find out if earthworms move more easily across rough or smooth surfaces.**

Suggested Time 20–30 minutes

Hints

- Remind students to handle the earthworms gently.

Safety Tip Have students wash their hands after handling the worms, for hygiene purposes. See page T21 for tips on disposing of animals appropriately.

Draw Conclusions Students should see that earthworms use body segments (muscles) and tiny bristles on their bodies to help them move. They move more quickly across rough surfaces because the bristles have a place to anchor the animal so that it can push forward.

ACTIVITIES FOR HOME OR SCHOOL

Reproducible copies of these activities are provided in the **Teaching Resources** book.

CHAPTER 3

Review and Test Preparation

VOCABULARY REVIEW (2 points each)

1. roundworms, parasites
2. cnidarians
3. Sponges
4. vertebrate, invertebrate
5. echinoderm
6. endoskeleton
7. segmented worms
8. arachnids
9. exoskeleton, arthropod
10. flatworm, mollusk
11. insects

CONNECT CONCEPTS (2 points each)

12. cnidarians
13. poison
14. insects
15. exoskeleton
16. echinoderms
17. tube feet
18. parasites
19. body walls
20. earthworm
21. circulatory system

CHECK UNDERSTANDING (4 points each)

22. D, protection
23. H, polyps and medusas
24. C, ink
25. F, six legs
26. B, echinoderms

CRITICAL THINKING (4 points each)

27. They both are filter feeders.
28. It disguises the insect, protecting it from predators.
29. The exposed surface area of the open-book structure is much greater, and the lung can absorb more oxygen.

CHAPTER 3 Review and Test Preparation

Vocabulary Review

Use the terms below to complete the sentences. The page numbers in ( ) tell you where to look in the chapter if you need help.

vertebrate (B70)
invertebrate (B70)
endoskeleton (B70)
exoskeleton (B71)
sponges (B76)
cnidarians (B78)
flatworm (B80)
roundworms (B81)
parasites (B81)
segmented worms (B82)
mollusk (B86)
arthropods (B88)
arachnids (B89)
insects (B89)
echinoderm (B91)

1. Many kinds of ___, the simplest worms that have digestive tracts with two openings, are ___, which harm their hosts.
2. All ___ live in water and have cells that can inject poison into other animals as a form of defense and attack.
3. ___ are filter feeders that live in water and have no tissues or organs.
4. A ___ has a backbone, while an ___ has no backbone.
5. An ___ has spines that project through its skin.
6. The system of connected bones within an animal's body is an ___.
7. Unlike roundworms, ___ have muscles that run in bands around their bodies.
8. Spiders and scorpions are ___.
9. An animal whose hard outer covering, or ___, is jointed is called an ___.
10. A planarian is a ___, and a clam is a ___.
11. The most numerous animals in the world are the ___.

Connect Concepts

Complete the chart by filling in terms from the Word Bank.

insects **body walls** **exoskeleton** **earthworm** **tube feet** **cnidarians** **parasites** **echinoderms** **poison** **circulatory system**

Invertebrate Adaptations

Animals	Adaptations of These Animals
Invertebrates that live only in water and have medusa and polyp forms are known as 12. ___.	These animals have cells that are specialized adaptations for defense and attack and that contain 13. ___.
The only invertebrates that can fly are 14. ___	Their bodies are supported by a hard 15. ___.
Animals with spiny skins but no exoskeleton are known as 16. ___.	Most of these animals have an unusual adaptation for movement, called 17. ___.
Flatworms and roundworms that feed on and harm other animals are 18. ___.	Some of these animals absorb food through their 19. ___.
An animal with two separate bands of muscles that lives in the soil is an 20. ___.	The characteristic that distinguishes this animal from other worms is its 21. ___.

B98

CHAPTER TEST

See **Assessment Guide pp. AG33–37** for the Chapter Test. Assessment Options are provided on p. B64e.

TAKE-HOME BOOK

Use the **Take-Home Book** (described on p. B64d) to provide more chapter content and activities.

Check Understanding

Write the letter of the best choice.

22. All animals need food, oxygen, water, and —
- A prey
- B lungs
- C exoskeletons
- D protection

23. Some species of cnidarians exist as both —
- F jellyfish and corals
- G larvae and bivalves
- H polyps and medusas
- J *Hydra* and comb jellies

24. An octopus can hide itself from predators with a cloud of —
- A poison
- B larvae
- C ink
- D parasites

25. An insect has —
- F six legs
- G eight legs
- H ten legs
- J no legs

26. Sand dollars and sea urchins are —
- A arthropods
- B echinoderms
- C mollusks
- D arachnids

Critical Thinking

27. In what way are sponges similar to baleen whales?

28. A praying mantis is an insect that looks like a twig. How does this adaptation help the animal survive?

29. The book lungs of arachnids look like open books with separated pages. Why is this "open-book" structure better than a "closed-book" structure would be?

30. Which is usually the larger animal—a parasite or its host? Explain the reasoning behind your answer.

Process Skills Review

31. If you **observe** that an animal has a jointed exoskeleton, what can you **infer** about the **classification** of the animal?

32. How would you **compare** three unidentified animals—an arthropod, a cnidarian, and a flatworm—to determine which one belongs to each group?

Performance Assessment

Crustacean Behavior

Work with a partner to design an experiment about the behavior of sow bugs. The equipment for your experiment can be everyday household objects such as paper towels, construction paper, a cookie sheet, and tape. Describe how you would find out whether sow bugs are more likely to be found in a dry environment or in a moist environment. Describe how you would find out whether they are more likely to live in light places or in dark places.

B99

WRITING ABOUT SCIENCE

Have students write a composition that involves writing for a specific purpose. You may wish to use the prompt that is provided in the **Workbook.** The prompt is accompanied by a graphic organizer that will help students organize their ideas before writing. Models for writing are provided in the **Teaching Resources** book. You may wish to reproduce those for students or display them on a transparency.

WRITING PRACTICE

A chapter writing prompt and a prewriting activity are provided on **Workbook p. WB85.**

TEST PREP Test-Taking Tips

Item type open-ended format in which the student is asked to write a short answer

To help students answer Question 30, share the following strategy with them. **Turn the question into a statement, such as "A parasite is usually smaller than (or larger than) its host because…" and then add facts to complete the sentence. Try to think of a parasite-host relationship to use as an example.**

30. The host is usually larger because it provides an environment and a food source for the parasite. The host is usually large enough to survive the harm caused by the parasite. Otherwise, the parasite would lose its supply of food.

PROCESS SKILLS REVIEW
(7 points each)

31. It is an arthropod.

32. Students' answers should include at least one distinguishing characteristic of each group.

Performance Assessment

Preparation Make sure that students realize that they are to design two experiments: one to test preference for moisture and one to test preference for light or dark.

Scoring Rubric
Performance Indicators

___ Includes in designs appropriate materials to simulate an environment with dry and moist sections.

___ Reflects in descriptions an understanding of the probability that the animals will move to the section (dry or moist) more suited to their needs.

___ Includes in designs appropriate materials to simulate an environment with dark and light sections.

___ Reflects in descriptions an understanding of the probability that the animals will move to the section (light or dark) more suited to their needs.

Observations and Rubric Score

3 2 1 0

CHAPTER 4 Planning Guide Vertebrates

LESSON	PACING	OBJECTIVES	MATERIALS
1 What Are Fish and Amphibians? B102–111 	3 days	▶ Distinguish between fish and amphibians. ▶ Describe the structure of fish and amphibians and how they meet their needs.	**Investigate** diagrams of frog and fish
2 What Are Reptiles, Birds, and Mammals? B112–121 	3 days	▶ Explain where reptiles, birds, and mammals live. ▶ List the characteristics of reptiles, birds, and mammals. ▶ Describe three different kinds of mammals.	**Investigate** diagrams of reptile, bird, and mammal skeletons, colored pencils
End of Chapter B122–127 **Science and Technology** **People in Science** **Activities for Home or School** **Chapter Review and Test Preparation**		▶ Evaluate the impact of research and technology on scientific thought, society, and the environment. ▶ Identify careers related to science. ▶ Connect chapter concepts with the contributions of scientists. ▶ Extend chapter concepts with hands-on investigations. ▶ Review chapter concepts.	

PROCESS SKILLS	VOCABULARY	RESOURCES AND TECHNOLOGY	REACHING ALL LEARNERS
Process Skill Tip observe, compare **Other Process Skills** record data, infer, plan and conduct a simple investigation	**fish** **amphibian** **cold-blooded**	**Workbook,** pp. WB86–91 **Vocabulary Cards,** pp. TR111–114 **Transparencies** B4, B4-1 **Harcourt Science Activity Video** **Harcourt Science Explorations CD-ROM**	**Visually Impaired and Kinesthetic Learners**, p. B108 **ESL Activity**, p. B105 **Investigation Challenge**, p. B110 **Advanced Learners**, p. B1g
Process Skill Tip observe, hypothesize **Other Process Skills** compare, predict	**reptile** **bird** **warm-blooded** **mammal** **mammary gland**	**Workbook,** pp. WB92–98 **Transparencies** B4, B4-2 **Harcourt Science Activity Video** **Harcourt Science Newsroom Video**	**Kinesthetic Learners**, p. B118 **ESL Activity**, pp. B115, B119 **Investigation Challenge**, p. B120
		Take-Home Book, pp. TH33–36 *Sea Turtles* **Activities for Home or School,** p. TR70 **Chapter Test**, pp. AG40–44	**Performance Assessment**, p. AG45 **Portfolio Evaluation Sheet**, p. AGxxiv

Prepare for Activities

Use this page to help you organize and prepare materials for the Investigations, which begin each lesson of this chapter.

LESSON INVESTIGATION	MATERIALS

1 Comparing Fish and Amphibians
B102–103

PURPOSE Observe and compare animals from these two groups of vertebrates to find their similarities and differences.

PREPARATION TIPS If you have the facilities available, obtain living specimens of the organisms shown. Provide students with hand lenses, and have them make observations of the living animals.

TIME 20–30 minutes

EXPECTED RESULTS Students should conclude that fish and amphibians have similarities and differences.

MATERIALS

- diagrams of frog and fish

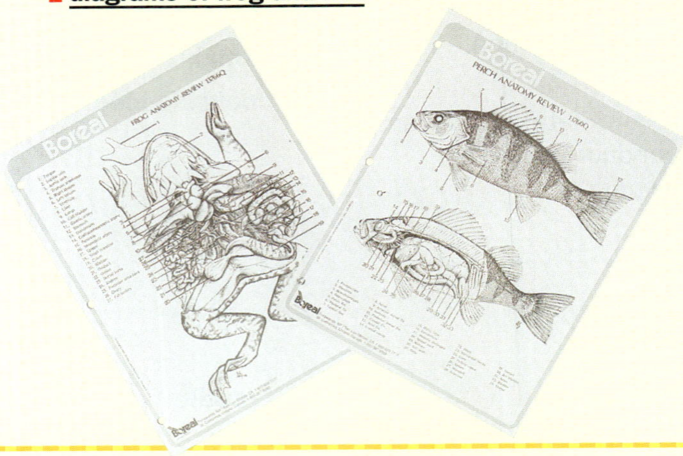

2 Comparing Skeletons
B112–113

PURPOSE Observe and compare the skeletons of a typical reptile, bird, and mammal.

PREPARATION TIPS If possible, set up a science station with examples of real bones from a reptile, bird, and mammal. Have students compare the bones in the diagrams with the real bones. For each pair of students, make a copy of the diagrams of animal skeletons available in Teaching Resources, pages TR167–168.

TIME 20–30 minutes

EXPECTED RESULTS Students should conclude that the skeletons of reptiles, birds, and mammals are similar.

MATERIALS

- diagrams of reptile, bird, and mammal skeletons
- colored pencils

Harcourt Science Activity Videos

Preview the Investigations The Activity Video for this unit shows students doing the Investigations. You may wish to view the video for classroom management ideas. The video can also be used to model the Investigation process for students.

Equipment Kits

Underlined items above are provided in the equipment kits (available separately).

Providing More Depth

. . . with Trade Books

These books provide in-depth information about certain vertebrates.

Do You Know the Difference? by Andrea and Michael Bischhoff-Miersch (North-South, 1995) pictures and discusses the differences between 12 pairs of look-alike mammals, such as the leopard and the jaguar. *NSTA Trade Book*

Hawk Highway in the Sky: Watching Raptor Migration by Caroline Arnold (Harcourt Brace, 1997) presents a close-up look at hawks, eagles, falcons, and other raptors through the work of HawkWatch International. *NSTA Trade Book*

Snakes by Eric S. Grace (Sierra Club, 1994) examines through photos and drawings the 10 families of snakes, their anatomy, behavior, and life cycle, dispelling myths with facts. *NSTA Trade Book*

Trade book titles are current at time of publication but may go out of print without notice.

. . . with More Hands-On Activities

Investigation Challenge

Activities for Home or School

The cross-curricular Links that appear at the end of each lesson in the student edition and throughout the teacher's edition also provide activities.

. . . with the Take-Home Book

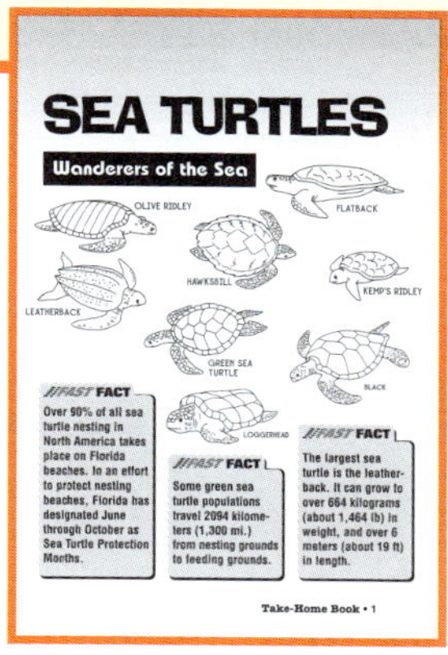

Sea Turtles

CONTENTS

. . . with Technology

Students can learn more about vertebrates through these components and special Internet sites. Look for Technology Links at the end of each lesson.

Harcourt Science Newsroom Video
Sea Turtles

Harcourt Science Explorations CD-ROM
Vertebrate Challenge

Visit The Learning Site for related links, activities, and resources: **www.harcourtschool.com**

WELCOME TO THE LEARNING SITE

Assessment Options ✓ Chapter 4

Formal Assessment

- **Chapter Review and Test Preparation**, PE pp. B126–127
- **Chapter Test**, pp. AG40–44

Performance Assessment

- **Chapter Review and Test Preparation**, PE p. B127
- **Chapter Performance Task**, pp. AG45–46
- **Project Evaluation Checklist**, p. AGxviii

Ongoing Assessment

- ✓ **Questions**, PE pp. B104–111, B114–121
- **Lesson Review**, PE pp. B111, B121
- **Informal Assessment**
  - **Classroom Observation**, TE p. B103
  - **Portfolio**, TE p. B121
- **Observation Checklist**, AGxiv

Student Self-Assessment

- **Self-Assessment–Investigate**, p. AGxvi
- **Self-Assessment–Learn About**, p. AGxvii
- **Project Summary Sheet**, p. AGxix

Portfolio Assessment

My Work

- **Science Experiences Record**, p. AGxxii
- **Guide to My Science Portfolio**, p. AGxxiii
- **Portfolio Evaluation**, p. AGxxiv

Chapter Test AG40

Vertebrates

Part I Vocabulary 1 point each

Use the letters of the terms in the Word Bank to complete the sentences.

A warm-blooded	D fish	G cold-blooded
B amphibians	E mammary glands	H reptiles
C bird	F mammals	

Animals with a backbone are called vertebrates. The classes of vertebrates that live entirely in water are called **1.** __D__. Another class of vertebrates, **2.** __B__, can live in water or on land but must return to the water to reproduce. Because both of these kinds of animals are **3.** __G__, their body temperatures change as the temperature of their surroundings changes. Lizards, snakes, alligators, and turtles are **4.** __H__. They have well-developed lungs, and most are cold-blooded. They reproduce by laying eggs on land. Another type of animal that reproduces by laying eggs is the **5.** __C__. This animal is **6.** __A__ and maintains its body temperature with energy from the food it eats and the feathers on its body. Warm-blooded animals that nurse their young are called **7.** __F__. They have milk-producing **8.** __E__ to nurse their young.

Part II Science Concepts and Understanding 2 points each

Write the letter of the best choice.

__B__ **9.** How are amphibians and fish alike?
- **A** Both have four to six fins.
- **B** Both reproduce in water.
- **C** Both can live in water and on land.
- **D** Both are warm-blooded.

Chapter Test AG41

__F__ **10.** How does a jawless fish differ from other kinds of fish?
- **F** It has a much simpler digestive system.
- **G** It has lungs.
- **H** It has a three-chamber heart.
- **J** It can live in salt water and in fresh water.

__D__ **11.** Why do bony fish lay thousands of eggs?
- **A** They are very tiny fish.
- **B** Many eggs are needed to build shelters for the hatchlings.
- **C** The eggs must combine to produce a fish.
- **D** Many of the eggs are eaten by other animals.

__H__ **12.** Why are cartilaginous fish such as sharks constantly moving?
- **F** They have big appetites and must find fish to eat.
- **G** Their skeletons are made of cartilage, not bone.
- **H** They have no swim bladder and would sink if they stopped.
- **J** The shape of their bodies does not allow them to sleep.

__D__ **13.** What adaptation allows adult amphibians to hibernate in mud during winter?
- **A** They can hide more easily because they no longer have a tail.
- **B** They don't need air when they are sleeping.
- **C** They can absorb water through their skin.
- **D** They can absorb oxygen through their skin.

__H__ **14.** The feathers of birds help them fly. What other purpose do feathers serve?
- **F** Feathers protect birds from sharp objects.
- **G** Feathers hide birds from predators.
- **H** Feathers keep birds warm on cold days.
- **J** Feathers take in oxygen while birds fly.

__A__ **15.** What do mammals have that no other animals have?
- **A** hair
- **B** teeth
- **C** lungs
- **D** beaks

Chapter Test AG42

__H__ **16.** Mammals that lay eggs outside their bodies are —
- **F** marsupials
- **G** placental mammals
- **H** monotremes
- **J** amphibians

__B__ **17.** With placental mammals, the mother and the embryo exchange —
- **A** genes, intelligence, and oxygen
- **B** nutrients, oxygen, and wastes
- **C** wastes, hormones, and size
- **D** nutrients, eye color, and DNA

__J__ **18.** Vertebrates that develop from embryos in jellylike eggs with no hard coverings are —
- **F** marsupials and reptiles
- **G** amphibians and birds
- **H** mammals and amphibians
- **J** fish and amphibians

__C__ **19.** Where do most pouched mammals, or marsupials, live?
- **A** Austria
- **B** the Amazon
- **C** Australia
- **D** Africa

__G__ **20.** Which of the following adaptations helps crocodiles and alligators stalk their prey?
- **F** rounded and pointed snouts
- **G** raised eyes and nostrils
- **H** tough skin covering
- **J** ability to give birth on land

__C__ **21.** What do the bodies of most mammals look like when they are born?
- **A** furry animal bodies
- **B** amphibian bodies
- **C** their parents' bodies
- **D** undeveloped embryos

Chapter Test AG43

Use the letters of the descriptions below to label the following pictures.

- **A** Embryo develops in an egg with a hard shell.
- **B** Embryo develops in an egg with a leathery shell.
- **C** Embryo develops in the mother's body before birth.
- **D** Eggs develop in water outside the parent's body.
- **E** Immature young develop in the mother's pouch.

22. __D__

23. __E__

24. __C__

25. __A__

26. __B__

Part III Critical Thinking 6 points each

Answer the following on a separate sheet of paper.

27. Describe the ways that the body structures and behaviors of bony fish enable them to escape from predators. Answers may include that bony fish have streamlined bodies that enable them to swim fast. Many also have a metallic color that makes them hard to see in water. Some have bright colors or patterns to warn away predators.

28. Explain how different types of adaptations enable amphibians to survive in different habitats. Possible answer: The color of their skin blends in with their surroundings, making them hard to see. Long, sticky tongues help frogs and toads catch insects for food. The shape of their feet helps them to swim (webbed) or climb (suction pad). They can dig into the mud and take in oxygen through their skin if it is too hot or too cold.

29. Why have scientists developed ways to raise and farm fish? The number of fish available for capture in the wild has decreased because of overfishing and pollution. Producing fish through farms helps to supply fish for people to eat.

30. In what ways are birds valuable to humans? Possible answer: Birds like chickens produce meat and eggs that humans eat. Birds eat insect pests and seeds of weeds. Birds kill rats and mice that attack grain crops.

Chapter Test AG44

Part IV Process Skills Application

Use the following chart to *compare* the vertebrates. Write your answers on the lines below. 2 points each

COMPARING VERTEBRATES

	Fish	Amphibians	(35)	Birds	Mammals
Chambers of heart	2	(33)	3 or 4	4	4
(31)	scales	none	scales or plates	feathers	(39)
Temperature regulation	(32)	cold-blooded	mostly cold-blooded	(37)	warm-blooded
Embryo development	mostly in jellylike eggs with no hard covering	mostly in jellylike eggs with ____ (34)	mostly in ____ (36) eggs	in ____ (38) eggs	mostly in ____ (40)

31. covering on skin or body
32. cold-blooded
33. 3
34. no hard covering
35. Reptiles
36. leathery
37. warm-blooded
38. hard-shelled
39. fur or hair
40. mother's body

6 **41.** If you **observed** a sea animal that had no gills or legs and it was nursing its offspring, what could you **hypothesize** about the animal?
It is a mammal because it is nursing its young.

6 **42.** **Compare** a cold-blooded animal such as a snake with a warm-blooded animal such as a rabbit. How would you **hypothesize** this difference would affect their behavior?
Answers may include that the snake would be dormant during very cold weather. It would come out to sun itself when the weather turns warm. The rabbit's body maintains the same temperature so it can be active all year long.

Assessment Continued

Performance Assessment AG45

A Table of Vertebrates

PERFORMANCE TASK

Materials

paper   pencil or pen

You want to make a study guide to help you and your friend study for a chapter test on vertebrates. To do this, you will make a table that compares and contrasts the features of fish, amphibians, reptiles, birds, and mammals.

To plan your table, answer the following questions. Then make a table to organize the information. Be sure to label the columns and rows of the table.

1. Where do the animals in each group live?

   ____

2. How do the animals in each group reproduce?

   ____

3. How do the animals in each group get oxygen?

   ____

4. What is the body covering of each group?

   ____

5. How do the animals in each group stay warm?

   ____

6. What are some of the animals in each group?

   ____

Performance Assessment AG46

PERFORMANCE TASK

A Table of Vertebrates

Materials paper, pencil or pen

Time 20–30 minutes

Suggested Grouping pairs

Science Processes record data, organize information, communicate

Preparation Hints You may wish to do this activity prior to the chapter test as preparation for Questions 31–40.

Introduce the Task Explain to students that tables are often used to organize information. This makes it easier to compare and contrast characteristics of groups. Tell them that they will use the information from the chapter to construct a table that compares fish, amphibians, reptiles, birds, and mammals. Remind students that a table consists of both columns and rows, and point out that both must be clearly labeled. The characteristics of each group can then be recorded in the appropriate sections.

Promote Discussion Display students' tables in the classroom. Encourage students to discuss the characteristics they included in their tables, summarizing the features that characterize each of the animal groups. Extend the discussion by comparing and contrasting the various animal groups.

Scoring Rubric

Performance Indicators

_____ Identifies the characteristics of each group.
_____ Constructs and labels a table to suit its use.
_____ Communicates information effectively in a table.
_____ Compares animal groups, noting similarities and differences.

Observations and Rubric Score

3   2   1   0

Workbook Support

Investigate Log WB87

Comparing Fish and Amphibians

Materials

diagrams of frog and fish

Activity Procedure

1. **Observe** the diagrams of the fish and the frog. **Compare** the animals' characteristics and behavior. Notice any traits that are unique to either animal.
2. Draw a picture of each animal. Include a size scale on each drawing. Write a brief description of each of the structures in your drawings.
3. Use the Venn diagram on this page to **record** how fish and amphibians are alike and how they are different. Add to the diagram any data that you already know about each kind of animal. **Infer** which characteristics are common to all fish and which are common to all amphibians.

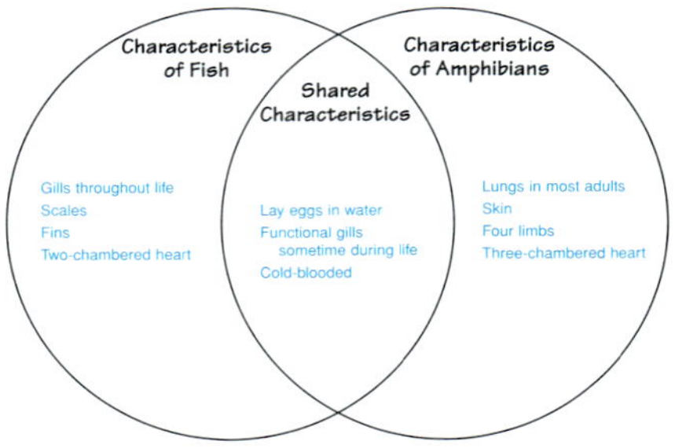

Investigate Log WB88

Draw Conclusions

1. From the information in your Venn diagram, in what ways are fish and amphibians alike? They both live in water for at least part of their lives. They both lay eggs to produce young. They are both cold-blooded.
2. In what ways are they different? Possible answers: Fish have gills; amphibians have lungs. Fish have external scales; amphibians do not. Fish have fins; amphibians have legs.
3. **Scientists at Work** *Comparative anatomy* is the study of the structures of different kinds of animals. It involves **observing** and **comparing** structures that may look different but have similar functions. What are two such structures of fish and amphibians? Answers will vary according to students' prior knowledge but are likely to include fins of fish and legs of amphibians, as well as gills of fish and lungs of amphibians.

Investigate Further **Plan and conduct a simple investigation** that will enable you to write good, complete definitions for *fish* and *amphibians.* Investigations may include observations of living creatures or research in the media center or on the internet. Studentdefinitions should mention that almost all fish have adaptations for living their entire lives in water and that amphibians have adaptations for living part of their lives in water and part on land.

Process Skills Practice WB89

Observe and Compare

Observing is the most basic of all scientific skills. When you compare things, you first must observe them to note similarities and differences between them.

Think About Observing and Comparing

Jose's class at school was studying growth patterns in living things. He learned in school that when plant roots grow longer, they grow at the tip and not at the base. He wondered how plant stems grew longer. He chose a growing plant and made a mark 1 centimeter below the top leaf on the stem. Then he drew four more marks down the stem, spaced 1 centimeter apart, as shown in the drawing. For five days Jose measured the distances between the marks. He recorded his observations in the table.

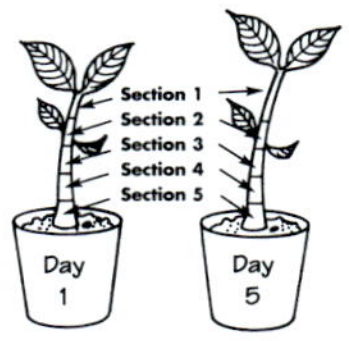

Growth (in centimeters)					
Day	**Section 1**	**Section 2**	**Section 3**	**Section 4**	**Section 5**
1	1 cm	1 cm	1 cm	1 cm	1 cm
2	1.25 cm	1 cm	1 cm	1 cm	1 cm
3	1.5 cm	1 cm	1 cm	1 cm	1 cm
4	1.75 cm	1 cm	1 cm	1 cm	1 cm
5	2 cm	1 cm	1 cm	1 cm	1 cm

1. What did comparing the changes in the plant tell Jose about the way plants grow taller? Plants grow taller only from their tips; the rest of the stem does not get longer.
2. How did Jose record his observations so they would be easy to compare? He made a table of the measurements.
3. How did observing and comparing help Jose understand how growth took place? Jose was able to compare the observations to see how the plant grew.

Reading Skills Practice WB90

Draw Logical Conclusions

Read the paragraph below. Then fill in the boxes with characteristics that describe the fish. Finally, draw a conclusion about what type of fish has been described, and write it in the circle.

Suppose that you have gone to a seafood market with an adult. On the ice in the glass cases you see many kinds of fish and other seafood. One fish in particular attracts your attention. You notice that it has knifelike fins on the side and a forked, narrow tail. The upper part of the fish's body is a metallic blue-green. The lower side of the fish is silvery. The body is very streamlined. What can you conclude about this fish?

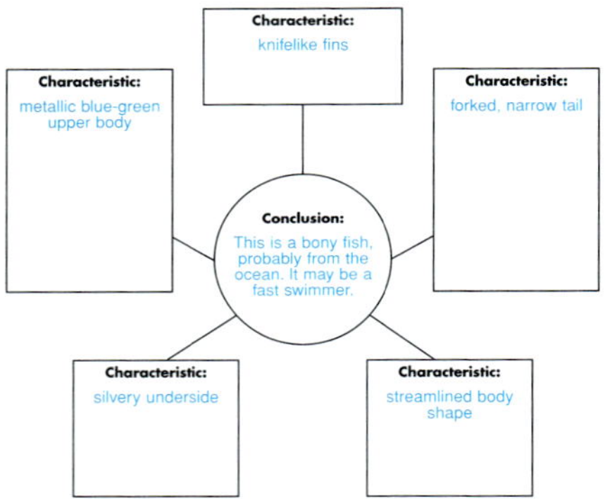

Lesson Concept Review/Assessment WB91

What Are Fish and Amphibians?

Lesson Concept

Fish and amphibians are cold-blooded vertebrates. Almost all fish live in water and breathe with gills throughout their lives. Most amphibians begin life in water as tadpoles with gills but then develop lungs and legs and live mostly on land. There are three kinds of fish—bony fish, cartilaginous fish, and jawless fish. Toads, frogs, newts, salamanders, and caecilians are all amphibians.

Vocabulary

fish (B104) **amphibians** (B104) **cold-blooded** (B104)

1. What is the most common characteristic feature of fish? Most spend all of their lives in water.
2. Both fish and amphibians are cold-blooded. Where geographically would you expect to find most species? warm tropical waters
3. Some fish have a substance like antifreeze in their blood. What does this tell you about these fish? They probably live in the Arctic or Antarctic where the water reaches freezing temperatures.
4. What are three characteristics of bony fish? Possibilities include swim bladder; caudal, anal, dorsal, and pelvic fins; eggs fertilized outside the body.
5. How are jawless fish different in additional ways from other fish? They have no fins or scales, and their digestive tract is much simpler.
6. Describe the typical life cycle of an amphibian. Eggs hatch into tadpoles that have gills, tails, and no legs. Legs and bodies develop and the gills are lost; tails may or may not be lost. Animal moves onto the land.
7. Do adult amphibians need a water environment? Yes; most must lay their eggs in water.

Investigate Log WB92

Comparing Skeletons

Materials

colored pencils

diagrams of reptile, bird, and mammal skeletons

Activity Procedure

1. **Observe** the diagrams. The skeletons can be divided into two parts. One part consists of the head, neck, and trunk, and the other consists of the limbs. Find these two parts on each diagram.
2. **Compare** the skeletons to find similarities. For example, the different skeletons have the same number of bones in corresponding parts of the limbs. Use the colored pencils to mark the similarities on each diagram.
3. **Observe** and **compare** the pictures of the bone cross sections of birds and mammals on page B113 of your science book. Find similarities and differences among the bones.

Workbook Support Continued

Investigate Log WB93

Draw Conclusions

1. What similarities did you **observe** in the head, neck, and trunk part of the three skeletons? Possible answers: a jointed backbone, a skull case, and ribs.
2. What similarities did you **observe** in the limb part of the skeletons? Each limb has two lower bones, one upper bone, and several smaller bones.
3. **Compare** the limbs of the skeletons. **Predict** how many limbs you would find on other animals in these groups. They have four limbs.
4. **Scientists at Work** Scientists use microscopes to study the structure of materials. From your **observations** of the bone cross sections, what can you **hypothesize** about the way a bird's bones help it fly? Students might notice that the bird's bones have air spaces, making them lighter than other animals' bones, and thereby helping the bird to fly.

Investigate Further **Observe** and **compare** the bones at the end of the limbs of the three animals. Discuss with your classmates how the adaptations in each animal are useful to it. Students should be able to observe differences and similarities in the number and arrangements of the bones. Have them find information about the adaptations of these animals and compare this to the structure of their digits.

Process Skills Practice WB94

Hypothesize

Hypothesizing is making a statement about a process, object, or event that you can support by an experiment or test. Usually the hypothesis is made after observing some natural occurrence.

Think About Hypothesizing

After reading letters to the editor written by local residents and observing local participation in recycling, a group of environmentalists hypothesized that a majority (more than 50 percent) of the people in their community cared enough about the endangered species in their state to become involved in conservation efforts. To test its hypothesis, the group hired you to do a mail survey.

1. List five questions you think would be important to ask in your survey. Assume that all the questions in your survey must be answered by indicating *Yes*, *No*, or *No opinion*. Answers will vary. Questions should be understandable and should test the hypothesis. Some possible questions: Are you aware that several species in our state are endangered? Do you know why certain species are endangered? Do you know if the ordinary citizen can do anything to help wildlife? Accept all reasonable responses.

Your survey must begin with an opening paragraph. Here are two possible choices.

A I am sending you this survey about an important issue—endangered species. Because I know you think it is important that plants and animals not become extinct, I know you will take the time to answer this survey.

B This survey is being sent to you to find out what you think about certain issues. Because your opinions are very important, please take the time to answer these questions.

2. Which of the above paragraphs, A or B, do you think would be better to use? Why? Paragraph B would probably be better because paragraph A already assumes that the individual thinks this issue is important. Paragraph B does not.
3. How did the hypothesis formed by the environmentalists help you design your survey? The hypothesis helped determine which questions to ask.

Reading Skills Practice WB95

Use Graphic Sources for Information

Use the chart below to develop at least two questions for other students to answer. Challenge a classmate to answer your questions based on the chart.

Flightless Birds

	Ostrich	Emu	Emperor Penguin
Habitat	**Africa**	**Australia**	**Antarctica**
Size	6–9 feet up to 300 lb.	5 feet up to 100 lb.	3–4 feet 60–100 lb.
Movement	can run 45 miles per hour	walks	swims 3–6 miles per hour; walks or toboggans on ice
Family life	males court females; shallow nests are in the ground	females court males; males incubate eggs and raise young; nest is a low platform of branches and twigs	only birds to breed in Antarctic winter; males incubate eggs
Food	seeds, leaves, small reptiles	insects, fruit, berries, grass	fish

Possible questions and answers are given.

Question:	Answer:
Which bird is the tallest?	Ostrich
Which bird eats fish?	Emperor penguin

Lesson Concept Review/Assessment WB96

What Are Reptiles, Birds, and Mammals?

Lesson Concept

Reptiles, birds, and mammals are vertebrates. A membrane that surrounds the reptile, bird, or mammal embryo contains water in which development takes place. Because of this, reptiles, birds, and mammals are able to live their entire lives on land. Most reptile embryos and all bird embryos develop inside eggs. Most mammal embryos develop inside the mother's body. Reptiles are cold-blooded and are covered with scales or plates. Birds are warm-blooded and have feathers. Mammals are warm-blooded and have hair and mammary glands.

Vocabulary

reptile (B115) **bird** (B116) **warm-blooded** (B116) **mammal** (B118) **mammary gland** (B118)

1. How are the lives of reptiles different from the lives of amphibians? Reptiles can spend all their lives on land. They do not need to reproduce in the water.
2. How do the eggs of reptiles, birds, and mammals allow the embryo to survive on land? The eggs have membranes that surround the embryo with water.
3. What are the three largest orders of reptiles? turtles, lizards and snakes, and alligators and crocodiles
4. List three characteristics of reptiles. They could include covering of horny scales or plates, well-developed lungs, cold-blooded, four legs except for snakes, and internal fertilization.
5. What is distinctive about birds? They have feathers; most can fly.
6. How does a bird get enough energy to fly? It eats an amount of food equal to about 30 percent of its body weight each day. To get enough oxygen, it has very efficient lungs connected to air sacs and a very efficient circulatory system with a four-chambered heart.
7. What skeletal and muscular adaptations does a bird have for flying? It has large chest muscles, a heavy breastbone, and lightweight bones with air spaces in them.

Vocabulary Review WB97

Recognize Vocabulary

Match the term on the left with its description on the right by writing the letter of the description in the blank next to the term.

Column A

E 1. fish
G 2. caecilian
M 3. amphibian
I 4. vertebrate
H 5. birds
L 6. bat
B 7. marsupial
K 8. placenta
C 9. mammals
A 10. reptiles
J 11. tuatara
D 12. monotremes
F 13. diaphragm

Column B

A These animals belong to four orders: turtles, lizards and snakes, crocodiles and alligators, and tuataras.
B The young of this animal group crawl into their mother's pouch and stay there until they are more fully developed.
C These are the only animals that have hair.
D These are the only mammals that lay eggs.
E There are three classes of this animal: bony, cartilaginous, and jawless.
F This is a muscle that only mammals have.
G This is an amphibian that has no legs.
H These are the only animals that have feathers.
I An animal that has a backbone made of cartilage or bone.
J This is the only remaining species of its order of reptiles.
K An organ through which nutrients, oxygen, and wastes are exchanged between the developing embryo and the mother.
L This is the only mammal that can fly.
M A salamander is an example of this type of animal.

Writing Practice WB98

Write a Folktale About a Vertebrate

Narrative Writing–Folktale

In his *Just-So Stories*, writer Rudyard Kipling made up fanciful folktales about the origins of certain animal characteristics. Two of his most famous tales are "How the Leopard Got Its Spots" and "How the Camel Got Its Hump." Write your own fanciful folktale about how a particular reptile, bird, or mammal developed a characteristic. Be inventive and playful. Use the story map below to help you plan your writing.

Title: How the ____________ **Got Its** ____________

Character (reptile, bird, or mammal):	Setting (habitat):
How the vertebrate originally looks:	
What happens to change the vertebrate:	
What characteristic develops:	
How life is different for the vertebrate:	

CHAPTER 4

Vertebrates

Generate Questions

Invite students to read and discuss the Fast Facts. Ask students if, after reading the interesting facts, they have any questions about vertebrates, or animals with backbones, for which they would like to find answers. Lead the class in brainstorming a list of questions, and record them on the board or on chart paper.

Ask students how they might go about finding answers to the questions they have. Suggest that the chapter will have investigations and reading material to help them answer their questions, but they may want to conduct other investigations or research on their own to answer some of their questions.

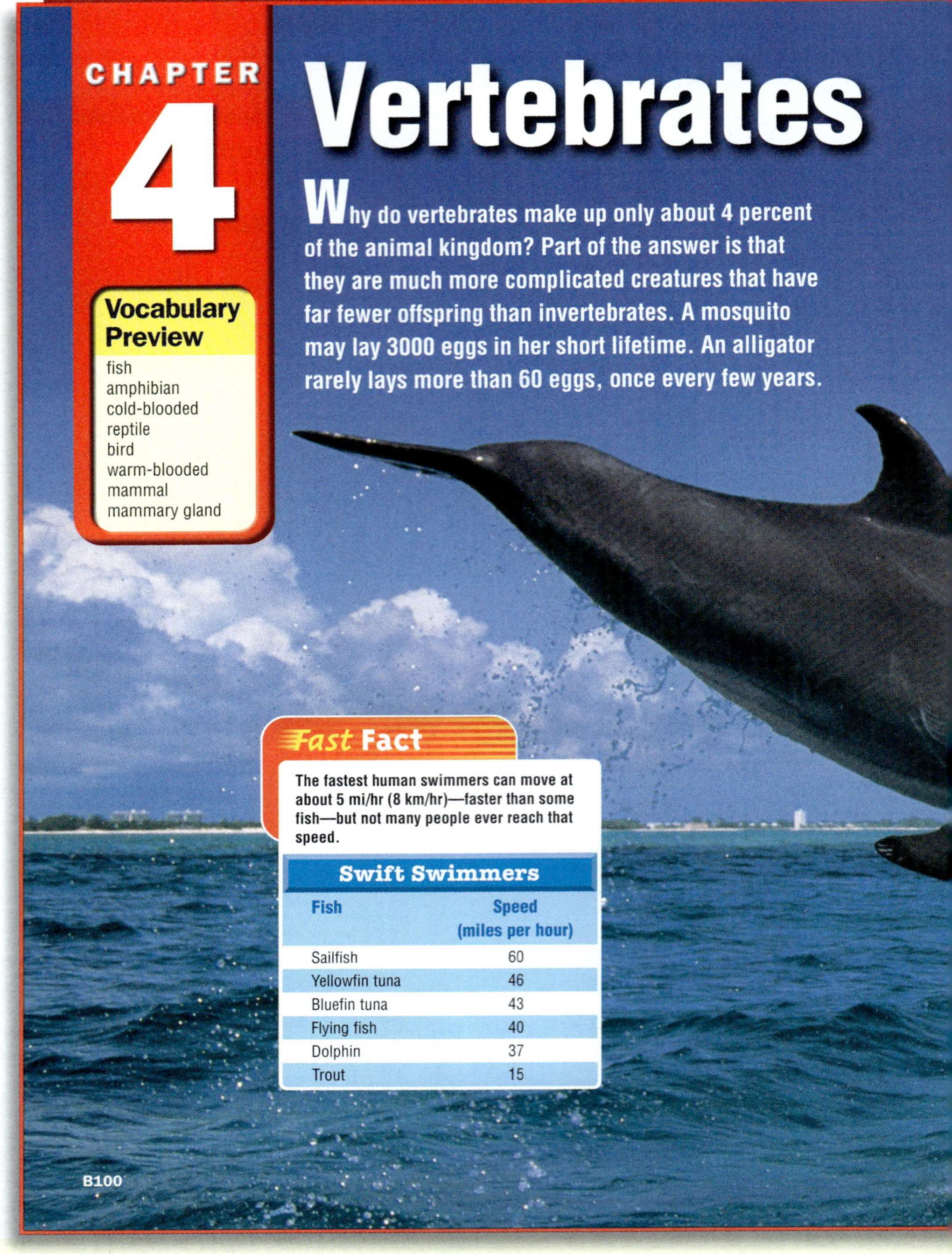

CHAPTER 4

Vertebrates

Why do vertebrates make up only about 4 percent of the animal kingdom? Part of the answer is that they are much more complicated creatures that have far fewer offspring than invertebrates. A mosquito may lay 3000 eggs in her short lifetime. An alligator rarely lays more than 60 eggs, once every few years.

Vocabulary Preview

fish
amphibian
cold-blooded
reptile
bird
warm-blooded
mammal
mammary gland

Fast Fact

The fastest human swimmers can move at about 5 mi/hr (8 km/hr)—faster than some fish—but not many people ever reach that speed.

Swift Swimmers

Fish	Speed (miles per hour)
Sailfish	60
Yellowfin tuna	46
Bluefin tuna	43
Flying fish	40
Dolphin	37
Trout	15

B100

Reading Skills Checklist

Strategies for developing the following reading skills are provided in this chapter.

- ❏ use prefixes and suffixes to determine word meanings
- ❏ use context to determine/confirm word meaning clues
- ☑ identify supporting facts and details *pp. B104, B106, B108, B110, B114, B116, B118, B121*
- ❏ arrange events in sequence
- ☑ identify the main idea *pp. B104, B106, B108, B110, B114, B116, B118*
- ❏ summarize and paraphrase a selection
- ❏ identify cause and effect
- ❏ predict probable future actions and outcomes
- ☑ use graphic sources for information *p. B120*
- ☑ draw logical conclusions *p. B109*
- ❏ compare and contrast
- ❏ make generalizations
- ☑ use reference sources *pp. B107, B122*
- ❏ distinguish fact and opinion

School-Home Connection

Distribute copies of the School-Home Connection, p. TR11. Have students take the page home to share with their families as you begin the chapter.

Follow Up Have volunteers compare the results of the activity they did at home. Additional School-Home Connections are provided by the **Activities for Home or School** (p. B125) and the **Take-Home Book** (p. B100d).

Teaching Resources, p. TR11

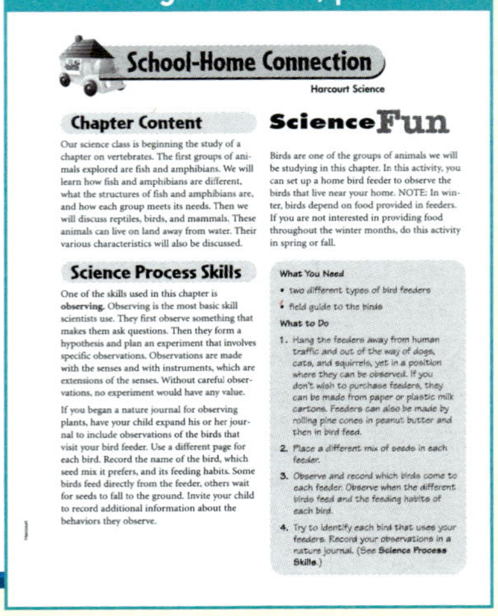

Prereading Strategies

K-W-L Chart

Have students make a four-column chart headed What I Know, What I Want to Know, What I Learned, and "+" for What I'll Learn More About. Ask them to look at the chapter title and fill in the first two columns. Ask them to predict what the chapter will be about. Have them fill in the third column as they study the chapter. At the end of the chapter, they can fill in the last column with additional things they'd like to learn.

K-W-L Chart

What I Know	What I Want to Know	What I Learned	+

Preview the Illustrations

Have students scan the pictures and captions in the chapter. Ask them to write questions raised by the illustrations and predict what the main idea of the chapter is.

Vocabulary Preview

Have students make a three-column chart for the words in the vocabulary list. In the first column, students should write each word and in the second column, what they think the word means. In the third column, students should write the definitions as they read through the chapter.

Vocabulary Chart

Word	What I Think It Means	Definition

BACKGROUND

Webliography

Keyword vertebrates

www.harcourtschool.com

Reptiles Millions of years ago, reptiles were the most numerous vertebrates. If dinosaurs are included, there were once 17 different orders. Now there are just four orders. The order **Squamata** includes about 3000 species of lizards, 2500 species of snakes, and 140 of worm lizards. **Chelonia** contains about 250 kinds of turtles and tortoises; **Crocodylia** has 21 species of crocodiles, alligators, gavials, and caimans. **Rhynochocephalia** today includes only the tuatara. Reptiles vary in size. The Indian gavial may be up to 9 m (30 ft) long and weigh 540 kg (1200 lb). The gecko lizard is about 13 cm (5 in.) long.

The Nile Crocodile was revered by certain ancient Egyptians. It has been hunted extensively. Although Nile crocodiles can still be found in the upper Nile and in other parts of Africa, it is considered an endangered species.

Fast Facts/Math Links

Use Logical Reasoning Discuss with students the voltage generated by electric eels. Ask students if they think this energy output could provide energy for lights or appliances. Encourage them to use logical reasoning to explain.

Interpret Data Have students analyze the Swift Swimmers table. The table shows top swimming speeds but does not show long-distance speeds. It also shows data only for fish. Data is difficult to obtain because of the practical problems of measuring the speed of a fish or a mammal underwater. Challenge students to make more tables about the swimming speeds of other vertebrates.

LESSON 1

What Are Fish and Amphibians?

Objectives

- Distinguish between fish and amphibians.
- Describe the structure of fish and amphibians and how they meet their needs.

Motivate

More Fast Facts

Spark interest with these facts and questions.

- The cornetfish is a very long fish. It can reach lengths of 1.8 m (about 6 ft), yet weighs only about 3.4 kg (7.5 lb). Instead of jaws it has a long, tubular mouth that it uses to suck in prey.

  How do you think the cornetfish and other fish differ from insects and other invertebrates? They have an internal skeleton.

- The Texas blind salamander lives in deep wells and underground streams in caves. It never sees daylight.

  How do you think this animal might get its food? It eats food that comes into the cave along with the water.

Access Prior Knowledge

- What makes fish different from other animals?
- What are some examples of amphibians? What are their characteristics?

When Minutes Count . . .

Whole-Class Investigate Option Complete the investigation as a class activity. Record observations on the board.

Using the Activity Video The video can be used to preview and model the investigate process with students. After viewing the video, students can conduct the investigation in small groups.

LESSON 1

What Are Fish and Amphibians?

In this lesson, you can . . .

- INVESTIGATE similarities and differences between fish and amphibians.
- LEARN ABOUT the structures and life processes of fish and amphibians.
- LINK to math, writing, language arts, and technology.

Both frog eggs and fish eggs are jellylike, without a hard or leathery covering. The eggs of both animals develop in water.

B102

INVESTIGATE

Comparing Fish and Amphibians

Activity Purpose Invertebrates are quite different from each other. But invertebrates have many similarities. Fish and amphibians are also different but similar. In this investigation you'll **observe** and **compare** animals from these two groups of vertebrates to find out their similarities and differences.

Materials

- diagrams of frog and fish

Activity Procedure

1. **Observe** the diagrams of the fish and the frog. **Compare** the animals' characteristics and behavior. Notice any traits that are unique to either animal. (Picture A)
2. Draw a picture of each animal. Include a size scale on each drawing. Write a brief description of each of the structures in your drawings.
3. Copy the Venn diagram, and **record** how fish and amphibians are alike and how they are different. Add to the diagram any data that you already know about each kind of animal. **Infer** which characteristics are common to all fish and which are common to all amphibians.

BACKGROUND

Fascinating Fish Several fish species have unusual characteristics that help them survive. When chased, a flying fish thrashes its tail until it bursts out of the water. It glides for about 488 m (1,600 ft) before falling back into the water. Some predators continue the chase and catch them when they return to the water. The warty frogfish is camouflaged and has a small, fishlike lure attached to its head. When another fish nears the lure, the frogfish extends its mouth and grabs its prey in milliseconds.

The Great Barrier Reef off the coast of Australia teems with a huge variety of fish. The scrawled file fish has scales so rough that it was once used as sandpaper. The leafy sea dragon disguises itself as floating seaweed, making it almost impossible to be detected. The weedy, or common, sea dragon is less flamboyant. The male carries the eggs under its tail.

Webliography GO ONLINE

Keyword vertebrates

www.harcourtschool.com

For student self-assessment, see Assessment Guide p. AGxvi.

Yellow perch

Picture A

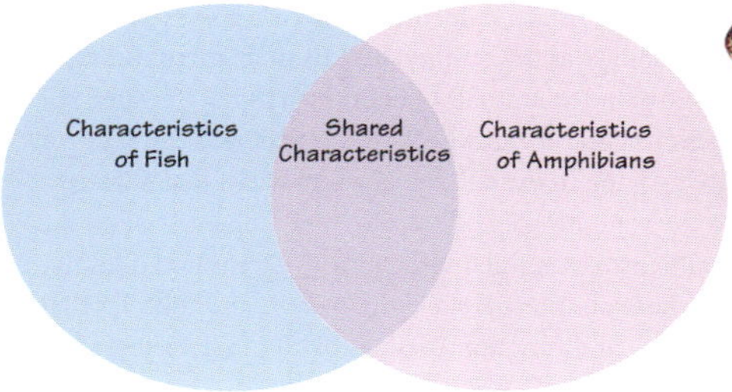

Leopard frog

Draw Conclusions

1. From the information in your Venn diagram, in what ways are fish and amphibians alike?
2. In what ways are they different?
3. **Scientists at Work** *Comparative anatomy* is the study of the structures of different kinds of animals. It involves **observing** and **comparing** structures that may look different but have similar functions. What are two such structures of fish and amphibians?

Investigate Further **Plan and conduct a simple investigation** that will enable you to write good, complete definitions for *fish* and *amphibians*.

Process Skill Tip

When you **compare**, you closely **observe** two or more objects to discover how they are alike and how they are different.

B103

Process Skill Tip

Observe and Compare Display the Process Skill Tip Transparency on an overhead projector or show the students similar drawings or photographs of a shark and a salamander. Have students follow the directions on the transparency.

Explain that an observation is information taken in through the senses. Point out that accurate observations are necessary in order to identify common and distinguishing characteristics among objects or events. Students should note that the Venn diagram they draw is similar to the one made in the investigation.

Process Skill Tip Transparency B4-1

Observe and Compare

- **Observe the shark and the salamander. What traits are the same in both? What is unique to each?** Both have a head and a tail. The shark has fins and gill slits. The salamander has feet and no gill slits.
- **Make a Venn diagram to show how the two animals are the same and different.**

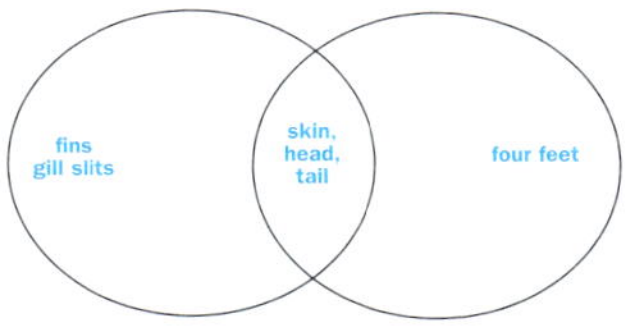

INVESTIGATE

Time 20–30 min **Preparation Tips** See p. B100c.

Grouping pairs

Process Skill Tip observe, compare

Other Process Skills record data, infer, plan and conduct a simple investigation

Expected Results Students should conclude that fish and amphibians have similarities and differences.

Teaching Tips

After students have completed the activity, draw on the board a Venn diagram that includes all the students' observations.

Draw Conclusions

1. Possible answers: Both live in water part of their lives, lay eggs to produce young, and are cold-blooded.
2. Possible answers: Fish have gills, scales, and fins; amphibians have lungs, no scales, and legs.
3. Answers will vary, but may include fins and gills of fish, and legs and lungs of amphibians.

Investigate Further Through their investigations, students should develop definitions that mention that fish have adaptations for living their entire lives in water and that amphibians have adaptations for living part of their lives in water and part on land.

Informal Assessment

Classroom Observation As students work in pairs to do the activity, observe how they interact. Encourage students to exercise patience with those who are slower at making observations.

INVESTIGATE LOG

Students can use **Workbook pp. WB87–88** to record the results of the investigation.

PROCESS SKILLS PRACTICE

To practice and apply process skills, see **Workbook p. WB89.**

LEARN ABOUT

1 Before Reading

SET A PURPOSE

FIND OUT Remind students that these statements will guide them in what to look for as they read.

VOCABULARY If students are doing the activity described in the Vocabulary Preview on page B101, remind them to add the definitions for these words to the chart.

Graphic Organizer for Chapter Concepts Display the transparency or have students use Workbook page WB86. Point out that by the end of the lesson, students should be able to fill in the information about fish and amphibians.

2 Guide the Learning

SCIENCE IDEAS

Help students focus on the supporting facts and details for this **main idea: Fish and amphibians are alike in some ways and different in others.**

- Why are fish and amphibians classified as vertebrates? They have an internal skeleton with a backbone that protects the spinal cord.
- How do fish and amphibians breathe? Most fish breathe through gills. Amphibians breathe through gills as young and live in water. As adults, they develop lungs, live on land, and breathe air.
- How do fish and amphibians reproduce? Both lay eggs without shells in water.
- How are the body temperatures of fish and amphibians affected by the temperature of their surroundings? They change as the temperature of their surroundings changes.
- What structures do fish and amphibians have for movement? Fish have fins. Amphibians have four limbs.
- How many chambers do their hearts have? Fish have two. Amphibians have three.

VOCABULARY CARDS AND ACTIVITIES

Students can use the Vocabulary Cards to make their own graphic organizers or to add to an ongoing file of science terms. The Vocabulary Cards and a variety of strategies and activities are provided beginning on p. TR85 in the **Teaching Resources** book.

LEARN ABOUT

Most fish have gills throughout their lives, while most amphibians begin life with gills and then develop lungs.

Fish and Amphibians

FIND OUT

- the differences between fish and amphibians
- the structure of fish and how they meet their needs
- the structure of amphibians and how they meet their needs

VOCABULARY

fish
amphibian
cold-blooded

Comparing Fish and Amphibians

Of the seven classes of vertebrates, three classes are **fish**, or vertebrates adapted to living their entire lives in water. One class is **amphibians**, or vertebrates that can live in water or on land but must return to the water to reproduce.

As you noticed in the investigation, fish and amphibians differ in important ways, but they share many characteristics. Both fish and amphibians lay their eggs in water. Most fish breathe through gills and spend their entire lives in water. Amphibian young also breathe through gills. But as amphibians mature, most develop lungs. As adults they spend much of their time on land.

Both fish and amphibians are **cold-blooded**. This means that their body temperature changes as the temperature of their surroundings changes. Many cold-blooded animals benefit from living in water because it changes temperature slowly. Some amphibians can control their temperature by lying in the sun or on warm rocks to get warm and returning to the water to cool off.

Fish and amphibians have similar circulatory systems and digestive systems. They also have an internal skeleton, nerves, and muscles that are similar in structure and function to those of all other vertebrates.

✓ **How do the respiratory systems of fish and amphibians differ?**

Comparison of Fish and Amphibians

Some shared characteristics of amphibians and fish are shown in the overlapping part of this Venn diagram.

Characteristics of Fish	Shared Characteristics	Characteristics of Amphibians
• Gills throughout life • Scales • Fins • Two-chambered heart	• Lay eggs in water • Jellylike eggs with no hard covering • Functional gills sometime during life • Cold-blooded	• Lungs in most adults • Skin • Four limbs • Three-chambered heart

B104

Graphic Organizer for Chapter Concepts

Transparency B4 • Workbook, p. WB86

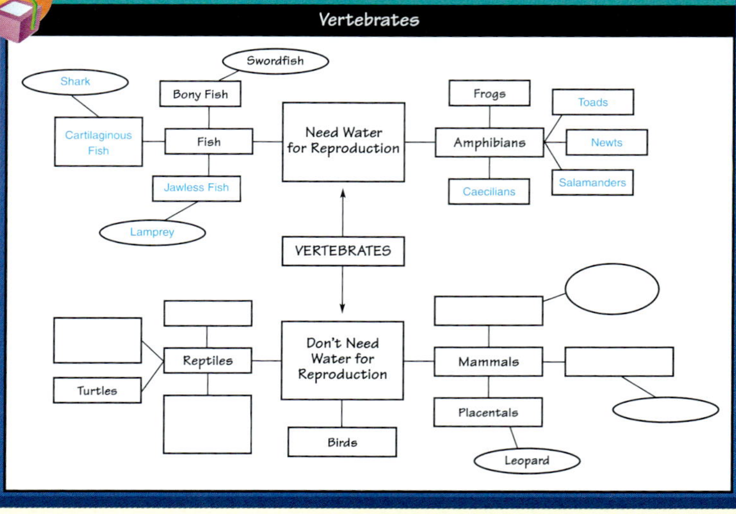

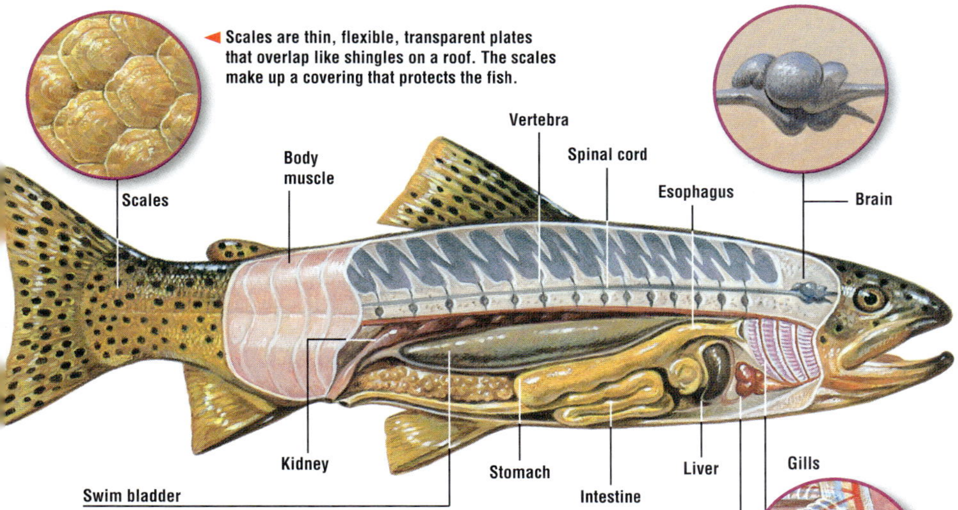

Scales are thin, flexible, transparent plates that overlap like shingles on a roof. The scales make up a covering that protects the fish.

A swim bladder is an organ that allows a bony fish to raise or lower itself in the water. Gas from the blood can move into the bladder, and gas from the bladder can move into the blood.

The heart of every fish has only two chambers. Blood enters the atrium of the heart and is pushed into the ventricle. The ventricle pushes the blood with enough force to send it through all of the fish's blood vessels.

A fish's gills are inside slits located behind its head. Water flows past the many blood vessels in the gills, and oxygen passes from the water into the blood.

Fish

A dolphin may look more like a fish than a sea horse does. But a dolphin isn't a fish, and a sea horse is. Dolphins do not have the characteristics of fish, and sea horses do.

The three classes of fish are bony fish, fish with skeletons made of cartilage (KART•uhl•ij)—called cartilaginous fish—and jawless fish. In place of the limbs of many other vertebrates, fish have fins for balance and movement. In a fish, blood is pumped through a network of arteries and veins by a two-chambered heart. The blood gets oxygen when it passes through the gills. Gills absorb oxygen from the water.

Most fish, except for the jawless fish, have hard scales covering their bodies. The scales of bony fish can be smooth, like those on carp and salmon, or rough, like those on bass and perch. The number of scales doesn't change during a fish's life. The scales that the fish started life with simply grow larger. Sharks and rays, which are cartilaginous fish, have scales that look like small teeth. These scales make the skin as rough as sandpaper.

The internal organs of all fish are similar. A fish has a brain that receives information about the outside world by way of sense organs. It has a digestive system to process food. The digestive systems of bony fish and cartilaginous fish include an esophagus, a stomach, and an intestine. A jawless fish has a simpler digestive system that is little more than a straight tube.

✓ **What can you tell about a sea horse's heart?** A sea horse's heart has two chambers because a sea horse is a fish.

B105

VISUAL LEARNING

Have students observe the illustration of the fish on page B105 and read the captions. Invite volunteers to name the various parts of the fish and to use their own words to describe the structure and function of each part.

DEVELOP SCIENCE VOCABULARY

fish, p. B104—Be sure students understand that living in water or swimming with a fishlike movement is not enough for an animal to be scientifically referred to as a fish. Have students list the characteristics of fish.

amphibian, p. B104—Tell students that the word *amphibian* comes from the Greek words *amphi*, meaning "on both sides," and *bios*, meaning "life." Help students see the connection between the word meaning and the animal it names: amphibians have a double life, in water and on land. Have students list other words that come from the Greek root *bio*.

cold-blooded, p. B104—Students may mistakenly think that *cold-blooded* means that the blood of these animals is always cold. Make sure they understand that cold-blooded animals are warmed by the sun or other heat sources. Ask them to describe what happens to the body temperature of fish and amphibians at night and during the day.

Ongoing Assessment

Use the ✓ questions in the pupil edition.

Cultural Connection

Sushi In Japan, fish is often eaten raw in **sushi**. Dried seaweed is pressed into thin sheets. The seaweed is wrapped around raw fish strips, vegetables, and cooked rice that is flavored with vinegar. The long rolls are then sliced, and each slice shows the bright colors of the vegetables and fish against the white rice and the dark roll of seaweed. Sushi may also include shellfish, such as shrimp, crab, octopus, and a garnish of fish eggs. It appears in many shapes and colors. Making sushi is an art form in Japan.

ESL ACTIVITY

Have students review the concepts in this lesson by making sentence-completion cards. Give students long strips of paper. Ask them to locate sentences that describe the characteristics of different animals in the lesson and copy them onto the strips of paper. For example, they might write, "Cartilaginous **fish** have no bones." Then ask students to cut the strips in half and, working in small groups, to mix up the half-strips and work together to rejoin them.

Guide the Learning continued

SCIENCE IDEAS

Help students focus on the supporting facts and details for this **main idea: Fish are grouped as bony fish, cartilaginous fish, and jawless fish.**

- What is the basic structure of a bony fish? How do bony fish differ? All bony fish have a bony skeleton and caudal, anal, dorsal, pelvic, and pectoral fins. Bony fish differ in overall shape, size of fins, and colors.
- How do most bony fish reproduce? Females lay eggs in the water, and then males of the same species fertilize the eggs.
- How are bony fish important to humans? They provide an important food source.
- What are the main characteristics of cartilaginous fish? a skeleton made of cartilage, no swim bladder, hinged jaws, eggs fertilized within the female
- What are the three main kinds of cartilaginous fish? What is a characteristic of each? Sharks, rays, and skates are cartilaginous fish. Sharks have several rows of sharp teeth; skates deposit fertilized eggs in leathery cases in the water; rays and skates have flattened bodies; many rays have poisonous stingers in their tails.
- In what ways are jawless fish similar to sharks, rays, and skates. In what ways are they different? All have skeletons made of cartilage. Jawless fish have no jaws, scales, or fins, and their digestive systems are basically straight tubes.

◀ Fish mouths and teeth differ greatly, depending on the kind of food they eat. The barracuda is a predatory fish that has a large mouth and long, sharp teeth.

Bony Fish

Most fish—more than 20,000 different species—are bony fish. Bony fish live in oceans, lakes, and streams from the equator to the polar seas. Because of their greatly different habitats, bony fish have many different forms and behaviors.

Fish that live in the open ocean have streamlined bodies. They have slim, knifelike side fins and narrow, forked tails. These allow them to swim fast to catch prey or to avoid predators. There is no place to hide in the open ocean, and many of these fish are well camouflaged. The upper parts of their bodies are a metallic blue-green, and their undersides are silvery. This coloring makes them difficult to see from above or below.

Fish that live in coral reefs are often brightly colored. Rather than making them stand out, this helps to hide them among the bright colors of the coral. Spots, stripes, and patches of color break up a fish's outline and confuse predators. Many reef fish have specially shaped fins and tails that help them maneuver around rocks and reefs.

All fish begin life as fertilized eggs. Eggs of most bony fish are fertilized in the water. Females lay thousands or even millions of eggs in the water. Males of the same species shed sperm over the eggs. Most of the eggs end up as food for other water animals, though some fish, such as sea horses, protect their eggs until they hatch.

Bony fish are an important source of food for people all around the world. Every year millions of tons of fish are taken from the ocean. The fishing industry has developed ways of catching large numbers of fish, and the populations of many ocean species have decreased. To stop the decrease, scientists and people who catch fish for a living have been developing ways to raise and farm fish, much as cattle are raised. This can help supply enough fish to eat and may prevent the loss of ocean fish.

✓ **What adaptations help fish survive in their habitats?** streamlining, camouflage, and specialized fins and body shapes

The cleaner wrasse swims right into the jaws of much bigger fish. There it scrapes out parasites and particles of food. As the cleaner wrasse gets fed, the larger fish gets dental care. ▼

The lionfish is one of the most venomous fish in the ocean. Glands in its spiny rays produce a powerful venom that can harm a predator and may even kill a human being. The brightly colored stripes of the lionfish are a warning to predators. ▼

Art Link

Mermaids are legendary creatures that are half woman, half fish. In Copenhagen, Denmark, the *Little Mermaid,* a statue of a mermaid sitting on a rock in the harbor, gazes out over the water searching for her true love. The *Little Mermaid* statue is based on the fairy tale by Hans Christian Anderson, which was also the basis for the movie by the same name. The statue was erected in 1913. In 1964, the mermaid's head was stolen. A new head was put on the statue, but the old one was never found.

▲ Rays vary in size from this blue spotted ray to manta rays. Manta rays can be up to 6 m (20 ft) across.

▲ Whale sharks are the largest of all sharks, reaching lengths of 8 m (about 26 ft) and weights of more than 17,000 kg (about 37,000 lb). Unlike predatory sharks, whale sharks get their food by sifting plankton from the water.

Fish Without Bones

Cartilaginous fish and jawless fish have no bones. Their skeletons are made of cartilage. Cartilage is a flexible tissue that forms the skeletons of most vertebrate embryos. It is also found as a kind of cushioning material in the adults of many vertebrates. For example, humans have cartilage in their joints, ears, and nose.

The cartilaginous fish include sharks, rays, and skates. Unlike bony fish, these fish have no swim bladders. Many of them cannot stop swimming because the weight of their bodies would pull them down to the bottom. Sharks, rays, and skates have hinged jaws that can open and close. Unlike the eggs of most other fish, the eggs of cartilaginous fish are fertilized within the female body. Many cartilaginous fish bear live young. Skates, however, deposit their fertilized eggs in leathery cases in the water.

More than 350 different kinds of sharks are known. Most have several rows of sharp teeth along their jaws. Although most sharks are fierce predators, they rarely attack humans.

Rays and skates have flattened bodies that allow them to glide along the ocean bottom near the coasts, where most of them live. They eat small fish and mollusks such as clams and oysters. The largest of the rays, the manta ray, lives in the upper waters of the open ocean. Many species of rays have poisonous stingers in their tails. They use the stingers for defense. These rays can be found in fresh water as well as salt water.

The hagfish and the lampreys are the only jawless fish still living. Many lampreys are parasites that attach themselves to their host with their suckerlike mouths. They scrape the host's flesh with their sharp teeth and then suck its blood. Jawless fish have the simplest structure of all fish. They have no jaws, scales, or fins, and their digestive systems are basically straight tubes.

✓ **In what way are jawless fish similar to sharks, rays, and skates?** Their skeletons are made of cartilage.

Most sharks have powerful jaws packed with many sharp teeth. Some sharks can bite hard enough to cut through a thick piece of steel. The great white shark is one of the ocean's fiercest predators, but usually it is not a danger to humans. ▼

USE PROCESS SKILLS

Compare Use a Venn diagram to help students compare the three groups of fish. On the board draw three circles that overlap in the middle. Make sure that the area of overlap is spacious. Label one circle Bony Fish, one Cartilaginous Fish, and one Jawless Fish. Invite volunteers to call out various characteristics of each type of fish and to identify where on the diagram to record the details. Write in the characteristics in the appropriate places. Encourage students to suggest characteristics of all fish as well as those of the various kinds of fish.

You may wish to challenge students to differentiate further among fish. Have them work together to make a more complex Venn diagram that includes various groups within each kind of fish, such as open-ocean-dwelling bony fish and coral-reef-dwelling bony fish, sharks, skates, rays, lampreys, and hagfish.

VISUAL LEARNING

Have students read the captions for each of the pictures on pages B106 and B107. Then have them identify specific structural or behavioral adaptations that can be found among specific species of fish.

Critical Thinking Guide students to better understand the relationship between structure and behavior by asking the following question.

► **What difference in structure do you think exists between the whale shark and other sharks? Why?** Since it obtains food by filtering plankton from the water, it probably does not have the many sharp teeth found in other sharks.

BACKGROUND

Webliography

Keyword vertebrates

www.harcourtschool.com

GO ONLINE

Sharks Most sharks are gray and are shaped like torpedoes. They have many characteristics of bony fish, but the small, pointed scales that cover their leathery skin differ in structure from the scales of bony fish. The upper part of a shark's tail fin is larger than the lower part, forcing the shark's head downward as it swims. Many species have rows of sharp teeth that may remain in the flesh of prey but that are quickly replaced by others. Contrary to popular belief, sharks' fins usually do not extend above water when the animals swim near the surface. Sharks tend to scavenge for injured fish, carrion, and garbage, as well as eat seals, turtles, birds, whales, crabs, and various fish. Shark young are referred to as **cubs**. A group of sharks is referred to as a **school** or a **shoal**.

THE QUESTIONS KIDS ASK!

Are electric eels really electric?

Address Misconceptions Electric eels and other electric fish, such as electric rays and skates, have specialized organs that produce electrical charges. Electric eels are capable of producing an electric jolt of more than 600 volts, which is strong enough to kill small animals and stun large animals and humans.

Encourage students to **use reference sources** to research different kinds of electric fish.

Guide the Learning continued

SCIENCE IDEAS

Help students focus on the supporting facts and details for this **main idea: Frogs, toads, newts, salamanders, and caecilians are classified as amphibians because of their characteristics.**

- **What are the main traits of amphibians?** They spend part of their lives in fresh water and part on land; they are vertebrates that have no covering of scales or hair; their skin is soft, thin, and moist and can absorb water; most of those with limbs have webbed feet with no claws; the adults can breathe in three ways: through lungs, skin, and roof of mouth.
- **Identify a trait of newts.** They have tails throughout their lives.
- **What are two main traits of salamanders?** They have tails throughout their lives, and they hatch in water as tadpole-like larva.
- **Talk about the two main traits of frogs.** They lose their tails as adults, and they have smooth, moist skin.
- **What are three main traits of toads?** They lose their tails as adults; they have rough, dry skin; and they visit ponds or lakes only to lay their eggs.
- **Describe an identifying trait of caecilians.** They have no limbs.

▲ The leopard frog is like most other amphibians without a tail. It has long hind legs and webbed feet adapted for swimming and hopping.

▲ There are about 4400 known species of amphibians. This Chinese salamander is an amphibian with a tail. It started life in the water as a larva and then developed into a land animal, which you see here.

▲ The caecilian is an amphibian that has no legs. Most caecilians live underground in the tropics and are almost blind.

Amphibians

Most amphibians spend part of their lives in fresh water and part on land. They are vertebrates that have no covering of scales or hair on their skin. The skin is usually soft, thin, and moist and can absorb water. If they have limbs, amphibians generally have webbed feet with no claws.

Frogs, toads, newts, salamanders, and caecilians (see•SIL•ee•uhnz) are all amphibians. Newts and salamanders have tails throughout their lives. Frogs and toads lose their tails when they reach maturity. Caecilians have no limbs.

The main difference between frogs and toads is in the texture of their skin. Toads have rough, dry skin and frogs have smooth, moist skin. Toads are able to live farther from water than frogs and may visit ponds or lakes only to lay their eggs.

Most frogs and toads hatch in water as tadpoles. Tadpoles swim freely, breathing through gills. They begin to develop lungs at the same time as they grow legs and lose their tails. They also change from having two-chambered hearts to having three-chambered hearts. Tadpoles eat

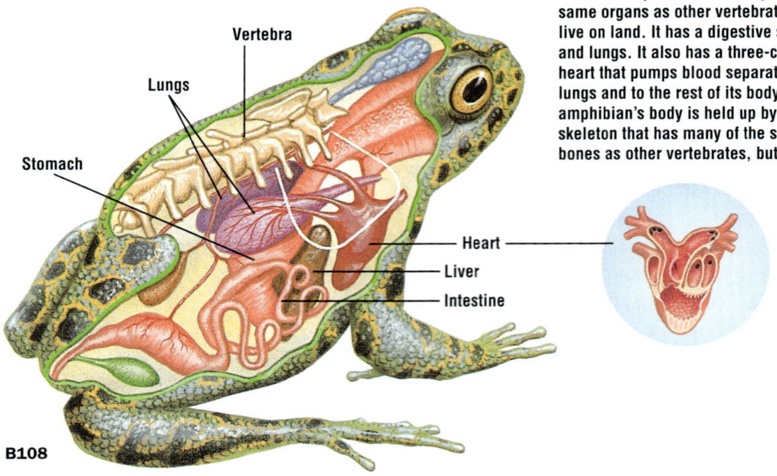

An adult amphibian has many of the same organs as other vertebrates that live on land. It has a digestive system and lungs. It also has a three-chambered heart that pumps blood separately to its lungs and to the rest of its body. The amphibian's body is held up by a bony skeleton that has many of the same bones as other vertebrates, but no ribs.

B108

Reaching All Learners

Visually Impaired and **Kinesthetic Learners**
Provide visually impaired and kinesthetic learners with plastic models of various fish and amphibians. Some models available from a scientific supply house clearly show features such as gills and scales. Provide students with the opportunity to touch the models, and then have them describe the structures that they feel. If possible, also provide live samples of fish and frogs or salamanders that students can observe.

BACKGROUND

Webliography GO ONLINE
Keyword vertebrates
www.harcourtschool.com

Amphibians come in a wide variety of colors and patterns of color. They may be blue, red, yellow, brown, or green, and may have stripes and spots of all sizes and shapes. Many are darker on their backs and lighter on their undersides. Like fish, their coloring and patterns either help them blend in with their surroundings or, in the case of brightly colored amphibians, warn predators of the poisons they contain. In some cases, the color may help the amphibian to absorb or reflect heat or to attract the opposite sex. The animals' colorings may change in responses to variations in humidity, light, or temperature. In warm, dry conditions, the colors become paler. In cool, moist conditions, they become darker.

plants, but most adult amphibians eat other animals, such as insects and worms.

Adult amphibians can breathe in three ways. First, their lungs exchange gases, though not as well as the lungs of many other vertebrates. Second, most amphibians can absorb oxygen through their skin. They get all their oxygen this way while they hibernate in mud during the winter. Third, a network of tiny blood vessels at the roof of an amphibian's mouth absorbs oxygen very well.

✔ **In what three ways can an adult frog breathe?** with its lungs, through its skin, and through the roof of its mouth

Life Cycle of a Salamander

The diagram here shows the stages in the life cycle of a Japanese giant salamander, the largest of the salamanders. It can reach a length of 1.5 m (about 5 ft) and can live up to 50 years. Like other amphibians, the Japanese giant salamander begins life in the water and goes through a series of changes before becoming an adult.

✔ **How are the life cycles of a salamander and a frog similar?** Both animals have a water-living stage with gills. Both live on land as adults.

The Life Cycle of a Japanese Giant Salamander

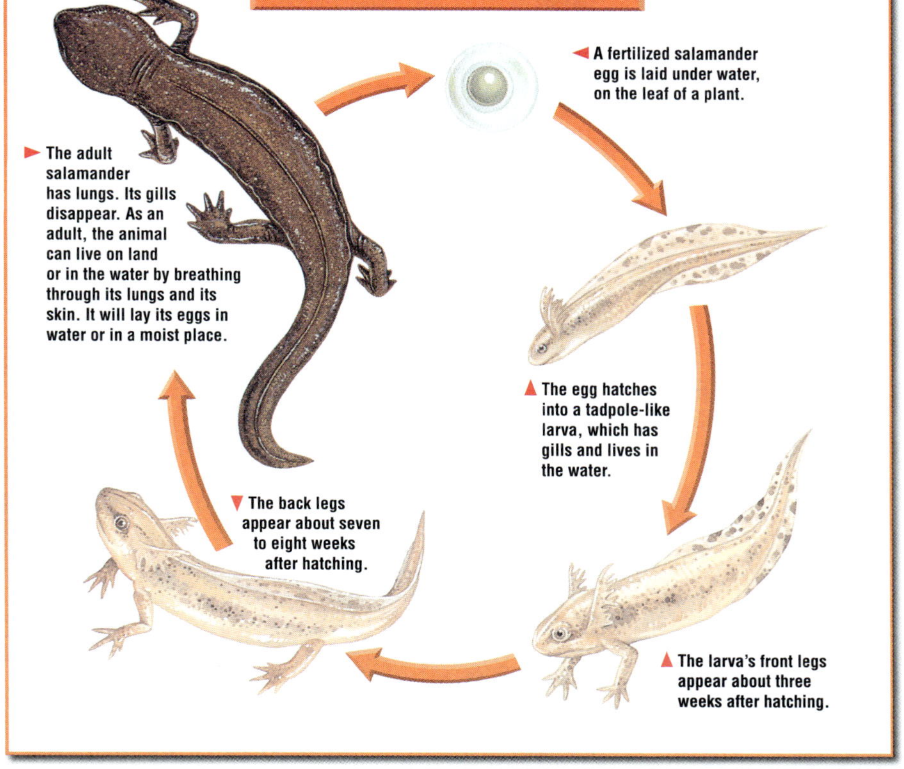

B109

VISUAL LEARNING

Invite students to thoroughly study the diagram of the life cycle of the Japanese giant salamander on page B109. Explain to them that this kind of life cycle is typical of amphibians, all of which undergo a change of body form and appearance as they develop into adult animals. To help students better understand the amphibian life cycle, ask the following questions:

- Where are a salamander's eggs fertilized? within the female
- Where does the salamander lay its eggs? under water, in this case on the leaf of a plant
- What is the appearance of the newly hatched salamander? It is a larva that has gills.
- How does the larva obtain oxygen? Its gills absorb oxygen from water.
- When do the larva's front legs appear? about three weeks after hatching
- When do the back legs appear? about seven to eight weeks after hatching
- How is the adult salamander different from the larva? The adult has lungs, fully developed legs, and no gills.
- How does the adult obtain oxygen when in the water? It takes in oxygen through its lungs and skin.

Reading Mini-Lesson

Draw Logical Conclusions Ask students to imagine they see an unusual animal while swimming in a pond. The animal has a fish-shaped body, gills, and no limbs, with a body covered in thick, gray skin and no scales. Display the transparency and have students draw a conclusion about what the animal is.

Reading Mini-Lesson Transparency B4-1

Draw Logical Conclusions

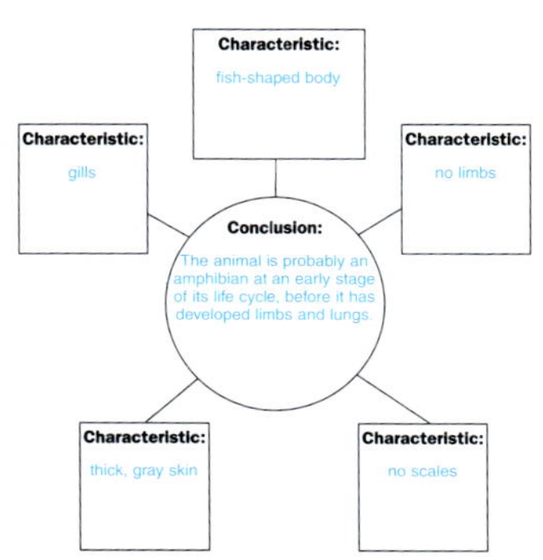

READING SKILLS PRACTICE

To practice and apply this reading skill, see **Workbook p. WB90.**

Guide the Learning continued

SCIENCE IDEAS

Help students focus on the supporting facts and details for this **main idea: Amphibians have adapted to live in many different environments.**

- **On what continents are amphibians found?** every continent except Antarctica
- **How does the color of amphibians help them?** Their coloring helps them blend into their surroundings to hide from predators.
- **What adaptation helps toads and frogs obtain food?** their long, sticky tongues
- **How are some amphibians' feet adapted to their environments?** Webbed feet aid in swimming; suction pads on toes help climbing frogs grip branches.
- **What behavior helps frogs and toads survive cold winters and hot summers?** The animals burrow into the mud and become inactive. They do nothing except absorb oxygen through their skin.
- **How do toads and poison-arrow frogs avoid being eaten by predators?** They are poisonous, and they often display startling colors to warn predators of the danger.

Amphibian Adaptations

Amphibians make up the smallest class of vertebrates. But they can be found on every continent except Antarctica, and they live in many different environments. Some frogs and toads, for example, despite their need for moisture, live far from riverbanks and ponds. They live on mountains, high up in rain-forest trees, and even in deserts.

Many amphibians are colored to blend in with their surroundings. Bullfrogs, for example, are green and live in ponds, which are often green. Other adaptations include long, sticky tongues that help frogs and toads catch insects. Their feet are also adapted to their environments. Frogs that swim have webbed feet. Climbing frogs have a suction pad on each toe that allows them to grip branches.

Amphibians are affected by the temperature. In winter, if their surroundings become too cold, frogs and toads dig into the mud and become inactive. They stay still and do nothing except absorb oxygen through their skin. In summer, frogs and toads burrow into the mud if the weather becomes too hot for them.

✔ **What are several amphibian environments?** Possible answer: almost any environment, including deserts, forests, and mountains

◄ This poison arrow frog from the rain forests of South America spends its life in water that collects in the cup of a flower blossom.

The giant toad is a native of the southwestern United States. It can reach a length of 23 cm (about 9 in.). Many toads are protected by a poison that spreads throughout their skin. It is produced in two organs located on the toad's head behind its eyes. These toads often swell up to discourage predators such as snakes. ▼

B110

INVESTIGATION CHALLENGE

Hands-On Activities: More About Amphibians

Use this idea for a hands-on activity to extend students' understanding of key lesson concepts.

Challenge students to observe and record a frog's life cycle. Collect frogs' eggs from a pond or obtain them from a biological supply house. Put the eggs in an aquarium or large jar filled with pond water, and keep the water at about 15–25°C (59–77°F). Feed the young tadpoles bits of boiled spinach or lettuce. Later, feed the growing tadpoles bits of hard-boiled egg yolk or raw liver. Ask volunteers to change the water twice a week. **CAUTION: Properly care for and dispose of the frogs.** Releasing non-native species could pose an environmental hazard or threaten local ecosystems. See page T21 for information.

For student self-assessment, see Assessment Guide p. AGxvii.

Summary

Fish and amphibians are cold-blooded vertebrates. Fish live in water and breathe with gills throughout their lives. Most amphibians begin life in water as tadpoles with gills but then develop lungs and legs and live mostly on land.

There are three kinds of fish—bony fish, cartilaginous fish, and jawless fish. Toads, frogs, newts, salamanders, and caecilians are all amphibians.

Review

1. What class of fish has the greatest number of species?
2. How are rays adapted to their environment?
3. In what form do most amphibians begin their lives after hatching from eggs?
4. **Critical Thinking** How do the hind legs of a frog help it survive?
5. **Test Prep** Fish and amphibians both have —
   - A gills throughout their lives
   - B a three-chambered heart
   - C fins throughout their lives
   - D a backbone

These long-tailed salamanders spend most of their lives in cool, dark, damp places. Since they have little protection against predators, they hide under logs or rocks. ▼

LINKS

MATH LINK

Write Percents A large number of fish species were recently classified by how close to extinction they are. Of these species, 1323 are not threatened with extinction, and 101 are becoming threatened. Another 443 species are vulnerable to extinction, and 291 are in immediate danger of extinction. What percent of all these fish are not threatened?

WRITING LINK

Informative Writing—Explanation Despite an old belief, a toad cannot give a person warts. Write an informative paragraph to some friends in which you explain this fact. Use reference books to find more information about this subject.

LANGUAGE ARTS LINK

Word Meanings Look up the word *salamander* in a dictionary or encyclopedia. You'll discover that one of the word's meanings is "a spirit that lives in fire." Find out what people once believed about salamanders.

TECHNOLOGY LINK

Learn more about types of vertebrates by investigating *Vertebrate Challenge* on **Harcourt Science Explorations CD-ROM.**

B111

3 Wrap Up and Assess

SUMMARY

Graphic Organizer for Chapter Concepts After reading the summary, have students complete the Lesson 1 section of the graphic organizer (shown on page B104).

REVIEW

1. bony fish
2. They have flattened bodies that allow them to glide over the bottom of the ocean.
3. tadpoles or larvae
4. They allow it to leap or swim away from predators very quickly.
5. D, a backbone

TEST PREP Test-Taking Tips

Item type multiple choice with a sentence fragment

To help students answer Question 5, share the following strategy with them. **Read the question and use facts you already know to help find the correct answer. Read each of the answer choices and think about which ones are true of fish and of amphibians. Select the answer that best fits.**

LESSON CONCEPT REVIEW/ASSESSMENT

Students can use **Workbook p. WB91** to review the lesson concepts.

LINKS TIPS

Links Objective Extend understanding of fish and amphibians.

Math Link If needed, show students how to find the percent. Find the total number of fish. To find what part the number of nonthreatened fish (1323) is of the total number of fish (2158), divide 1323 by 2158. Then multiply the quotient by 100 to find the percent. about 61 percent.

Writing Link Show students different ways to organize the information they have gathered. Explain that one approach is to compile related facts onto a single note card. Then have students arrange the note cards in the order they wish to present their information. Writing models are provided in the **Teaching Resources** book.

LESSON 2

What Are Reptiles, Birds, and Mammals?

Objectives

- ▶ **Explain where reptiles, birds, and mammals live.**
- ▶ **List the characteristics of reptiles, birds, and mammals.**
- ▶ **Describe three different kinds of mammals.**

Motivate

More Fast Facts

Spark interest with these facts and questions.

- The Marion's tortoise can live to more than 152 years—longer than any other animal on Earth. Three other kinds of tortoises can live to be more than 100.

  What kind of animals are tortoises? reptiles

- The largest bird is the North African ostrich, at 2.5 m (about 8 ft) tall and 156 kg (about 344 lb). The bee hummingbird is the smallest, at 5.7 cm (about $2\frac{1}{4}$ in.) long and 1.6 g ($\frac{6}{100}$ oz).

  Name one trait these birds have in common. Both have feathers.

Access Prior Knowledge

- ▶ What are some examples of birds, reptiles, and mammals?
- ▶ What characteristics do birds, reptiles, and mammals have in common?

When Minutes Count . . .

Whole-Class Investigate Option Display on an overhead projector transparencies of the diagrams to guide students as they complete each step of the investigation.

Using the Activity Video The video can be used to preview and model the investigate process with students. After viewing the video, students can conduct the investigation in small groups.

LESSON 2

What Are Reptiles, Birds, and Mammals?

In this lesson, you can . . .

- INVESTIGATE the skeletons of reptiles, birds, and mammals.
- LEARN ABOUT how reptiles, birds, and mammals carry out life functions.
- LINK to math, writing, art, and technology.

INVESTIGATE

Comparing Skeletons

Activity Purpose Vertebrates can be divided into two groups by comparing the kinds of eggs they have. In one group are the fish and amphibians. In the other group are the reptiles, birds, and mammals. There are clear differences among these last three kinds of animals. But what makes them similar to each other? In this investigation you'll **observe** and **compare** the skeletons of a typical reptile, bird, and mammal.

Materials

- diagrams of reptile, bird, and mammal skeletons
- colored pencils

Activity Procedure

1. **Observe** the diagrams. The skeletons can be divided into two parts. One part consists of the head, neck, and trunk, and the other consists of the limbs. Find these two parts on each diagram. (Picture A)
2. **Compare** the skeletons to find similarities. For example, the different skeletons have the same number of bones in corresponding parts of the limbs. Use the colored pencils to mark the similarities on each diagram.
3. **Observe** and **compare** the pictures of the bone cross sections of birds and mammals. Find similarities and differences among the bones.

◀ The golden lion tamarin is an endangered relative of monkeys and apes. All these types of animals are mammals.

BACKGROUND

Webliography
Keyword vertebrates
www.harcourtschool.com

The Komodo dragon is one of a group of lizards characterized by snakelike mouths. They have a long, forked tongue that retracts into a sheath. Their lower jaw comes unhinged so they can swallow large prey whole.

The brown pelican, like other pelicans, stores the fish it catches in the pouch of its bill. Later, the pelicans bring out the food either to eat it or to feed it to their young. Pelicans live in large colonies and build nests of twigs and branches.

The wombat has teeth that grow continuously, so it must constantly use them. Its long claws are adapted for digging the burrows in which it lives. A female wombat bears one offspring each year and carries it in her pouch for six months. Young wombats are protected from dirt and debris when the mother is digging because her pouch faces to the rear.

For student self-assessment, see Assessment Guide p. AGxvi.

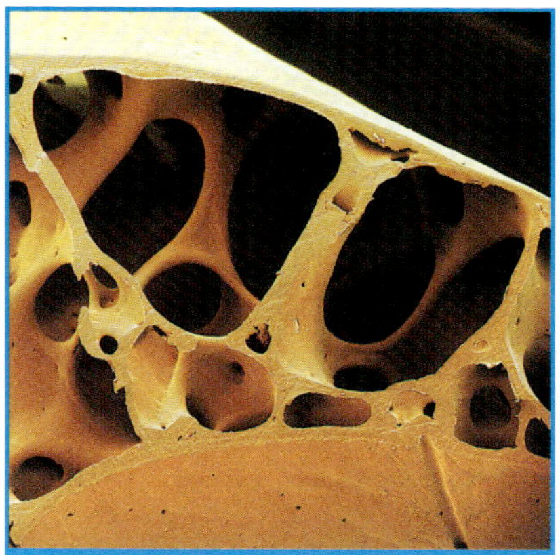

Cross section of bird bone (10×)

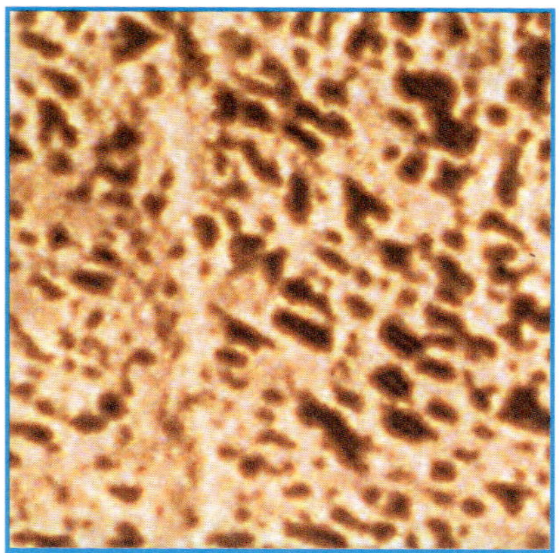

Cross section of mammal bone (10×)

Draw Conclusions

1. What similarities did you **observe** in the head, neck, and trunk part of the three skeletons?
2. What similarities did you **observe** in the limb part of the skeletons?
3. **Compare** the limbs of the skeletons. **Predict** how many limbs you would find on other animals in these groups.
4. **Scientists at Work** Scientists use microscopes to study the structure of materials. From your **observations** of the bone cross sections, what can you **hypothesize** about the way a bird's bones help it fly?

Investigate Further **Observe** and **compare** the bones at the end of the limbs of the three animals. Discuss with your classmates how the adaptations in each animal are useful to it.

Picture A

Process Skill Tip

When you **hypothesize** about an object or event, you use **observations** to help develop an explanation for it.

B113

INVESTIGATE

Time 20–30 min **Preparation Tips** See p. B100c.

Grouping pairs

Process Skill Tip observe, hypothesize

Other Process Skills compare, predict

Expected Results Students should conclude that the skeletons of reptiles, birds, and mammals are similar.

Teaching Tips

For reproducible masters of the animal skeletons, see Teaching Resources pages TR167–168. Encourage students to find as many similarities and differences as possible and to organize their observations in a chart or a Venn diagram.

Draw Conclusions

1. Possible answers: a jointed backbone, a skull case, and ribs
2. Each limb has two lower bones, one upper bone, and several smaller bones.
3. They have four limbs.
4. Students might notice that the bird's bone has air spaces, making it lighter than other animals' bones and thereby making it possible for the bird to fly.

Investigate Further Students should be able to observe differences and similarities in the number and arrangements of the bones. Have them find information about the adaptations of these animals and compare this to the structure of their digits.

INVESTIGATE LOG

Students can use **Workbook pp. WB92–93** to record the results of the investigation.

PROCESS SKILLS PRACTICE

To practice and apply process skills, see **Workbook p. WB94.**

Process Skill Tip

Observe and Hypothesize Display the Process Skill Tip Transparency on an overhead projector. Point out that a hypothesis is an educated guess about the relationships between variables, based on observation, prior knowledge, and experimental outcomes. Have students first observe and then hypothesize about the duck's structure. Then discuss the strengths and weaknesses of their hypotheses.

Process Skill Tip Transparency B4-2

Observe, Hypothesize

Closely observe the physical characteristics of the duck.

- Why do you think this animal could not survive in the desert or in a rocky, waterless area? The duck's webbed feet are suited for swimming and its flat bill is suited for scooping up fish.
- State a hypothesis you could test to determine if your observations and conclusions are correct. Animals with webbed feet would have difficulty walking on rocky terrain.

LEARN ABOUT

1 Before Reading

SET A PURPOSE

FIND OUT Remind students that these statements will guide them in what to look for as they read.

VOCABULARY If students are doing the activity described in the Vocabulary Preview on page B101, remind them to add the meanings of these words to the chart.

Graphic Organizer for Chapter Concepts
Display the transparency (shown below) or have students use Workbook page WB86. Point out that by the end of the lesson, students should be able to fill in the information about the remaining vertebrates on the transparency.

2 Guide the Learning

SCIENCE IDEAS

Help students focus on the supporting facts and details for this **main idea: Reptiles are divided into four orders: turtles, lizards and snakes, crocodiles and alligators, and tuataras.**

- **What distinguishes reptiles from other animals?** a dry, protective covering of horny scales or plates; eggs fertilized within the female
- **What are the largest reptiles?** alligators and crocodiles
- **How are alligators and crocodiles similar and different?** Similar: live much of life in water; use tails for swimming; covered with horny plates that are strengthened with bone; eyes and nostrils are raised so they can be above water while the rest of the body is below water. Different: alligators have rounded snouts, whereas crocodiles have triangular heads with pointed snouts.
- **What are the characteristics of turtles?** a domed upper shell, which contains the fused backbone and ribs, and a flatter lower shell; hard, toothless beak and strong jaws; may live either on land or in water; eggs always laid on land

Ongoing Assessment

Use the ✓ questions in the pupil edition.

LEARN ABOUT

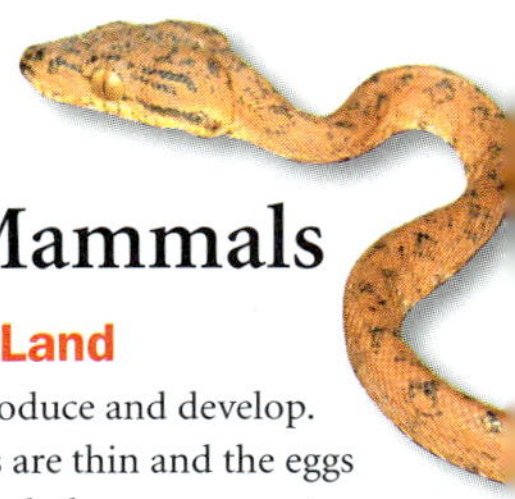

Reptiles, Birds, and Mammals

FIND OUT

- why reptiles, birds, and mammals can live on land, away from water
- the characteristics of reptiles, birds, and mammals
- three different kinds of mammals

VOCABULARY

reptile
bird
warm-blooded
mammal
mammary gland

Reproducing and Developing on Land

Fish and amphibians depend on water to reproduce and develop. Their eggs have membranes, but the membranes are thin and the eggs must be laid in water. Reptiles, birds, and mammals, however, can stay on land throughout their lives. They do not have to lay their eggs in water. Each developing embryo is protected by a membrane that encloses it. Within this membrane is the water the embryo needs in order to develop. In addition, the egg has food and nutrients just as fish and amphibian eggs do. The stage in which reptiles, birds, and mammals have gill slits takes place within the membrane.

In birds and most reptiles, the membrane is inside an egg that has a shell. In mammals, the embryo usually develops enclosed within a membrane inside the uterus of the mother. For most reptiles, birds, and mammals, when development inside the membrane is complete, the young hatch or are born having basically the same form as their parents.

✓ **How does the membrane that encloses an embryo allow animal species to survive on land?** It surrounds a developing embryo in the water that it needs.

An egg contains its own water supply, allowing the embryo to develop on land. ▼

The Komodo dragon is the largest living lizard, reaching a length of 3 m (about 10 ft). It is found on Komodo and other small islands of Indonesia, where it hunts other animals during the day. ►

▲ Most reptiles, like this turtle, hatch from eggs.

B114

Graphic Organizer for Chapter Concepts

Transparency B4 • Workbook, p. WB86

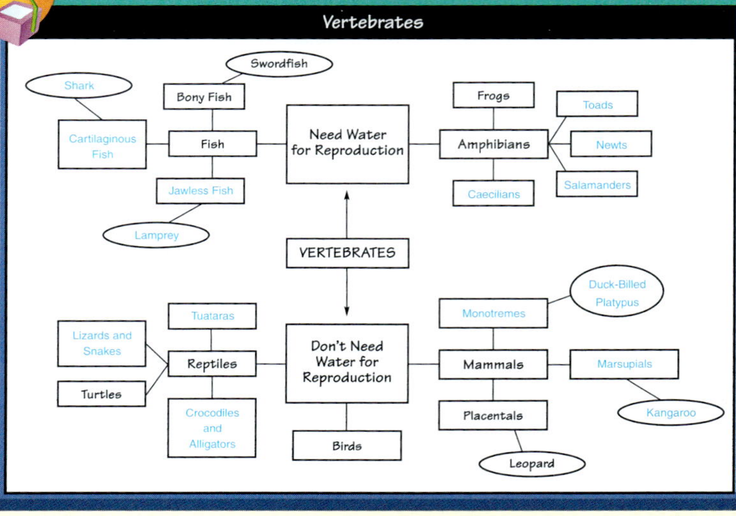

Reptile eggs usually have a leathery covering and are laid on land. Amphibian eggs are jellylike with no hard covering and are laid in water.

◀ Many snakes are brown, black, or dull-colored. Bright coloring often serves as a warning to other animals that the snake is venomous, although that is not the case with this tree snake.

The Nile crocodile of Africa is one of the largest living reptiles. It can reach a length of 6 m (about 20 ft). It has a long, low body, short legs, and a long, powerful tail. The Nile crocodile can live in salt water as well as fresh water.

Reptiles

Reptiles have a dry protective covering of scales or plates. They usually have four legs with clawed toes. However, snakes and some lizards don't have any legs. And turtles that live in water have flippers instead of clawed toes. Reptiles have well-developed lungs. Most reptiles are cold-blooded.

There are about 6000 species of reptiles. They belong to four different orders: tuataras (too•uh•TAH•ruhz), turtles, crocodiles and alligators, and lizards and snakes. The tuatara is the only remaining species of its order. Its closest relatives became extinct more than 100 million years ago. Tuataras reach a length of about 60 cm (2 ft) and look like large lizards. They live on islands off the coast of New Zealand.

The largest reptiles are the crocodiles and alligators. Crocodiles have triangular heads with pointed snouts. Alligators have more rounded snouts. Crocodiles and alligators spend much of their lives in water. They use their powerful, flattened tails for swimming. They are covered with plates that are strengthened with bone. The eyes and nostrils are raised and stick out above the water while the rest of the animal is under the water. Crocodiles and alligators can stay hidden in the water as they approach prey.

Turtles have a domed upper shell and a flatter lower shell. Some turtles can completely withdraw their tail, legs, and head into the shell. Turtles have a hard, toothless beak and strong jaws. Tortoises, or turtles that live on land, are often plant eaters. Even turtles that live in water, such as the giant sea turtle and the snapping turtle, lay their eggs on land.

Lizards and snakes are members of the same order. Both have scales, and both can unhinge their jaws. This is easier to see in snakes, which can swallow fairly large animals whole. As lizards and snakes grow larger, they shed their skins. Some snakes shed several times a year.

Most lizards eat other animals—usually insects and worms—but some are plant eaters. All snakes are meat eaters. Very few snakes are venomous. Gila (HEE•luh) monsters and beaded lizards are the only venomous lizards. The Gila monster lives in the southwestern United States.

Reptile eggs are always fertilized within the female. Most reptiles lay eggs, which usually have a leathery shell. Hard-shelled eggs are laid by tortoises, crocodiles, and gecko lizards. In some lizards and snakes, the young develop within the female and are born alive.

✔ **How do reptile eggs differ from amphibian eggs?**

A turtle is a reptile that is easy to recognize because of its shell. The top part of the shell contains the backbone and ribs, fused into the plates. ▼

B115

Have students continue to compare and contrast the orders of reptiles.

▶ **How are lizards and snakes similar? How are they different?** Both have scales and can unhinge their jaws, and both shed their skins as they grow larger. Some lizards are plant eaters, but all snakes are meat eaters.

▶ **What other reptiles does the tuatara resemble?** large lizards

VISUAL LEARNING

Invite students to study the egg illustration on page B114. Ask volunteers to list the parts, characteristics, and functions of a reptile egg.

Critical Thinking Guide students to better understand the function of a reptile egg.

▶ **How do each of the structures of the egg relate to the fact that the egg is laid on land, rather than in the water?** Explain that the membrane enclosing the embryo allows it to survive on land by surrounding the embryo in the water that it needs.

DEVELOP SCIENCE VOCABULARY

reptile, p. B115—Explain to students that the word *reptile* comes from the Latin word *repere,* meaning "to crawl." Invite students to explain the connection between the two words. Challenge them to list those reptiles that best fit this description.

BACKGROUND

Sea turtles include the green, loggerhead, hawksbill, ridley, and leatherback turtles. The leatherback is the largest living turtle. It weighs up to 540 kg (about 1200 lb). Sea turtles live primarily in warm, tropical waters. The leatherback feeds mainly on jellyfish, the green turtle eats seaweed and sea grasses, and the hawksbill consumes crabs and other shellfish.

Sea turtles have long flippers that propel them through the water. Many sea turtles migrate long distances. The green turtle migrates 3340 km (about 2071 mi) from Brazil to the island of Ascension in the Mid-Atlantic, where it breeds.

Female sea turtles lay their eggs in a large hole above the high-tide line. Shoreline development has destroyed some sea turtle nesting grounds, and many turtles drown in fishing trawls. These environmental changes have caused some turtles to be endangered.

Webliography GO ONLINE

Keyword vertebrates

www.harcourtschool.com

ESL ACTIVITY

Have students draw or paste animal pictures onto index cards to make decks of animal cards. Ask them to make an equal number of bird, **reptile**, and mammal cards. Then have them devise and play a game using the cards and facts they know about the three animal groups. If they need help, suggest they design a game like "Hearts," in which the designated suit (a type of vertebrate) always trumps the other cards.

Guide the Learning continued

SCIENCE IDEAS

Help students focus on the supporting facts and details for this **main idea: Birds have feathers and can fly.**

- **What are the functions of down feathers and contour feathers?** Down feathers trap air and hold in a bird's body heat; contour feathers help a bird fly.
- **What are three ways that flying is useful for birds?** to escape predators, to find food, and to migrate
- **Why do birds eat a lot of food and use a lot of oxygen?** to get the energy they need to fly and maintain body temperature
- **How do birds get the large amounts of oxygen they need?** Birds have both lungs and air sacs that fill up with oxygen when they breathe.
- **Which muscles enable a bird to fly?** chest muscles that are attached to the largest bone in a bird's body, the breastbone
- **What adaptations help keep a bird's body weight light so the bird can fly?** bones with tiny air spaces; feathers that are light and that compress during flight to give the bird a more streamlined shape; lightweight beak; no teeth

Birds

Birds are the only animals that have feathers. Unlike the vertebrates that you've read about so far, birds are **warm-blooded**. Their bodies stay at the same temperature regardless of the temperature of their surroundings. Warm-blooded animals get heat from the energy released in their cells as they burn food. Being warm-blooded allows birds to do more active work than cold-blooded animals. It also allows birds to be active all the time because they can keep warm in cold weather.

A bird's feathers not only enable it to fly but also keep it warm. Down feathers, which are soft and finely divided, are found next to the bird's skin. They trap a layer of air that holds in body heat. The larger contour feathers are used for flying.

Almost all birds can fly. Flight allows them to escape from predators that cannot fly. It helps birds catch insects in the air and search for food over a large area. Flight also allows birds to migrate as the seasons change.

However, flight and maintaining body temperature use up a lot of energy. Many birds eat as much as 30 percent of their body weight in food each day to get the energy they need. They also need a great deal of oxygen to release the energy from the food. Birds have very efficient lungs that are connected to air sacs. These sacs, as well as the lungs, fill with air when a bird breathes, giving the bird more oxygen. The sacs also help keep the bird cool as it flies. Oxygen is carried through a very efficient circulatory system that is powered by a four-chambered heart.

Owls have large, broad heads with feathers around the eyes. They usually live alone and hunt at night. Their soft feathers muffle the swishing sound that most birds make when they fly. So owls can swoop down on their prey without being heard. ▼

▲ Birds' eggs are similar in structure to those of reptiles. But the shells of birds' eggs are brittle rather than leathery.

BACKGROUND

Young birds vary in the amount of care they need. **Altricial** (al•TRISH•uhl) young are blind and have little or no down. They cannot stand and are totally dependent on their parents. **Precocial** (pree•KOH•shuhl) young have open, functional eyes and dense down. Soon after hatching they can walk, and within a few days they begin to find their own food. Songbirds have altricial young, as do woodpeckers, swifts, and pelicans. Young fowl, such as ducks, chickens, and pheasants, are precocial. Some young birds are in an intermediate state of development. Young gulls and terns are downy, have open eyes, and run within a day or two, but depend on their parents for food for several weeks.

Webliography GO ONLINE

Keyword vertebrates

www.harcourtschool.com

◄ Not all birds fly. Penguins have wings adapted for swimming. Rheas and ostriches are too heavy to fly, but they can run very fast. The cassowary lives in the forests of Australia and New Guinea. It stands 1.5 m (about 5 ft) tall and can run at speeds of 64 km/hr (about 40 mi/hr).

Flying also takes enormous muscle power. The muscles that enable a bird to fly are not in its wings, but in its chest. These muscles are attached to the largest bone in a bird's body, the breastbone. They account for more than one-third of the bird's weight.

To be able to fly, a bird cannot weigh very much. A bird has adaptations that keep it very light for its size. Its bones are not solid but have tiny air spaces in them. The structure of the bones with all these air spaces makes them very strong. The feathers, too, are light and strong. During flight the feathers compress so that the bird's body takes on a more streamlined shape.

A bird also has feet and a beak that are suited for its behavior. A duck, for example, has webbed feet for swimming and a broad, flat bill that it uses to strain water or mud for food. Birds that hunt have long, sharp claws and a hooked beak for tearing.

The arrangement of the toes is different for different birds. Climbing birds such as woodpeckers have two toes in front and two at the back. Perching birds have a single toe at the back. Running birds have all their toes at the front of their feet.

All birds hatch from eggs that have brittle, hard shells. The embryo inside each egg must be kept warm in order to grow and develop. Parent birds sit on their eggs to keep the embryos warm and to protect the eggs from predators.

Birds' eggs have a brittle, hard shell; reptiles' eggs have a leathery shell.

Many newly hatched birds are helpless. Their eyes are closed, and the few feathers that they have are small down feathers. The parents must keep their young warm, protect them from predators, and bring them food. But young birds grow very fast, and many are ready to leave the nest within a few weeks.

Birds are valuable to humans in several ways. Chickens and other poultry birds provide people with meat and eggs. Birds also help farmers by eating weed seeds and insect pests. Some of the larger birds, such as hawks and owls, kill rats and mice that eat grain crops.

Last but not least, people like birds. Listening to their songs and observing their colors and activities are experiences that many people enjoy.

✓ **What is the difference between birds' eggs and reptiles' eggs?**

A bird's skeleton has adaptations that enable the bird to fly. The wing muscles are attached to the keel, a large extension of the breastbone. Most birds also have a lightweight beak, no teeth, and bones that are light and porous. ▼

B117

USE PROCESS SKILLS

Communicate Allow students time to read page B117 and to study the illustrations on pages B116 and B117. Then elicit responses to the following:

► **Describe some adaptations of birds that help them live in their particular environments.** Possible responses: Ducks have webbed feet they use for swimming and a flat bill they use to strain water or mud for food; birds that hunt have long, sharp claws and a hooked beak for tearing meat; birds such as woodpeckers have two toes in front and two toes in back to help them climb; perching birds have a single toe in back to help them stay perched; running birds have all toes at the front to help them stay balanced; pelicans have bills with pouches to hold a large catch of fish.

DEVELOP SCIENCE VOCABULARY

bird, p. B116—Write the heading *Birds* on the board. As volunteers list the characteristics of birds, write their suggestions under the heading. Guide students to include the following traits: warm-blooded; have feathers; most can fly; air sacs connected to lungs; bones have air spaces; eggs fertilized in females; and eggs have an interior membrane and a hard shell.

warm-blooded, p. B116—Write the words *warm-blooded* and *cold-blooded* on the board. Help students compare the two terms. Make sure students understand that, unless they are experiencing fevers, the bodies of warm-blooded animals stay at the same temperature at all times.

Music Link

Camille Saint-Saëns was a French composer, pianist, and organist, who was born in Paris in 1835 and died in Algiers, Algeria, in 1921. At age $4\frac{1}{2}$, he could play Beethoven sonatas on the piano, and at age 18 he composed his first symphony. One of his famous works is a suite called *The Carnival of the Animals*, composed in 1886. Some of the movements include *Tortoises*, *The Elephant*, *Kangaroos*, *Creatures with Long Ears*, and *Swan*.

Play a recording of *The Carnival of the Animals* for the class. Encourage students to list each animal and describe how the music illustrates the animal's characteristics.

Guide the Learning continued

SCIENCE IDEAS

Help students focus on the facts and supporting details for this **main idea: Mammals are warm-blooded vertebrates that have mammary glands, hair, and a diaphragm.**

- What are mammary glands? What is their function? They are milk-producing glands. Female mammals use them to nurse their young.
- How does hair help a mammal survive? It keeps the animal warm and helps conserve body heat.
- Since whales have little hair, how do they stay warm? A thick layer of fat under the skin keeps them warm.
- How do the body systems of mammals differ from those of other vertebrates? Mammals have more highly developed body systems, including a four-chambered heart, a diaphragm, and a sophisticated brain and nervous system.
- What is the diaphragm and what is its function? It is a sheet of muscle in the chest cavity that helps a mammal breathe efficiently.

Mammals

Mammals are warm-blooded vertebrates that have milk-producing glands called **mammary glands**. Because of this adaptation, female mammals can nurse their young without leaving them to find food. Nursing also creates a bond between a mother and her helpless off-spring. It ensures that the mother will take care of them. The pictures on these pages show some of the ways in which mammals bear their young. Generally the young are born alive in a form like that of their parents.

Most mammals have hair and teeth. In fact, mammals are the only animals that have hair. It keeps them warm and helps them save the body heat they produce as warm-blooded animals. Hair can have different forms, such as the quills of a porcupine and the fur of a bear. Some mammals, such as whales and porpoises, have little hair. Armadillos also have little hair and no teeth. They are the only mammals covered with an "armor" of bony plates.

Most types of mammals live on land. Only one type—bats—can fly. They have leathery

THE INSIDE STORY

Mammals are classified into three groups—monotremes (MAHN•uh•treemz), marsupials (mar•SOOP•ee•uhlz), and placentals (pluh•SEN•tuhlz). These groups of mammals differ in how their young develop before and shortly after birth.

Monotremes

There are only three species of monotremes—two species of spiny anteaters and the duckbilled platypus. They lay eggs in burrows in the ground, and the mother keeps the eggs warm until they hatch. After hatching, the young depend on milk from the mother's mammary glands.

A young echidna hatches from its leathery-shelled egg. It will feed on the milk from its mother's mammary glands. ▶

▲ The spiny echidna is a kind of anteater from Australia. It lays eggs but it is a mammal.

Marsupials

Marsupials, or pouched mammals, bear live young, but the newborn is very immature. A newborn marsupial crawls into a pouch on its mother's abdomen, where it nurses on milk from the mammary glands. It stays in the pouch until it has grown and developed more fully. Marsupial mammals are found mainly in Australia. They include kangaroos, koalas, and Tasmanian wolves. The only marsupial that is native to North America is the opossum.

A newborn kangaroo is tiny, blind, and helpless. It crawls to its mother's pouch, where it will stay while it develops. ▶

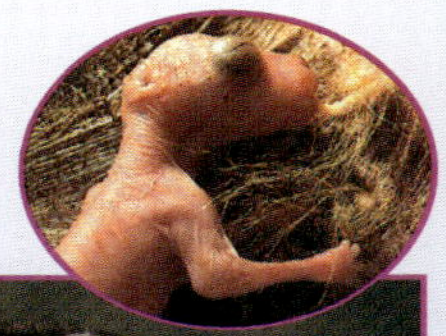

▲ The kangaroo is a stout marsupial that carries its young in a pouch. Kangaroos are plant eaters that live mostly in Australia.

B118

Reaching All Learners

Kinesthetic Learners Have students place their hands just below their rib cage and feel the actions of their diaphragm as they exhale sharp, quick breaths. Explain that when we breathe in, the diaphragm flattens and the chest cavity expands; when we breathe out, the diaphragm relaxes, the chest cavity gets smaller, and air is forced out of the lungs.

wings without feathers. Whales and their relatives live in the ocean. Like all mammals, they have mammary glands and feed milk to their young. Rather than hair, whales have thick layers of fat under their skin to protect them from the cold.

The body systems of mammals are highly developed. Their brains and nervous systems are the most developed of all vertebrates. Their circulatory systems are powered by a four-chambered heart. It pumps blood to the lungs to pick up oxygen. Then the oxygen-rich blood returns to the heart and is pumped to the other parts of the body. Mammals are the only animals with a sheet of muscle called a *diaphragm* (DY•uh•fram) within the chest cavity. This muscle is important in enabling them to breathe very efficiently.

Human beings have used other mammals for thousands of years. Animals such as cattle, pigs, and sheep are used as food. Horses and dogs are sometimes used for work. Mammals also provide wool and leather for clothing. And mammals such as cats and dogs are commonly kept as pets.

✔ **What characteristics of mammals make them different from other animals?** Mammals have hair, mammary glands, and a diaphragm.

A pig's offspring grow and develop within the mother pig. They are born with the same form as their parents.

Placentals

Placental mammals develop within the female more completely than marsupials do. They are attached by an umbilical cord to a structure called the placenta. Through this organ nutrients, oxygen, and wastes are exchanged between the developing embryo and the mother.

Placental mammals take a long time to develop before birth. Even after birth they depend on their parents until their nervous systems are fully developed and they have learned to take care of themselves. Some placental mammals, such as horses, can stand and walk minutes after birth. But others, such as cats, are not nearly so mature.

Pigs are placental mammals that have stout, heavy bodies covered with coarse hair. The adult females give birth to large groups of young, which they feed with milk from their bodies.

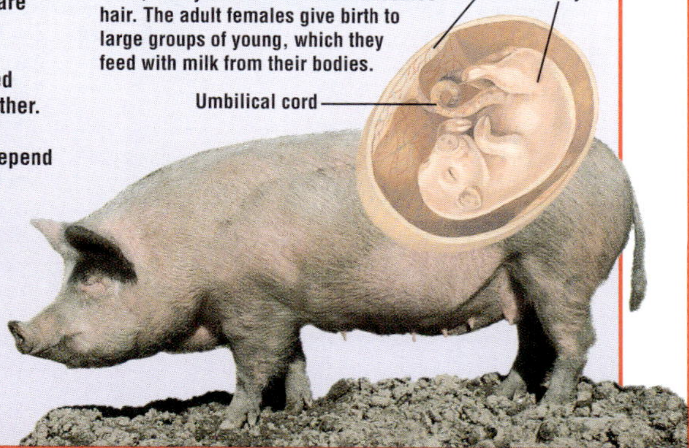

VISUAL LEARNING

Have students read *The Inside Story.*

- **What is characteristic of monotremes?** They lay eggs, which the mother keeps warm until they hatch.
- **What are the three living species of monotremes?** two species of spiny anteaters, including the spiny echidna and the duck-billed platypus
- **Describe a characteristic of marsupials.** The newborns are very immature. They crawl into the mother's pouch to nurse until fully developed.
- **Name examples of marsupials.** kangaroos, koalas, Tasmanian wolves, opossums, wombats
- **What is characteristic of placental mammals?** Placental mammals develop within the female more completely than marsupials do. The young are connected to the placenta by an umbilical cord, allowing nutrients, oxygen, and wastes to be exchanged between the embryo and the mother.

DEVELOP SCIENCE VOCABULARY

mammal and **mammary gland**, p. B118—Explain that the term *mammal* comes from the Latin word *mamma*. Write the two vocabulary words on the board for students to compare. Be sure they understand that a mammal is an animal with mammary glands and that mammary glands produce milk to nourish the young.

Cultural Connection

Camels are native to the desert regions of Asia and northern Africa. The Arabian camel has one hump, and the Bactrian camel has two. Both have been domesticated since ancient times and are adapted to the desert. They can live without water for several days. Their humps store fat that provides nutrients when food is scarce. Camels can carry heavy loads and have been used as beasts of burden. Nomadic people of Southwest Asia and North Africa depend on camels. Their meat and milk provide food, their hide is made into leather, and their long hair is woven into cloth. In these cultures, the camel breeders are highly regarded.

ESL ACTIVITY

Invite pairs of students to practice reading names of common **mammals** that are familiar to them. Ask students to say the names of those animals using their first language. Explain how in some instances if a word sounds similar to an English word, it may be because it comes from the same Latin root. Many of the words we use in English and in other languages originally came from another language. The words have changed over the centuries to become the words we are familiar with today.

Guide the Learning continued

ANALYZE AND INTERPRET DATA

Allow students time to study the table on page B120. Then have small groups of students select a particular group of animals, such as amphibians, and compare it to the other four groups. For each feature, instruct the students to find out which other groups of vertebrates exhibit the same characteristic and which are different. Have a volunteer from each group report the findings to the class. Then ask the following questions:

- **Which vertebrates have a four-chambered heart and lay hard-shelled eggs?** birds
- **Identify the vertebrates that have scales and need water for reproduction.** fish

Have students write additional questions of this nature. Provide time for small groups of students to quiz each other.

Tell students to analyze the circle graph on page B120. Help students interpret the data displayed in the circle graph by asking the following questions.

- **Which vertebrate group makes up almost half of all vertebrate species?** fish
- **Which vertebrate group has the most species? the second most?** fish; birds
- **Name the vertebrate group that has the fewest species.** amphibians
- **What percent of vertebrate species are mammals?** 10.7 percent

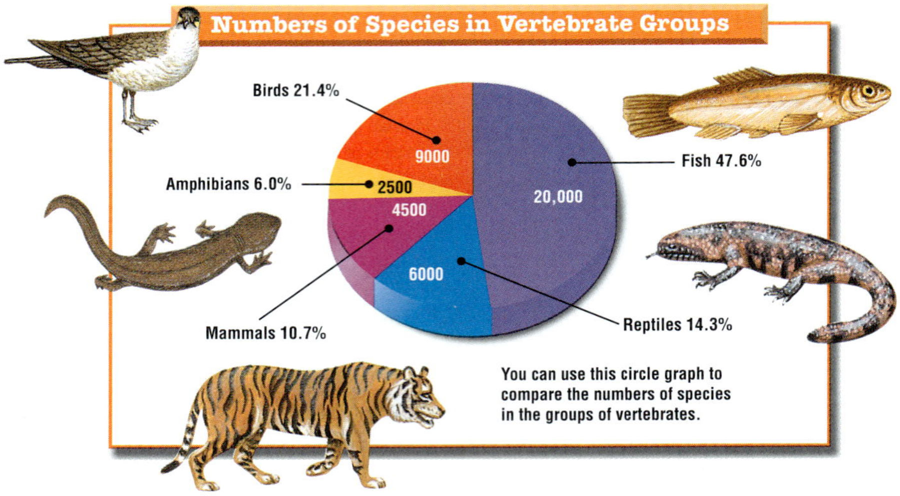

You can use this circle graph to compare the numbers of species in the groups of vertebrates.

Review of Vertebrates

Animals as different as fish, frogs, snakes, birds, and cats are all vertebrates. This group of animals is much smaller than the invertebrates, both in number of species and number of individuals.

A vertebrate has a bony or cartilaginous skeleton that includes a jointed backbone. A vertebrate's brain is enclosed in a bony skull. Most vertebrates are large compared to invertebrates and have sharp senses and a large brain. They all have similar body systems that include many of the same organs.

The vertebrates include the classes of fish, amphibians, reptiles, birds, and mammals. The characteristics of each of these groups are summarized in the chart below. The percent of species of each type of vertebrate is shown in the circle graph above.

✓ **Which group of vertebrates shown on the circle graph contains the most species?** fish

Comparing Vertebrates

	Fish	Amphibians	Reptiles	Birds	Mammals
Chambers of heart	2	3	3 or 4	4	4
Covering on skin	scales	none	scales or plates	feathers	hair
Number of limbs	varies	4 (most common)	4 or none	4	4
Temperature regulation	cold-blooded	cold-blooded	mostly cold-blooded	warm-blooded	warm-blooded
Environment for reproduction	needs water	needs water	doesn't need water	doesn't need water	doesn't need water
Embryo development	mostly in jellylike eggs with no hard covering	mostly in jellylike eggs with no hard covering	mostly in leathery eggs	in hard-shelled eggs	mostly in mother's body

B120

INVESTIGATION CHALLENGE

Hands-On Activities: More About Birds

Use these hands-on activities to extend students' understanding of key lesson concepts.

Have students investigate the aerodynamics of birds in flight, and challenge them to construct a mobile model (a modified model airplane) that simulates a bird in motion.

Provide students with pictures of birds and have them listen to a tape of corresponding bird calls. Challenge students to match the sights with the sounds.

Reading Mini-Lesson

Use Graphic Sources for Information

Tell students to use the graphic source on the bottom of page B120 for additional information on vertebrates. Display the transparency and ask each student to write two questions that could be answered from the information on the table. Have students exchange questions and use the table to find the answers.

Reading Mini-Lesson Transparency B4-2

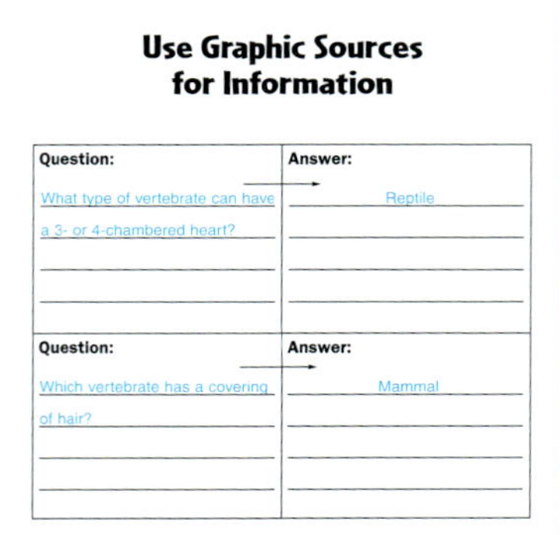

READING SKILLS PRACTICE

To practice and apply this reading skill, see **Workbook p. WB95.**

For student self-assessment, see Assessment Guide p. AGxvii.

Summary

Reptiles, birds, and mammals are vertebrates. Within the membrane that surrounds their embryos is water needed for development. Because of this, reptiles, birds, and mammals are able to spend their entire lives on land. Most reptile embryos and all bird embryos develop inside eggs. Most mammal embryos develop inside the mother's body. Reptiles are cold-blooded and are covered with scales or plates. Birds are warm-blooded and have feathers. Mammals are warm-blooded and have hair and mammary glands.

Review

1. How can snakes swallow large animals?
2. Which is the largest bone of a bird's body? Why is it so large?
3. What are the three kinds of mammals?
4. **Critical Thinking** In cold climates some mammals and reptiles hibernate during the winter. What can birds do instead of hibernating?
5. **Test Prep** How are birds similar to reptiles?
   A Both lay eggs.
   B Both have membranes that protect the developing embryo.
   C Both are warm-blooded.
   D Both **A** and **B**

LINKS

MATH LINK

Solve Problems The smallest living mammals are shrews, which weigh only about 4 g. They eat as much as two times their own weight in food every day. About how many grams of food can one shrew eat in a growing season that is 4 months long?

WRITING LINK

Persuasive Writing—Opinion Most snakes are harmless to people, and they eat many rodent and insect pests. Write a brief newspaper editorial about how people benefit from wild animals. Use snakes as an example.

ART LINK

Unicorns and Camelopards The Greek historian Herodotus described the strange and wonderful animals that he saw on a trip to Africa. Find out about his observations of the animals he called unicorns and camelopards. Draw pictures of these animals. What are their current names?

TECHNOLOGY LINK

Find out more about one type of vertebrate by watching *Sea Turtles* on the **Harcourt Science Newsroom Video.**

CNN Turner Learning

B121

LINKS TIPS

Links Objective Extend understanding of reptiles, birds, and mammals.

Math Link Guide students through the following calculations: multiply 4 g × 2 to get the amount a shrew eats in one day 8 g; multiply the daily amount by 120 to determine the food intake for four months, or 120 days. about 960 g

Writing Link Encourage students to begin by researching specific ways in which snakes are helpful to people. Show students how to use a graphic organizer, such as a word web, to organize their facts. Then have them **identify supporting facts and details** they will use in their editorials. Remind students to write a strong opening sentence that will grab the reader's interest. Writing models are provided in the **Teaching Resources** book.

3 Wrap Up and Assess

SUMMARY

Graphic Organizer for Chapter Concepts After reading the summary, have students complete the Lesson 2 section of the graphic organizer (shown on page B114).

REVIEW

1. by unhinging their jawbones
2. The breastbone; the muscles used for flight are attached to it.
3. monotremes, marsupials, and placentals
4. They can migrate to warmer regions.
5. D, both A and B

TEST PREP Test-Taking Tips

Item type multiple choice with a question stem

To help students answer Question 5, share the following strategy with them. **Read the question carefully and find key words that are clues to the correct answer choice, such as *birds* and *reptiles*. Try to recall some facts that will help narrow the choices. Eliminate any answers that are obviously wrong. Then select the best answer to the question.**

Informal Assessment

Portfolio Have students place in their portfolios their research regarding specific ways in which snakes are helpful to people.

LESSON CONCEPT REVIEW/ASSESSMENT

Students can use **Workbook p. WB96** to review the lesson concepts.

VOCABULARY REVIEW

Students can use **Workbook p. WB97** to review the chapter vocabulary.

HEALTH HANDBOOK

To reinforce or extend the science concepts in this lesson, you may want to use the content and activities in the Health Handbook, pp. R24–R37 (T55–T61).

SCIENCE AND TECHNOLOGY

Objectives

- **Evaluate the impact of research and technology on scientific thought, society, and the environment.**
- **Identify careers related to science.**

1 Before Reading

SET A PURPOSE

Ask students to examine the title and pictures for the article and then to predict what the article is about. Then ask students to list questions they would like answered based on their predictions. Write these on the board and check them off as they are answered. Encourage students to use reference sources to find the answers to any unanswered questions.

2 Guide the Learning

SCIENCE IDEAS

Help students gather information about this non-fiction science article by asking these questions:

- **Why is it important to know the location of manatees?** Since they are an endangered species, any manatees that are in trouble can be helped out and possibly saved from danger or death.
- **What is the main function of GPS?** to precisely identify locations on Earth
- **What is the role of satellites in GPS?** Twenty-four car-sized satellites orbit 19,000 km above the equator. Each satellite stays above a certain spot on Earth's surface by traveling at the same speed as Earth's rotation. Each one also sends out a radio signal that identifies it. The signals are picked up by transmitters on the object under study.
- **How might you someday use GPS in your daily life?** to keep track of children and family pets

SCIENCE AND TECHNOLOGY

Tracking Animals by Satellite

Biologists can use satellite technology to track animals in the wild and help those in danger.

A GPS system

Why do biologists track the locations of manatees?

Years ago, manatees were hunted heavily. As a result, the manatee is now an endangered species. Often biologists capture manatees that are sick or have been injured, get them healthy again, and then release them back into the wild. To follow manatees' movements and to make sure they can survive on their own after living in captivity, biologists place radio tags on some

Manatee, or sea cow

B122

Reading Mini-Lesson

Use Reference Sources There are many uses for GPS technology. Have students use a variety of reference sources to find specific examples of other uses of GPS. Then challenge students to write or deliver oral reports on, or to make posters of, the different ways this technology can be used.

manatees. The radio tags enable biologists to track the manatees by using a satellite system called the Global Positioning System, or GPS.

How does GPS work?

GPS uses 24 car-sized satellites orbiting 19,000 km (about 12,000 mi) above the equator. Each satellite stays above a certain spot on Earth's surface by traveling at the same speed as Earth's rotation. Each one also sends out a radio signal that identifies it. The location of a GPS receiver on Earth's surface is determined by the length of time it takes radio signals to reach it from four of the satellites. Because each satellite sends out a unique signal and stays above one location, mathematical calculations done by a computer in the receiver can pinpoint the latitude and longitude of any GPS receiver.

How does GPS help biologists track manatees?

In addition to the GPS receiver, the manatee's tag also contains a radio transmitter that continuously sends information about the animal's location to biologists who are tracking it. The position of the animal can be shown on a video map display. If a manatee seems to be in trouble, scientists can go to it and try to help.

How has GPS saved a manatee's life?

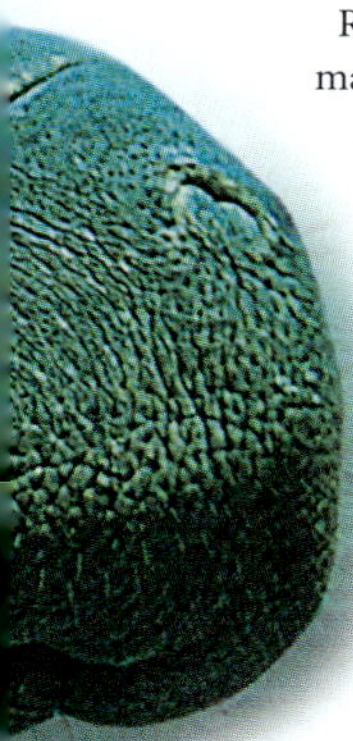

Recently GPS was used to save a manatee named Mo. After being raised in captivity, Mo was released off the coast of Florida. Unfortunately, he wandered far offshore and was carried south about 454 km (280 mi) by ocean currents. Biologists checked on Mo's location and found that he was far from areas where he could get food and fresh water. Biologists quickly captured Mo and returned him to Miami for treatment that saved his life.

What are other uses of GPS?

GPS has many other uses. Some of them might even be familiar to you. It is used to help ships, boats, and commercial aircraft to navigate. Some new cars have GPS-run systems that pinpoint a car's location on a computer map and show the driver the best route to a nearby location. One GPS developer thinks that in the future, families could use GPS devices the size of a wristwatch to keep track of their children and pets.

THINK ABOUT IT

1. How is GPS used to identify the location of things on Earth's surface?
2. How has GPS helped biologists save the life of a manatee?

CAREERS
MARINE BIOLOGIST

What They Do
Marine biologists study ocean species and their environments. They gather specimens from various marine habitats to follow the overall health of ocean organisms and ocean habitats.

Education and Training
Marine biologists must have a background in biology, oceanography, chemistry, and environmental studies. They should also have an interest in plants and animals and should enjoy being outdoors and on the ocean. Marine biologists must have at least a bachelor's degree in biology, but many also have a master's degree or doctorate so they can do advanced research.

WEB LINK
For Science and Technology updates, visit the Harcourt Internet site.
www.harcourtschool.com

B123

BACKGROUND

Webliography
Keyword careers
www.harcourtschool.com

Marine Biologist Marine biologists play an important part in our environment. Their primary goal is to classify and describe marine organisms. They study all aspects of organisms from reproduction to food habits. They also study how people can use these organisms for food without depleting their populations. More and more these scientists are trying to determine how the substances that organisms produce can be used to treat human illnesses such as cancer and arthritis. They usually perform much of their research in laboratories near oceans throughout the world.

DEVELOP SCIENCE VOCABULARY

Global Positioning System, or GPS, p. B122—Have students explain why each word in this phrase is an appropriate word. For example, GPS is global because it can locate a receiver anywhere on the globe. It is involved in positioning because it tells the position or location of an object. It is a system because it is made up of many parts.

manatee, p. B122—Another name for the manatee is *sea cow*. Have students explain why this is an appropriate name.

CHAPTER CONCEPTS

This article focuses on the technology scientists are using to keep track of individuals of an endangered species.

- **How is GPS helping scientists to preserve an endangered species?** Scientists track their movements and make sure that they can survive on their own after being cared for in captivity due to an illness or injury.
- **Why is GPS especially useful for tracking marine animals?** Marine animals cannot always be seen and are difficult to follow because they live in the water.

3 Wrap Up and Assess

THINK ABOUT IT

1. GPS finds the location of a radio tag on Earth by calculating the time it takes radio signals from four satellites to travel from a known position in orbit to the receiver.
2. The manatee was wearing a GPS receiver and a radio transmitter that allowed biologists to track its location and rescue the manatee when it was in need of help.

PEOPLE IN SCIENCE

Objectives

- ▶ Connect chapter concepts with the contributions of scientists.

1 Before Reading

SET A PURPOSE

Conservation biologists study populations of plants and animals with an eye toward preserving these populations. Have students read to find out how Dr. Archie Carr has helped save sea turtles from extinction.

2 Guide the Learning

THE WORK OF SCIENTISTS

Help students focus on the methods and processes scientists use and the impact of their work.

- ▶ **How did the U.S. Navy's Operation Green Project help save sea turtle populations?** It distributed green sea turtle eggs around the Caribbean and Gulf of Mexico.
- ▶ **If you were a conservation biologist, what sorts of endangered plants or animals would you study, and why?**

3 Wrap Up and Assess

THINK ABOUT IT

1. Some sea turtle populations were decreasing, and their species were in danger of extinction.
2. They study populations of plants and animals and try to discover what is causing problems for these populations. Then they seek solutions to the problems and work hard to save those populations.

PEOPLE IN SCIENCE

Archie Carr

CONSERVATION BIOLOGIST/NATURALIST

Archie Carr was a writer, teacher, and naturalist who was the world's leading expert on sea turtles. For a long time, Dr. Carr taught students and conducted research at the University of Florida. That was where he started his campaign to preserve nesting grounds for sea turtles. Because of his work, several sea turtle species were saved from extinction.

Dr. Carr grew up with a love of nature. The back yard of his home in Georgia was filled with caged snakes, lizards, and, of course, turtles. It was no surprise when he decided to become a biologist, and in 1937 Carr earned his Ph.D. in zoology, the study of animals. He went on to teach biology, continuing his special interest in sea turtles. When he wasn't teaching, he traveled in Central America and around the world doing research on sea turtles and their place in various ecosystems. His discovery that sea turtles in some places were in danger of extinction led him to his work as a conservationist.

Dr. Carr decided to devote himself to saving endangered sea turtle species. For 20 years he directed the international movement for the conservation of sea turtles. In this role he fought the killing of endangered sea turtles for turtle products. He also campaigned to continue protection of once-endangered species, even when their numbers began to increase. In the 1960s he started and directed the U.S. Navy's Operation Green Turtle. The project distributed green-turtle eggs and young turtles to nesting beaches around the Caribbean and Gulf of Mexico. Biologists wanted to reestablish or increase sea turtle populations at these places.

Scientists who study sea turtles today have concluded that some endangered sea turtle populations are now on the road to recovery. Dr. Carr played a large part in that success.

THINK ABOUT IT

1. Why was it important for Dr. Carr to fight for the protection of sea turtles?
2. What else do conservation biologists such as Dr. Carr do?

B124

BACKGROUND

Webliography
Keyword biographies
www.harcourtschool.com
GO ONLINE

More About Dr. Archie Carr
Dr. Archie Carr, who lived from 1909 to 1987, was a leader of the Caribbean Conservation Corporation. He put tags on many sea turtles as a way to learn more about their patterns of migration. Dr. Carr helped establish Tortuguero National Park in Costa Rica to preserve that country's sea turtle hatching grounds and to provide a place for research. He opposed commercial sea turtle "ranching" and the removal of sea turtles from the endangered or threatened list even when specific populations seemed to be doing better. Carr's knowledge of world ecosystems was legendary, and he was well known for his sense of humor. He once attended a party dressed as a witch and carrying a live vulture. He was a teacher whose students remembered his classes and field trips long after they were over.

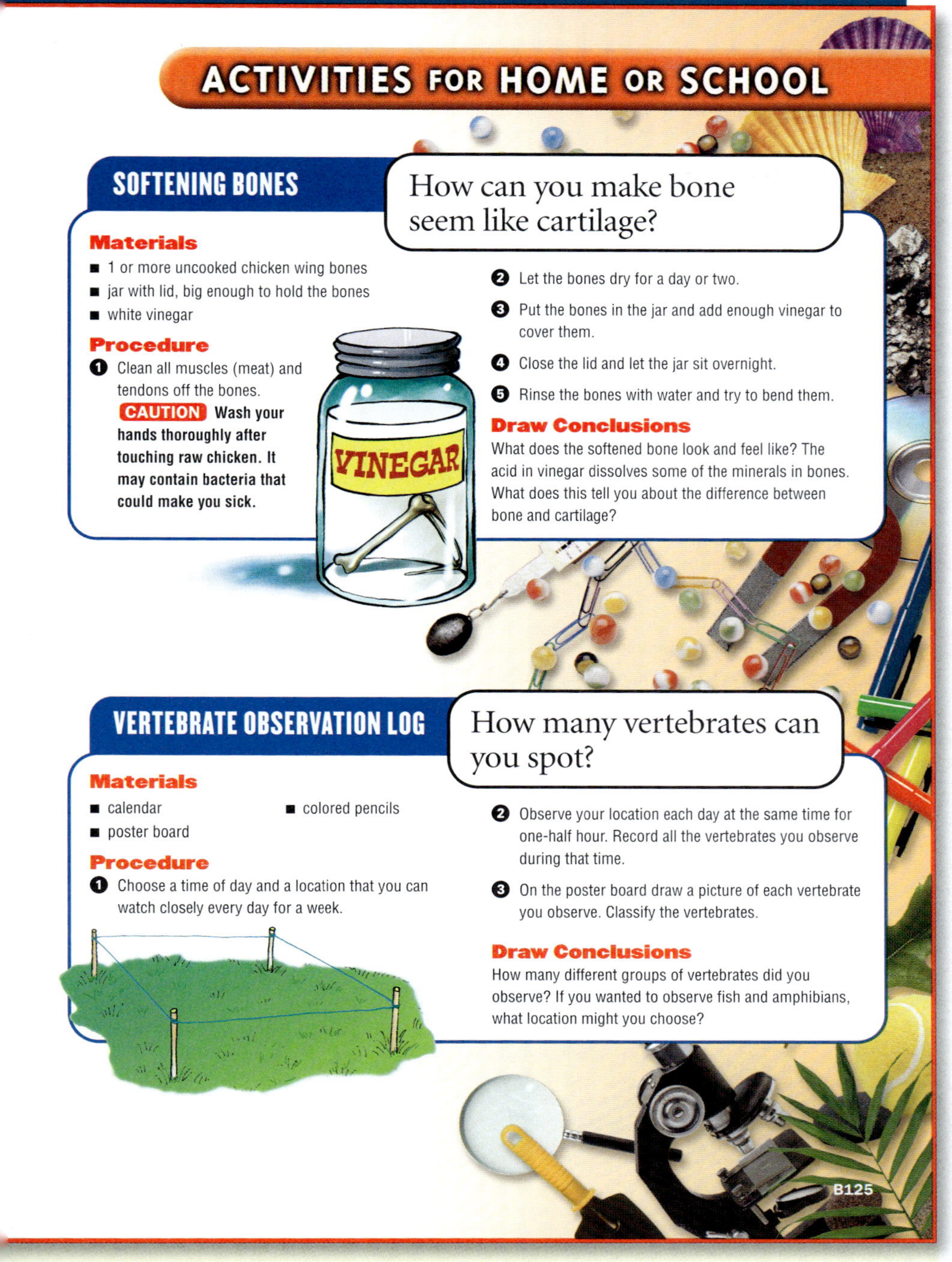

ACTIVITIES FOR HOME OR SCHOOL

SOFTENING BONES

How can you make bone seem like cartilage?

Materials

- 1 or more uncooked chicken wing bones
- jar with lid, big enough to hold the bones
- white vinegar

Procedure

1. Clean all muscles (meat) and tendons off the bones. **CAUTION** **Wash your hands thoroughly after touching raw chicken. It may contain bacteria that could make you sick.**
2. Let the bones dry for a day or two.
3. Put the bones in the jar and add enough vinegar to cover them.
4. Close the lid and let the jar sit overnight.
5. Rinse the bones with water and try to bend them.

Draw Conclusions

What does the softened bone look and feel like? The acid in vinegar dissolves some of the minerals in bones. What does this tell you about the difference between bone and cartilage?

VERTEBRATE OBSERVATION LOG

How many vertebrates can you spot?

Materials

- calendar
- poster board
- colored pencils

Procedure

1. Choose a time of day and a location that you can watch closely every day for a week.
2. Observe your location each day at the same time for one-half hour. Record all the vertebrates you observe during that time.
3. On the poster board draw a picture of each vertebrate you observe. Classify the vertebrates.

Draw Conclusions

How many different groups of vertebrates did you observe? If you wanted to observe fish and amphibians, what location might you choose?

B125

School-Home Connection

These activities provide an excellent opportunity to assign hands-on activities for the students and their families. The materials are often easy to obtain and are safe to work with. For those activities that you do not wish to do in the classroom, encourage students to complete them at home. Remind them that communicating their results is always important, but especially so for home activities since you will not be there to observe their activities. Ask students to use diagrams, sketches, graphs, or even photographs to show their results.

ACTIVITIES FOR HOME OR SCHOOL

Softening Bones

Objectives

- **Demonstrate how bone can be made to seem like cartilage.**

Suggested Time 15–20 minutes on Day 1; 10–15 minutes on Day 2 or 3; 15–20 minutes on Day 3 or 4

Hints

- Provide students with utensils for removing the meat and tendons from the bones.
- If necessary, allow the bones to sit for more than one night in the vinegar until they are very soft.

Safety Tip Remind students to wash their hands thoroughly with soap and hot water after touching the bones.

Draw Conclusions Students should realize that the softened bones are similar to cartilage and that bone is mineralized and cartilage is not.

Vertebrate Observation Log

Objectives

- **Observe the vertebrates in an area of one square meter over the course of a week.**

Suggested Time 45–60 minutes every day for one week

Hints

- Review with students what kinds of organisms are considered vertebrates.
- Encourage students to choose locations that include a variety of plant life.
- Remind students that they are observing all kinds of vertebrates, including humans.
- Encourage students to use reference books to help them classify the vertebrates.

Safety Tip Students should choose a spot that is easily observed from a safe place.

Draw Conclusions Depending on the location, the season, and the region of the country, students should observe between one and five classes of vertebrates. They would probably choose a pond or lake to observe fish and amphibians.

ACTIVITIES FOR HOME OR SCHOOL

Reproducible copies of these activities are provided in the **Teaching Resources** book.

CHAPTER 4

Review and Test Preparation

VOCABULARY REVIEW (2 points each)

1. Reptiles
2. mammals
3. fish
4. mammary glands
5. birds
6. cold-blooded, warm-blooded
7. amphibians

CONNECT CONCEPTS (3 points each)

8. fish
9. reptiles
10. birds
11. gills
12. tadpoles
13. eggs with leathery shells
14. feathers
15. hair
16. bear live young
17. in water
18. on land

CHECK UNDERSTANDING (3 points each)

19. A, sharks
20. H, skin and the roof of its mouth
21. B, reptile
22. G, bird
23. B, marsupial
24. H, oxygen
25. A, lays eggs

CHAPTER 4 Review and Test Preparation

Vocabulary Review

Use the terms below to complete the sentences. The page numbers in ( ) tell you where to look in the chapter if you need help.

cold-blooded (B104)
fish (B104)
amphibians (B104)
reptiles (B115)
birds (B116)
warm-blooded (B116)
mammals (B118)
mammary glands (B118)

1. ___ are cold-blooded vertebrates that lay eggs with leathery coverings.
2. Warm-blooded vertebrates that produce milk are ___.
3. Vertebrates that live their entire lives in the water are ___.
4. The main identifying characteristic of mammals is the presence of ___.
5. ___ are the only animals that have feathers.
6. Animals that take on the temperature of their surroundings are ___. Those that maintain their body temperature despite the temperature of the surroundings are ___.
7. Animals that spend part of their lives in the water and part on land are ___.

Connect Concepts

Complete the chart using terms and phrases from the Word Bank.

in water **hair** **on land** **eggs with leathery shells** **gills** **bear live young** **feathers** **birds** **reptiles** **fish** **tadpoles**

Vertebrates				
Cold-blooded			Warm-blooded	
8. ___	Amphibians,	9. ___	10. ___	Mammals
have 11. ___ all their lives.	such as frogs, have gills as 12. ___	have scales. They lay 13. ___	have 14. ___. They lay eggs with brittle shells.	have 15. ___. They 16. ___.
They reproduce 17. ___.		They reproduce 18. ___.		

B126

CHAPTER TEST

See **Assessment Guide pp. AG40–44** for the Chapter Test. Assessment Options are provided on p. B100e.

TAKE-HOME BOOK

Use the **Take-Home Book** (described on p. B100d) to provide more chapter content and activities.

Check Understanding

Write the letter of the best choice.

19. Cartilaginous fish that have jaws include —
- **A** sharks
- **B** lampreys
- **C** whales
- **D** barracudas

20. An adult frog can breathe with its lungs and through its —
- **F** gills
- **G** skin
- **H** skin and the roof of its mouth
- **J** skin and gills

21. A cold-blooded vertebrate that lays eggs with leathery shells is —
- **A** an amphibian
- **B** a reptile
- **C** a fish
- **D** a bird

22. One example of a warm-blooded vertebrate is —
- **F** an amphibian
- **G** a bird
- **H** a reptile
- **J** a fish

23. A mammal with a pouch is called a —
- **A** monotreme
- **B** marsupial
- **C** whale
- **D** placental

24. In order to fly, birds need a lot of —
- **F** weight
- **G** bone mass
- **H** oxygen
- **J** humidity

25. A monotreme is a mammal that —
- **A** lays eggs
- **B** has a placenta
- **C** lives in the water
- **D** bears live young

Critical Thinking

26. Most birds lay many fewer eggs at one time than reptiles do. Why might this be so?

27. Some animals have feathers or hair as insulation. Why don't reptiles or amphibians have any insulation?

28. How can you use the structure of the skeleton of a vertebrate to infer its habitat?

Process Skills Review

29. What **observations** would you make to **compare** the early development of marsupials and monotremes?

30. If you **observed** an animal that had gills and four limbs, what could you **hypothesize** about the animal?

Performance Assessment

Comparing Animals

Your teacher has photographs of five different animals. Make a chart listing the characteristics of each animal. Circle in red those characteristics that all five animals have in common. Circle in blue a characteristic that is unique to each animal. Describe how each characteristic circled in blue helps the animal that has it.

CRITICAL THINKING (6 points each)

26. Birds take care of their young once they hatch, whereas most reptiles do not. More offspring of birds are likely to survive. Also, the parents of birds could not take care of more young.

27. Reptiles and amphibians are cold-blooded and do not need to save heat. Their temperature changes with that of the environment.

TEST PREP Test-Taking Tips

Item type open-ended format in which the student is asked to write a short answer

To help students answer Question 28, share the following strategy with them. **Turn the question into a statement to begin your answer, such as "The structure of the skeleton of a vertebrate can be used to infer its habitat by" and then add the facts you know to complete the sentence.**

28. Forelimbs that look like wings suggest the ability to fly. Those that look like paddles suggest a predominantly ocean-dwelling animal. Those that look like legs suggest a land-dwelling animal.

PROCESS SKILLS REVIEW (6 points each)

29. Possible answers: Observe the birth of the immature marsupial that then grows in its mother's pouch and the leathery monotreme egg that hatches as a smaller copy of its parents.

30. Students should hypothesize that the animal was a young amphibian.

Performance Assessment

Preparation Cut out or copy photographs of five different vertebrates, including a fish, an amphibian, a reptile, a bird, and a mammal. Also provide red and blue pencils or markers for students to use.

Scoring Rubric

Performance Indicators

___ Understands characteristics that all five animals have in common.

___ Observes characteristics unique to each animal.

___ Describes how adaptations enable animals to follow their ways of life.

Observations and Rubric Score

3 2 1 0

WRITING ABOUT SCIENCE

Have students write a composition that involves writing for a specific purpose. You may wish to use the prompt that is provided in the **Workbook.** The prompt is accompanied by a graphic organizer that will help students organize their ideas before writing. Models for writing are provided in the **Teaching Resources** book. You may wish to reproduce those for students or display them on a transparency.

WRITING PRACTICE

A chapter writing prompt and a prewriting activity are provided on **Workbook p. WB98.**

UNIT B EXPEDITIONS

Objectives

- Identify different types of plants and animals.
- Classify living things as either vertebrates or invertebrates.

Hints

A field trip to Historic Bartram's Garden, the Exhibit Museum of Natural History, or a similar destination may not be feasible for your class. If so, consider visiting The Learning Site on the Internet. This site provides links to sites appropriate for student learning.

GO ONLINE www.harcourtschool.com

UNIT B EXPEDITIONS

There are many places to go to learn about how living things grow and respond. You can learn about types of plants and animals by visiting the places below. You'll also have fun while you learn.

Historic Bartram's Garden

WHAT An ornamental garden displaying a wide assortment of plants
WHERE Philadelphia, Pennsylvania
WHAT CAN YOU DO THERE? Tour the Bartram home, explore the garden, see the unique *Franklinia alatamaha* and the oldest gingko tree in the country, and observe the adaptations of the many plants.

Exhibit Museum of Natural History

WHAT One of a number of museums maintained by the University of Michigan
WHERE Ann Arbor, Michigan
WHAT CAN YOU DO THERE? Tour the museum, visit the many life science exhibits, and see the Zoology section.

GO ONLINE Plan Your Own Expeditions

If you can't visit Historic Bartram's Garden or the Exhibit Museum of Natural History, visit a garden or museum near you. Or log on to The Learning Site at **www.harcourtschool.com** to visit these science sites and see what more you can learn about plant and animal diversity.

B128

UNIT TEST

See **Assessment Guide pp. AG47–AG50** for the Unit Assessment. The Unit Assessment includes items for all the chapters in this Unit.

HARCOURT SCIENCE

•References
•Teacher Resources
•Index

This Section Contains:

- **Pupil Edition References**
  - **Science Handbook**
  - **Health Handbook**
  - **Using Science Reading Strategies**
  - **Building a Science Vocabulary**
  - **Glossary**
- **Bibliography of Science Trade Books**
- **Soaring into Space**
- **Tips and Trashpile Treasures**
- **Teacher's Edition Index**

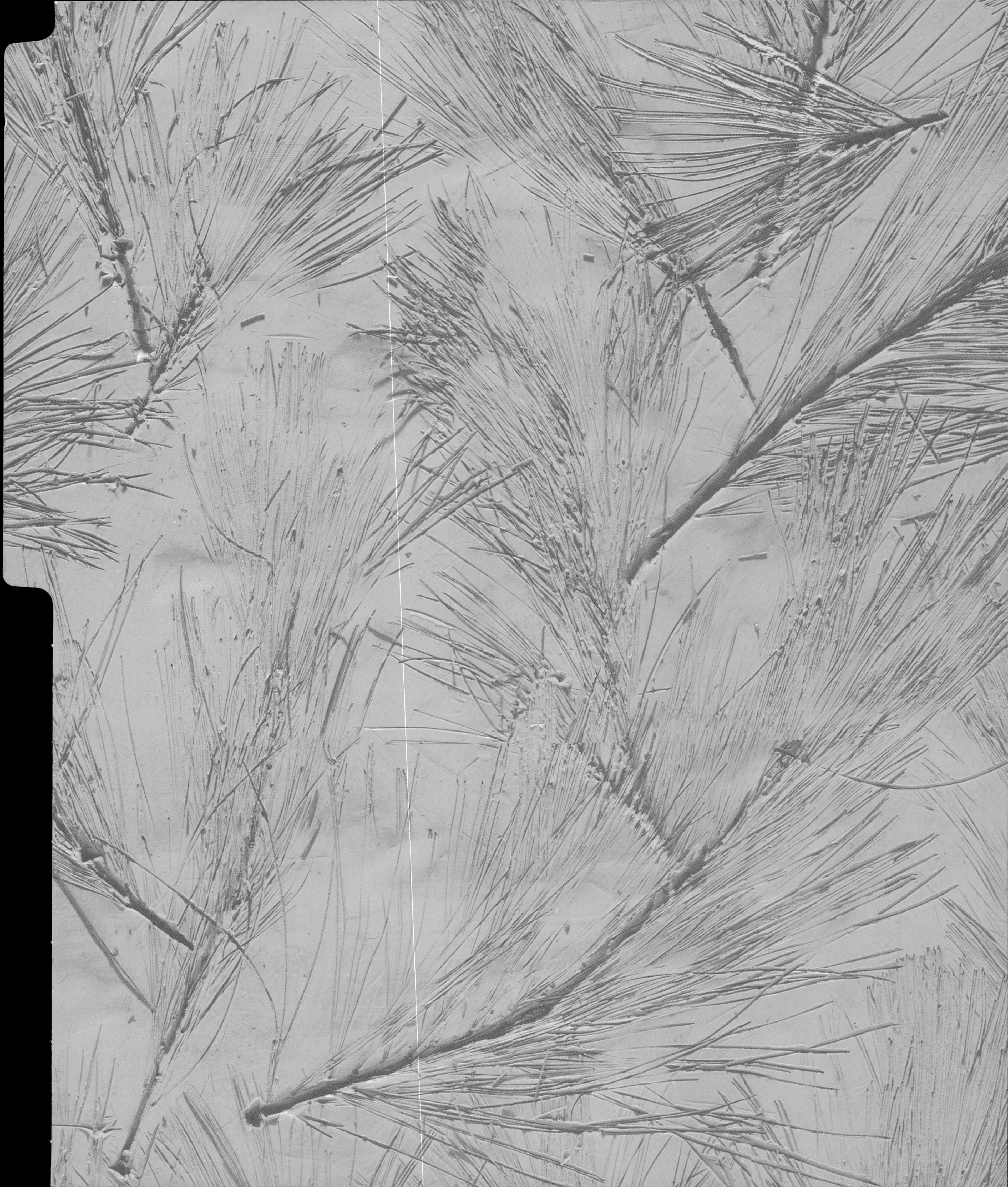

References

Using Science Tools

Using a Hand Lens

A hand lens magnifies objects, or makes them look larger than they are.

1. Hold the hand lens about 12 cm (5 in.) from your eye.
2. Bring the object toward you until it comes into focus.

Using a Thermometer

A thermometer measures the temperature of air and most liquids.

1. Place the thermometer in the liquid. Don't touch the thermometer any more than you need to. Never stir the liquid with the thermometer. If you are measuring the temperature of the air, make sure that the thermometer is not in line with a direct light source.
2. Move so that your eyes are even with the liquid in the thermometer.
3. If you are measuring a material that is not being heated or cooled, wait about two minutes for the reading to become stable. Find the scale line that meets the top of the liquid in the thermometer, and read the temperature.
4. If the material you are measuring is being heated or cooled, you will not be able to wait before taking your measurements. Measure as quickly as you can.

R2

SCIENCE HANDBOOK

Caring for and Using a Microscope

A microscope is another tool that magnifies objects. A microscope can increase the detail you see by increasing the number of times an object is magnified.

Caring for a Microscope

- Always use two hands when you carry a microscope.
- Never touch any of the lenses of a microscope with your fingers.

Using a Microscope

1. Raise the eyepiece as far as you can using the coarse-adjustment knob. Place your slide on the stage.
2. Always start by using the lowest power. The lowest-power lens is usually the shortest. Start with the lens in the lowest position it can go without touching the slide.
3. Look through the eyepiece, and begin adjusting it upward with the coarse-adjustment knob. When the slide is close to being in focus, use the fine-adjustment knob.
4. When you want to use a higher-power lens, first focus the slide under low power. Then, watching carefully to make sure that the lens will not hit the slide, turn the higher-power lens into place. Use only the fine-adjustment knob when looking through the higher-power lens.

You may use a Brock microscope. This is a sturdy microscope that has only one lens.

1. Place the object to be viewed on the stage.
2. Look through the eyepiece, and begin raising the tube until the object comes into focus.

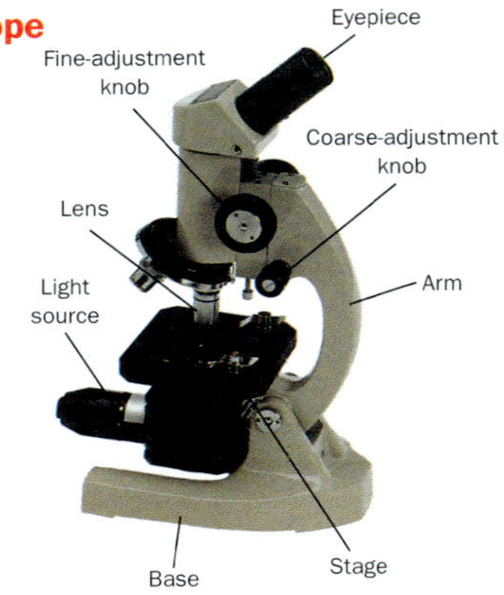

A Light Microscope

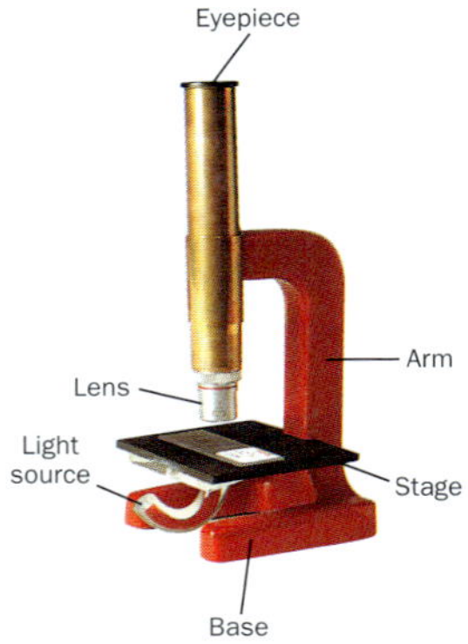

A Brock Microscope

R3

Science Handbook pages R2–R3

Background

Thermometer The design of a thermometer is based on the principle that matter expands as it gains heat and contracts as it loses heat. The liquid in a thermometer tube and bulb expands as it gains heat and moves upward in the tube. The liquid contracts when it loses heat and moves downward in the tube.

Microscope Most historians believe that Zacharias Janssen of Holland discovered the principle of the compound microscope about 1590.

The amateur scientist Anton van Leeuwenhoek, also of Holland, made lenses that could magnify up to 270x. He also built microscopes more powerful than others available at that time. Leeuwenhoek was the first person to observe microscopic life and record his observations.

Activities

Making a Water-Drop Lens Students can make a water-drop lens to use when a hand lens is not available. A water-drop lens is best used to make flat objects, such as leaves and pieces of paper, appear larger.

Materials 2 erasers about 1 cm thick, piece of acetate, water, dropper

1. Place the object you want to observe on a table between two erasers that are the same size.
2. Place a sheet of acetate on top of the erasers, about 1 cm above the object.
3. Use the dropper to place one drop of water on the sheet over the object.
4. Move the object if you need to until you can see it well.

Making a Thermometer Students can make this simple thermometer easily. It won't give an exact temperature reading, but students can tell if the temperature is going up or down.

Materials plastic soda bottle, clear plastic straw, ruler, clay, colored water, dropper, marker

1. Add colored water to the bottle until it is nearly full.
2. Place the straw in the bottle. Finish filling the bottle with water, but leave about 1 cm of space at the top.
3. Lift the straw until 10 cm of it sticks up out of the bottle. Use the clay to seal the mouth of the bottle.
4. Use the dropper to add colored water to the straw until it is a little more than half full.
5. Mark the level of the water on the straw. This is the *start* level.

Students can get an idea of how the thermometer works by placing it in a bowl of hot water. They will see the level of water in the straw rise. If they put the bottle in a bowl of ice, they will see the level of water in the straw fall.

Using a Balance

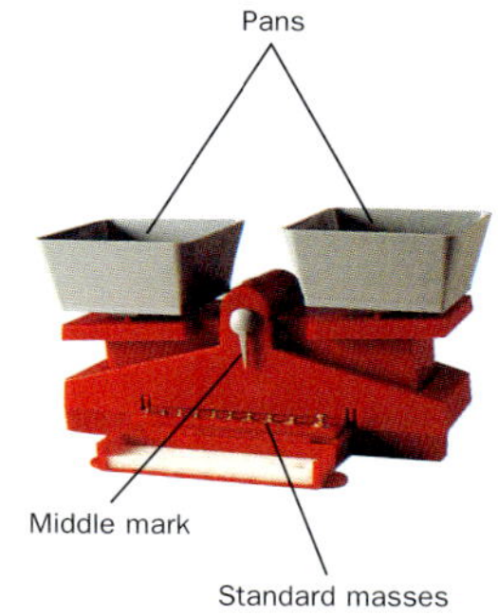

Use a balance to measure an object's mass. Mass is the amount of matter an object has.

1. Look at the pointer on the base to make sure the empty pans are balanced.
2. Place the object you wish to measure in the left pan.
3. Add the standard masses to the other pan. As you add masses, you should see the pointer move. When the pointer is at the middle mark, the pans are balanced.
4. Add the numbers on the masses you used. The total is the mass in grams of the object you measured.

Using a Spring Scale

Use a spring scale to measure forces such as the pull of gravity on objects. You measure weight and other forces in units called newtons (N).

Measuring the Weight of an Object

1. Hook the spring scale to the object.
2. Lift the scale and object with a smooth motion. Do not jerk them upward.
3. Wait until any motion of the spring comes to a stop. Then read the number of newtons from the scale.

Measuring the Force to Move an Object

1. With the object resting on a table, hook the spring scale to it.
2. Pull the object smoothly across the table. Do not jerk the object.
3. As you pull, read the number of newtons you are using to pull the object.

R4

Beaker Graduate

Measuring Liquids

Use a beaker, a measuring cup, or a graduate to measure liquids accurately.

1. Pour the liquid you want to measure into a measuring container. Put your measuring container on a flat surface, with the measuring scale facing you.
2. Look at the liquid through the container. Move so that your eyes are even with the surface of the liquid in the container.
3. To read the volume of the liquid, find the scale line that is even with the surface of the liquid.
4. If the surface of the liquid is not exactly even with a line, estimate the volume of the liquid. Decide which line the liquid is closer to, and use that number.

Using a Ruler or Meterstick

Use a ruler or meterstick to measure distances and to find lengths of objects.

1. Place the zero mark or end of the ruler or meterstick next to one end of the distance or object you want to measure.
2. On the ruler or meterstick, find the place next to the other end of the distance or object.
3. Look at the scale on the ruler or meterstick. This will show the distance you want or the length of the object.

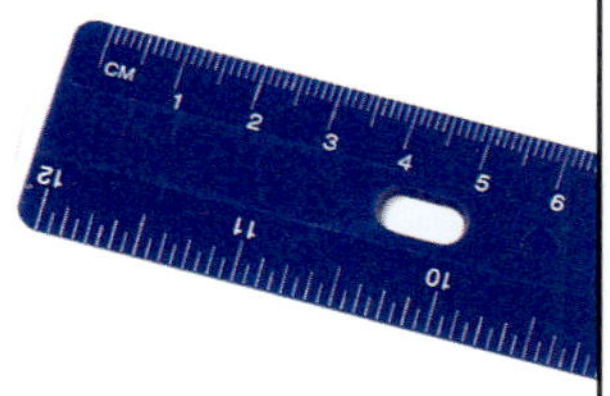

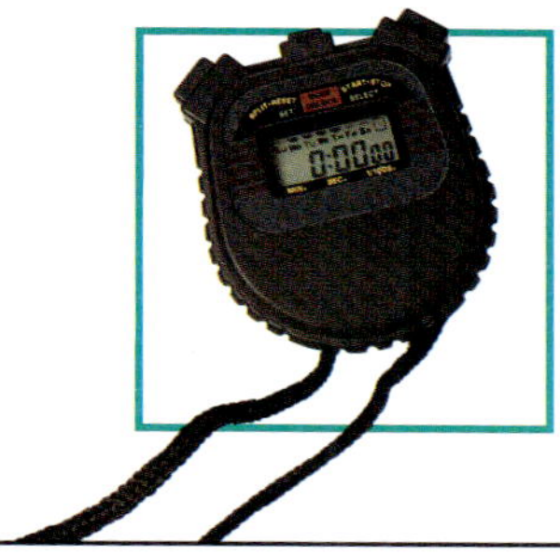

Using a Timing Device

Use a timing device such as a stopwatch to measure time.

1. Reset the stopwatch to zero.
2. When you are ready to begin timing, press start.
3. As soon as you are ready to stop timing, press stop.
4. The numbers on the dial or display show how many minutes, seconds, and parts of seconds have passed.

R5

Science Handbook pages R4–R5

Background

Mass and Weight *Mass* is the amount of matter an object has, for example 5 kg. The pull of gravity gives objects their *weight*. If an object weighs 50 newtons, it is because Earth's gravity is pulling on it with a force of 50 newtons.

Activities

Making a Balance Students can make this balance using a few simple materials. It is best used to measure small masses.

Materials ruler, string, tape, 2 paper cups, 2 large paper clips

1. If the ruler has holes in it, tie the string through the middle hole. If it does not have holes, tie the string around the middle of the ruler.
2. Tape the other end of the string to a table. Let the ruler hang down from the side of the table. Make sure the ruler hangs level.
3. Unbend the end of each paper clip a little. Push these ends through the sides of the paper cups. Hang the cups on the ruler using the paper clips.
4. Move the cups until the ruler is level.

Making a Spring Scale Students can make this spring scale from common materials. Advise them not to pull on the rubber band any more than necessary.

Materials heavy cardboard (10 cm × 30 cm), large rubber band, stapler, marker, large paper clip, paper strip (about 1 cm × 3 cm), 100-g masses (about 1 N each)

1. Staple one end of the rubber band (the part with the sharp curve) to the middle of one end of the cardboard so that the rubber band hangs down the length of the cardboard. Color the loose end of the rubber band with a marker to make it easy to see.
2. Bend the paper clip so that it is slightly open and forms a hook. Hang the paper clip by its unopened end from the rubber band.
3. Put the narrow paper strip across the rubber band, and staple the strip to the cardboard. The rubber band and hook must be able to move easily.
4. While holding the cardboard upright, hang one 100-g mass from the hook. Allow the mass to come to rest, and mark the position of the bottom of the rubber band on the cardboard. Label this position on the cardboard 1 N. Add another 100-g mass for a total of 200 g.
5. Continue to add masses and mark the cardboard. Each 100-g mass adds a force of about 1 N.

Measurement Systems

SI Measures (Metric)

Temperature
Ice melts at 0 degrees Celsius (°C)
Water freezes at 0°C
Water boils at 100°C

Length and Distance
1000 meters (m) = 1 kilometer (km)
100 centimeters (cm) = 1 m
10 millimeters (mm) = 1 cm

Force
1 newton (N) = 1 kilogram × meter/second/second (kg-m/s^2)

Volume
1 cubic meter (m^3) = 1m × 1m × 1m
1 cubic centimeter (cm^3) = 1 cm × 1 cm × 1 cm
1 liter (L) = 1000 milliliters (mL)
1 cm^3 = 1 mL

Area
1 square kilometer (km^2) = 1 km × 1 km
1 hectare = 10,000 m^2

Mass
1000 grams (g) = 1 kilogram (kg)
1000 milligrams (mg) = 1 g

Rates (Metric and Customary)
kmh or km/hr = kilometers per hour
m/s = meters per second
mph or mi/hr = miles per hour

Customary Measures

Volume of Fluids
8 fluid ounces (fl oz) = 1 cup (c)
2 c = 1 pint (pt)
2 pt = 1 quart (qt)
4 qt = 1 gallon (gal)

Temperature
Ice melts at 32 degrees Fahrenheit (°F)
Water freezes at 32°F
Water boils at 212°F

Length and Distance
12 inches (in.) = 1 foot (ft)
3 ft = 1 yard (yd)
5,280 ft = 1 mile (mi)

Weight
16 ounces (oz) = 1 pound (lb)
2,000 pounds = 1 ton (T)

R6

HEALTH HANDBOOK

Health Handbook

R7

Science Handbook pages R6–R7

Using the Health Handbook

This section of the student edition provides information that addresses important health concerns of students such as nutrition, physical fitness, and safety and first aid. It also provides interesting and detailed information about the systems of the human body. The structure and function of each system is identified. Student activities and tips for caring for each system are also included.

The Health Handbook can be used as a reference section to support concepts presented in other chapters. It can also be used for a unit of study about health and the human body.

The Food Guide Pyramid

No one food or food group supplies all the nutrients you need. That's why it's important to eat a variety of foods from all the food groups. The Food Guide Pyramid can help you choose healthful foods in the right amounts. By choosing more foods from the groups at the bottom of the pyramid, and few foods from the group at the top, you will eat nutrient-rich foods that provide your body with energy to grow and develop.

Understanding Serving Size

The Food Guide Pyramid suggests a number of servings to eat each day from each group. But a serving isn't necessarily the amount you eat at a meal. A plate full of macaroni and cheese may contain three or four servings of pasta (macaroni) and three servings of cheese. That's about half your bread group servings and all your milk servings at one sitting! The table below can help you estimate the number of servings you are eating.

Food Group	Amount of Food in One Serving	Easy Ways to Estimate Serving Size
Bread, Cereal, Rice, and Pasta Group	$\frac{1}{2}$ cup cooked pasta, rice, or cereal 1 ounce ready-to-eat (dry) cereal 1 slice bread, $\frac{1}{2}$ bagel	ice-cream scoop large handful of plain cereal or a small handful of cereal with raisins and nuts
Vegetable Group	1 cup of raw, leafy vegetables $\frac{1}{2}$ cup other vegetables, cooked or chopped raw $\frac{3}{4}$ cup vegetable juice $\frac{1}{2}$ cup tomato sauce	about the size of a fist ice-cream scoop
Fruit Group	medium apple, pear, or orange $\frac{1}{2}$ large banana, or one medium banana $\frac{1}{2}$ cup chopped or cooked fruit $\frac{3}{4}$ cup of fruit juice	about the size of a baseball
Milk, Yogurt, and Cheese Group	$1\frac{1}{2}$ ounces of natural cheese 2 ounces of processed cheese 1 cup of milk or yogurt	about the size of two dominoes $1\frac{1}{2}$ slices of packaged cheese
Meat, Poultry, Fish, Dried Beans, Eggs, and Nuts Group	3 ounces of lean meat, chicken, or fish 2 tablespoons peanut butter $\frac{1}{2}$ cup of cooked dried beans	about the size of your palm
Fats, Oils, and Sweets Group	1 teaspoon of margarine or butter	about the size of the tip of your thumb

Health Handbook pages R8–R9

Background

Eating Disorders (ED) Five million women in the United States have an ED. Each year approximately 1,000 die from the disorder, some within months, others after decades of pain. People who develop an ED tend to be achievement-oriented perfectionists, have poor self-respect, high-stress familial settings, unstable intimate relationships, and a history of sexual, physical, or mental abuse or other life trauma. Danger signs of EDs include preoccupation with weight and food, weight loss or fluctuation, exercising compulsively, bingeing, purging, depression or mood swings, dizziness, and poor circulation.

Discussion

How is an equal-arm balance like a balanced diet? You need the right foods to balance a diet just as you need the right masses on the equal-arm balance to make the two sides balance. Explain that the Food Guide Pyramid is a tool that can be used to achieve a balanced diet.

Why is it unnecessary to eat a lot of foods rich in fats and sugars? Because fats and sugars are contained in many foods from other groups, it is not necessary to intentionally eat foods rich in these nutrients. Also, the body requires less of these nutrients than it does of other nutrients.

Activities

Music Link - Food Group Songs Organize the class into six groups, one for each food group. Have each group write a short song, using their own or a familiar melody. They should tell about what kinds of foods are in their food group, what nutrients the foods contain, and how many servings a person should eat from the group each day. Have them present their songs to the class.

Social Studies Link - Deficiency Diseases Have students research diseases caused by nutrient deficiencies. Some examples are rickets, beriberi, pellagra, and kwashiork. Have students find out what causes each disease, where the disease is most common, and how the disease is treated or cured.

Preparing Foods Safely

Fight Bacteria

You probably already know to throw away food that smells bad or looks moldy. But food doesn't have to look or smell bad to make you ill. To keep your food safe and yourself from becoming ill, follow the procedures shown in the picture below. And remember—when in doubt, throw it out!

Food Safety Tips

Tips for Preparing Food

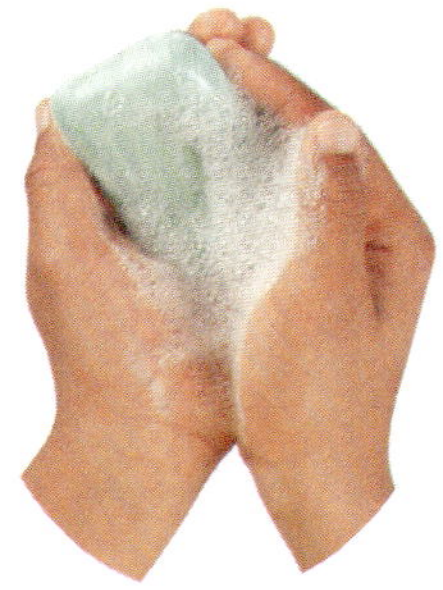

- Wash hands in hot, soapy water before preparing food. It's also a good idea to wash hands after preparing each dish.
- Defrost meat in the microwave or the refrigerator. Do NOT defrost meat on the kitchen counter.
- Keep raw meat, poultry, fish, and their juices away from other food.
- Wash cutting boards, knives, and countertops immediately after cutting up meat, poultry, or fish. Never use the same cutting board for meats and vegetables without thoroughly washing the board first.

Tips for Cooking Food

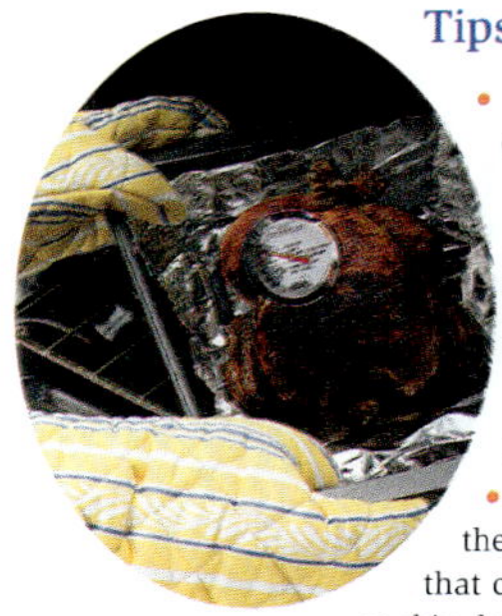

- Cook all food thoroughly, especially meat. Cooking food completely kills bacteria that can make you ill.
- Red meats should be cooked to a temperature of 160°F. Poultry should be cooked to 180°F. When done, fish flakes easily with a fork.
- Eggs should be cooked until the yolks are firm. Never eat food that contains raw eggs. Never eat cookie dough made with raw eggs.

Tips for Cleaning Up the Kitchen

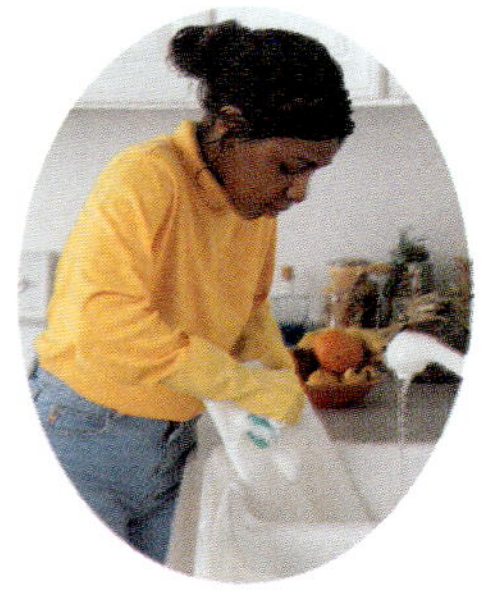

- Wash all dishes, utensils, and countertops with hot, soapy water. Use a disinfectant soap, if possible.
- Store leftovers in small containers that will cool quickly in the refrigerator. Don't leave leftovers on the counter to cool.

Health Handbook pages R10–R11

Background

Pathogens Foods can be contaminated by bacteria, fungi, parasites, and viruses. The most common bacterial pathogens causing food illness are Salmonella bacteria, Staphylococcus aureus, Clostridium perfringens, Clostridium botulinum, Escherichia coli, and the Monocytogenes. Parasites include Trichinella spiralis and Toxoplasmosa gondii; these are linked to undercooked meat and poultry, but the latter can also be transferred to humans from cat feces. Hepatitis A is a virus that can get into shellfish from water contaminated with sewage or into foods from infected persons who handle these foods without proper hygiene.

Discussion

Why does the freshness date on dairy products provide the day and month but not the year? The products have such a short shelf life and spoil so quickly that the year is not needed.

Would a pack date most likely appear on a food that spoils quickly or on a food that has a long shelf life? Pack dates are most common on foods that have long shelf lives.

Activities

Math Link - Cooking in the Mountains Have students find out why cooks who live at an altitude of 1,000 feet or more need to boil foods longer to kill bacteria than do those living at sea level. Also have them find out what altitude cooking times are based on, what other changes in recipes high-altitude cooks need to make, and why.

Social Studies Link - Research Local Pollution Have students find out if water pollution or factory-dumped chemicals have been a problem in or near their community. Have them describe the problem and steps that have been taken to prevent pollution of drinking water, farmland, and homes.

Being Physically Active

Planning Your Weekly Activities

Being active every day is important for your overall health. Physical activity strengthens your body systems and helps you manage stress and maintain a healthful weight. The Activity Pyramid, like the Food Guide Pyramid, can help you choose a variety of activities in the right amounts to keep your body strong and healthy.

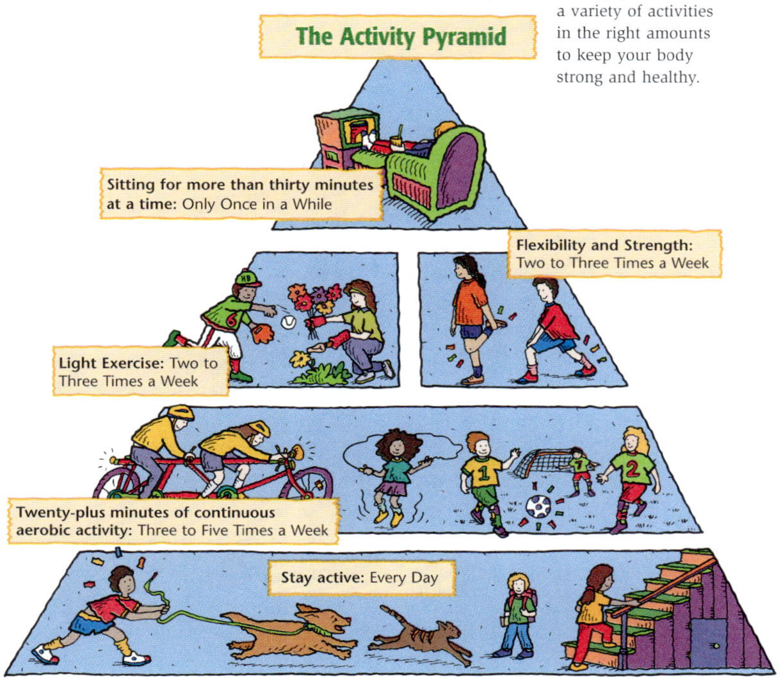

R12

HEALTH HANDBOOK

Guidelines for a Good Workout

There are three things you should do every time you are going to exercise—warm up, work out, and cool down.

Warm-Up When you warm up, your heart rate, respiration rate, and body temperature gradually increase and more blood begins to flow to your muscles. As your body warms up, your flexibility increases, helping you avoid muscle stiffness after exercising. People who warm up are also less prone to exercise-related injuries. Your warm-up should include five minutes of stretching, and five minutes of a low-level form of your workout exercise. For example, if you are going to run for your primary exercise, you should spend five minutes stretching, concentrating on your legs and lower back, and five minutes walking before you start running.

Workout The main part of your exercise routine should be an aerobic exercise that lasts twenty to thirty minutes. Some common aerobic exercises include walking, bicycling, jogging, swimming, cross-country skiing, jumping rope, dancing, and playing racket sports. You should choose an activity that is fun for you and that you will enjoy doing over a long period of time. You may want to mix up the types of activities you do. This helps you work different muscle groups, and provides a better overall workout.

Cool-Down When you finish your aerobic exercise, you need to give your body time to return to normal. You also need to stretch again. This portion of your workout is called a cool-down. Start your cool-down with three to five minutes of low-level activity. For example, if you have been running, you may want to jog and then walk during this time. Then do stretching exercises to prevent soreness and stiffness.

R13

Health Handbook pages R12–R13

Background

Balancing Calories You don't have to exercise to use all the Calories you eat. Your body uses a good deal of energy just to keep your heart beating and performing other necessary body functions. This energy is called your basal metabolism. An adult's basal metabolism uses up about 10 Calories per day per pound of body weight. Thus, a 100-pound person's body would use 1,000 Calories per day just to keep itself going. The basal metabolism of adolescents varies depending on the person's growth rate, degree of physical maturation, and body composition.

Discussion

Why is it important to consider height when determining healthy weights? People who are taller need to weigh more.

Why do people who weigh more burn more Calories doing the same activity than people who weigh less? It takes more energy to move their bodies because their bodies are larger.

Activities

Math Link - Target Heart Rate Your target heart rate is the rate at which your heart must beat so that you get maximum benefit from exercise without putting an unnecessary strain on your heart. Your heart rate should remain at or slightly above your target heart rate during a workout. To find your target heart rate, begin with 220. Subtract your age. Multiply this amount by 0.70 for the target heart rate. For a 20-year-old, the target heart rate is $220 - 20 = 200 \times 0.70 = 140$ beats per minute.

Art Link - A Favorite Exercise Have students draw, color, or paint a series of pictures of themselves taking part in a favorite fitness activity or exercise. The series should include components of the activities that represent the three parts of fitness.

Using a Computer Safely

Good Posture at the Computer

Good posture is important when using the computer. To help prevent eyestrain, muscle fatigue, and injuries, follow the posture tips shown below. Remember to grasp your mouse lightly, keep your back straight, avoid facing a monitor toward a window, and take frequent breaks for stretching.

R14

HEALTH HANDBOOK

Safety on the Internet

The Internet is a remarkable tool. You can use it for fun, education, research, and more. But like anything else, there are some downsides to the Internet. Some people compare the Internet to a real city; not all the people there are people you want to meet and not all the places you can go are places you want to be. Just like in a real city, you have to use common sense and follow safety guidelines to protect yourself. Below are some easy rules to follow to ensure that you are safe on-line.

Rules for Online Safety

- Talk with an adult family member to set up rules for going online. Decide what time of day you can go online, how long you can be online, and appropriate places you can visit. Do not access other areas or break the rules you establish.
- Don't give out personal information like your address, telephone number, or the name or location of your school.
- If you find any information online that makes you uncomfortable, tell an adult family member right away.
- Never agree to meet with anyone in person. If you want to get together with someone you meet online, check with an adult family member first. If a meeting is approved, arrange to meet in a public place and bring an adult with you.
- Don't send your picture or anything else to a person you meet online without first checking with an adult.
- Don't respond to any messages that are mean or make you feel uncomfortable. If you receive a message like that, tell an adult right away.

R15

Health Handbook pages R14–R15

Background

Repetitive Strain Injuries The carpal tunnel is the structure of bones in the wrist through which the median nerve runs. Tendons in the carpal tunnel can swell and become inflamed, pinching the median nerve, which connects to the thumb and first three fingers. Symptoms of carpal tunnel syndrome (CTS) include pain, weakness, and tingling, especially in the thumb and fingers. For mild cases, rest and varied movement lead to improvement. For more serious cases, treatment involves steroid drugs or surgery. If not treated, CTS can lead to permanent paralysis.

Discussion

How does good posture affect breathing? It's easier to take a deep breath when sitting up. Explain that when you breathe deeply, your body gets more oxygen and you stay more relaxed. **Even if you don't use a computer much now, why should you develop good habits working at the computer?** You might use a computer more later and the good habits you develop now will be helpful then.

Activities

Language Arts Link - Learning the Keyboard Encourage students to set up a list of simple words or a sentence that they can type to learn the positions of letters on the computer keyboard. Tell them their words or sentences should include all the letters of the alphabet. For example: A quick brown fox jumps over the lazy dog.

Being Safe at Home

Safety Tips for Baby-Sitters

Baby-sitting is a very important job. As a sitter you are responsible for the safety of the children in your care. Adults depend on you to make good decisions. Here are some tips to help you be a successful and safe baby-sitter.

When you accept a baby-sitting job, ask . . .

- what time you should arrive.
- how long the adults will be away.
- what your responsibilities are.
- the amount of pay you will receive.
- what arrangements will be made for your transportation to and from the home.

When you arrive to start a job . . .

- Arrive several minutes early so that the parents or guardians have time to give you information about caring for the children.
- Write down the place and phone number of where the parents are going and when they will be home.
- Know where emergency phone numbers are listed. The list should have numbers for the police, the fire department, and the children's doctor.
- Know where first-aid supplies are kept and if the children need any special medication. Remember: Never give the children any medication unless directed to do so by the parents or the doctor.
- Ask what and when the children should eat.
- Ask what activities the children can do.
- Ask when the children should go to bed and what their bedtime routine is.

R16

While you are baby-sitting . . .

- Never leave a baby alone on a changing table, sofa, or bed.
- Never leave a child alone, even for a short time.
- Check children often when they are playing and when they are sleeping.
- Never leave a child alone near a pool or in the bathtub.
- Never let a child play with a plastic bag.
- Keep dangerous items out of a child's reach.
- Know where all the doors are, and keep them locked. Do not let anyone in without permission from the parents.
- If the phone rings, take a message. Do not tell the caller that you are the baby-sitter or that the parents are out.
- If there is an injury or if a child gets sick, don't try to take care of it by yourself. Call the parents, or if they are unreachable, call the emergency numbers on the list.

◄ Never leave children playing alone.

▼ Never leave children alone near a pool or in the bathtub.

▼ Never leave a child to eat alone.

R17

Health Handbook pages R16–R17

Background

Cordless Phones Cordless phones are convenient but probably should not be relied on to use when dialing 911 in a life-threatening emergency. Remind students of the possibility of having the battery in the cordless phone go dead, of traveling beyond the range of the phone, and of receiving interference from other cordless or cellular phones or from television or radio stations. Since it is vital to maintain communication with the 911 dispatcher until help arrives, a "hard-wired" telephone should be used if one is easily available.

Discussion

Give the telephone to a student. Explain that you will describe a situation and the student must decide whether or not to call 911 for help. The student can pretend to dial 911 or can explain whom to call.

Some communities have telephone numbers for people to call for emergencies that are not life-threatening. Why do you think communities would do this? There are too many emergency calls. A second number makes it easier for the operator to respond. Non-life-threatening emergencies don't need the same kind of response that life-threatening emergencies do.

Activities

Art Link - Emergency Numbers Bookmarks Have students make bookmarks that list emergency numbers on one side and non-emergency numbers on the other. These numbers can be found in the white pages of the telephone book.

Drama Link - Emergency Role-Play Have pairs of students role-play calling 911 (or the local fire or police department in areas where 911 is not available) for different types of emergencies.

Safety in Emergencies

Fire Safety

Fires cause more deaths than any other type of disaster. But a fire doesn't have to be deadly if you prepare your home and follow some basic safety rules.

- Install smoke detectors outside sleeping areas and on any additional floors of your home. Be sure to test the smoke detectors once a month and change the batteries in each detector twice a year.
- Keep a fire extinguisher on each floor of your home. Check monthly to make sure each is properly charged.
- Work with your family to make a fire escape plan for each room of your home. Ideally, there should be two routes out of each room. Sleeping areas are most important, because most fires happen at night. Plan to use stairs only; elevators can be dangerous in a fire.
- Pick a place outside for everyone to meet. Designate one person to call the fire department or 911 from a neighbor's home.
- Practice crawling low to avoid smoke. If your clothes catch fire, follow the three steps listed below.

R18

Earthquake Safety

An earthquake is a strong shaking of the ground. The tips below can help you and your family stay safe in an earthquake.

Before an Earthquake	During an Earthquake	After an Earthquake
• Secure tall, heavy furniture, such as bookcases, to the wall. Store the heaviest items on the lowest shelves. • Check for potential fire risks. Bolt down gas appliances, and use flexible hosing and connections for both gas and water utilities. • Reinforce and anchor overhead light fixtures to help keep them from falling.	• If you are outdoors, stay outdoors and move away from buildings and utility wires. • If you are indoors, take cover under a heavy desk or table or in a doorway. Stay away from glass doors and windows and heavy objects that might fall. • If you are in a car, drive to an open area away from buildings and overpasses.	• Continue to watch for falling objects as aftershocks shake the area. • Adults should have the building checked for hidden structural problems. • Check for broken gas, electric, and water lines. If you smell gas, an adult should shut off the gas main and leave the area. Report the leak.

Thunderstorm Safety

Thunderstorms are severe storms. Lightning associated with thunderstorms can injure or kill people, cause fires, and damage property. Here are some thunderstorm safety tips.

- **If you are inside, stay there.** The best place to take cover is inside a building.
- **If you are outside, try to take shelter.** If possible, get into a closed car or truck. If you can't take shelter, get into a ditch or low area, if possible.
- **If you are outside, stay away from tall objects.** Don't stand under a lone tree, in an open field, on a beach, or on a hilltop. Find a low place to stay.
- **Stay away from water.** Lightning is attracted to water, and water conducts electricity.
- **Listen for weather bulletins and updates.** The storms that produce lightning may also produce tornadoes. Be ready to take shelter in a basement or interior hallway away from windows and doors.

R19

Health Handbook pages R18–R19

Background

Fire Escape Plan When making an escape plan, family members should check to see that all doors and windows can be easily opened from the inside. If someone does become trapped in a room during a fire, being able to open and close a window without a struggle can help that person remain calm and possibly prevent injury or death. Family members should also be reminded that once they have escaped, they should not go back into a burning building for any reason.

Discussion

Why are regular fire drills important? They provide an opportunity for the family to learn if their fire escape plan works and to make sure everyone understands it.

What types of fires could people attempt to put out with a fire extinguisher? small wood, paper, trash, plastic, cloth, or cardboard fires; kitchen grease fires; small electrical fires

Activities

Language Arts Link - Be Prepared Have students write scripts for public service messages that encourage people to be prepared for natural disasters common to their region. Students can record or videotape their messages and present them to the class. If public access television is available, students might inquire about having their messages televised.

Social Studies - Interview Firefighters Have students contact the local fire department to ask about ideal locations for fire extinguishers in the home. They should also find out the minimum number of extinguishers needed in a home and what type of extinguisher is most appropriate for each room.

First Aid

The tips on the next few pages can help you provide simple first aid to others or yourself. Always tell an adult about any injuries that occur.

For Choking . . .

If someone else is choking . . .

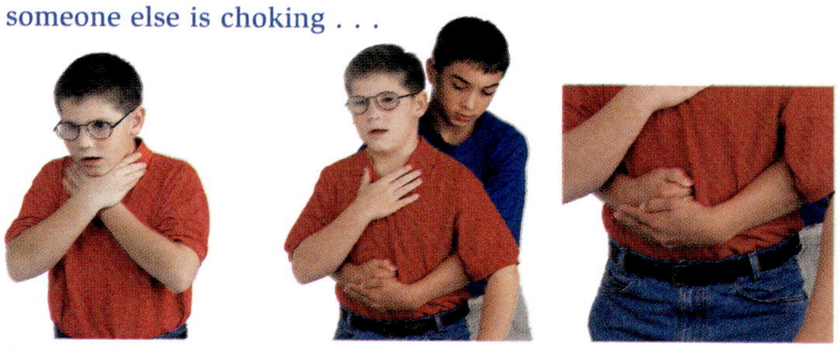

1. Recognize the Universal Choking Sign—grasping the throat with both hands. This sign means a person is choking and needs help.
2. Stand behind the choking person, and put your arms around his or her waist. Place your fist above the person's belly button. Grab your fist with your other hand.
3. Pull your hands toward yourself, and give five quick, hard, upward thrusts on the person's stomach.

If you are choking when alone . . .

1. Make a fist, and place it above your belly button. Grab your fist with your other hand. Pull your hands up with a quick, hard thrust.

2. Or, keep your hands on your belly, lean your body over the back of a chair or over a counter, and shove your fist in and up.

R20

For Bleeding . . .

If someone else is bleeding . . .

Wash your hands with soap, if possible.

Put on protective gloves, if available.

Wash small wounds with soap and water. Do *not* wash serious wounds.

Place a clean gauze pad or cloth over the wound. Press firmly for ten minutes. Don't lift the gauze during this time.

If you don't have gloves, have the injured person hold the gauze or cloth in place with his or her hand for ten minutes.

If after ten minutes the bleeding has stopped, bandage the wound. If the bleeding has not stopped, continue pressing on the wound and get help.

If you are bleeding . . .

- Wash your wound if it is a small cut. If it is a serious wound, do *not* wash it.
- Place a gauze pad or clean cloth over the wound, and hold it firmly in place for ten minutes. Don't lift the gauze or cloth until ten minutes has passed.
- If you have no gauze or cloth, apply pressure with your hand.
- If after ten minutes the bleeding has stopped, bandage the wound. If the bleeding has not stopped, continue pressing on the wound and get help.

R21

Health Handbook pages R20–R21

Background

Objects that Cause Choking Objects that are rounded and smooth, such as hot dogs, are most likely to cause choking. The nonfood item that most frequently causes choking deaths is balloons.

Discussion

The sight of gushing blood is frightening. A few reassuring words delivered in a calm controlled voice can help put an injured person at ease and make it possible to give first aid while waiting for help to arrive. People who are bleeding profusely are in danger of fainting from loss of blood flow to the brain.

Why might it be a good idea to have a bleeding person lie down? Possible answers: to increase the blood flow to the brain; if the person does faint, it will help prevent further injury; to prevent shock

Activities

Abdominal Thrusts Plastic soda bottles can also be used for practicing abdominal thrusts. Have students find empty 3-liter plastic soda bottles and fit the tops with plugs made of tightly wrapped gauze or some other material. Have them tie the plugs to the necks of the bottles with 2–3-in. pieces of string, and put the plugs firmly in the bottles. Next, students should hold the bottles in front of themselves as if the bottles were choking victims, and apply thrusts to pop the plugs out of the bottles.

Drama Link - Rescue Play Have students work together to write and perform a play that incorporates the first-aid techniques explained in the lesson, including treatment for choking and severe bleeding. Students might adapt the technique used in Thornton Wilder's play *Our Town* in which the character of the Stage Manager narrates and explains the actions that occur in a series of flashbacks.

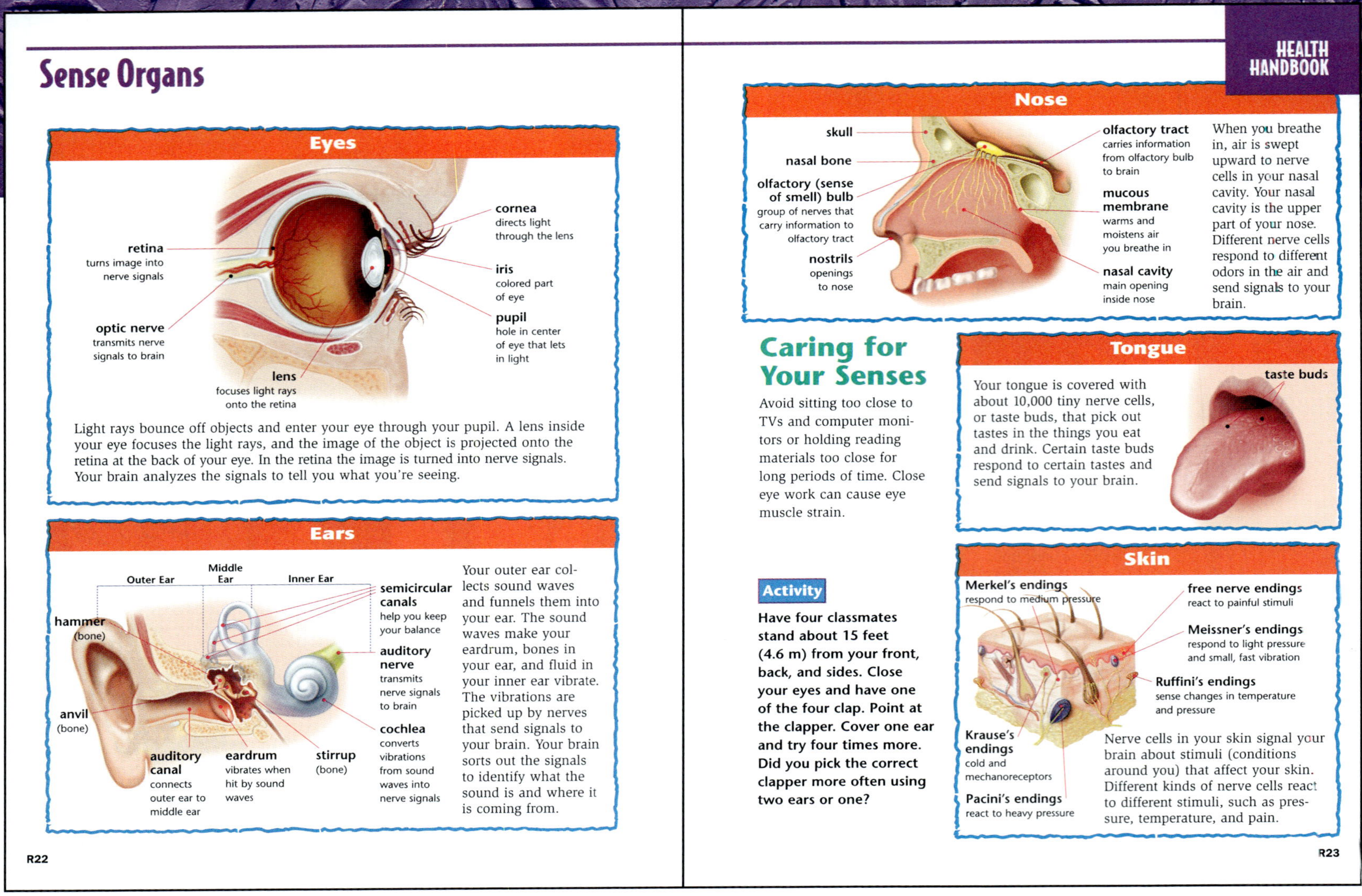

HEALTH HANDBOOK

Sense Organs

Eyes

retina
turns image into nerve signals

optic nerve
transmits nerve signals to brain

lens
focuses light rays onto the retina

cornea
directs light through the lens

iris
colored part of eye

pupil
hole in center of eye that lets in light

Light rays bounce off objects and enter your eye through your pupil. A lens inside your eye focuses the light rays, and the image of the object is projected onto the retina at the back of your eye. In the retina the image is turned into nerve signals. Your brain analyzes the signals to tell you what you're seeing.

Ears

Outer Ear

Middle Ear

Inner Ear

hammer (bone)

anvil (bone)

auditory canal
connects outer ear to middle ear

eardrum
vibrates when hit by sound waves

stirrup (bone)

semicircular canals
help you keep your balance

auditory nerve
transmits nerve signals to brain

cochlea
converts vibrations from sound waves into nerve signals

Your outer ear collects sound waves and funnels them into your ear. The sound waves make your eardrum, bones in your ear, and fluid in your inner ear vibrate. The vibrations are picked up by nerves that send signals to your brain. Your brain sorts out the signals to identify what the sound is and where it is coming from.

R22

Nose

skull

nasal bone

olfactory (sense of smell) bulb
group of nerves that carry information to olfactory tract

nostrils
openings to nose

olfactory tract
carries information from olfactory bulb to brain

mucous membrane
warms and moistens air you breathe in

nasal cavity
main opening inside nose

When you breathe in, air is swept upward to nerve cells in your nasal cavity. Your nasal cavity is the upper part of your nose. Different nerve cells respond to different odors in the air and send signals to your brain.

Caring for Your Senses

Avoid sitting too close to TVs and computer monitors or holding reading materials too close for long periods of time. Close eye work can cause eye muscle strain.

Tongue

taste buds

Your tongue is covered with about 10,000 tiny nerve cells, or taste buds, that pick out tastes in the things you eat and drink. Certain taste buds respond to certain tastes and send signals to your brain.

Activity

Have four classmates stand about 15 feet (4.6 m) from your front, back, and sides. Close your eyes and have one of the four clap. Point at the clapper. Cover one ear and try four times more. Did you pick the correct clapper more often using two ears or one?

Skin

Merkel's endings
respond to medium pressure

Krause's endings
cold and mechanoreceptors

Pacini's endings
react to heavy pressure

free nerve endings
react to painful stimuli

Meissner's endings
respond to light pressure and small, fast vibration

Ruffini's endings
sense changes in temperature and pressure

Nerve cells in your skin signal your brain about stimuli (conditions around you) that affect your skin. Different kinds of nerve cells react to different stimuli, such as pressure, temperature, and pain.

R23

Health Handbook pages R22–R23

Background

Acne Some sixth-grade students may have acne. Students with acne might be self-conscious when discussing it. Direct all discussions away from individual students. Point out that acne can be caused by hormones and genetics. Emphasize that a person with acne can help alleviate the problem by practicing good hygiene but that acne is not necessarily caused by poor hygiene or unhealthful eating. Tell students that a doctor can suggest a regimen to reduce acne, and that with age the problem usually disappears.

Discussion

Direct students to the illustration of the ear.

What structure separates the outer ear from the middle ear? eardrum

Name the three bones of the middle ear. hammer, anvil, stirrup

What part of the ear controls your balance? semicircular canals

Recall what it's like to spin in circles. Why do you feel dizzy after you've stopped spinning? The fluid in the semicircular canals continues to whirl for a short time, causing the spinning sensation.

Activities

Science Link - The Better to See You With, My Dear! Provide students with old eyeglasses for both nearsighted and farsighted individuals. Have students use the lenses to observe a small object such as a coin, noting the distance each lens must be held from the coin in order to produce a clear view.

Multicultural Link - Acupuncture Have students find out about the Asian procedure of acupuncture, which is used to relieve pain, cure certain diseases, anesthetize a person, or improve general health.

Acupuncture involves inserting one of more small needles into the skin and underlying tissues at precise points, which correspond to nerve endings.

Skeletal System

Your bones fit together and attach to your muscles at your joints, allowing you to move in many directions. Each joint is designed to do a certain type of job.

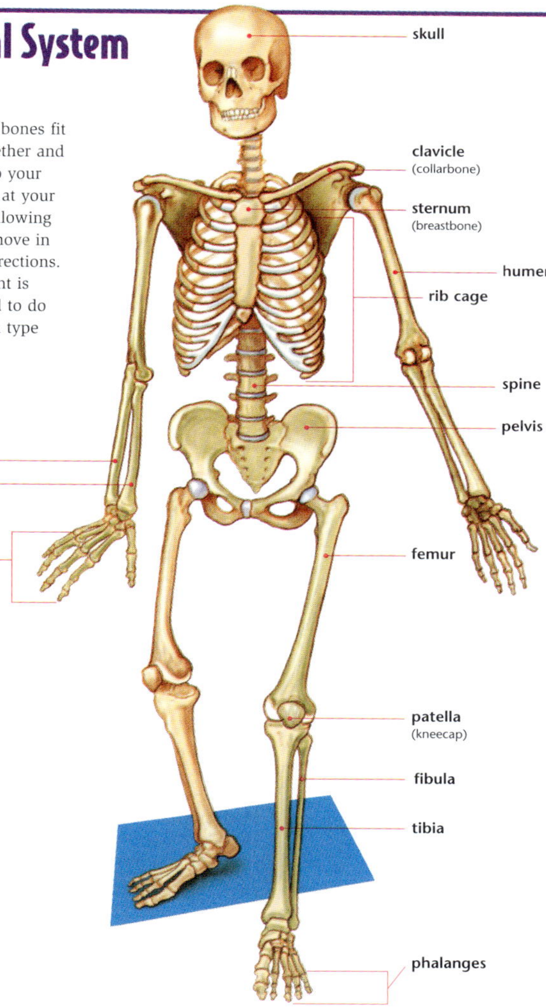

R24

HEALTH HANDBOOK

Some Types of Joints

Ball-and-socket joints allow rotation and movement in many directions. Your hips and shoulders are ball-and-socket joints.

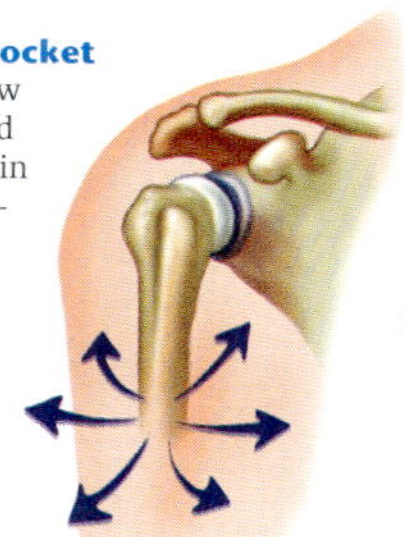

Hinge joints move only back and forth. Your elbows and fingers have hinge joints.

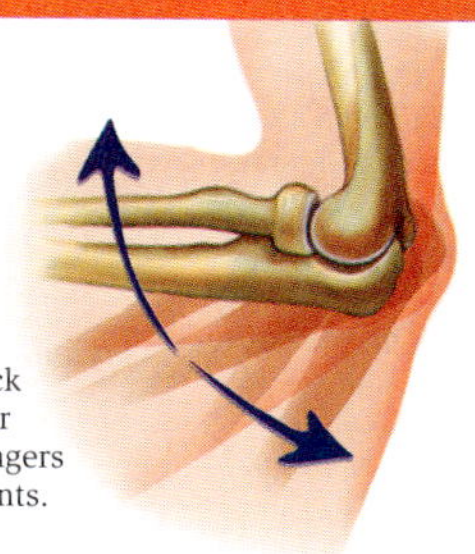

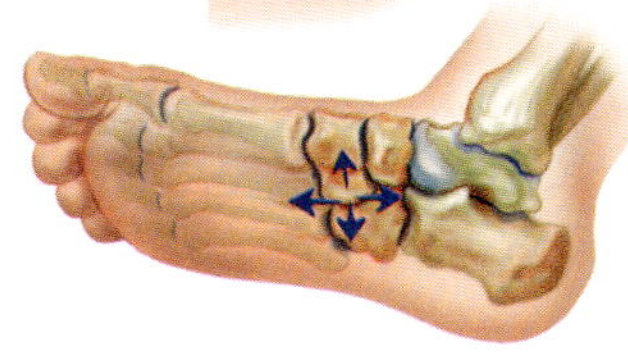

Gliding joints allow only side-to-side and back-and-forth movements. The vertebrae of your spine form gliding joints. Your wrist is a gliding joint. Your feet also contain gliding joints.

Caring for Your Skeletal System

- Move and flex your joints regularly through exercise. Strong muscles can protect your joints from injuries.
- Be sure to warm up and cool down whenever you exercise.
- You can injure a joint by using it too much at one time or without warming up or by moving it in a way it is not designed to move.

Activities

1. Curl your index finger. Point it and move it in a circle. Which joint is a hinge joint? Which joint is a gliding joint?
2. Make a model of a joint. Cut out two strips of cardboard and join them together using a round metal fastener. What type of joint does the model most nearly represent?
3. Try to run without bending your knees. Try to write without bending any of your fingers. Can you do either of these things?

R25

Health Handbook pages R24–R25

Background

Ligaments, Cartilage, and Tendons Joints are held together and often enclosed by tough bands of connective tissue called ligaments. Joints that permit a wide range of motion usually have more ligaments associated with them than joints that provide limited movement. Cartilage, the tissue from which bones develop, covers the ends of bones. Cartilage reduces friction between adjacent bones. Tendons, like cartilage, are thick bands of connective tissue that attach muscles to bones.

Discussion

What are two major functions of your skeletal system? The skeletal system gives support and shape to the body and it protects internal organs.

Name two places on your body that contain immovable joints. skull and pelvis

What is an example of a body movement that involves a pivot joint? turning the head

Activities

Activity 1 Students discover that curling the finger involves the use of a hinge joint. Moving the finger in a circular motion involves a gliding joint.

Activity 2 Students make a simple hinge joint that allows back and forth movement. Models will vary.

Music Link - Them Bones Have students use the tune of this folk song to describe the relationships among the various sets of bones shown on page R24. For example, "the skull bone's connected to the spine bone; the spine bone's connected to the pelvis bone..."

Muscular System

The muscles that make your body move are attached to bones. When one of these muscles contracts, or gets shorter, it pulls on the bone it's connected to and the bone moves. Your digestive and circulatory systems also have muscles.

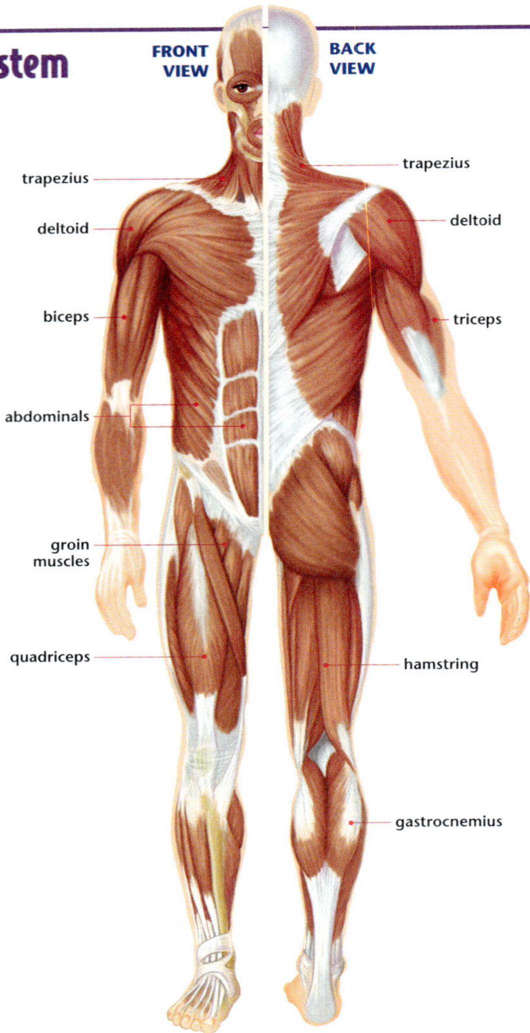

R26

Types of Muscle

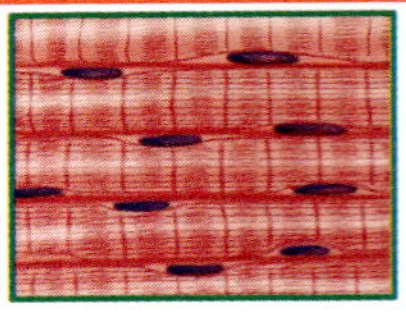

Striated Muscle Striated means striped. If you could look at your muscles without any covering of skin, you could see the thin stripes. All the muscles you use to walk and move your body are striated muscles.

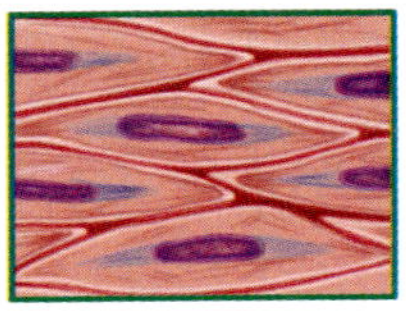

Smooth Muscle Smooth muscle is part of other body systems. The walls of your esophagus, stomach, and intestines are lined with smooth muscle. These muscles squeeze and relax to move food through your digestive system.

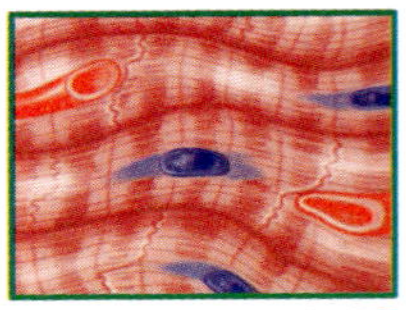

Cardiac Muscle Your heart is made of a special kind of muscle called cardiac muscle. It squeezes and relaxes every second of every day to pump blood through your body.

Caring for Your Muscular System

- An easy jog and gentle stretches before you start to play a sport loosen your tendons, ligaments, and muscles and help prevent injuries. This is called a warm-up.
- You should cool down after exercise too. Because your muscles contract during exercise, they need to be stretched when you finish.

Activities

1. With one hand, pull up on your desk. Feel the muscle on the front side of your upper arm. This is your biceps. It is a striated muscle.
2. Close your hand around a partly empty, closed tube of toothpaste. Gently squeeze starting with your little finger and ending with your index finger. Your fingers move toothpaste through the tube like smooth muscles move food through your digestive system.
3. Each second for a minute squeeze one hand into a fist and relax it. This is like the work your cardiac muscle does all the time.

R27

Health Handbook pages R26–R27

Background

Steroid Use Among Adolescent Females A 1997 study published in the Archives of Pediatric and Adolescent Medicine indicates that steroid use among girls is rising. In fact, as many as 175,000 high school girls, or 2.4 percent across the United States, have used anabolic steroids to build more muscle. Another study has shown that the percentage of eighth grade females who have used steroids nearly doubled between 1991 and 1996. Stress the dangers of the use of steroids by both girls and boys, which include female breast shrinkage and menstrual problems; testicle shrinkage and development of breast tissue in males; and hypertension, extreme mood swings, aggression, excessive acne, and liver problems in both sexes.

Discussion

Challenge students not to "move a muscle." While they probably think they are sitting quite still, explain to students that some of their muscles (involuntary muscles) stretch and contract on their own.

What are some involuntary muscles? muscles in the eyelids, the heart, certain stomach muscles, and the diaphragm

Activities

Activity 1 Students should feel the biceps contract as they exert force on the desk. Have students use the diagram on page R26 to identify the other muscle in the pair, the triceps.

Activity 2 This activity simulates the slow, progressive contractions of the muscles in the digestive tract. Have students contrast this with the relatively quick contraction of cardiac and striated muscles.

Activity 3 Using 60 beats per minute as an average, how many times does your heart beat in an hour? 3600 **in a typical day?** 86,400 **in a year?** 31,536,000 **since you were born?** If students are 11 years old, their hearts have beat an average of 346,896,000 times.

Digestive System

Digestion is the process of breaking food into tiny pieces that are absorbed by your blood and carried to all parts of your body. Each part in your digestive system does a different job.

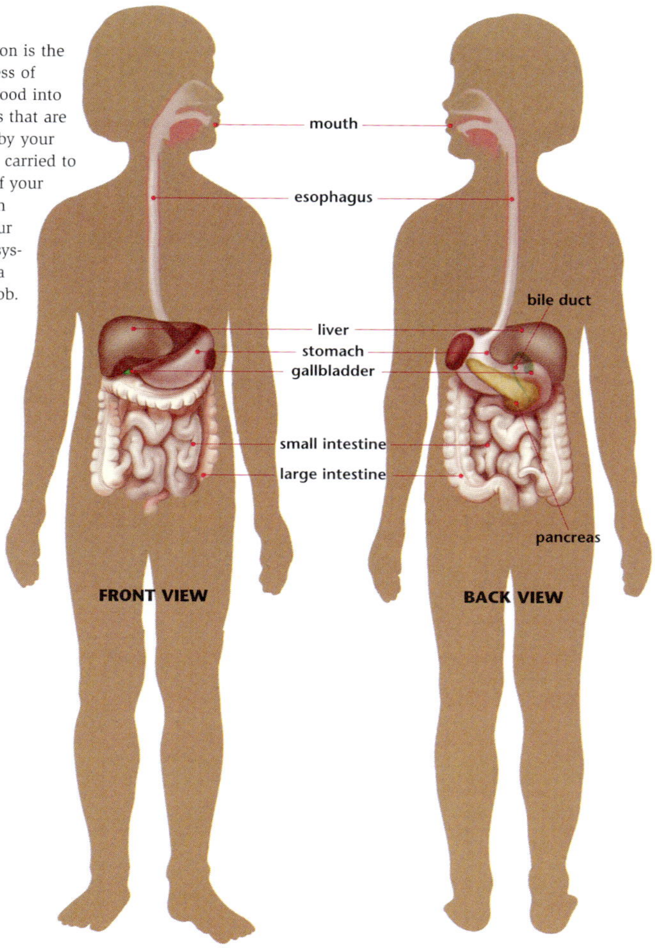

R28

HEALTH HANDBOOK

Some Digestive Organs

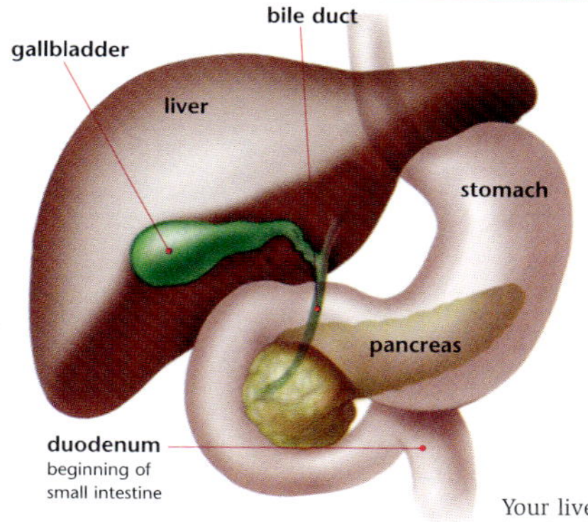

Liver, Gallbladder, Bile Duct, and Pancreas Your liver receives blood full of freshly digested nutrients from your small intestine. It stores some substances, such as vitamins and certain nutrients that supply your body with energy. It lets other nutrients pass through to other parts of your body. The liver also destroys poisonous waste products created during the digestion process. It breaks down chemicals such as medicines and alcohol too.

Your liver also makes bile, a fluid that breaks fats into smaller pieces. Bile is stored in your gallbladder. When it is needed, the bile flows through the bile duct into your small intestine to help with the digestive process.

Digestive juices made in your pancreas flow along a narrow tube into your small intestine, where they aid in digesting fats and proteins. Your pancreas also makes a special substance called insulin. Insulin helps your body move sugar from blood into cells.

Caring for Your Digestive System

- Drink a lot of water. Water helps food move through your digestive system.
- Don't drink alcohol. It's illegal for people your age. Drinking too much alcohol can damage your liver and pancreas.
- Keep a healthy body weight. Overweight people are more likely to have difficulty producing enough insulin.

Activities

1. Vinegar is a kind of acid. Soak a potato chip in vinegar for 30 minutes. What happens? What process is demonstrated?
2. Foods with fat are harder to digest. Look at the nutrition labels on foods in your home. Which ones have the most fat per unit of mass (gram or g)?
3. Pretend you are a piece of food and write a story about your journey through the digestive system.

R29

Health Handbook pages R28–R29

Background

Obesity Obesity is a complex issue with varied causes. A tendency toward obesity seems to occur in some families, though researchers are uncertain about genetic causes versus environmental effects. Besides overeating and lack of exercise, weight gain may result from taking certain medications, from some types of glandular disorders, and from repeated "yo-yo" dieting. Hyperplastic obesity results when too many fat cells form in the body. These cells may never disappear, and thus may create a long-term tendency toward obesity.

Discussion

Digestion involves both chemical reactions and mechanical processes.

Where do mechanical processes involved in digestion take place? in the mouth (chewing and grinding) and the stomach (mixing of food with digestive juices)

Where do the chemical processes occur? in the mouth (food reacts with saliva), the stomach (food reacts with digestive juices), and the small intestine

Activities

Activity 1 The digestion process is demonstrated.

Activity 2 Foods high in fats include ice cream, chocolate, rich desserts, fried foods, foods containing saturated oils, marbled red meats, cheeses and other whole milk dairy products, and many processed food items.

Activity 3 Encourage creativity but insist on accuracy. In their stories, students should include the physical and chemical processes food undergoes as it passes through each organ of the digestive tract.

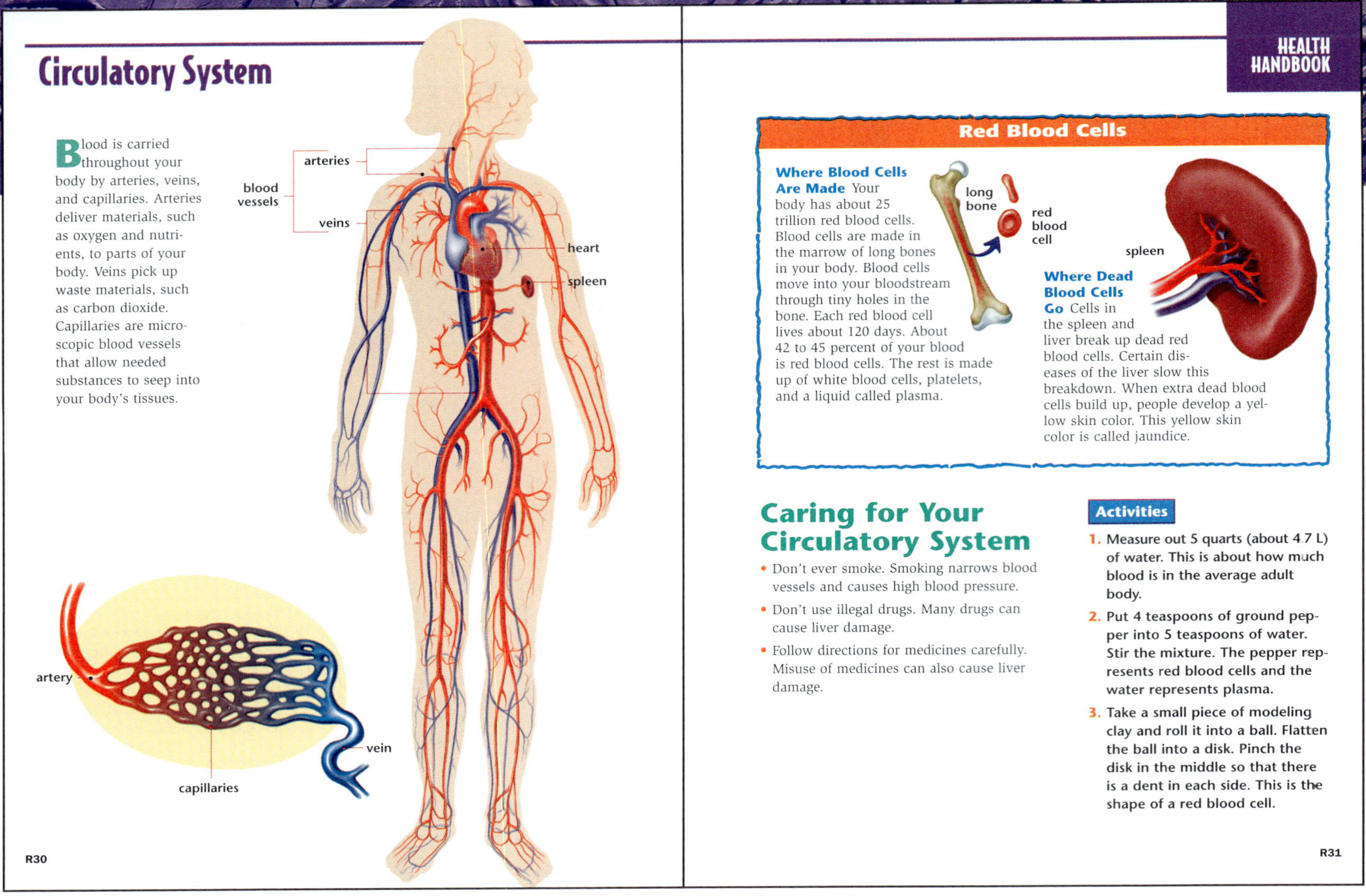

Circulatory System

Blood is carried throughout your body by arteries, veins, and capillaries. Arteries deliver materials, such as oxygen and nutrients, to parts of your body. Veins pick up waste materials, such as carbon dioxide. Capillaries are microscopic blood vessels that allow needed substances to seep into your body's tissues.

R30

HEALTH HANDBOOK

Red Blood Cells

Where Blood Cells Are Made Your body has about 25 trillion red blood cells. Blood cells are made in the marrow of long bones in your body. Blood cells move into your bloodstream through tiny holes in the bone. Each red blood cell lives about 120 days. About 42 to 45 percent of your blood is red blood cells. The rest is made up of white blood cells, platelets, and a liquid called plasma.

Where Dead Blood Cells Go Cells in the spleen and liver break up dead red blood cells. Certain diseases of the liver slow this breakdown. When extra dead blood cells build up, people develop a yellow skin color. This yellow skin color is called jaundice.

Caring for Your Circulatory System

- Don't ever smoke. Smoking narrows blood vessels and causes high blood pressure.
- Don't use illegal drugs. Many drugs can cause liver damage.
- Follow directions for medicines carefully. Misuse of medicines can also cause liver damage.

Activities

1. Measure out 5 quarts (about 4.7 L) of water. This is about how much blood is in the average adult body.
2. Put 4 teaspoons of ground pepper into 5 teaspoons of water. Stir the mixture. The pepper represents red blood cells and the water represents plasma.
3. Take a small piece of modeling clay and roll it into a ball. Flatten the ball into a disk. Pinch the disk in the middle so that there is a dent in each side. This is the shape of a red blood cell.

R31

Health Handbook pages R30–R31

Background

Atherosclerosis and Smoking In a study of atherosclerosis in more than 10,000 middle-aged adults published in January 1998, researchers found that on the average, arteries of people who smoke harden 50 percent faster than do those of nonsmokers. Ex-smokers' arteries harden about 25 percent faster than the arteries of nonsmokers. The most striking finding was the effect of secondhand smoke on nonsmokers. Arteries of nonsmokers exposed to about 20 hours of secondhand smoke each week harden 20 percent faster than arteries of nonsmokers not exposed to the smoke.

Discussion

Explain that arteries carry blood away from the heart; veins carry blood toward the heart. **How are veins and arteries similar?** Both are blood vessels that move blood through the body. **What is the relationship among capillaries, veins, and arteries?** Capillaries connect veins and arteries.

Activities

Activity 2 For best results, use fine-grained pepper. Clear, glass custard bowls provide good containers for the mixture. This activity allows students to visualize the percentage of red blood cells in blood.

What percentage of blood is plasma, white blood cells, and platelets? 55 to 58 percent

Activity 3 Encourage students to make scale models of red blood cells. Tell them that a red cell is about three and a half to four times as wide (diameter) as it is thick. Thus, a scale model would have a diameter of about 8 cm and would be about 2 cm high.

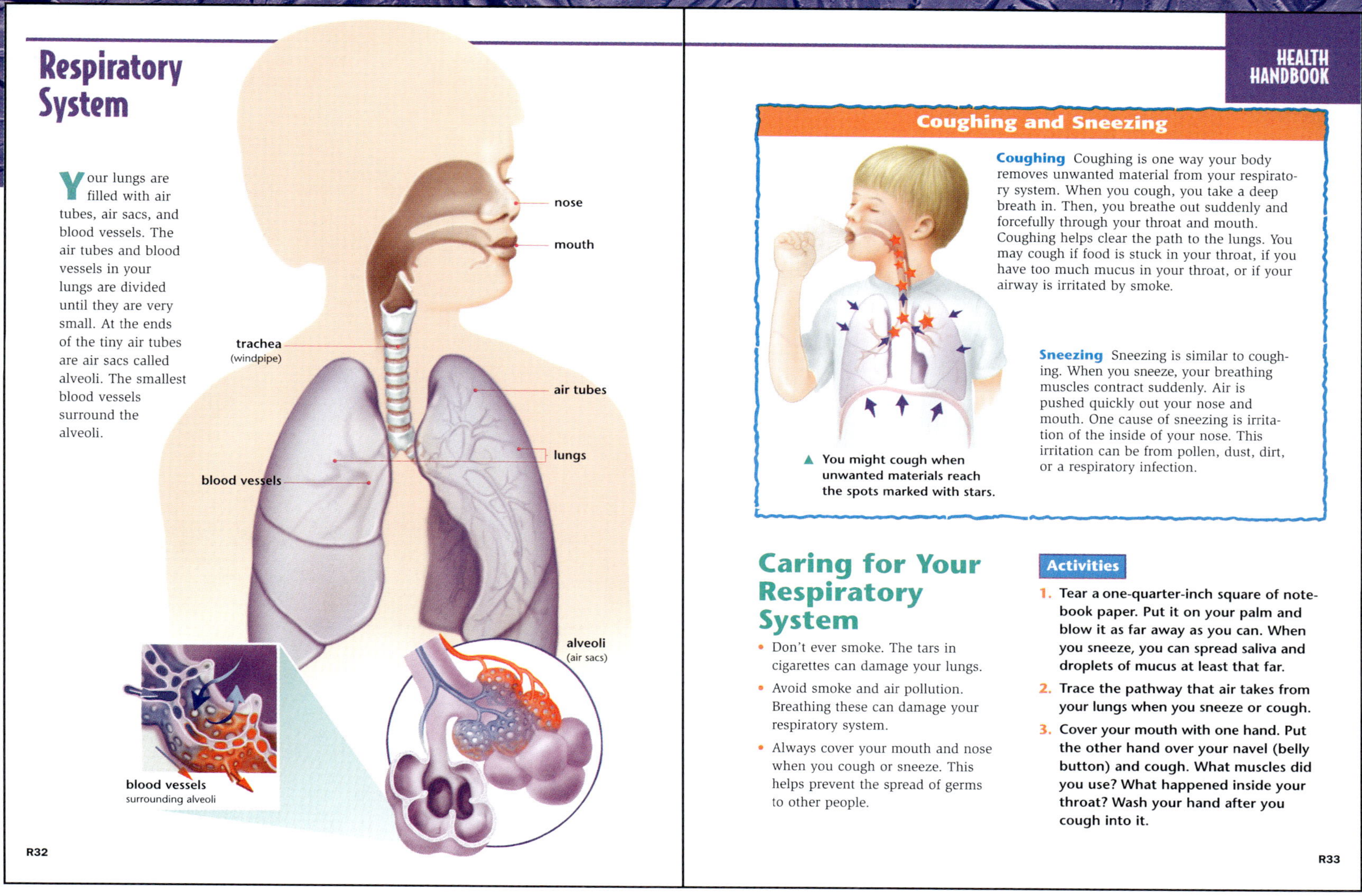

Respiratory System

Your lungs are filled with air tubes, air sacs, and blood vessels. The air tubes and blood vessels in your lungs are divided until they are very small. At the ends of the tiny air tubes are air sacs called alveoli. The smallest blood vessels surround the alveoli.

R32

HEALTH HANDBOOK

Coughing and Sneezing

Coughing Coughing is one way your body removes unwanted material from your respiratory system. When you cough, you take a deep breath in. Then, you breathe out suddenly and forcefully through your throat and mouth. Coughing helps clear the path to the lungs. You may cough if food is stuck in your throat, if you have too much mucus in your throat, or if your airway is irritated by smoke.

▲ You might cough when unwanted materials reach the spots marked with stars.

Sneezing Sneezing is similar to coughing. When you sneeze, your breathing muscles contract suddenly. Air is pushed quickly out your nose and mouth. One cause of sneezing is irritation of the inside of your nose. This irritation can be from pollen, dust, dirt, or a respiratory infection.

Caring for Your Respiratory System

- Don't ever smoke. The tars in cigarettes can damage your lungs.
- Avoid smoke and air pollution. Breathing these can damage your respiratory system.
- Always cover your mouth and nose when you cough or sneeze. This helps prevent the spread of germs to other people.

Activities

1. Tear a one-quarter-inch square of notebook paper. Put it on your palm and blow it as far away as you can. When you sneeze, you can spread saliva and droplets of mucus at least that far.
2. Trace the pathway that air takes from your lungs when you sneeze or cough.
3. Cover your mouth with one hand. Put the other hand over your navel (belly button) and cough. What muscles did you use? What happened inside your throat? Wash your hand after you cough into it.

R33

Health Handbook pages R32–R33

Background

Lung Structure The lungs are two spongelike organs inside the chest cavity. They are made of millions of air sacs. If the walls of the sacs that form the lungs were spread out flat, they would have a surface area about 40 times that of the outside skin. The left lung, next to the heart, is slightly smaller than the right lung.

The Trachea The trachea, or windpipe, carries air from the nose and mouth through the pharynx (throat) and past the larynx (voice box) to the chest.

Discussion

Over 150,000 people die each year from lung cancer caused by smoking cigarettes. Exposure to environmental tobacco smoke also increases a person's risk of developing lung cancer. **Why are these statistics especially tragic?** Unlike many other factors that affect health, smoking is a choice.

Activities

Activity 1 This activity reinforces how easily pathogens are spread via sneezes and coughs. **What are some ways to reduce the spread of pathogens when you have a cold or other infectious respiratory illness?** covering the mouth with disposable tissues, washing hands often with warm water and soap, and keeping hands away from the nose, mouth, and eyes

Activity 2 Have students refer to the figure on page R33 to sequence the events involved in coughing. During a cough, air moves from the trachea to the air tubes and into the air sacs. Then the air flow reverses and air exits the mouth. A sneeze is similar except that during a sneeze, air is pushed out of the nose as well as the mouth.

Activity 3 If necessary, refer students to the diagram of the muscular system on page R26 to help them with Activity 3. Muscles used during coughing include abdominal muscles, the diaphragm, and muscles in the throat. When a person coughs, air and sometimes foreign matter are expelled from the body via the throat.

Nervous System

Your nervous system is responsible for all of your thoughts and your body's activity. It makes your heart beat and your lungs work. It allows you to see, hear, smell, taste, and touch. It lets you learn, remember, and feel emotions. It moves all of your muscles.

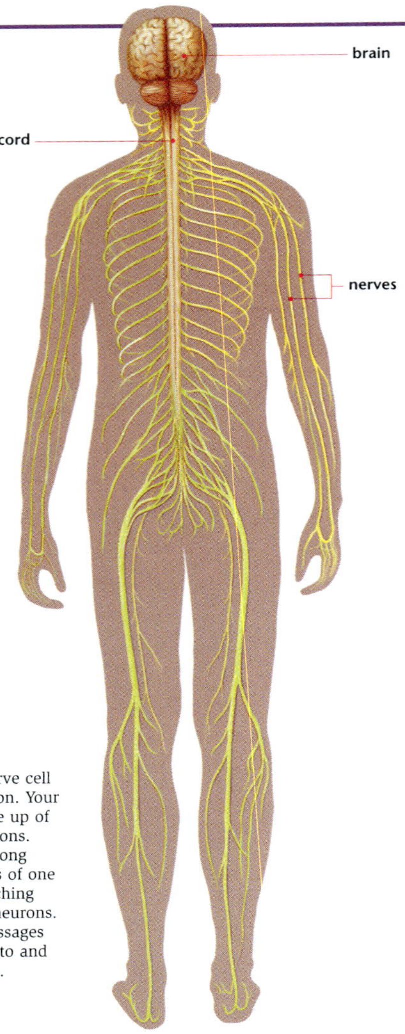

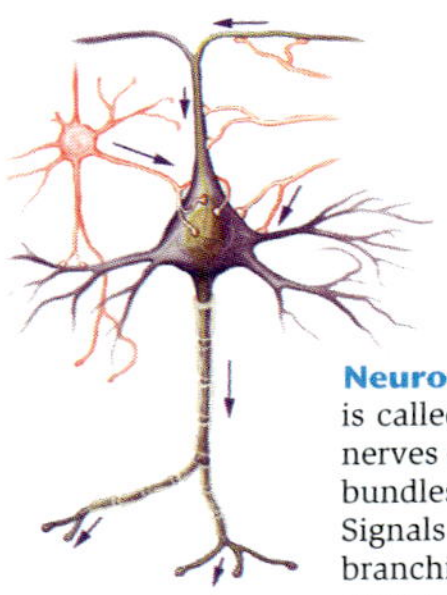

Neurons A nerve cell is called a neuron. Your nerves are made up of bundles of neurons. Signals travel along branching fibers of one neuron to branching fibers of other neurons. This is how messages are transmitted to and from your brain.

R34

HEALTH HANDBOOK

The Brain

Central Nervous System Your brain and spinal cord make up your central nervous system. Different areas of your brain control different activities.

Cerebrum The cerebrum is the top layer of your brain. It begins behind your forehead and fills the space inside the top and sides of your skull to where your neck begins. The cerebrum splits down the middle from front to back. The left and right parts are called hemispheres.

Four Lobes Each hemisphere of the cerebrum is divided into four smaller parts called lobes. The lobes are named for the skull bones that cover them.

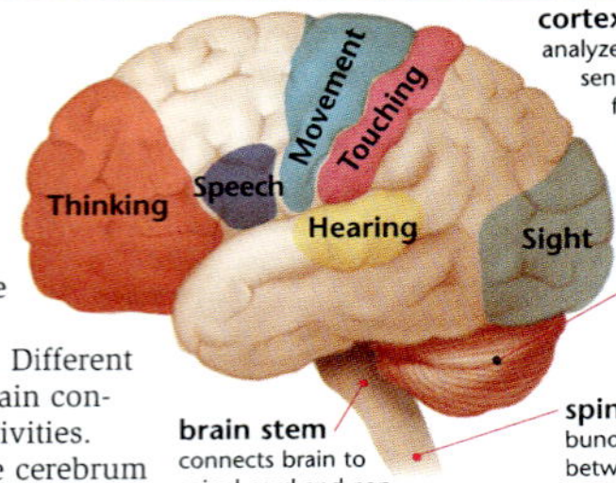

cortex analyzes information from your sense organs and is responsible for thinking abilities, such as reasoning, planning, and problem solving.

cerebellum helps with your coordination and balance

brain stem connects brain to spinal cord and controls vital functions, such as breathing

spinal cord bundle of nerves that relays messages between your brain and the rest of the nerves in your body

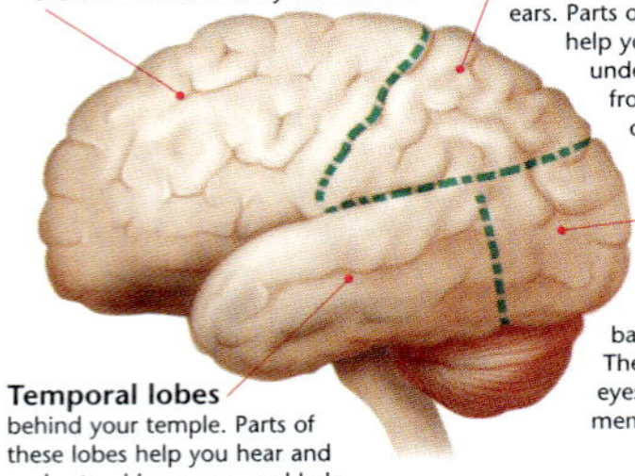

Frontal lobes behind your forehead. This part of the brain controls body movement.

Parietal lobes above and just behind your ears. Parts of these lobes help you combine and understand signals from your sense organs.

Occipital lobes behind the bulge on the back of your head. These lobes control eyesight and visual memory.

Temporal lobes behind your temple. Parts of these lobes help you hear and understand language, and help you speak.

Caring for Your Nervous System

- Do not take any drugs unless chosen by your parents or guardian or a doctor. Some drugs can affect your brain cells.
- Eat a well-balanced diet. Your nervous system cannot work properly without certain nutrients.

Activity

Have a friend hold a ruler so that the 0-inch mark is at the bottom. Put your hand just under the ruler. Ask your friend to drop the ruler while you catch it. Note the inch mark at which you caught the ruler. Try it nine more times. What happens?

R35

Health Handbook pages R34–R35

Discussion

There are two major divisions of the human nervous system: the central nervous system and the peripheral nervous system.

Think about what the terms central and peripheral mean. Which two organs shown are parts of the central nervous system? the brain and the spinal cord

A person who is paralyzed has experienced damage to which lobes? frontal lobes

Which lobes control your sense of hearing? temporal lobes

Which part of the brain controls coordination and balance? the cerebellum

Which part of the brain controls the heartbeat? the brain stem

Immune System and Endocrine System

Your immune system defends your body from harmful invaders, such as organisms that cause infection. White blood cells, which are your immune system's primary infection fighters, are produced in your bone marrow, thymus, lymph nodes, and spleen.

Lymph Nodes Lymph nodes are bean-sized organs that help filter germs from your blood. The plasma in your blood moves from your capillaries into your lymph ducts and through these ducts to your lymph nodes. At the lymph nodes the plasma is cleaned and then drained back into the blood. Lymph nodes also produce plasma cells. Plasma cells produce antibodies, an essential part of your immune system.

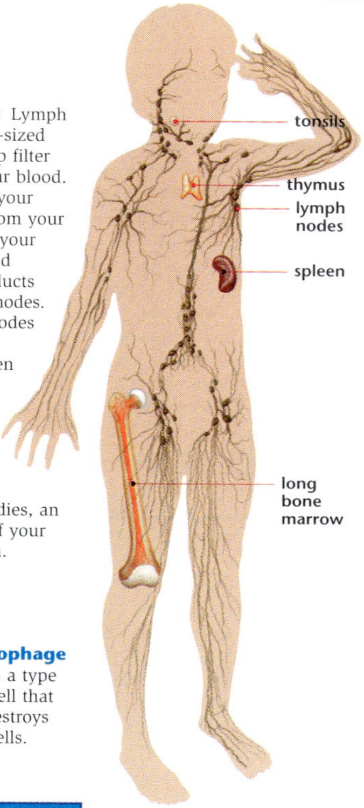

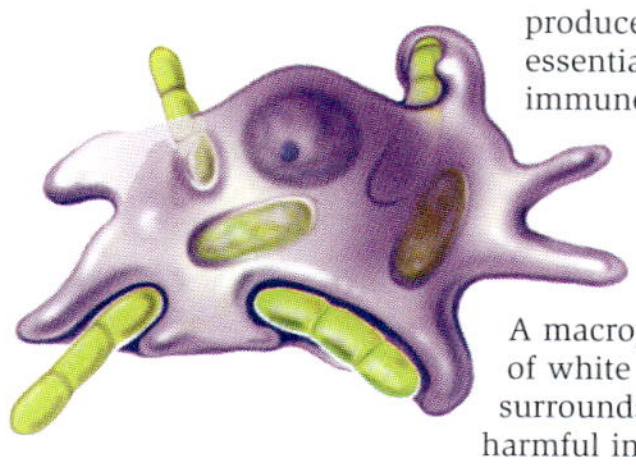

Macrophage A macrophage is a type of white blood cell that surrounds and destroys harmful invader cells.

Caring for Your Immune System

One way to help your body fight disease is through immunization. You probably were immunized before you began school. More immunizations (boosters) may be required as you grow, if you travel to other countries, or when you are an adult.

Activity

An average disease-causing organism grows and divides into two new organisms about every thirty minutes. Calculate how many organisms there would be at the end of five hours.

R36

HEALTH HANDBOOK

Fight or Flight When you are threatened or under stress, your body prepares to fight or to run away. Your pupils dilate. Your heartbeat and breathing rates increase. Also, your adrenal glands release adrenaline into your blood to help your body keep up this reaction.

Your endocrine system works with your nervous system to help your body grow normally and work and react as it should. The endocrine system does this by releasing hormones into your bloodstream. The hormones deliver instructions to various organs and tissues.

Caring for Your Endocrine System

- Doctors can sometimes give medicine to treat an endocrine system that produces too little of any chemical.

Activity

The next time you are startled, feel your heartbeat. How is it different from normal?

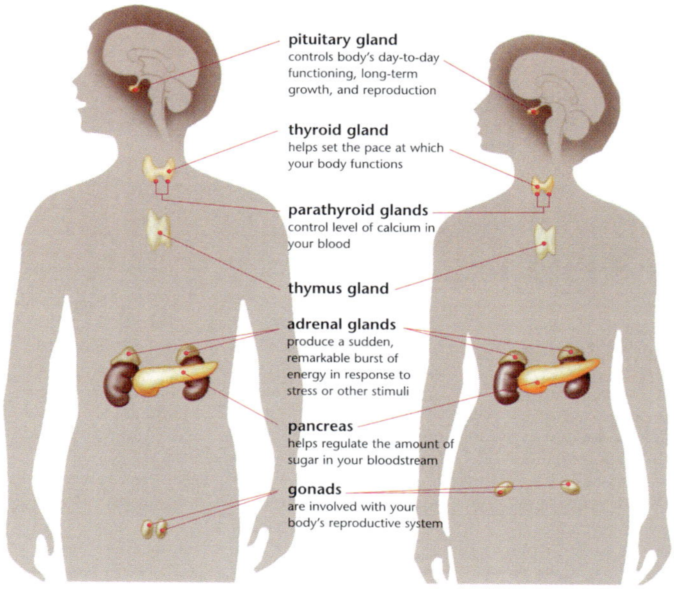

R37

Health Handbook pages R36–R37

Background

White Blood Cells Phagocytes are white blood cells that ingest and destroy pathogens by surrounding them and eventually engulfing the disease-causing microbes. A macrophage is a type of phagocyte that migrates out of capillaries and attacks anything it recognizes as foreign.

Discussion

Locate some of the lymph nodes in your body. Lymph nodes are found in the neck, in the armpits, in the groin area, and behind the ears, among other places.

Why does the doctor feel the lymph nodes behind your ears and under your neck when you complain of an earache and a sore throat? Swollen lymph nodes can indicate infection.

What are situations in which you have experienced the fight or flight response? Students' answers might include when they were scared; under pressure to perform well before an exam, sporting event, or concert; or placed in an unfamiliar situation.

Activities

Art Link - Visual Learners Demonstrate the relationship between a foreign substance in the body (an antigen) and an antibody by giving students several pieces from a jigsaw puzzle. Make sure only two of the pieces interlock correctly. Use the interlocking pieces to explain that a unique antibody is formed in response to each antigen the body encounters. The antibody binds to an antigen, making it harmless.

Using Science Reading Strategies

Some strategies work better than others when it comes to reading science text. Below are effective strategies to use and ways you might use them.

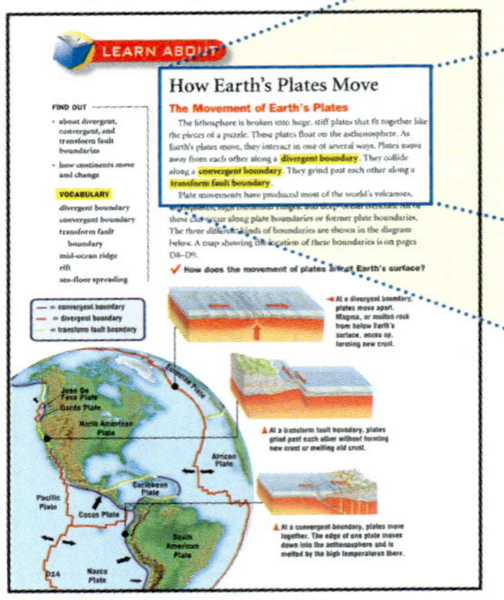

When reading a text like this...

> ## How Earth's Plates Move
> ### The Movement of Earth's Plates
> The lithosphere is broken into huge, stiff plates that fit together like the pieces of a puzzle. These plates float on the asthenosphere. As Earth's plates move, they interact in one of several ways. Plates move away from each other along a **divergent boundary.** . . .
>
> page D14

A good strategy to use is...

REREAD

Think about whether what you're reading makes sense to you. It's always OK to go back and read a section again. Maybe the text involves a scientific concept from an earlier chapter. You may have to look back at that chapter to remind yourself of information you need to understand the new passage.

What you might say to yourself...

"Wait. I'm getting lost. I'd better not read more about how the plates move until I know what's going on. I'll reread the beginning of this paragraph . . .

"Now that I've reread it, I can figure out the problem. I'm confused by two terms that were already covered: *lithosphere* and *asthenosphere.* I'd better look back at the section on Earth's structure to remind myself what these terms mean. Then I'll reread the beginning of the paragraph to see if I understand the text . . .

"All right, I'm clear on what the text is saying. I'm ready to move on."

When reading a text like this...

> Plants can't move around as most animals do, so plants can't respond to their environments in the same way. However, tropisms help plants get the things they need to survive.
>
> Plant stems as well as leaves show positive phototropism—they grow toward the light. The leaves of plants turn so that they get the largest amount of sunlight possible.
>
> Because of positive gravitropism, the roots of a plant grow down into the soil. . . .
>
> page B15

A good strategy to use is...

USE PRIOR KNOWLEDGE

As you read, think about questions like these:

- What do you already know about the topic?
- What new information are you learning?
- How does the information fit in with what you already know?

What you might say to yourself...

"I've seen plants at my house with their leaves turned toward sunlight. I've also seen plant roots coming from the bottom of pots, so I know they're growing down. I knew that this was what plants did when they grew.

"But I never thought about why plants *had* to do these things—because they can't move around on their own to get the things they need to survive and grow."

When reading a text like this...

> Glaciers also change Earth's surface. When Earth was cooler, a dense blanket of ice spread over much of North America. Remains of that glacier still cover most of Greenland. Today most glaciers form slow-moving rivers of ice.
>
> A glacier can erode the land and deposit rock, sand, and mud. As the ice is pulled downhill by gravity, it drags rocks and pebbles along with it, acting as a sort of sandpaper that grinds away the land beneath. When the ice melts, the material carried by the glacier is deposited to form ridges, hills, and gravel-covered plains.
>
> pages D10–D11

A good strategy to use is...

CREATE MENTAL IMAGES

Think about how an artist might draw a picture to appear on the page with the text you're reading. If a passage is describing a process, try to picture how it would look in the frames of a cartoon. That way, you can tell if you're understanding what you're reading.

What you might say to yourself...

"To get a clear idea of this passage, I'm going to picture each of the things described:

"First, I picture a mountain that is covered on top with a thick layer of ice. I see the ice slowly moving down the mountainside.

"The text says that the ice acts like sandpaper, so I picture the moving ice scraping up rocks, pebbles, and soil as it goes.

"Now I picture the ice melting. Everything the ice collected either clumps together in mounds or spreads out as the ice disappears."

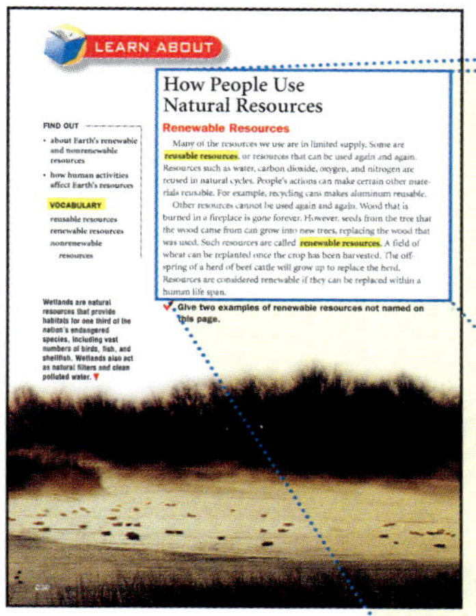

When reading a text like this...

How People Use Natural Resources

Renewable Resources

Many of the resources we use are in limited supply. Some are reusable resources, or resources that can be used again and again. Resources such as water, carbon dioxide, oxygen, and nitrogen are reused in natural cycles. People's actions can make certain other materials reusable. For example, recycling cans makes aluminum reusable.

Other resources cannot be used again and again. Wood that is burned in a fireplace is gone forever. However, seeds from the trees that the wood came from can grow into new trees, replacing the wood that was used. Such resources are called renewable resources. A field of wheat can be replanted once the crop has been harvested. The offspring of a herd of beef cattle will grow up to replace the herd. Resources are considered renewable if they can be replaced within a human life span.

page C30

A good strategy to use is...

SELF-QUESTION

Think about questions you could ask yourself. If you can answer your questions, you are understanding what you're reading. In your science book, one easy way to do this is to turn headings into questions that you answer as you read.

What you might say to yourself...

"I can turn both of these headings into questions. I'll try to answer them as I read. My first question would be, 'How do people use natural resources?' I should probably wait to answer that question until I finish the whole section.

"The question I can answer first, after I read a few paragraphs, comes from the second heading: 'What are renewable resources?' . . . Now that I've read most of the second paragraph, I can answer the question. I know that renewable resources are resources that can't be reused, but they can be replaced—within a human life span. I think that's about 70 years."

When reading a text like this...

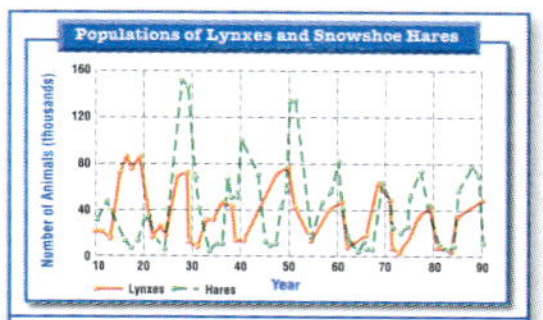

Snowshoe hares are the main food for lynxes. The graph shows the populations of lynxes and hares over time. The population of hares reaches a peak every 9 or 10 years. Then it decreases. The same changes can be seen with the lynx population.

There are several ways to explain the repeating pattern shown in the graph. As the prey population decreases . . .

page C47

A good strategy to use is...

USE GRAPHIC AIDS

Look at any pictures, graphs, charts, diagrams, or time lines that are included with the text. Think about why the author included these graphic aids.

- Are they there to help you organize the information that is presented in the text? If so, you can use them as you read.
- Are they there to add new information? If so, take time to study them carefully, to learn the new information.

What you might say to yourself...

"This graph gives more information than the text includes. As I read the text, I'll refer to the graph, going back and forth.

"The text refers to the graph, so I'll study the graph. I notice that the patterns aren't exact matches. The hare population decreases more than the lynx population. There is a delay between a decrease in the hare population and a decrease in the lynx population.

"Now I can go back to the text, to see why there are changing patterns on the graph."

When reading a text like this...

Why is a DVD better than a CD or videocassette tape?

A DVD can store the data for an entire movie and provide a better image than a videotape can. A DVD also produces sharp images. . . .

A DVD has a longer playing time than a CD because it can hold seven times as much information. . . . Because it can hold much more data, a DVD has many more samples of the music. With more samples, the DVD provides a richer sound . . .

page E131

A good strategy to use is...

SUMMARIZE AND PARAPHRASE

Think about how you could shorten a passage without leaving out any key ideas. Telling the key points of a section in your own words will help you know if you understood it. When you finish a chapter, look for a chapter summary to help you recall the main points. If there is no summary, come up with one of your own.

What you might say to yourself . . .

"When I summarize this section, I want to include all the important points but I want to keep it as brief as possible. Here is my summary:

"A DVD holds much more information than a videotape or CD, so it has a longer playing time, sharper images, and a richer sound."

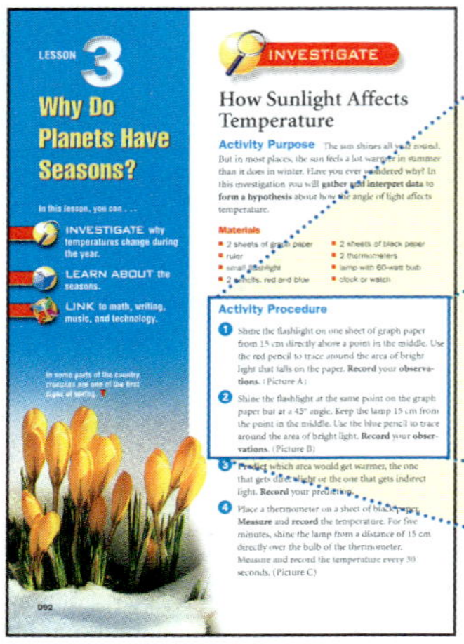

When reading a text like this...

Activity Procedure

1. Shine the flashlight on one sheet of graph paper from 15 cm directly above a point in the middle. Use the red pencil to trace around the area of bright light that falls on the paper. **Record** your **observations**.
2. Shine the flashlight at the same point on the graph paper but at a 45° angle. Keep the lamp 15 cm from the point in the middle. Use the blue pencil to trace around the area of bright light. **Record** your **observations**.

page D92

A good strategy to use is...

ADJUST READING RATE

Think about the kind of science text you are reading. You have to read some kinds of text more slowly and carefully than others.

- Introductions or material you have read before can be read quickly.
- New information or a set of directions should be read more slowly so you can be sure to understand what you read.

What you might say to yourself...

"I see three kinds of text on this page: The first paragraph tells me what the activity is about. The next kind of text lists what I'll need for the investigation. The last kind of text tells me step by step what I need to do.

"I can probably give the description a quick read—just to get an idea of what the activity is about.

"I'll read through the list quickly, but I may look back at it more than once—to make sure I have everything I need.

"I'll have to slow down and read all the directions carefully. I'll probably have to reread and check them a few times. I want to make sure I'm doing the investigation the right way. If I don't, the information I gather might not be useful."

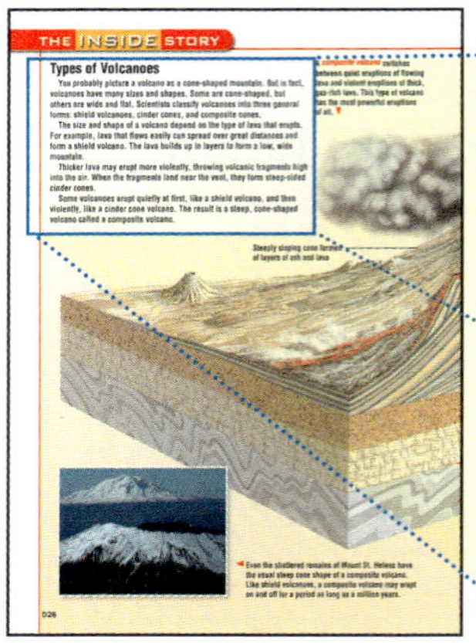

When reading a text like this...

Types of Volcanoes

You probably picture a volcano as a cone-shaped mountain. But in fact, volcanoes have many sizes and shapes. Some are cone-shaped, but others are wide and flat. Scientists classify volcanoes into three general forms: shield volcanoes, cinder cones, and composite cones.

The size and shape of a volcano depend on the type of lava that erupts. For example, lava that flows easily can spread over great distances and form a shield volcano. The lava builds up in layers to form a low, wide mountain.

Thicker lava may erupt more violently, throwing volcanic fragments high into the air. When the fragments land near the vent, they form steep-sided cinder cones.

Some volcanoes erupt quietly at first, like a shield volcano, and then violently, like a cinder cone volcano. The result is a steep, cone-shaped volcano called a composite volcano.

page D26

A good strategy to use is...

USE TEXT STRUCTURE AND FORMAT

Think about how the headings and captions can help you figure out how the author organized information.

- Do the paragraphs tell about causes and effects?
- Do they show the sequence in a process?
- Do they compare two or more things?
- Do they classify concepts?

Recognizing how the information is arranged helps you better understand it.

What you might say to yourself...

"The heading makes me think that this material is organized to show how scientists classify types of volcanoes.

"As I read this section, I see I'm right. The paragraph tells me how scientists classify volcanoes: They use the shape of the volcano to decide whether it is a shield volcano, a cinder cone, or a composite cone. I see that the shape depends on the way the lava flows when the volcano erupts.

"As I read on, I have to remember that the author can switch patterns from section to section—even from paragraph to paragraph."

Building a Science Vocabulary

Reading and understanding scientific terms can take some thinking! Science has its own language—words that many people hardly ever use. Learning how some scientific terms came about will help you remember many of them. Once you understand how the language of science works, you will be better able to understand and use what you read.

Eponyms—Words from People's Names

Eponyms are words that come from people's names. When you reach for a large sheet of warm cloth, think of Thomas Blanket. He was a British weaver who made the first blankets more than seven hundred years ago. When you enjoy the thrills and chills of a ride on a giant wheel at a carnival or fair, think of G. W. Gale Ferris. He was an American engineer who built the first Ferris wheel in 1893.

Many scientific terms are eponyms—words based on scientists' names. These words honor scientists for their work, their breakthroughs, their inventions, or their discoveries. Consider the examples on the next page.

R44

Scientific Eponyms

Scientific Term	Named After	Definition of Term
Richter scale	its inventor, Charles Francis Richter, an expert in the study of earthquakes	A scale that measures the energy released during an earthquake
Linnaean system	its developer, Linnaeus, an expert in the study and classification of plants	A classification system for living things
Punnett square	its developer, Reginald Punnett, a biologist who studied genetics	A diagram used to find the possible combinations of factors in the offspring of two parents
Celsius scale	its inventor, Anders Celsius, a Swedish astronomer	A scale, used to measure temperature, on which water freezes at 0° and boils at 100°
Fahrenheit scale	its inventor, Gabriel Daniel Fahrenheit, a German meteorologist	A scale, used to measure temperature, on which water freezes at 32° and boils at 212°

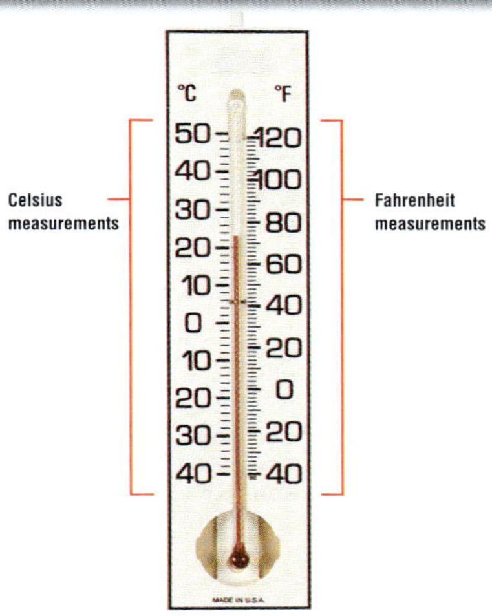

THINK ABOUT IT

As a meteorologist, Fahrenheit studied the weather. Can you see how a way to measure the temperature became connected to his name? What about you? If a scientific term were formed from your name, what would the term be and what would it mean?

R45

Greek and Latin Word Parts

Scientific terms often combine roots—or word parts—to create a word. For example, the word part *thermo* comes from the ancient Greek word *therme*, which means "heat." The word *meter* comes from the ancient Greek word *metron*, which means "to measure." Put those word parts together, and you have the English word *thermometer*, an instrument for measuring thermal energy. Knowing the meanings of Greek and Latin roots will help you to understand the meanings of many scientific terms.

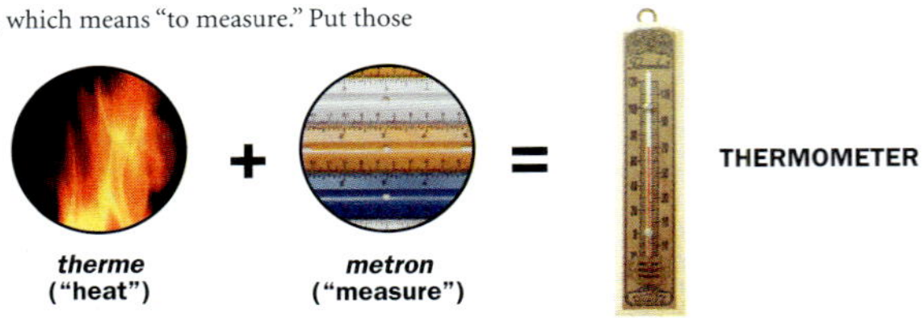

Common Greek and Latin Roots in Scientific Words

Root	Meaning	Science Vocabulary Words	Other English Words
dens (Latin)	compact	condensation, density	condense, condensed book, dense
nuc (Latin)	nut; a hard center	nuclear energy, nuclear membrane, nucleus	nuclear reactor, nuclear fission, nuclear fusion
photo (Greek)	light	phototropism	photograph, photocopy
plast (Greek)	formed	chloroplast	plaster, plastic
radi (Latin)	ray	radiant energy, radiation	radar, radio, radius
trans (Latin)	across	translucent, transparent, transpiration	transatlantic, transfer, transform

THINK ABOUT IT

The Greek root *chloro* means "green." Think about the meaning of *plast*, as shown in the chart. How does knowing about the Greek roots help you to remember the definition of *chloroplast*?

Using Base Words

Many words in the English language are formed by adding word parts to base words. When you come to an unfamiliar scientific term in your reading, look for clues that will help you figure out its meaning.

- First, look for a familiar base word inside the unfamiliar word.
- Then, think about what you already know about the base word and other word parts.

Think about the vocabulary word *axial tilt*. It contains the base word *axis*. Notice that the spelling was changed slightly before the suffix *-al* was added. You know that *axis* means "an imaginary line through Earth's center, from the North Pole to the South Pole." You also know that the word *tilt* means "standing at an angle." Therefore, you could use these two clues to figure out what *axial tilt* means. It means "the angle at which a planet's axis is tilted."

To make sure that your conclusions about the meanings of new words are correct, always check a dictionary or science glossary. Here are more examples of science words containing base words you may know.

Using Base Words

Science Vocabulary Word	Base Word + New Word Part	Clues to Use	Meaning of Science Vocabulary Word
abyssal plain	abyss + al plain	An abyss is a very deep hole. A plain is a large, flat area.	the vast floor of the deep oceans
cementation	cement + ation	Cement is a material made from particles of sand and rock. The suffix *-ation* means "the act of."	the gluing together of particles to form rock
consumer	consume + er	*Consume* means "to eat." The suffix *-er* means "one who."	a living thing that eats other living things
galactic cluster	galaxy + tic cluster	A galaxy is a large system of stars. A cluster is a group.	a group of galaxies

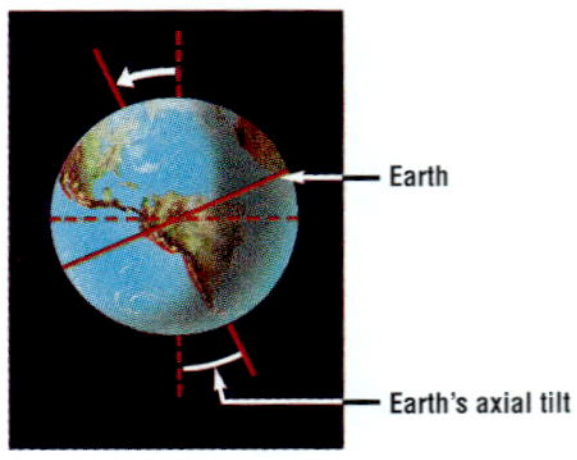

THINK ABOUT IT

Have you ever gone on a scavenger hunt? In this game, you are given a list of items to find. You must use your wits to find the items, sometimes in unusual places. Use what you know about a *scavenger hunt* to figure out what *scavengers* are. Hint: Think about what these living things might eat. Check your answers in a dictionary or a science glossary.

Phonics: Sounds and Spellings in Scientific Terms

Your knowledge of letters and letter sounds helps you to read and pronounce new words when you read stories. Those same skills can help when you come to unfamiliar scientific terms. For example, you can use your knowledge of the sounds of *ph* and *th* to read and pronounce such words as *phase, phototropism, thermometer,* and *thermosphere.*

Think about the rules: Knowing some of the rules of phonics can help you with new words. For example, the letters *c* and *g* can be hard or soft, depending on the letters that follow them. Think of all the science words that these rules will help you pronounce!

Rules for Pronouncing Letters

The Letter *C*	The Letter *G*
Rule: *C* is soft when it is followed by *e, i,* or *y.*	**Rule:** *G* is soft when it is followed by *e* or *y.* Sometimes it is soft when it is followed by *i.*
Sound: Soft, as in *cent*	**Sound:** Soft, as in *gentle*
Science Vocabulary: cell wall, cytoplasm, frequency	**Science Vocabulary:** gene, oxygen, gymnosperm, fungi
Rule: *C* is hard when it is followed by *a, o, u,* or most consonants.	**Rule:** *G* is hard when it is followed by *a, o, u,* or most consonants. Sometimes it is hard when it is followed by *i.*
Sound: Hard, as in *cat*	**Sound:** Hard, as in *goal*
Science Vocabulary: carbon dioxide, fulcrum, comet	**Science Vocabulary:** galaxy, gas, gravity, gill

Think about the possibilities: Some letter combinations can stand for more than one sound.

- The letter pair *ch* can have three different sounds, as in *chair, echo,* and *parachute.* When you come to an unfamiliar word containing *ch,* try out each sound. If you are unsure, check a dictionary.

Which sound does *ch* stand for in each of these vocabulary words: *arachnid, chloroplast, chromosome, echinoderm, machine, niche, Richter scale*? Check your answers in your science glossary.

- The vowel pair *ei* can have the long *a* sound as in *neighbor,* or the long *i* sound as in *height.* Examples of scientific terms and names containing this vowel pair include:

  Long *a* sound: *atomic weight, reindeer, vein*

  Long *i* sound: *meiosis, Albert Einstein*

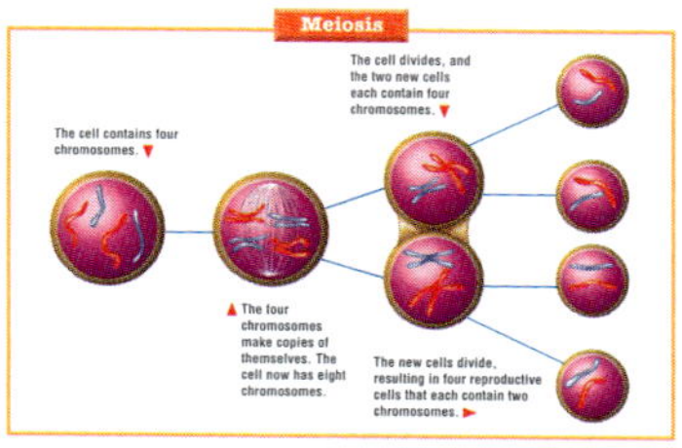

Think about whether a letter stands for any sound at all: Remember that some letters are silent. Sometimes, the consonants *c, k,* and *p* are silent when they come before *n.* Examples of words containing such silent letters include *cnidarian, knowledge,* and *pneumonia.*

cnidarian

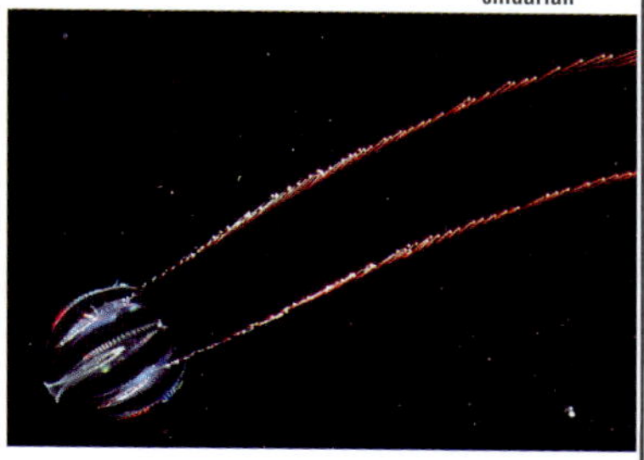

THINK ABOUT IT

Use what you know about letter sounds and the rules of phonics to figure out how to pronounce the following scientific terms.

knee joint
seismology
circuit
dichotomous key
convergent boundary
electromagnet
precipitation
thermal energy

Can you match your pronunciations with any of the phonics rules presented on these pages? Feel free to check a dictionary or science glossary if you're not sure.

Visit the Multimedia Science Glossary to see illustrations of these words and to hear them pronounced.
www.harcourtschool.com/science

Glossary

As you read your science book, you will notice that new or unfamiliar words have been respelled to help you pronounce them quickly while you are reading. Those respellings are *phonetic respellings*. In this Glossary you will see a different kind of respelling. Here, diacritical marks are used, as they are used in dictionaries. *Diacritical respellings* provide a more precise pronunciation of the word.

When you see the ′ mark after a syllable, pronounce that syllable with more force than the other syllables. The page number at the end of the definition tells where to find the word in your book. The boldfaced letters in the examples in the Pronunciation Key that follows show how these letters are pronounced in the respellings after each glossary word.

PRONUNCIATION KEY

a	**a**dd, m**a**p	m	**m**ove, see**m**	u	**u**p, d**o**ne
ā	**a**ce, r**a**te	n	**n**ice, ti**n**	û(r)	b**ur**n, t**er**m
â(r)	c**a**re, **a**ir	ng	ri**ng**, so**ng**	yōō	f**u**se, f**ew**
ä	p**a**lm, f**a**ther	o	**o**dd, h**o**t	v	**v**ain, e**v**e
b	**b**at, ru**b**	ō	**o**pen, s**o**	w	**w**in, a**w**ay
ch	**ch**eck, ca**tch**	ô	**o**rder, j**aw**	y	**y**et, **y**earn
d	**d**og, ro**d**	oi	**oi**l, b**oy**	z	**z**est, mu**s**e
e	**e**nd, p**e**t	ou	p**ou**t, n**ow**	zh	vi**s**ion, plea**s**ure
ē	**e**qual, tr**ee**	ŏŏ	t**oo**k, f**u**ll	ə	the schwa, an
f	**f**it, hal**f**	ōō	p**oo**l, f**oo**d		unstressed vowel
g	**g**o, lo**g**	p	**p**it, sto**p**		representing the
h	**h**ope, **h**ate	r	**r**un, poo**r**		sound spelled
i	**i**t, g**i**ve	s	**s**ee, pa**ss**		*a* in **a***bove*
ī	**i**ce, wr**i**te	sh	**s**ure, ru**sh**		*e* in *sick***e***n*
j	**j**oy, le**dg**e	t	**t**alk, si**t**		*i* in *poss***i***ble*
k	**c**ool, ta**k**e	th	**th**in, bo**th**		*o* in *mel***o***n*
l	**l**ook, ru**l**e	t͟h	**th**is, ba**th**e		*u* in *circ***u***s*

Other symbols:

- • separates words into syllables
- ′ indicates heavier stress on a syllable
- ′ indicates light stress on a syllable

A

abiotic [ā′bī•ot′ik] Relating to the nonliving parts of the environment **(C6)**

abyssal plain [ə•bis′əl plān′] The vast floor of the deep oceans **(C71)**

acid [as′id] A compound that reacts easily with other substances and turns blue litmus paper red **(E32)**

adaptation [ad′əp•tā′shən] A structure or feature of an organism that helps it meet a particular need in its natural habitat **(A45)**

air mass [âr′mas′] A large body of air that has the same characteristics throughout **(C96)**

air pressure [âr′ presh′ər] The weight of air pressing down on an area **(C100)**

amphibians [am•fib′ē•ənz] Vertebrates that can live in water or on land but must return to the water to reproduce **(B104)**

amplitude [am′plə•tōōd′] The distance that any point on a wave has moved from its resting position; equal to half the wave height **(E109)**

angiosperm [an′jē•ō•spûrm′] A vascular plant that produces flowers **(B44)**

Animalia [an′i•māl′yə] The name of the kingdom that is made up of animals **(A44)**

anther [an′thər] The part of the stamen of a flower that produces pollen grains **(B46)**

arachnid [ə•rak′nid] An arthropod with eight legs and no antennae, such as a spider, scorpion, tick, or mite **(B89)**

arthropod [är′thrə•pod′] An animal that has a jointed exoskeleton and jointed limbs **(B88)**

asexual reproduction [ā•sek′shōō•əl rē′prə•duk′shən] The form of reproduction in which a new organism is produced without the joining of a sperm cell and an egg cell **(B29)**

asteroid [as′tə•roid] A chunk of rock and metal, smaller than a planet and larger than a meteoroid, that orbits the sun **(D80)**

asthenosphere [as•then′ə•sfir′] A zone of partially melted rock in the upper mantle of the Earth **(D7)**

astronomical unit (AU) [as′trə•nom′i•kəl yōō′nit] A measurement equal to the distance between Earth and the sun, a distance of about 150 million km (93 million miles) **(D82)**

atoll [at′ôl] A ring of islands around a shallow central lagoon **(C75)**

atom [at′əm] The smallest unit of a pure substance that still has the properties of that substance **(E6)**

atomic number [ə•tom′ik num′bər] The number of protons in an atom **(E8)**

average speed [av′rij spēd′] The measure of the total distance an object has moved divided by the total time elapsed **(F15)**

axial tilt [ak′sē•əl tilt′] The angle that a planet's axis is tilted from vertical **(D95)**

axis [ak′sis] An imaginary line through the Earth's center, from the North Pole to the South Pole **(D86)**

B

balanced forces [bal′ənst fôrs′əz] The opposite forces acting on an object that cancel each other out **(F26)**

base [bās] A compound that reacts easily with other substances and turns red litmus paper blue **(E32)**

biome [bī′ōm] A region of the world that is defined by its climate and the unique plants and animals that live there **(C14)**

biotic [bī•ot′ik] Relating to the living parts of the environment **(C6)**

bird [bûrd] A warm-blooded animal that has feathers and lays eggs with hard shells **(B116)**

buoyancy [boi′ən•sē] The ability to float in a liquid or rise in a gas **(E47)**

C

carbon cycle [kär′bən sī′kəl] The flow of carbon dioxide and oxygen through Earth's ecosystems **(C24)**

cell membrane [sel′ mem′brān′] A thin layer that surrounds the cell and keeps it together, and controls the substances passing into and out of the cell **(A8)**

cell wall [sel′ wôl′] The stiff outer layer that protects a plant cell and gives it shape **(A8)**

cementation [sē′men•tā′shən] The gluing together of particles of sediment **(D49)**

chemical energy [kem′i•kəl en′ər•jē] A form of potential energy stored in reactants **(E92)**

chemical property [kem′i•kəl prop′ər•tē] A property that describes the ability of a substance to react with other materials and form new substances **(E53)**

chemical rock [kem′i•kəl rok′] One of a group of sedimentary rocks that is made from minerals that were dissolved in water, came out of solution, and were then deposited **(D49)**

chloroplast [klôr′ə•plast′] A kind of organelle in plant cells that contains chlorophyll, which enables the plant to make its own food **(A9)**

chromosome [krō′mə•sōm′] A structure in the cell that provides instructions for all the activities and traits of the cell and the organism **(A9)**

circuit [sûr′kit] The connection needed for current to travel between the poles of two electric energy sources **(E86)**

classification [klas′ə•fi•kā′shən] The arrangement of things in groups of similar items **(A42)**

clastic rock [klas′tik rok′] One of a group of sedimentary rocks that is made of sediments that were weathered, transported, and deposited in layers **(D49)**

cnidarian [nī•dar′ē•ən] An animal with tentacles that have the ability to sting its prey or predators **(B78)**

cold-blooded [kōld′blud′id] Relating to an animal whose body temperature changes as the temperature of the surroundings changes **(B104)**

colloid [kol′oid] A suspension that isn't easily separated, because of the small size of the particles **(E64)**

combustibility [kəm•bus′tə•bil′ə•tē] The ability to burn **(E53)**

comet [kom′it] A ball of ice, rock, and frozen gases that orbits the sun **(D81)**

commensalism [kə•men′səl•iz′əm] A symbiotic relationship that benefits one organism and doesn't harm or help the other organism **(C58)**

community [kə•myōō′nə•tē] The populations living in the same environment at the same time **(C7)**

compound [kom′pound] A substance formed of the atoms of two or more elements **(E30)**

compound machine [kom′pound mə•shēn′] A device that is made up of two or more simple machines **(F58)**

condensation [kon′dən•sā′shən] The process by which water vapor changes from a gas to a liquid **(C23)**

conduction [kən•duk′shən] The transfer of thermal energy that results from the collision of particles **(E79)**

conifer [kon′ə•fər] A plant that grows cones that produce seeds **(B36)**

consumer [kən•sōōm′ər] An organism that eats other organisms **(C46)**

continental shelf [kon′tə•nen′təl shelf′] The gently sloping part of a continent that is under water and rings each continent **(C71)**

continental slope [kon′tə•nen′təl slōp′] The edge of a continent that drops steeply down to the deep-ocean floor **(C71)**

convection [kən•vek′shən] The transfer of thermal energy through liquids and gases **(E79)**

convergent boundary [kən•vûr′jənt boun′də•rē] A boundary where crustal plates collide with each other **(D14)**

coral reef [kôr′əl rēf′] A structure built by living creatures **(C81)**

core [kôr] The Earth's innermost layer **(D6)**

crust [krust] The Earth's outermost layer **(D6)**

cytoplasm [sī′tə•plaz′əm] A watery, jellylike substance that fills a cell and contains the other cell parts **(A8)**

D

deciduous forest [di•si′jə•wəs fôr′əst] A biome in which the dominant plants are broad-leaved trees that shed their leaves each fall **(C16)**

decomposer [dē′kəm•pōz′ər] A type of consumer that gets its food by breaking down animal wastes and remains of dead plants and animals **(C46)**

density [den′sə•tē] The amount of mass in a given volume **(E47)**

deposition [dep′ə•zish′ən] The dropping of sediment from wind or water that slows down, or from ice that melts **(D49)**

dichotomous key [dī•kot′ə•məs kē′] A guide used to identify an organism based on its characteristics **(A56)**

dicotyledon [dī′kot•ə•lēd′ən] A seed that has two seed leaves **(B52)**

divergent boundary [di•vûr′jənt boun′də•rē] A boundary where crustal plates move away from each other **(D14)**

DNA [dē′en•ā′] A complex chemical that contains information about every part of an organism **(A10)**

dominant [dom′ə•nənt] Relating to a factor that will determine whether a trait is shown **(A26)**

ductility [duk•til′ə•tē] The ability of a substance to be pulled into thin threads without breaking **(E47)**

E

earthquake [ûrth′kwāk′] The sudden movement of rock along a fault which releases energy vibrations **(D22)**

echinoderm [ē•kī′nə•dûrm′] An invertebrate that has an internal skeleton and spines that are part of its skin **(B91)**

ecosystem [ek′ō•sis′təm] A community and the abiotic parts of its environment **(C7)**

efficiency [i•fish′ən•sē] The ratio of the work that comes out of a machine to the work put into that machine **(F62)**

effort arm [ef′ərt ärm′] The part of a lever on which a person or another machine applies a force **(F42)**

electric energy [i•lek′trik en′ər•jē] The movement of electrons from one place to another; current electricity **(E86)**

electricity [i•lek′tris′i•tē] The interaction of electric charges—usually classified as static electricity or current electricity **(E86)**

electromagnet [i•lek′trō•mag′nit] A device made when electric current travels through a wire coiled around a piece of iron **(E87)**

electromagnetic waves [i•lek′trō•mag•net′ik wāvz] Waves of vibrating electric and magnetic fields **(E124)**

electron [i•lek′tron′] A negatively charged subatomic particle **(E7)**

element [el′ə•mənt] A pure substance that is made up of only one kind of atom **(E14)**

ellipse [i•lips′] An oval-shaped path **(D87)**

embryo [em′brē•ō] The tiny plant that forms from the fertilized egg **(B53)**

endoskeleton [en′dō•skel′ə•tən] A system of connected bones or plates that lie within the body of an organism **(B70)**

energy [en′ər•jē] The ability to cause changes in matter; the ability to do work **(E76)**

environment [in•vī′rən•mənt] The surroundings in which an organism lives, which provide the specific things an organism needs **(C6)**

epicenter [ep′ə•sent′ər] The point on the Earth's surface directly above the focus of an earthquake **(D22)**

erosion [i•rō′zhən] The movement or transportation of sediment to new locations **(D49)**

estuary [es′chōō•er′ē] A place where a freshwater environment and a saltwater environment meet **(C82)**

evaporation [i•vap′ə•rā′shən] The change of water from liquid water to water vapor **(C22)**

exoskeleton [ek′sō•skel′ə•tən] A skeleton that is outside of the body of an invertebrate **(B71)**

F

fern [fûrn] A vascular plant that has feathery fronds and reproduces without seeds **(B34)**

fertilization [fûr′təl•ə•zā′shən] The process in which a sperm cell joins with an egg cell to produce a new organism **(B53)**

first quarter [fûrst′ kwôr′tər] The waxing phase of the moon viewed from Earth where half the moon's lit side is visible **(D102)**

fish [fish] A vertebrate adapted to living its entire life in the water **(B104)**

fission [fish′ən] The reaction that releases nuclear energy when a nucleus breaks into two parts **(E95)**

flatworm [flat′wûrm′] An animal, such as a planarian, that has a flattened body, a digestive system with only one opening, and a simple nervous system **(B80)**

focus [fō′kəs] The point in the Earth's crust where the first major movement of an earthquake fault occurs **(D22)**

force [fôrs] A push or a pull **(F6)**

forecast [fôr′kast′] A prediction of what the weather will be like in the future **(C108)**

frequency [frē′kwən•sē] The number of vibrations a wave has in a given time **(E111)**

friction [frik′shən] A force that opposes motion whenever two surfaces rub against each other **(F25)**

front [frunt] The border between two air masses that collide **(C102)**

fulcrum [fōōl′krəm] The fixed point on a lever **(F42)**

Fungi [fun′jī′] The name of the kingdom of plant-like organisms that do not make their own food but take the nutrients that they need from the environment **(A44)**

fusion [fyōō′zhən] A process in which nuclei of atoms are joined **(D117)**

Multimedia Science Glossary: www.harcourtschool.com/science

G

galactic cluster [gə•lak′tik klus′tər] A grouping of galaxies **(D124)**

galaxy [gal′ək•sē] A large system of stars **(D124)**

gas [gas] A state of matter that has no definite shape or volume **(E23)**

gene [jēn] A section of DNA that controls specific cell activities and the characteristics of every organism **(A15)**

genus [jē′nəs] The first part of an organism's scientific name; a group of organisms that share major characteristics and are therefore closely related **(A53)**

grafting [graft′ing] A form of asexual reproduction in which parts of two plants are joined together to make a single plant **(B55)**

grassland [gras′land] A biome where the dominant plants are grasses **(C17)**

gravitropism [grav′i•trō′piz•əm] A tropism in which the roots of a plant grow downward, in the direction of gravity **(B14)**

gravity [grav′ə•tē] The force that pulls everything toward the center of Earth; the force of gravitation on Earth **(F7)**

groundwater [ground′wô′tər] Water that soaks into the ground **(C23)**

gymnosperm [jim′nə•spûrm′] A seed-bearing plant whose seeds are not surrounded by a container **(B36)**

H

habitat [hab′ə•tat′] The part of an environment in which an organism lives **(C7)**

heat [hēt] The transfer of thermal energy **(E78)**

hurricane [hûr′ə•kān′] A large, spiraling storm system that can be as much as 600 km across **(C118)**

I

igneous rock [ig′nē•əs rok′] A rock formed when magma or lava cools and hardens **(D41)**

inclined plane [in′klīnd plān′] A flat, sloping surface **(F50)**

insect [in′sekt] An arthropod with six jointed legs, such as a grasshopper **(B89)**

intertidal zone [in′tər•tīd′əl zōn′] The area between the high-tide mark and the low-tide mark **(C80)**

invertebrate [in•vûr′tə•brit] An animal that does not have a backbone **(B70)**

K

kinetic energy [ki•net′ik en′ər•jē] The energy of motion **(E76)**

L

lava [lä′və] Magma that reaches the Earth's surface **(D41)**

law of conservation of energy [lô′ uv kon′ser•vā′shən uv en′ər•jē] A law that states that the total amount of energy in a system is always the same, and that energy cannot be created or destroyed **(E76)**

lever [lev′ər] A simple machine having an arm that moves about a fixed point **(F42)**

Linnaean system [lə•nē′ən sis′təm] A classification system useful for storing and finding information about living things **(A43)**

liquid [lik′wid] A state of matter that has a definite volume but no definite shape **(E23)**

lithosphere [lith′ə•sfir′] The cool, solid portion of Earth that includes all of the crust and part of the upper mantle **(D7)**

long-day plant [lông′dā′ plant′] A plant that needs short periods of darkness before it can bloom **(B16)**

longitudinal wave [lon′jə•tōōd′ən•əl wāv′] A wave in which the particles vibrate back and forth in the same direction as the wave is moving **(E110)**

loudness [loud′nis] The perception of the amplitude of a sound wave **(E119)**

M

machine [mə•shēn′] A device that changes the amount of a force applied to an object, the direction of a force, or both **(F41)**

magma [mag′mə] Melted rock within the Earth **(D41)**

magnet [mag′nit] An object which has all the domains pointing in the same direction and magnetic fields that work together **(E84)**

main sequence [mān′ sē′kwəns] The band on the Hertzsprung-Russell diagram where ninety percent of stars are found **(D117)**

malleability [mal′ē•ə•bil′ə•tē] The property of a metal that indicates that it can be bent and rolled into sheets **(E15)**

mammal [mam′əl] A warm-blooded vertebrate that has milk-producing glands **(B118)**

mammary gland [mam′ə•rē gland′] A milk-producing gland on a female mammal used for nursing young **(B118)**

mantle [man′təl] The layer directly beneath the Earth's crust **(D6)**

meiosis [mī•ō′sis] The process that forms reproductive cells **(A17)**

mesosphere [mes′ō•sfir′] The layer of atmosphere where air is thin and most meteors burn up **(C95)**

metal [met′əl] An element that conducts heat and electricity well **(E14)**

metamorphism [met′ə•môr′fiz′əm] The process by which any kind of rock is changed into metamorphic rock **(D56)**

meteoroid [mēt′ē•ə•roid′] A chunk of material smaller than an asteroid that orbits the sun **(D81)**

mid-ocean ridge [mid′ō′shən rij′] The place where plates of the Earth's crust along the ocean floor are being split apart and molten rock pushes up to form new ocean floor and a mountain range **(C72, D15)**

mineral [min′ər•əl] A natural, solid substance that has a definite chemical composition and physical structure **(D40)**

mitochondria [mīt′ə•kon′drē•ə] Bean-shaped organelles that break down sugar to produce energy **(A8)**

mitosis [mī•tō′sis] The process of cell division that produces new body cells with complete sets of chromosomes **(A16)**

mixture [miks′chər] A combination of two or more pure substances that are not chemically combined with each other **(E60)**

mollusk [mol′əsk] An animal with a soft body and no bones **(B86)**

Monera [mə•nir′ə] The kingdom that includes bacteria **(A44)**

monocotyledon [mon′ə•kot′ə•lēd′ən] A seed that has only one seed leaf **(B52)**

moss [môs] A very short green plant that does not have true leaves, stems, and roots **(B28)**

mutualism [myōō′chōō•əl•iz′əm] The symbiotic relationship that benefits both organisms involved **(C56)**

N

nebula [neb′yə•lə] A huge, cool, dark cloud of gases and dust where stars are formed **(D118)**

neutron [nōō′tron] A subatomic particle that has the same mass as a proton but no electric charge **(E7)**

newton [nōōt′ən] The metric unit for forces including weight **(F10)**

niche [nich] The role of an organism in an ecosystem **(C9)**

nitrogen cycle [nī′trə•jən sī′kəl] The movement of nitrogen through ecosystems **(C26)**

nonmetal [non′met′əl] An element that does not conduct electricity and is not shiny or malleable **(E15)**

nonrenewable resources [non′ri•nōō′ə•bəl rē′sôrs′ez] Resources that cannot be replaced within a human life span **(C31)**

nonvascular plant [non•vas′kyə•lər plant′] A plant that does not have tissues to transport food and water **(B10)**

nova [nō′və] A star that briefly becomes thousands of times brighter than normal **(D119)**

nuclear energy [nōō′klē•ər en′ər•jē] The energy that holds protons and neutrons together in an atom's nucleus **(E95)**

nuclear membrane [nōō′klē•ər mem′brān′] A membrane that surrounds the nucleus of a cell and controls the materials passing into and out of it **(A9)**

nucleus [nōō′klē•əs] 1 *(cell)* The control center of the cell, which directs all the cell's activities. 2 *(atom)* The center of an atom, which contains protons and neutrons. **(A9, E7)**

O

opaque [ō•pāk′] Not allowing light to pass through **(E128)**

orbit [ôr′bit] The path an object follows as it revolves around another object **(D87)**

ovary [ō′və•rē] The base of the pistil of a flower, containing ovules **(B47)**

P

parasite [par′ə•sīt′] An organism that survives by living on or in another animal and feeding on that animal **(B81)**

parasitism [par′ə•sīt′iz•əm] A symbiotic relationship in which one organism benefits and the other organism is harmed **(C55)**

periodic table [pir′ē•od′ik tā′bəl] A table of elements arranged according to their atomic number **(E17)**

petal [pet′əl] The part of the plant that is usually the largest and most visible part of a flower and helps protect the other parts of the flower **(B46)**

phase [fāz] A stage in the changing shape that the moon seems to have when it is viewed from Earth **(D102)**

phloem [flō′əm] The vessels in a vascular plant that carry food from the leaves to the other parts of the plant for use and storage **(B9)**

phototropism [fō′tō•trō′piz•əm] The movement of a plant toward or away from light **(B14)**

physical properties [fiz′i•kəl prop′ər•tēz] Properties that can be observed or measured without changing an object, or any part of the materials it is made of, into something else **(E46)**

pistil [pis′təl] The female reproductive part of a flower **(B46)**

pitch [pich] The level of a sound's perceived highness or lowness **(E118)**

planetary system [plan′i•ter′ē sis′təm] A system of planets revolving around a star **(D79)**

Plantae [plan′tē] The name of the kingdom that is made up of plants **(A44)**

plate tectonics [plāt′ tek•ton′iks] The theory scientists use to explain the movements of plates on the Earth's surface **(D8)**

pollen [pol′ən] The powderlike male spores that develop into two male sex cells **(B37)**

pollination [pol′ə•nā′shən] The process by which pollen from the anthers of a flower lands on the stigma of a flower **(B47)**

population [pop•yə•lā′shən] Organisms of the same species living together in the same environment **(C7)**

potential energy [pō•ten′shəl en′ər•jē] The energy that exists due to the position or condition of an object **(E76)**

precipitation [pri•sip′ə•tā′shən] The process by which water returns to Earth's surface, usually as rain or snow **(C23)**

predator [pred′ə•tər] An animal that feeds on other living animals **(C47)**

prey [prā] The animals predators eat **(C47)**

producer [prə•do͞os′ər] An organism that makes its own food **(C46)**

Protista [prə•tist′ə] The kingdom of classification in which members often have traits of both plants and animals **(A44)**

proton [prō′ton] A subatomic particle that has a positive electric charge **(E7)**

pulley [po͝ol′ē] A wheel that has a groove along its edge **(F45)**

Punnett square [pun′ət skwer′] A checkerboard-type diagram used to find the possible combinations of factors in the offspring of two parents **(A29)**

P wave [pē′ wāv′] A primary wave or "push-pull" wave of released energy in the Earth's surface that causes a back-and-forth vibration in the same direction that the wave moves **(D23)**

R

radiant energy [rā′dē•ənt en′ər•jē] The kind of energy carried by electromagnetic waves **(E124)**

radiation [rā′dē•ā′shən] The transfer of thermal energy as waves **(E79)**

reactivity [rē′ak•tiv′ə•tē] The ability of a substance to go through a chemical change **(E53)**

recessive [ri•ses′iv] Relating to a factor not shown when one of a pair is dominant **(A27)**

reflecting telescope [ri•flekt′ing tel′ə•skōp′] A telescope that uses mirrors to gather light and magnify images **(D130)**

reflection [ri•flek′shən] The bouncing of light waves off a surface **(E126)**

refracting telescope [ri•frakt′ing tel′ə•skōp′] A telescope that uses glass lenses to gather light and magnify objects **(D130)**

refraction [ri•frak′shən] The bending of a light wave when it passes from one medium to another **(E127)**

relative humidity [rel′ə•tiv hyo͞o•mid′ə•tē] A comparison of the actual amount of moisture in the air to the greatest possible amount that could be in the air at the same temperature and pressure **(C101)**

renewable resources [ri•no͞o′ə•bəl rē′sôrs•əz] Resources that can be replaced within a human life span **(C30)**

reptile [rep′til] A cold-blooded animal that has a dry, protective covering of horny scales or plates, and lays eggs with leathery shells **(B115)**

resistance arm [ri•zis′təns ärm′] The part of a lever that applies a force on an object **(F42)**

reusable resources [rē•yo͞o′zə•bəl rē′sôrs•əz] Resources that can be used again and again **(C30)**

revolution [rev′ə•lo͞o′shən] The movement of one object in an orbit around another object **(D87)**

Richter scale [rik′tər skāl′] A scale that measures the amount of energy released during an earthquake **(D24)**

rift [rift] A deep ocean valley formed where two crustal plates move apart **(D15)**

rock cycle [rok′ sī′kəl] A process by which rocks are formed from one another **(D64)**

rotation [rō•tā′shən] The turning of an object on an axis **(D86)**

roundworm [round′wûrm′] An animal, with a round tubelike body, that has a digestive system with two openings **(B81)**

runner [run′ər] A stem that grows close to the ground and puts out roots and shoots **(B55)**

S

scavenger [skav′in•jər] An animal that eats the remains of animals that have died **(C47)**

screw [skro͞o] An inclined plane wrapped around a cylinder or cone **(F53)**

sea-floor spreading [sē′flôr′ spred′ing] A process in which magma is slowly pushed up through cracks in a rift and then cools to form new sea floor **(D15)**

seamount [sē′mount′] A steep-sided volcanic mountain under the ocean **(C72)**

season [sē′zən] A period of the year with a certain level of temperature and type of weather **(D94)**

sedimentation [sed′ə•men•tā′shən] The process of building up layers of sediment over millions of years **(D49)**

segmented worm [seg′mənt•id wûrm′] An animal, such as the earthworm, whose body is made up of connected sections, or segments **(B82)**

sexual reproduction [sek′sho͞o•əl rē′prə•duk′shən] The form of reproduction by the joining of a male reproductive cell and a female reproductive cell **(A24)**

short-day plant [shôrt′dā′ plant′] A plant that needs long periods of darkness before it can bloom **(B16)**

solid [sol′id] A state of matter that has definite shape and volume **(E23)**

solution [sə•lo͞o′shən] A mixture in which the particles of each substance are mixed evenly **(E62)**

species [spē′shēz] The second part of an organism's scientific name that identifies one specific organism in the genus **(A53)**

speed [spēd] A measure of how far something moves during a period of time **(F15)**

sponge [spunj] A type of animal that filters the water it lives in to get food **(B76)**

spore [spôr] A structure that contains cells that can grow into a new plant without joining with other cells **(B29)**

stability [stə•bil′ə•tē] The condition where a substance does not go through chemical changes easily **(E53)**

stamen [stā′mən] The male reproductive part of a flower **(B46)**

station model [stā′shən mod′əl] An arrangement of symbols and numbers that show the weather conditions recorded at a weather station **(C108)**

stigma [stig′mə] The tip of the pistil of a flower where pollen grains land and stick **(B46)**

stratosphere [strat′ə•sfir′] The layer of the atmosphere in which ozone is present **(C95)**

style [stīl] The stemlike part of the pistil of a flower that connects the stigma to the ovary **(B47)**

supernova [sōō′pər•nō′və] A supergiant star that explodes **(D119)**

surface map [sûr′fis map′] A map that includes station models and information about fronts and about centers of high pressure and low pressure **(C108)**

surface wave [sûr′fis wāv′] An energy wave from an earthquake that travels only at the surface and moves less quickly than P waves and S waves but makes the ground roll and sway **(D23)**

suspension [sə•spen′shən] A mixture that contains particles that are large enough to be seen and can be separated from the mixture by using filter paper **(E64)**

S wave [es′ wāv′] A slower kind of energy wave released by an earthquake that causes vibrations at right angles to the wave's direction of travel **(D23)**

symbiosis [sim′bī•ō′sis] A relationship between two organisms of different species that benefits one or both of the organisms **(C54)**

T

taiga [tī′gə] A biome where winters are very cold and long and the dominant plants are conifers **(C18)**

texture [teks′chər] An observable physical property; how the surface of a substance feels **(E47)**

thermal energy [thûr′məl en′ər•jē] The kinetic energy of moving particles of a substance or object **(E78)**

thermosphere [thûr′mō•sfir′] The upper layer of the atmosphere where temperatures are extremely high **(C95)**

third quarter [thûrd kwôr′tər] The waning phase of the moon viewed from Earth where half of the moon's lit side is visible **(D103)**

thunderstorm [thun′dər•stôrm′] A very strong storm with a lot of rain, thunder, and lightning **(C116)**

tornado [tôr•nā′dō] An intense windstorm that often forms within a severe thunderstorm **(C120)**

transform fault boundary [trans′fôrm′ fôlt′ boun′də•rē] A boundary where crustal plates grind past each other **(D14)**

translucent [trans•lōō′sənt] Relating to a substance that allows light to pass through it but scatters the light **(E129)**

transparent [trans•par′ənt] Relating to a medium or substance that light can travel through **(E128)**

transpiration [tran′spə•rā′shən] The process by which plants lose water to the air **(C23)**

transverse wave [trans′vûrs wāv′] A wave in which the particles vibrate at right angles to the direction of the wave **(E110)**

trench [trench] A place where two plates of the Earth's crust hit each other, forming a deep part of the ocean **(C72)**

tropical storm [trop′i•kəl stôrm] The stage of hurricane development when the winds of a tropical depression reach a constant speed of 63 km/hr (39 mph) **(C118)**

tropism [trō′piz′əm] The response of a plant to turn toward or away from something in its environment **(B14)**

troposphere [trō′pə•sfir′] The bottom layer of the Earth's atmosphere **(C95)**

U

unbalanced forces [un•bal′ənst fôrs′əz] The forces acting on an object that cause a change in velocity, making the object speed up, slow down, or change direction **(F27)**

V

vacuole [vak′yōō•ōl′] A storage space in a cell enclosed by a membrane **(A9)**

vascular plant [vas′kyə•lər plant′] A plant that has tissues that carry water and dissolved materials **(B8)**

velocity [və•los′ə•tē] The speed of an object in a particular direction **(F16)**

vertebrate [vûr′tə•brit] An animal that has a backbone **(B70)**

volcano [vol•kā′nō] A mountain that may form around an opening in the Earth's surface where an eruption of molten rock occurs **(D25)**

W

warm-blooded [wôrm•blud′id] Relating to an animal whose body stays the same temperature regardless of the temperature of its surroundings **(B116)**

water cycle [wôt′ər sī′kəl] The movement of water through Earth's ecosystems **(C23)**

wave [wāv] A disturbance that travels through matter or space, carrying energy from one place to another **(E108)**

wavelength [wāv′length′] The measurement of the distance between the crests and troughs in a series of waves **(E109)**

weather balloon [weth′ər bə•lōōn′] A balloon released into the atmosphere that carries a package of instruments that records data about temperature, air pressure, and humidity **(C109)**

weathering [weth′ər•ing] The process of breaking rock into smaller pieces **(D48)**

weather map [weth′ər map′] A map that shows data about recent weather conditions across a large area **(C110)**

wedge [wej] A simple machine composed of one or two inclined planes **(F52)**

wheel and axle [hwēl′ and ak′səl] A simple machine that has a central pivot around which a wheel moves **(F46)**

work [wûrk] The result when a force causes an object to move **(F40)**

X

xylem [zī′ləm] The vessels in a vascular plant that carry water and nutrients from the roots to the leaves **(B8)**

Teacher Notes and Resources

Bibliography

of Science Trade Books

Adventures with Atoms and Molecules: Chemistry Experiments for Young People, Book IV by Robert C. Mebene and Thomas R. Rybolt (Enslow, 1992) includes 30 experiments designed to clarify atomic structure, along with the explanation for the outcome of each.

American Environmental Heroes by Phyllis M. Stanley (Enslow, 1996) presents biographies of 10 American environmentalists, including Thoreau, Carson, and Carver. *NSTA Trade Book*

Ancient Machines by Michael Woods and Mary B. Woods (Runestone Press, 2000) presents ancient uses of machines such as levers, wheels, and pulleys.

Antoni Van Leeuwenhoek: First to See Microscopic Life by Lisa Yount (Enslow, 1994) tells of the scientist's many discoveries examining microscopic life. *NSTA Trade Book*

The Big Rock by Bruce Hiscock (Atheneum, 1988) tells the story of a rock's history and describes the formation of mountains and seas. *NSTA Trade Book*

Black Holes by Heather Couper and Nigel Henbest (Dorling Kindersley, 1996) discusses how we know about black holes, how they affect matter around them, and what would happen if you got inside one. *NSTA Trade Book*

Caves by Stephen Kramer (Carolrhoda, 1995) describes how caves form and the creatures and strangely shaped rocks that can be found there. *NSTA Trade Book*

Cell Wars by Frank Balkwill (Carolrhoda, 1993) explains how cells fight off diseases and viruses in the human body.

The Cell Works by Patrick A. Bauerle and Norbert Landa (Barron's, 1997) explores a cell's nucleus and the double-helix structure of DNA, how new cells are made, and how they form organisms based on their gene patterns.

Chemistry **(Eyewitness Science)** by Ann Newmark (Dorling Kindersley, 1993) explores the subject of chemical reactions and shows the role that chemistry plays in our world. *NSTA Trade Book*

Do You Know the Difference? by Andrea and Michael Bischhoff-Miersch (North-South, 1995) pictures and discusses the differences in 12 pairs of look-alike mammals, such as the leopard and the jaguar. *NSTA Trade Book*

"The Dog of Pompeii" from ***The Donkey of God*** by Louis Untermeyer (iuniverse.com, 1999) is a story of friendship between a homeless boy and a dog in the city of ancient Pompeii.

Down to a Sunless Sea: The Strange World of Hydrothermal Vents by Kate Madin (Steck-Vaughn, 1999) recounts man's early encounters with this unique ecosystem and examines the organisms that thrive in these warmer, "smelly" waters on the ocean floor.

Down to Earth by Michael J. Rosen (Harcourt, 1998) features recipes and activities related to plants and gardening from several authors and illustrators, including Lynne Cherry and Lois Ehlert.

Earth Alive! by Sandra Markel (Lothrop, Lee & Shepard 1991) discusses changes in Earth, including how rocks erode, geysers erupt, and sinkholes swallow the land.

Earthquakes by Sally M. Walker (Carolrhoda, 1996) explains what earthquakes are, how they occur, and their effects. *NSTA Trade Book*

Ecology **(Eyewitness Science)** by Steve Pollock (Dorling Kindersley, 1993) examines the interrelationships between plants and animals and the energy and materials from the environment they all need to survive.

Ecosystems & Environment by Ann Fullick (Heinemann Library, 2000) discusses ecosystems and the environment, including habitats, food chains, adaptation, and genetic engineering.

Electricity **(Science Projects)** by Simon de Pinna (Steck-Vaughn, 1998) provides answers to many intriguing questions about electricity and includes several simple experiments that can be done both at home and at school.

Energy by Jane Featherstone (Steck-Vaughn, 1999) discusses power generated by sun, wind, and water, and ways energy can be conserved.

An Extraordinary Life: The Story of a Monarch Butterfly by Laurence Pringle (Orchard, 1997) chronicles the life cycle, feeding habits, migration, predators, and mating of a monarch butterfly named *Danaus*. *NSTA Trade Book*

Forces and Motion by Simon de Pinna (Steck-Vaughn, 1998) examines forces, such as magnetic, electrostatic, and gravitational, and various aspects of motion, followed by simple experiments that apply these laws of physics.

Hawk Highway in the Sky: Watching Raptor Migration by Caroline Arnold (Harcourt, 1997) presents a close-up look at hawks, eagles, falcons, and other raptors through the work of HawkWatch International. *NSTA Trade Book*

How Did We Find Out About Neptune? by Isaac Asimov (Walker, 1990) presents an account of astronomers' observations over the years, leading to the discovery of Neptune's existence. *NSTA Trade Book*

How to Make a Chemical Volcano and Other Mysterious Experiments by Alan Kramer (Franklin Watts, 1991) allows the student detective to solve the mysteries of chemistry with experiments using household materials.

Hurricanes: Earth's Mightiest Storms by Patricia Lauber (Scholastic, 1996) tells how hurricanes form, how scientists study them, and how the storms affect people. *NSTA Trade Book*

I Didn't Know That Some Plants Grow in Midair by Claire Llewellyn (Copper Beach Books, 1998) discusses amazing facts about rain forests and contains hands-on fun projects.

Inside Guides: Incredible Plants by Barbara Taylor (Dorling Kindersley, 1997) reveals amazing facts about plants and uses three-dimensional models to show how the parts of a plant operate.

Invertebrate Zoology by Ellen Doris (Thames and Hudson, 1993) examines seven phyla of invertebrates and includes ideas for projects, field trips, and other fun suggestions for investigating these "spineless" creatures.

Investigating Solids, Liquids, and Gases with Toys: States of Matter and Changes of State by Jerry Sarquis, Lynn Hogue, Mickey Sarquis, and Linda Woodward (McGraw-Hill, 1997) explores the three states of matter and changes of state with 24 experiments using toys and everyday items.

Isaac Newton: The Greatest Scientist of All Time by Margaret J. Anderson (Enslow, 1996) not only recounts his life and scientific discoveries, but concludes with experiments related to these scientific principles. *NSTA Trade Book*

Janice VanCleave's Machines by Janice Pratt VanCleave (John Wiley, 1993) consists of easy experiments that demonstrate scientific properties of machines and include roller bearings, conveyors, gears, and compound machines.

Janice VanCleave's Physics for Every Kid: 101 Easy Experiments in Motion, Heat, Light, Machines, and Sound by Janice Pratt VanCleave (John Wiley, 1991) includes a special *Why?* section that explains each result scientifically in easily understood terms.

Journey to the Planets by Patricia Lauber (Harcourt, 1990) describes in detail our solar system, including the sun. *ALA Notable*

Just a Dream by Chris Van Allsburg (Houghton Mifflin, 1990) tells of Ben's trip to the future where energy resources have been treated carelessly and the environment has been dangerously threatened by pollution. *NSTA Trade Book*

Light Action!: Amazing Experiments with Optics by Vicki Cobb and Josh Cobb (HarperCollins, 1993) examines the basic principles of light and how it is used in lasers and fiber optics.

The Lorax by Dr. Seuss (Random House, 1971) employs the author's delightful characters and rhymes in an ecological warning of environmental destruction.

***Marie Curie: Discovery of Radium* (Great Minds of Science)** by Margaret Poynter (Enslow, 1994) recounts Marie Curie's life and research in the fields of chemistry, physics, and medicine and includes activities and a copy of the Periodic Table of Elements. *NSTA Trade Book*

Marine Biology by Ellen Doris (Thames and Hudson, 1993) encourages students to explore the rich variety of ocean life, from plankton to blue whales, through field trips and activities such as setting up a saltwater aquarium.

Matter by Christopher Cooper (Dorling Kindersley, 1992) explains the science of matter, from the early ideas of philosophers to the latest scientific discoveries. *Children's Choice*

Mirrors: Finding Out About the Properties of Light by Bernie Zubrowski (Morrow Junior Books, 1992) includes many simple experiments that demonstrate the behavior of light and the laws of reflection. *NSTA Trade Book*

The Most Beautiful Roof in the World: Exploring the Rainforest Canopy by Kathryn Lasky (Harcourt, 1997) examines this unique home to previously unknown species of plants and animals. *NSTA Trade Book*

The Mysterious Ocean Highway by Benjamin Franklin (Raintree/Steck-Vaughn, 2000) travels with two centuries of oceanographers to find out about the Gulf Stream—the amazing river in the sea.

Nature Cross-Sections by Moira Butterfield and Richard Orr (Dorling Kindersley, 1995) illustrates the variety of life in various habitats through cross-section views. *NSTA Trade Book*

Nature Got There First by Phil Gates (Kingfisher, 1995) describes several human-made inventions that parallel natural solutions, such as the jet engine and its squid precursor. *NSTA Trade Book*

***Nuclear Energy* (Inventors & Inventions)** by Gini Holland (Marshall Cavendish, 1996) provides an introduction to nuclear energy: its history, sources, uses, and future, as well as the dangers it poses.

Oceans by Seymour Simon (Turtleback, 1997) describes the physical characteristics, life forms, and the fragile nature of the world's oceans.

Our Patchwork Planet: The Story of Plate Tectonics by Helen Roney Sattler (Lothrop, Lee & Shepard, 1995) discusses plate tectonics and the forces that cause motion and change in our planet. *NSTA Trade Book*

Our Solar System by Seymour Simon (William Morrow, 1992) presents the origin and characteristics of the planets, moons, meteoroids, asteroids, comets, and the sun in our solar system. *NSTA Trade Book*

Our Universe: A Guide to What's Out There by Russell Stannard (Kingfisher, 1995) takes students on a tour through space with a physicist who explores atoms, planets, galaxies, black holes, and strange dark matter.

Plant Survival: Adapting to a Hostile World by Brian Capon (Timber Press, 1994) illustrates fascinating mechanisms that plants use to survive.

***Plants* (The Kingdoms of Life)** by Alvin, Virginia, and Robert Silverstein (Twenty-First Century Books, 1996) discusses specific kinds of plants after describing the plant kingdom and its classification.

Pond Water Zoo: An Introduction to Microscopic Life by Peter Loewer (Atheneum, 1996) examines the microscopic life in a drop of pond water.

Sand and Fog: Adventures in Southern Africa by Jim Brandenburg (Walker, 1994) depicts the specially adapted lifeforms in the Namib Desert through the eyes of an award-winning nature photographer. *NSTA Trade Book*

The Science Book of Gravity by Neil Ardley (Harcourt, 1992) invites students to explore the force of gravity with photographs of easy-to-follow demonstrations and explanations of the point behind each activity.

Simon & Schuster Children's Guide to Insects & Spiders by Jinny Johnson (Simon & Schuster, 1996) describes family characteristics and habits of more than 100 insects and arachnids. *NSTA Trade Book*

***Smart Survivors* (World of Discovery)** by Sneed B. Collard (Northwood, 1994) describes "smart things" that plants and animals do to assure their survival.

Snakes by Eric S. Grace (Sierra Club, 1996) examines through photographs and drawings the 10 families of snakes, their anatomy, behavior, and life cycle, dispelling myths with facts. *NSTA Trade Book*

Sound Fundamentals: Funtastic Science Activities for Kids by Robert W. Wood (McGraw-Hill, 1997) explores the physics of sound, such as the Doppler effect, amplification, pitch, and resonation, with 29 simple activities and explanations of expected results.

"Stopping by Woods on a Snowy Evening" from ***The Poetry of Robert Frost*** (Henry Holt, 1987) is a poem that tells the story of a man who stops to watch snow falling in the woods.

The Story of Oxygen by Karen Fitzgerald (Franklin Watts, 1996) explores the chemical element oxygen: its history, its chemistry, how it works in the body, and its importance in our lives. *NSTA Trade Book*

Summer Ice: Life Along the Antarctic Peninsula by Bruce McMillan (Houghton Mifflin, 1995) describes this continent's active summer life from the tiny plants on the ice cliffs to the scavenging birds. *NSTA Trade Book*

Surtsey: The Newest Place on Earth by Kathryn Lasky (Hyperion, 1992) describes the formation and colonization of the volcanic island and how the first plants and animals became established there. *NSTA Trade Book*

A Tree in a Forest by Jan Thornhill (Simon & Schuster, 1991) traces the life of a maple tree and its vital role in the well-being of its forest home.

***The Visual Dictionary of Plants* (Eyewitness Visual Dictionaries)** by Deni Brown (Dorling Kindersley, 1992) not only presents the spectrum of plants from mosses to angiosperms, but discusses plant parts and processes as well, concluding with classification information.

***The Visual Dictionary of the Universe* (Eyewitness Visual Dictionaries)** by Sue Becklake (Dorling Kindersley, 1993) depicts the wonders of the universe and explains the solar system, galaxies, and nebulae in a photographic dictionary. *NSTA Trade Book*

Waves: The Electromagnetic Universe by Gloria Skurzynski (National Geographic Society, 1996) examines the spectrum of electromagnetic waves, from radio to infrared to ultraviolet to gamma rays, as well as their uses. *NSTA Trade Book*

Weather by John Farndon (Dorling Kindersley, 1992) discusses such elements of weather as clouds, wet air, frost, ice, wind, and air pressure. *NSTA Trade Book*

What Is a Reptile? by Robert Snedden (Sierra Club, 1994) answers numerous questions about reptiles with many illustrations and photographs. *NSTA Trade Book*

When Earth Moves by Sandra Downs (Millbrook, 2000) describes and explains the geological phenomena of earthquakes, landslides, sinkholes, quicksand, and avalanches.

Wild, Wet, and Windy by Clair Llewellyn (Candlewick, 1997) describes all kinds of weather, including hurricanes, tornadoes, fog, and lightning, and discusses the effects of weather on people and animals. *Children's Choice*

Wings, Stings, and Wriggly Things by Martin Jenkins (Candlewick, 1996) introduces readers to the world of such invertebrates as bees, snails, and butterflies.

The Young Astronomer by Harry Ford (Dorling Kindersley, 1998) provides readers with an introduction to the basic concepts of astronomy.

Soaring into Space

Bernard A. Harris, Jr., M.D.

Dr. Bernard Harris, a **Harcourt Science** *advisor, tells how his boyhood dream of exploring space became a lifelong adventure. Through his efforts to reach the stars, he became a medical doctor and an astronaut. He served on two space shuttle missions and holds the distinction of being the first African American to "walk" in space. In this article, Dr. Harris tells his story of pursuing his dream, and he invites students to participate in space exploration so they can share in the excitement and challenges it brings. He also stresses how important science education is to the future of our nation.*

Pursuing My Dream

I was born in Texas, but at the age of seven I was transported to the Navajo Indian reservation in New Mexico and Arizona. My mother was an educator who taught at the boarding schools on the reservation. She was a wonderful teacher who instilled in her students and her own children the value of a good education.

It was out there, under starry skies on the great plains of the Navajo nation, that I conceived my dream of becoming an astronaut. On television I watched the American space program develop before my eyes and was fascinated by what I saw. Young men were risking their lives for their country in order to win the crown of being the first person on the moon. The space program provided an outlet for my imagination. I watched every television program and read every book about space that I could. I explored the future through the creative writings of authors like H.G. Wells, Isaac Asimov, and Jules Verne.

Science is a natural vehicle for inspiring children to learn. For me as a child, studying how humans transcended earthly bounds to get into space inspired me to learn more about many fields of science. When I took imaginary trips through space, I was inspired to learn about astronomy. I sought answers to questions such as "How do the astronauts use the positions of the stars to navigate their vessels?" I learned about the human body after asking questions such as "How do people survive in the harsh environment of space?" I learned about advances in technology through questions such as "How do the basic requirements for survival in space lead to scientific discoveries?" These questions and others inspired me to look deeper into the science of the universe.

While I was in middle school, I set my goal to become an astronaut. In high school, I discovered that I could pursue two of my interests—medicine and spaceflight. I chose to follow in the footsteps of medical doctors who worked in space medicine. I obtained a bachelor's degree and an M.D. degree, completed a residency in internal medicine at the Mayo Clinic, and then set my sights on space.

I joined NASA as a flight surgeon in 1987 and became an astronaut in 1990. I am a veteran of two spaceflights, having logged more than 438 hours in space and traveled more than 7.2 million miles.

The Role of Educators

The role of educators has been critical in my life. They assisted me in my personal development and in my preparation for a career in medicine and space. Just as the teachers of my past helped me to reach my goals, the

During both of my missions, I conducted research in life science, materials science, and astronomy. The shuttle crews and I completed over 150 investigations during three weeks in space. The results of our investigations contributed to the ever-growing pool of knowledge about space and the world in which we live. ▼

NASA photo

My view from space was incredibly beautiful. I could see Earth as our home planet and appreciate its position in our galaxy. The unique vantage point of space enhances our understanding of who we humans are and helps us learn about our place in the universe. ►

NASA photo

teachers of today have the power to shape the scientific future of our nation. Among today's students are the scientists, physicians, and astronauts of tomorrow. For those students, education is the passport to their future.

As an astronaut who is concerned about the education of our youth, I utilize every opportunity to highlight the importance of space and science education. I continue to rely on education to reach my goals by seeking additional advanced college degrees. My belief in education has come from the educators in my life. A teacher has the power to open a child's mind to life's possibilities. Once a child's mind is opened, the sky is the limit!

The STARS Academy™

One way to open children's minds to science and all areas of learning is to get them involved in a hands-on, interactive education program. The STARS Academy, presenting space-based and earth-based multidisciplinary adventures via the Internet, is such a program.

Students participate in experiments on the International Space Station and the Space Shuttle through pictures, interactive activities, and video. Comparing information from the weightless environment of space, they conduct classroom experiments and form hypotheses about what they think will happen in space. STARS Academy students have studied orb spiders, butterflies, ladybugs, and most recently, an entire aquatic ecosystem.

Academy members also explore problems here on Earth—tracking and protecting endangerd species, for example—and the role space technology plays. By exchanging research ideas and cultural information online, STARS Academy students throughout the world work together to create solutions for the future. The STARS Academy online program also makes today's technologies more accessible and exciting for students and teachers alike.

For more information, visit the STARS Academy website.

GO ONLINE **www.starsacademy.com**

Students from Beijing, China, study silkworms in preparation for the STS-107 space shuttle mission.

Students from Albany, Georgia, prepare their butterfly experiment for the STS-93 space shuttle mission.

Student photos courtesy of SPACEHAB, Inc.

Tips and Trash-Pile Treasures

By Carol Valenta

When materials for science activities are difficult to obtain or too expensive for your budget, you can find acceptable and less expensive substitutes in unlikely places. You may want to include the following list of tips, activity ideas, and trash-pile "treasures" as a supplement to your file of materials and ideas.

Activity Place Mats

Have students make large place mats out of newsprint to help them organize and account for materials during activities. Provide each team with a large sheet of newsprint and a marker. As you show students a sample of each item they will receive, have them draw and label on the newsprint a box where they can place the item.

Plastic Bottles

Make terrariums from 2-liter bottles by removing the label and cutting off the top. Fill a clean butter tub with soil, and plant seeds or small plants. Cover the plants by inserting the clear plastic "dome" into the soil.

Other ways to use 2-liter bottles include these:

Funnel	Save the top that was cut off while making the terrarium.
Aquarium	Place the bottle on its side, and stabilize with clay or cutout boxes.
Planter	Place the bottle on its side, stabilize with clay or cutout boxes, and cut out the new "top half."
Bird Feeder	Cut out windows around the side, and push disposable chopsticks through the bottle to serve as perches.

Disposable Microscope Slides

Cut an index card into five strips that are each 3 inches long. Use a paper punch to make a hole in the middle of each strip. Place transparent tape over each hole. The adhesive on the opposite side can be used to hold small crystals, tiny insects, hairs, and so on. Place these "slides" on the microscope or hold them up for viewing with a hand lens.

Disposable Spoons

When you need to conserve materials and avoid contamination of substances, use large wooden tongue depressors as disposable spoons. Mark a line to indicate how far to "fill."

Substitute Droppers

Oil, soap, and acids are often hard to remove from droppers and leave the droppers contaminated. When precision in measuring is not necessary, substitute drinking straws for droppers. Put one end of the straw into the liquid, and cover the other end with a finger. Move the straw to the desired location, and remove the finger. Students need practice to perfect these movements.

Standard Masses

When multiple standard masses are needed for weighing, use the following: paper clips (plastic or metal), individual staples (ends folded), toothpicks, small uniform tiles, full aluminum soft drink cans, milliliters of water (1 mL to 1 g), or coins.

Saturated Solutions

Many household substances are fine crystals or salts that can be dissolved in water to form saturated solutions. Some examples are table salt, Epsom salts, alum crystals, and sugar.

To make a saturated solution, heat water in a coffee maker and store in a thermos until ready to use. Pour the hot water into a large plastic foam cup until it is half full. Using a small spoon, stir in one spoonful of crystals at a time until no more will dissolve. The solution is saturated at that temperature when no more will dissolve. Since hot water dissolves more of the solid than cool water, crystals form as the solution cools.

Soap-Bubble Solutions

Soap-bubble solutions may need to be used with care in the classroom but can be explored with great abandon outdoors. In a large plastic dishpan, mix the following:

4 parts water

2 parts liquid detergent

1 part corn syrup

1 part glycerine

Use plastic coat hangers, small plastic-mesh fruit and vegetable baskets, or the plastic holders from six-packs of canned drinks as frames for making bubbles.

Crystal Snow Scenes

Make a saturated solution of Epsom salts (magnesium sulfate) to create crystal snow scenes that form while you watch. Paint snowflakes and snow scenes on dark-colored paper with an old paintbrush dipped in the solution. Set the drawings aside to dry. Although the crystals that form are easily visible, it is better to view them through a hand lens.

Shapely Crystals

Make a saturated alum solution. (Alum can be purchased at a pharmacy.) Pour the solution into a glass jar. Use a 15-cm length of colored chenille stem to create an interesting shape, and suspend it from a pencil with a piece of sewing thread. Then completely submerge the shape in the saturated alum solution. As the solution cools, crystals will form. A faint tint of the color from the chenille stem will show through the crystals.

Salt-Crystal Fantasies

Make a saturated solution of table salt. Place the solution in a small open plastic tub, and set it aside where it will be exposed to air. Check frequently. Fluffy salt crystals form as evaporation takes place. These crystals are similar to those that form on wood near the ocean.

Amorphous Starch Stuff

There are many ways to make this amazing goo that acts like a liquid one minute and a solid the next. The following version allows students to observe the changes.

Fill paper cups or empty yogurt containers approximately one-third full with cornstarch. Have students slowly add water colored with food coloring to the cornstarch, one small spoonful at a time, and stir the mixture. Have them stop adding water when the mixture first has liquid properties. When their mixture is about 2 parts cornstarch and 1 part colored water, it should be runny but thick enough to roll into a ball.

Kitchen Indicators

Tea is an indicator for acids and bases. Adding lemon juice, an acid, to tea lightens the color of tea. Adding any base, such as baking soda, darkens the color.

Red cabbage juice and grape juice can also be used as indicators for acids and bases. Make an indicator with cabbage juice by chopping up several red cabbage leaves and soaking them for an hour in very hot water. The juice turns red in an acid and blue in a base.

Testing for Starch

A dilute iodine solution can be used to test for starch. Slowly add tincture of iodine, available at drugstores, to about 250 mL of water. Stop adding drops when the solution is about the tint of filtered apple juice. This works with any strength of iodine. Store this solution in dark dropper bottles or away from the light. Have students wash their hands after use.

Starchy Papers and Vanishing Art

Use a dilute iodine solution to test and classify paper scraps for starch. Collect a wide variety of papers—notebook and typing paper, napkins, and so on.

Place a small sample of paper on foil or wax paper and test with a drop of dilute iodine solution. A change to bluish-black indicates the presence of starch.

Make two more iodine solutions of different strengths that are stronger than the first solution. Paint them on paper containing starch. Hang this monochromatic "art work" in a brightly lighted place.

Substitutes for Iron Filings

Plain, unsoaped steel wool can be used as a substitute for iron filings. Cover the table with newspaper, and cut the steel wool into fine shreds using yard clippers or old scissors. Pour these shreds into storage containers.

For larger-size pieces, use a stapler to punch out individual staples until you have a large supply. These have a small enough mass to be attracted by a magnet and align with the magnetic field.

Magnet Sources

Small magnets, such as refrigerator magnets, are used in electromagnets. Electronics supply stores are excellent sources for small, inexpensive magnets.

Lemon Lights

Lemons can function as chemical cells. Push a probe made from a steel paper clip into the lemon. About 1 cm away from the first probe, push in a second probe made from a brass thumbtack. Attach a very fine insulated copper wire to the probes to complete the circuit. Students can try other combinations of metal to explore the changes in potential energy. To test the circuit, use a holiday cool light or bring a compass close to it.

Little Lights

Strings of holiday lights are a good source of cool lights for activities. An old string of lights can be cut apart to provide enough lights for the whole class. This is an easy item for families to donate.

To separate the bulbs, simply cut the wire midway between each light. Then use wire cutters or scissors to carefully remove about 1 cm of insulation from the ends of the wires coming from each bulb.

Circuit Stand-Ins

The following items can be used to substitute for parts of circuits:

Wire	Place a narrow transparent tape down the center of a 2.5-cm by 30-cm strip of aluminum foil. Fold the foil over the tape to form as narrow a strip as possible.
Bulb Holder	Wooden spring-clip clothespins can be used to attach wire and bulbs together and to hold the bulbs.
Batteries	Wide rubber bands make good battery holders, securing wires to the cells and offering flexibility.
Clips	Small metal clips for holding hair can be used in place of alligator clips in making circuits.

Great Gears

Glue strips of corrugated paper, with the "bumps" out, around the outside edges of plastic lids from jars or plastic tubs to model gears for experimentation. Other strips of corrugated paper can be used as pulleys when the ends are fastened together to form a ring. Old-fashioned soft-drink bottle tops can also be used as gears.

Pendulum Patterns

A liquid detergent bottle half-filled with sand and suspended can be used as a pendulum to demonstrate patterns of motion.

"Eggciting" CO_2

Place several eggshells in a zip-type clear plastic bag. Use a can, rolling pin, or dowel to crush the shells. Pour the shells into a 1-liter plastic bottle. Add vinegar, and cover the bottle opening with a balloon. Carbon dioxide will form and inflate the balloon. Seashell chips and small pieces of broken chalk will also work.

Terrific Timers

Place an aluminum pie plate in a plastic dishpan. From a string over the pie plate, suspend a large plastic bottle filled with water. Find the lowest part of the bottle. Use a thumbtack or T-pin to make a hole at this low point. Enlarge the hole as necessary until one drop of water falls to hit the pan. Count the number of drops in a given period of time to establish a benchmark unit of time, such as 10 drops in 15 seconds.

Colored Light Gels

Gelatin can be used to create brightly colored gels for light experiments. Dissolve one envelope of unflavored gelatin in 3 tablespoons of boiling water in a heat-resistant glass bowl. Stir until the gelatin is dissolved, and then add 2 or 3 drops of food coloring. Pour into margarine tub lids or other similar lids, to a depth of about 1 cm. Allow the gels to dry several days until hard.

Index

Note: Page numbers in boldface indicate pages in the *Teacher's Edition*. All other page numbers refer to pages in the *Pupil Edition,* which are reproduced in this guide.

A

B

D

E

M

N

Q

R

S

X

Y

Z

HARCOURT SCIENCE

Correlations

This Section Contains Correlations to:

- **National Science Education Standards**
- **Stanford Achievement Test**

National Standards/SAT Correlations

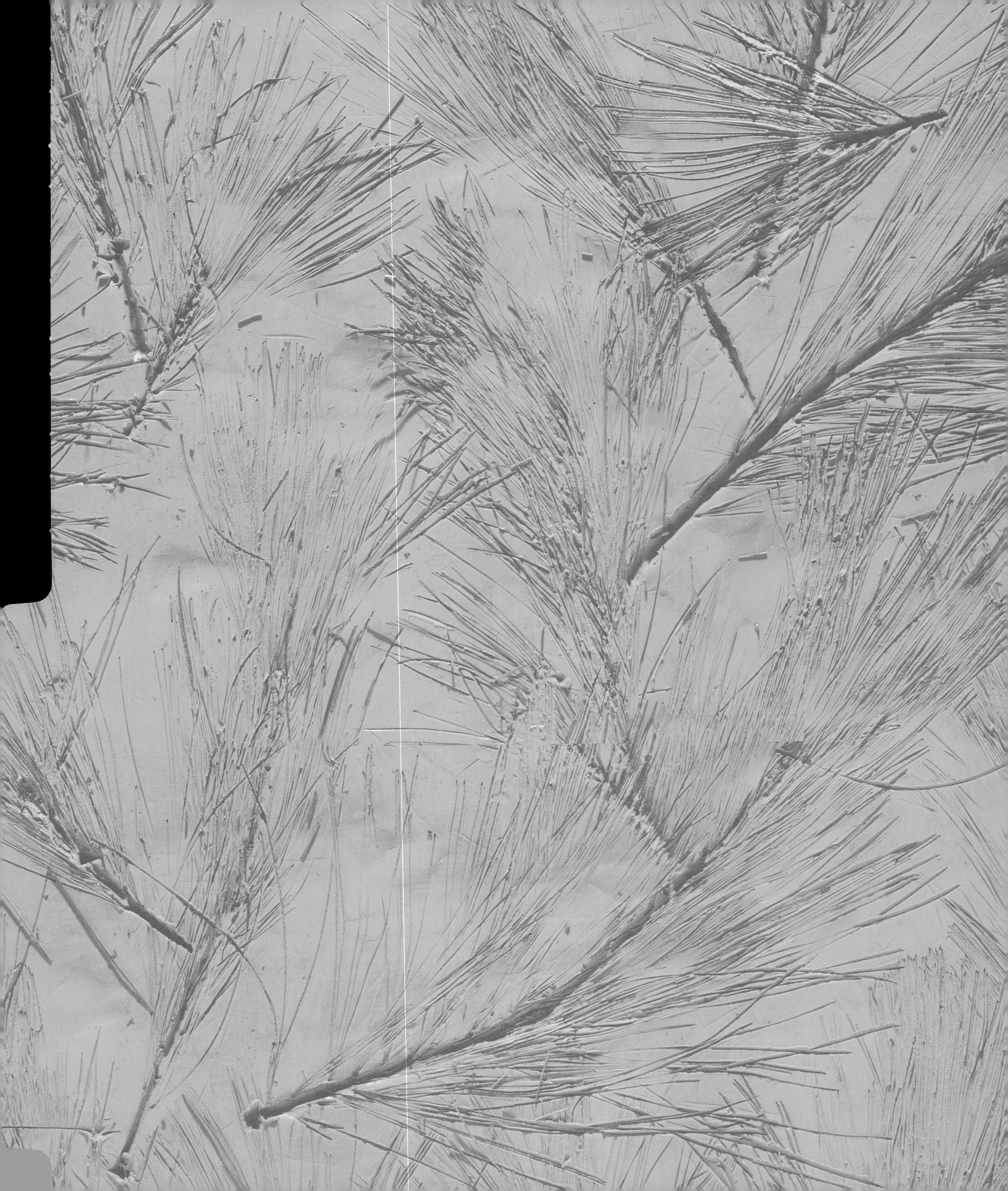

National Science Education Standards

The Science Content Standards, which are within the National Science Education Standards, outline what students should know, understand, and be able to do in the natural sciences for grades Kindergarten through 8. The Standards are broken into two grade-level sets: standards for grades K–4, and standards for grades 5–8. The following chart shows where in *Harcourt Science* Standards are met at grades 5 and 6. Since Harcourt Science is a K–6 program, some of the standards would be met in grades 7 and 8 of a middle school science program. Please consult the grades K–4 Teacher's Editions for correlations to the Standards for other grade levels of *Harcourt Science*. For more information about the Standards, please see pages T8–T9.

National Science Education Standards, Grades 5-8		Harcourt Science Teacher Edition pages, Grades 5-6
PHYSICAL SCIENCE		
Properties and Changes of Properties in Matter	A substance has characteristic properties, such as density, a boiling point, and solubility, all of which are independent of the amount of the sample.	**Grade 5:** C34-C37, E6-E11, E41-E43 **Grade 6:** E20-E21, E44-E45
	A mixture of substances often can be separated into the original substances using one or more of the characteristic properties.	**Grade 5:** E10-E11 **Grade 6:** E60-E65
	Substances react chemically in characteristic ways with other substances to form new substances (compounds) with different characteristic properties.	**Grade 5:** E44-E45, E48-E49, F20-F23 **Grade 6:** E30-E35, E50-E56
	In chemical reactions, the total mass is conserved.	**Grade 5:** E26-E27
	Substances often are placed in categories or groups if they react in similar ways; metals is an example of such a group.	**Grade 5:** E38, E40-E43, E46-E47 **Grade 6:** E16-E19
	Chemical elements do not break down during normal laboratory reactions involving such treatments as heating, exposure to electric current, or reaction with acids.	**Grade 5:** E12-E19 **Grade 6:** E32-E35
	There are more than 100 known elements that combine in a multitude of ways to produce compounds, which account for the living and nonliving substances that we encounter.	**Grade 5:** E46-E49 **Grade 6:** E16-E19
Motions and Forces	The motion of an object can be described by its position, direction of motion, and speed. That motion can be measured and represented on a graph.	**Grade 5:** F32-F35, F48-F51, F60-F61 **Grade 6:** E74-E75, F12-F13
	An object that is not being subjected to a force will continue to move at a constant speed and in a straight line.	**Grade 5:** F12, F41 **Grade 6:** F12-F13, F18
	If more than one force acts on an object along a straight line, then the forces will reinforce or cancel one another, depending on their direction and magnitude.	**Grade 5:** F12-F15 **Grade 6:** F22-F23, F24-F29
	Unbalanced forces will cause changes in the speed or direction of an objects motion.	**Grade 5:** F12-F15, F42, F48-F51 **Grade 6:** F19-F21, F27-F29
Transfer of Energy	Energy is a property of many substances and is associated with heat, light, electricity, mechanical motion, sound, nuclei, and the nature of a chemical.	**Grade 5:** E102-E107, F40-F45, F62-F63, F68-F73, F76-F81, F96-F97, F108-F109, F110-F113 **Grade 6:** E76-E79, E92-E97, E106-E113, E116-E129
	Energy is transferred in many ways.	**Grade 5:** C102-C107, F62-F65, F102-F103, F108-F109, F110-F113 **Grade 6:** E76-E77, E80, E92-E97, E108-E113, E116-E129
	Heat moves in predictable ways, flowing from warmer objects to cooler ones, until both reach the same temperature.	**Grade 5:** F82-F83, F85 **Grade 6:** E78-E79
	Light interacts with matter by transmission (including refraction), absorption, or scattering (including reflection).	**Grade 5:** F74-F78, F90 **Grade 6:** E122-E129
	To see an object, light from that object—emitted by or scattered from it—must enter the eye.	**Grade 5:** F74-F75, R22 **Grade 6:** E12-E123
	Electrical circuits provide a means of transferring electrical energy when heat, light, sound, and chemical changes are produced.	**Grade 5:** F66-F73 **Grade 6:** E86-E90, E94
	In most chemical and nuclear reactions, energy is transferred into or out of a system. Heat, light, mechanical motion, or electricity might all be involved in such transfers.	**Grade 5:** F85, F96-F97 **Grade 6:** E92-E97
	The sun is a major source of energy for changes on the Earth's surface.	**Grade 5:** D38-D39, F98 **Grade 6:** E95
	The sun loses energy by emitting light.	**Grade 5:** D38-D39, F78

National Science Education Standards Grades 5-8		Harcourt Science Teacher Edition pages, Grades 5-6
	A tiny fraction of the sun's light reaches the Earth, transferring energy from the sun to the Earth.	**Grade 5:** D38-D39
	The sun's energy arrives on Earth as light with a range of wavelengths, consisting of visible light, infrared, and ultraviolet radiation.	**Grade 5:** D38-D39, F78 **Grade 6:** E124-E125, E127-E128
LIFE SCIENCE		
Structure and Function in Living Systems	Living systems at all levels of organization demonstrate the complementary nature of structure and function.	**Grade 5:** A9, A19-A27 **Grade 6:** A4-A11, A42-A45, B76-B79
	Important levels of organization in living systems for structure and function include cells, organs, tissues, organ systems, whole organisms, and ecosystems.	**Grade 5:** A12, A17-A27 **Grade 6:** A4-A11, A43-A45, B76-B79
	All organisms are composed of cells—the fundamental unit of life.	**Grade 5:** A6-A12 **Grade 6:** A4-A11, B78
	Most organisms are single cells; other organisms, including humans, are multicellular.	**Grade 5:** A5, A12, A17-A27 **Grade 6:** A44-A48
	Cells carry on the many functions needed to sustain life. They grow and divide, thereby producing more cells. This requires that they take in nutrients, which they use to provide energy for the work that cells do and to make the materials that a cell or an organism needs.	**Grade 5:** A8-A9, A90-A94 **Grade 6:** A12-A19, B7-B17
	Specialized cells perform specialized functions in multicellular organisms.	**Grade 5:** A7, A17-A27 **Grade 6:** A14, B7-B17, B46-B49
	Groups of specialized cells cooperate to form a tissue, such as a muscle.	**Grade 5:** A14-A15, A22-A25 **Grade 6:** B78
	Different tissues are in turn grouped together to form larger functional units, called organs.	**Grade 5:** A12, A14-A27, R22-R37 **Grade 6:** B105
	Each type of cell, tissue, and organ has a distinct structure and set of functions that serve the organism as a whole.	**Grade 5:** A12, A90-A99 **Grade 6:** B7-B17, B42-B43, B52-B57, B105
	The human organism has systems for digestion, respiration, reproduction, circulation, excretion, movement, control, and coordination, and for protection from disease. These systems interact with one another.	**Grade 5:** A14-A27, R22-R37 **Grade 6:** R22-R37
	Disease is a breakdown in structures or functions of an organism. Some diseases are the result of intrinsic failures of the system. Others are the result of damage by infection by other organisms.	**Grade 5:** A28-A29, R10-R11
Reproduction and Heredity	Reproduction is a characteristic of all living systems; because no individual organism lives forever, reproduction is essential to the continuation of every species.	**Grade 5:** A62-A63, A68-A69 **Grade 6:** A14-A21, B46, B52-B57, B114-B119
	Some organisms reproduce asexually. Other organisms reproduce sexually.	**Grade 5:** A67-A68 **Grade 6:** A15-A19, B52-B57
	In many species, including humans, females produce eggs and males produce sperm.	**Grade 6:** A20
	Plants also reproduce sexually—the egg and sperm are produced in the flowers of flowering plants.	**Grade 5:** A103 **Grade 6:** A19, A24, B46-B49, B52-B57
	An egg and sperm unite to begin development of a new individual. That new individual receives genetic information from its mother (via the egg) and its father (via the sperm).	**Grade 5:** A80 **Grade 6:** A20
	Sexually produced offspring never are identical to either of their parents.	**Grade 5:** A77 **Grade 6:** A20-A21
	Every organism requires a set of instructions for specifying its traits. Heredity is the passage of these instructions from one generation to another.	**Grade 5:** A78-A81 **Grade 6:** A24-A31
	Hereditary information is contained in genes, located in the chromosomes of each cell.	**Grade 5:** A65, A80-A81 **Grade 6:** A15

National Science Education Standards Grades 5-8		Harcourt Science Teacher Edition pages, Grades 5-6
	Each gene carries a single unit of information. An inherited trait of an individual can be determined by one or by many genes, and a single gene can influence more than one trait.	**Grade 5:** A80-A81 **Grade 6:** A30
	A human cell contains many thousands of different genes.	**Grade 5:** A80-A81 **Grade 6:** A28-A31
	The characteristics of an organism can be described in terms of a combination of traits. Some traits are inherited and others result from interactions with the environment.	**Grade 5:** A62-A69 **Grade 6:** A28-A31, B66-B67
Regulation and Behavior	All organisms must be able to obtain and use resources, grow, reproduce, and maintain stable internal conditions while living in a constantly changing external environment.	**Grade 5:** A92, B42-B47, B54-B55 **Grade 6:** B27-B31, B38-B41, B72, B109
	Regulation of an organism's internal environment involves sensing the internal environment and changing physiological activities to keep conditions within the range required to survive.	**Grade 6:** B104
	Behavior is one kind of response an organism can make to an internal or environmental stimulus.	**Grade 5:** B42-B47, B54-B55 **Grade 6:** C56-C57
	A behavioral response requires coordination and communication at many levels, including cells, organ systems, and whole organisms.	**Grade 5:** A27
	Behavioral response is a set of actions determined in part by heredity and in part from experience.	**Grade 5:** B46
	An organism's behavior evolves through adaptation to its environment. How a species moves, obtains food, reproduces, and responds to danger are based in the species' evolutionary history.	**Grade 5:** B54-B55 **Grade 6:** B38-B42, B88, B110-B111, C18-C19
Populations and Ecosystems	A population consists of all individuals of a species that occur together at a given place and time.	**Grade 5:** B18-B19, B26-B31 **Grade 6:** C78-C83
	All populations living together and the physical factors with which they interact compose an ecosystem.	**Grade 5:** B18-B19, B26-B31, B62-B80 **Grade 6:** C6-C11, C14-C19, C78-C83
	Populations of organisms can be categorized by the function they serve in an ecosystem.	**Grade 5:** B24-B25, B34-B39 **Grade 6:** C7, C48-C51
	Plants and some micro-organisms are producers—they make their own food.	**Grade 5:** A39, B34-B39 **Grade 6:** B7-B9, C81
	All animals, including humans, are consumers, which obtain food by eating other organisms.	**Grade 5:** A110-A111, B34-B39 **Grade 6:** C44-C45
	Decomposers, primarily bacteria and fungi, are consumers that use waste materials and dead organisms for food.	**Grade 5:** B35 **Grade 6:** C25-C26, C46
	Food webs identify the relationships among producers, consumers, and decomposers in an ecosystem.	**Grade 5:** B36-B37 **Grade 6:** C44-C45, C48-C51
	For ecosystems, the major source of energy is sunlight. Energy entering ecosystems as sunlight is transferred by producers into chemical energy through photosynthesis. That energy then passes from organism to organism in food webs.	**Grade 5:** A96, B38 **Grade 6:** C24, C78-C79
	The number of organisms an ecosystem can support depends on the resources available and abiotic factors, such as quantity of light and water, range of temperatures, and soil composition.	**Grade 5:** B30, B62-B81 **Grade 6:** C10, C32-C35, C78-C83
	Given adequate biotic and abiotic resources and no disease or predators, populations (including humans) increase at rapid rates. Lack of resources and other factors, such as predation and climate, limit the growth of populations in specific niches in the ecosystem.	**Grade 5:** B29-B30 **Grade 6:** C32-C35, C78-C83
Diversity and Adaptations of Organisms	Millions of species of animals, plants, and microorganisms are alive today.	**Grade 5:** A34-A35
	Although different species might look dissimilar, the unity among organisms becomes apparent from an analysis of internal structures, the similarity of their chemical processes, and the evidence of common ancestry.	**Grade 5:** A36-A54

National Science Education Standards Grades 5-8		Harcourt Science Teacher Edition pages, Grades 5-6
	Biological evolution accounts for the diversity of species developed through gradual processes over many generations.	
	Species acquire ma ny of their unique characteristics through biological adaptation, which involves the selection of naturally occurring variations in populations.	**Grade 5:** B42-B46 **Grade 6:** B33, B38-B41
	Biological adaptations include changes in structures, behaviors, or physiology that enhance survival and reproductive success in a particular environment.	**Grade 5:** A116, B40-B41, B54-B55 **Grade 6:** B33, B38-B41, C76-C77
	Extinction of a species occurs when the environment changes and the adaptive characteristics of a species are insufficient to allow its survival.	**Grade 5:** B48-B51 **Grade 6:** C10
	Fossils indicate that many organisms that lived long ago are extinct.	**Grade 5:** C24-C25
	Extinction of species is common; most of the species that have lived on the Earth no longer exist.	**Grade 5:** B51
EARTH AND SPACE SCIENCE		
Structure of the Earth System	The solid Earth is layered with a lithosphere; hot, convecting mantle; and dense, metallic core.	**Grade 5:** C2-C19 **Grade 6:** D6-D7, E80
	Lithospheric plates on the scales of continents and oceans constantly move at rates of centimeters per year in response to movements in the mantle.	**Grade 5:** C15-C16, C20-C25 **Grade 6:** C72, D8-D9, D14-D19
	Major geological events, such as earthquakes, volcanic eruptions, and mountain building, result from plate motions.	**Grade 5:** C2-C19 **Grade 6:** D8-D19, D22-D29, E80
	Landforms are the result of a combination of constructive and destructive forces. Constructive forces include crustal deformation, volcanic eruption, and deposition of sediment, while destructive forces include weathering and erosion.	**Grade 5:** B92-B95, C6-C10, C16-C19 **Grade 6:** C70-C75, D8-D19, D22-D29, D46-D53, E80
	Some changes in the solid Earth can be described as the "rock cycle." Old rocks at the Earth's surface weather, forming sediments that are buried, then compacted, heated, and often recrystallized into new rock. Eventually, those new rocks may be brought to the surface by the forces that drive plate motions, and the rock cycle continues.	**Grade 5:** C36, C50-C52 **Grade 6:** D46-D53, D56-D61, D64-D67, E80
	Soil consists of weathered rocks and decomposed organic material from dead plants, animals, and bacteria.	**Grade 5:** B92-B95, C7 **Grade 6:** D10-D11, D46-D53
	Soils are often found in layers, with each having a different chemical composition and texture.	
	Water, which covers the majority of the Earth's surface, circulates through the crust, oceans, and atmosphere in what is known as the "water cycle." Water evaporates from the Earth's surface, rises and cools as it moves to higher elevations, condenses as rain or snow, and falls to the surface where it collects in lakes, oceans, soil, and in rocks underground.	**Grade 5:** B12-B19, C67-C69 **Grade 6:** C22-C23
	Water is a solvent. As it passes through the water cycle it dissolves minerals and gases and carries them to the oceans.	**Grade 5:** C97, E10 **Grade 6:** D51
	The atmosphere is a mixture of nitrogen, oxygen, and trace gases that include water vapor.	**Grade 5:** C67-C69 **Grade 6:** C94
	The atmosphere has different properties at different elevations.	**Grade 5:** C64-C66 **Grade 6:** C95
	Clouds, formed by the condensation of water vapor, affect weather and climate.	**Grade 5:** B15, C67 **Grade 6:** C102-C105
	Global patterns of atmospheric movement influence local weather.	**Grade 5:** C73-C74, C80-C83 **Grade 6:** C100-C101, C110-C113
	Oceans have a major effect on climate, because water in the oceans holds a large amount of heat.	**Grade 5:** C73-C74, C80-C83

National Science Education Standards Grades 5-8		Harcourt Science Teacher Edition pages, Grades 5-6
	Living organisms have played many roles in the Earth system, including affecting the composition of the atmosphere, producing some types of rocks, and contributing to the weathering of rocks.	**Grade 5:** C84 **Grade 6:** C10, C24-C25, D49, D51
Earth's History	The Earth processes we see today, including erosion, movement of lithospheric plates, and changes in atmospheric composition, are similar to those that occurred in the past.	**Grade 5:** C15-C25, C108-C109
	Earth history is also influenced by occasional catastrophes, such as the impact of an asteroid or comet.	**Grade 5:** D10 **Grade 6:** E112
	Fossils provide important evidence of how life and environmental conditions have changed.	**Grade 5:** C23-C25, F98-F99 **Grade 6:** D118
Earth in the Solar System	Earth is the third planet from the sun in a system that incudes the moon, the sun, eight other planets and their moons, and smaller objects, such as asteroids and comets.	**Grade 5:** D6-D11, D12-D20 **Grade 6:** D78-D81, D135
	The sun, an average star, is the central and largest body in the solar system.	**Grade 5:** D16 **Grade 6:** D78-D81
	Most objects in the solar system are in regular and predictable motion. Those motions explain such phenomena as the day, the year, phases of the moon, and eclipses.	**Grade 5:** D6-D8, D14-D19, F50 **Grade 6:** D78-D81, D86-D91, D94-D99, D102-D105
	Gravity is the force that keeps planets in orbit around the sun and governs the rest of the motion in the solar system.	**Grade 5:** F48-F51 **Grade 6:** D89
	Gravity alone holds us to the Earth's surface and explains the phenomena of the tides.	**Grade 5:** F6-F9 **Grade 6:** D89
	The sun is the major source of energy for phenomena on the Earth's surface, such as growth of plants, winds, ocean currents, and the water cycle.	**Grade 5:** D36-D43 **Grade 6:** D94-D95
	Seasons result from variations in the amount of the sun's energy hitting the surface, due to the tilt of the Earth's rotation on its axis and the length of the day.	**Grade 5:** D15 **Grade 6:** D89, D94-D99
SCIENCE IN PERSONAL AND SOCIAL PERSPECTIVES		
Personal Health	Natural environments may contain substances (for example, radon and lead) that are harmful to human beings. Maintaining environmental health involves establishing or monitoring quality standards related to use of soil, water, and air.	**Grade 5:** B98-B101
Populations, Resources, and Environments	When an area becomes overpopulated, the environment will become degraded due to the increased use of resources.	**Grade 5:** B48-B51, B100-B102 **Grade 6:** C31-C35
	Causes of environmental degradation and resource depletion vary from region to region and from country to country.	**Grade 5:** B48-B49, C84 **Grade 6:** C32-C35
Natural Hazards	Internal and external processes of the Earth system cause natural hazards, events that change or destroy human and wildlife habitats, damage property, and harm or kill humans. Natural hazards include earthquakes, landslides, wildfires, volcanic eruptions, floods, storms, and even possible impacts of asteroids.	**Grade 5:** C1-C12, R18-R19 **Grade 6:** C34-C35, C74-C75, C114-C121, E112
	Human activities also can induce hazards through resource acquisition, urban growth, land-use decisions, and waste disposal. Such activities can accelerate many natural changes.	**Grade 5:** B18-B19, B96-B102, C84 **Grade 6:** C32-C35
	Natural hazards can present personal and societal challenges because misidentifying the change or incorrectly estimating the rate and scale of change may result in either too little attention and significant human costs or too much cost for unneeded preventive measures.	**Grade 6:** E112-E113
Risks and Benefits	Risk analysis considers the type of hazard and estimates the number of people that might be exposed and the number likely to suffer consequences. The results are used to determine the options for reducing or eliminating risks.	**Grade 5:** B10-B11, B56, B84, B110-B111, F54 **Grade 6:** C32-C35, C119
	Students should understand the risks associated with natural hazards (fires, floods, tornadoes, hurricanes, earthquakes, and volcanic eruptions), with chemical hazards (pollutants in air, water, soil, and food), with biological hazards (pollen, viruses, bacterial, and parasites), social hazards (occupational safety and transportation), and with personal hazards (smoking, dieting, and drinking).	**Grade 5:** A28-A29, B10-B11, B98-B101, F54, R14-R15, R18-R19 **Grade 6:** C74, C116-C120, D24

National Science Education Standards Grades 5-8		Harcourt Science Teacher Edition pages, Grades 5-6	
	Individuals can use a systematic approach to thinking critically about risks and benefits. Examples include applying probability estimates to risks and comparing them to estimated personal and social benefits.	**Grade 5:** B56, B84, B104-B107, B110-B111, C120, F54, F100, F114-F115	**Grade 6:** C32-C35
	Important personal and social decisions are made based on perceptions of benefits and risks.	**Grade 5:** B56, B84, B104-B107, B110-B111, C120, F54, F114-F115	**Grade 6:** B18-B19, C32, C35
Science and Technology in Society	Science influences society through its knowledge and world view. Scientific knowledge and the procedures used by scientists influence the way many individuals in society think about themselves, others, and the environment.	**Grade 5:** A30, C122-C123, D22-D27, D58-D59, F24, F52-F53, F90, F114-F115, F116	**Grade 6:** A6, A24, D130-D131, E56
	The effect of science on society is neither entirely beneficial nor entirely detrimental.	**Grade 5:** A30, D58-D59	**Grade 6:** B18-B19, C32-C35
	Societal challenges often inspire questions for scientific research, and social priorities often influence research priorities through the availability of funding for research.	**Grade 5:** B56, C120, C122-C123, D22-D27, D58-D59, F25-F26, F90, F114-F115,	
	Technology influences society through its products and processes.	**Grade 5:** A6, A28-A29, B82-B83, B104-B105, C116-C121, D22-D27, E28-E29, F25, F53, F90, F114-F115	**Grade 6:** A6, E56, F40-F57
	Technology influences the quality of life and the ways people act and interact.	**Grade 5:** A30, B104-B105, C116-C117, F52-F54, F90, F114-F116	**Grade 6:** A32-A33, A58-A59, B18-B19, B94-B95, C60-C61, F30-F31, F66-F67
	Technological changes are often accompanied by social, political, and economic changes that can be beneficial or detrimental to individuals and to society.	**Grade 5:** A116, B10-B11, B104-B105, C116-C117, D22-D27, D28-D29, E28-E29, F54, F114-F116	
	Social needs, attitudes, and values influence the direction of technological development.	**Grade 5:** B82-B83, B104-B105, D22-D29, F54, F90, F100, F114-F116	
	Science and technology have advanced through contributions of many different people, in different cultures, at different times in history.	**Grade 5:** A6, A30, C116-C119, C122, C125, D22-D27, D30, D58-D59, F26, F54, F90, F114-F116	**Grade 6:** A6, A24, D78-D79, D130-D131, D134,
	Science and technology have contributed enormously to economic growth and productivity among societies and groups within societies.	**Grade 5:** B82-B83, C122-C123, C125, D58-D59, F114-F116	
	Scientists and engineers work in many different settings, including colleges and universities, businesses and industries, specific research institutes, and government agencies.	**Grade 5:** A30, A116, B56, C116-C123, C125, D30, E29-E30, F224-F26, F54, F90, F114-F116	**Grade 6:** A34, A60, B20, B60, B94, B124, C38, C62, C86, C124, D32, D70, D108, D140, E38, E68, E100, E132, F32, F68
	Scientists and engineers have ethical codes requiring that human subjects involved with research be fully informed about risks and benefits associated with the research before the individuals choose to participate. This ethic extends to potential risks to communities and property. In short, prior knowledge and consent are required for research involving human subjects or potential damage to property.	**Grade 5:** F54	
	Science cannot answer all questions and technology cannot solve all human problems or meet all human needs. Students should understand the difference between scientific and other questions. They should appreciate what science and technology can reasonably contribute to society and what they cannot do. For example, new technologies often will decrease some risks and increase others.	**Grade 5:** A30, B56, B110B111, C88, D28-D29, E52	
HISTORY AND NATURE OF SCIENCE			
Science as a Human Endeavor	Women and men of various social and ethnic backgrounds—and with diverse interests, talents, qualities, and motivations—engage in the activities of science, engineering, and related fields such as the health professions. Some scientists work in teams, and some work alone, but all communicate extensively with others.	**Grade 5:** A30, A56, A79, A84, A116, B19, B55-B56, B84, C27-C28, C88, C123-C125, D30, D60, E30, E52, F24, F26, F54, F90, F115-F116	**Grade 6:** A34, A60, B20, B60, B94, B124, C38, C62, C86, C124, D32, D70, D108, D140, E38, E68, E100, E132, F32, F68
	Science requires different abilities, depending on such factors as the field of study and type of inquiry.	**Grade 5:** A83, A115, B19, B55-B56, B83-B84, C27-C28, C88, C123-C125, D30, D60, F29-F30, E52, F24, F54, F90, F115	**Grade 6:** A59, B123, C37, C123, D69, D107, D139, E67, E99, F31

National Science Education Standards Grades 5-8		Harcourt Science Teacher Edition pages, Grades 5-6
	Science is very much a human endeavor, and the work of science relies on basic human qualities, such as reasoning, insight, energy, skill, and creativity—as well as on scientific habits of mind, such as intellectual honesty, tolerance of ambiguity, skepticism, and openness to new ideas.	**Grade 5:** A84, A116, B56, C88, C122-C125, D30, D60, E30, E52, F54, F90, F116 **Grade 6:** A34, A60, B20, B60, B94, B124, C38, C62, C86, C124, D32, D70, D108, D140, E38, E68, E100, E132, F32, F68
Nature of Science	Scientists formulate and test their explanations of nature using observation, experiments, and theoretical and mathematical models. Although all scientific ideas are tentative and subject to change and improvement in principle, for most major ideas in science, there is much experimental and observational confirmation. Those ideas are not likely to change greatly in the future.	**Grade 5:** A79, B56, C122, E30, E52, F54, F116 **Grade 6:** E6-E7
	Scientists do and have changed their ideas about nature when they encounter new experimental evidence that does not match their existing explanations	**Grade 5:** A54-A55, B114-B115, C86-C87, C116-C123, D59-D59, D60, E30, F90 **Grade 6:** D78-D79, E6-E7
	In areas where active research is being pursued and in which there is not a great deal of experimental or observational evidence and understanding, it is normal for scientists to differ with one another about the interpretation of the evidence or theory being considered. Different scientists might publish conflicting experimental results or might draw different conclusions from the same data. Ideally, scientists acknowledge such conflict and work towards finding evidence that will resolve their disagreement.	**Grade 5:** B114-B115, C27, E52, F26 **Grade 6:** A30, A32-A33, A34, A42-A43, D30, C86
	It is part of scientific inquiry to evaluate the results of scientific investigations, experiments, observations, theoretical models, and the explanations proposed by other scientists. Evaluation includes reviewing the experimental procedures, examining the evidence, identifying faulty reasoning, pointing out statements that go beyond the evidence, and suggesting alternative explanations for the same observations. Although scientists may disagree about explanations of phenomena, about interpretations of data, or about the value of rival theories, they do agree that questioning, response to criticism, and open communication are integral to the process of science. As scientific knowledge evolves, major disagreements are eventually resolved through such interactions between scientists.	**Grade 5:** A55, A79, B114-B115, C124, E52, F26, F54, F116 **Grade 6:** D78-D79
History of Science	Many individuals have contributed to the traditions of science. Studying some of these individuals provides further understanding of scientific inquiry, science as a human endeavor, the nature of science, and the relationships between science and society.	**Grade 5:** A30, A54-A55, A79, B20, B84, B116, C88, C116-C117, C124, D28-D29, D30, D60, E50-E52, F26, F54, F116 **Grade 6:** A6, A24, A42-A43, D78-D79, D131-D132, E6-E7
	In historical perspective, science has been practiced by different individuals in different cultures. In looking at the history of many peoples, one finds that scientists and engineers of high achievement are considered to be among the most valued contributors to their culture.	**Grade 5:** A30, A56, A79, B116, C86-C88, C116-C117, C124, D30, D60, E50-E52, F26, F54, F116 **Grade 6:** A42-A43, D78-D79, D131-D132, E6-E7
	Tracing the history of science can show how difficult it was for scientific innovators to break through the accepted ideas of their time to reach the conclusions that we currently take for granted.	**Grade 5:** A54-A55, B114-B115, C86-C87, C116-C119, D28-D29, E50-E52, F26 **Grade 6:** A24, A42-A43, D78-D79, E6-E7

Stanford Achievement Test

Harcourt Science prepares students to perform successfully on a variety of standardized tests, especially the Stanford Achievement Test. The following charts show how *Harcourt Science* correlates to the objectives of the various levels of the Stanford Achievement Test. Please consult the Teacher's Editions for other grades of *Harcourt Science* for correlations to other levels of the Stanford Achievement Test.

Stanford Achievement Test Objectives for INTERMEDIATE 3		Harcourt Science Teacher Edition pages
Physical Science	Use observations to determine the relative amounts of energy in a substance	**Grade 4:** E40-E41, E46-E47, E63, E68-E69, F4-F5, F23, F63 **Grade 5:** F60-F61, F98 **Grade 6:** D92-D93, E74-E75, F12-F13
	Use a data chart to predict the motion of objects	**Grade 4:** D68-D69, F42, F54-F55, F63 **Grade 5:** F16-F17, F32-F33, F38-F39, F60-F61 **Grade 6:** E75, F12-!3
	Predict the effects of heat on objects	**Grade 4:** E8, E40-E41, E47, E63 **Grade 5:** E12-E13, E33, F96-F97, F108-F109 **Grade 6:** D94- D96
	Make a prediction about the effect of force on an object	**Grade 4:** D4-D7, E5, E16-E17, E22, F5, F17, F23, F44-F45, F55 **Grade 5:** F4-F5, F10-F11, F24, F32-F33, F38-F39, F61 **Grade 6:** F6-F11, F12-F13, F19-F29
	Deduce the relative amounts of energy used in a simple machine	**Grade 4:** F60-F61, F66-F75, F76-F81, F84-F89, F93 **Grade 5:** F16-F23 **Grade 6:** F40-F58
	Interpret data about the physical properties of matter	**Grade 4:** D57, E6-E8, E11, E14, E35 **Grade 5:** C34-C35, E4-E5, E12-E13, E31 **Grade 6:** E14-E19, F6-F11
	Apply an understanding of energy changes	**Grade 4:** D93, E42-E44, E49-E52, E68-E69, E70-E74, E76-E78, E80, F23, F33 **Grade 5:** F38-F39, F60-F61, F76-F81, F97, F104-F107 **Grade 6:** D92-D93, E82-E83, E95-E96, E108-E129
	Apply an understanding of changes in states of matter	**Grade 4:** E8, E24-E30, E63 **Grade 5:** B12-B13, C94-C95, E12-E13 **Grade 6:** E20-E21, E24-E27
	Predict the relative effect of variables on a force experiment	**Grade 4:** D5, F45, F56-F58 **Grade 5:** F4-F5, F10-F11, F24, F38-F39 **Grade 6:** E75, F19-F29
Earth and Space Science	Identify sources of energy on Earth	**Grade 4:** C7, C16, C21-C23, C29, C50-C57, D12-D15, D40-D45, D52-D53, E57 **Grade 5:** F58-F59, F64-F65, F69-F73, F76-F81, F84-F87, F98-F99, F100, F104-F107 **Grade 6:** D22-D29, D41-D44, E80
	Use observations to determine the source of Earth products	**Grade 4:** C21-C23, C29, C50-C51, C61, D52-D53 **Grade 5:** B8-B9, C36-C37, C40-C45, F98-F113 **Grade 6:** D26-D29, D40-D45
	Understand causes and effects of weather phenomena	**Grade 4:** D12-D16, D20-D23, D32-D33 **Grade 5:** B14-B17, C60-C89 **Grade 6:** C102-C105, C114-C121, D94-95
	Read a weather map	**Grade 4:** D13-D14, D22-D23 **Grade 5:** C74-C77 **Grade 6:** C107-C113

Stanford Achievement Test Objectives for INTERMEDIATE 3		Harcourt Science Teacher Edition pages
	Deduce the relative motion of objects in the sky	**Grade 4:** D62-D63, D64-D66, D68-D73 **Grade 5:** D6-D8, D12-D19, D32-D33 **Grade 6:** D84-D91, D102-D105
	Apply an understanding of celestial events	**Grade 4:** D60-D67, D68-D70, D72-D73, D80 **Grade 5:** D6-D8, D12-D15 **Grade 6:** D102-D105, D118-D121
	Apply an understanding of precipitation	**Grade 4:** D35 **Grade 5:** B15, C60-C69 **Grade 6:** C103-C104
Life Science	Use observations to classify organisms	**Grade 4:** A33, B10-B11 **Grade 5:** A8, A9, A36-A54 **Grade 6:** A40-A57, B28-B41, B52-B57, B84-B85, B102-B103
	Apply an operational definition to animal models	**Grade 4:** A7, A47, B26-B27 **Grade 5:** A42-A43, B40-B41 **Grade 6:** B84-B85, B102, B112-B113
	Associate organisms with their characteristics	**Grade 4:** A14-16, A20-A22, A23, A26-A28, A73, A78-A79 **Grade 5:** A5, A8-A9, A12, A17-A27, A90-A97 **Grade 6:** A20-A31, B7-B17, B50-B57, B84-B85
	Analyze a model of biological process	**Grade 4:** A47, A68-A69, A73, A78-A79, B2-B3, B54-B57 **Grade 5:** A10, A17-A27, A62-A85, A105-A106, B6-B9, B48-B49, B59 **Grade 6:** A4-A5, A12-A21
	Make a prediction based on a food web	**Grade 4:** A42-A43, B23, B45 **Grade 5:** B36-B37, B59 **Grade 6:** C48-C51
	Use observations to sequence the growth of organisms	**Grade 4:** A27, B14-B15, B74-B75, B78 **Grade 5:** A64-A66, A70-A75 **Grade 6:** A14-A19, B4-B5
	Associate a product with its source	**Grade 4:** A73, B32, D26-D27 **Grade 5:** A73, A108-A113, B32, C36-C37, F98-F106 **Grade 6:** A18-A19, B40, B56
	Associate an activity with a career	**Grade 4:** A32, A63, B43, B44, D25-D26, D48, D55, D84, D92, E33-E34, E61-E 62, E91, F62, F92 **Grade 5:** A29, A30, A56, A83, A115, B19, B20, B55-B56, B83, C27, C55, C123, D54, E29-E30, F25, F53, F115 **Grade 6:** A59, B19, B95, B122, D32, D69, D107, D139, E37, E67, E99. E131, F31
Science Processes	Interpret/analyze a graph	**Grade 4:** B22, B32, B70, D36, B70, D46-D47, D48-D49, E55, E85 **Grade 5:** A60, A102, B8, B30-B31, C83-C85, E10, F94, F98 **Grade 6:** C4-C5, C9, C31, C47, C93, E20-21
	Use a model to make a prediction	**Grade 4:** A4-A5, B4-B5, B10-B11, B34-B35, B50-B51, B59, C4-C5, C29, C34-C35, C61, D27, D47-D47, D57, D69, D75, D83, F82-F83, F93 **Grade 5:** A36-A37, A43, B59, B91, C4-C5, C29, C100-C101, C114-C115, F32-F33 **Grade 6:** A4-A5, C20-C21, C28-C29, D68-D69, C99, C106-C107, D4-D5, D12-D13, D46-D47, D54-D55, D76-D77, D100-D101, D122-D123, D128-D129
	Use a chart to make a prediction	**Grade 4:** A19, A39, A96-A97, A102-A103, A108-A109, B50-B51, D11, D69, F11, F55, F69 **Grade 5:** B41, B48, B57, C33, C63, C78, C115, E9, E18, E45, F97 **Grade 6:** C13, D4-D5, D20-D21, E13, E20-E21

Stanford Achievement Test Objectives for INTERMEDIATE 3		Harcourt Science Teacher Edition pages
	Evaluate the design of an experiment	**Grade 4:** A24-A25, A71, B50-B51, B59, D11, D69, D75, D83, E25, E55, E83, F93
		Grade 5: A1, A43, A71, B1, B13, B109, C1, C5, C21, C33, D1, D13, D20-D21, E1, E13, E45, F1, F47, F102-F103 , F109
		Grade 6: A1, A5, B1, C1, D1, D47, D128-D129, E1, E5, E21, E83, E91, E115, E123, F1, F4-F5
	Read a chart	**Grade 4:** A3, A38-A39, A47, A69, A96-A97, A102-A103, A108-A109, B2, B48, B50-B51, B62, B72, B78, C3, D3, D11, D22, D31, D61, D69, D72, D75, E14, E38, E67, E96, F3, F11, F37, F42, F55, F57, F66, F69
		Grade 5: B40-B41, B48-B49, B57, B63, C35, C115, E46-E47, F34, F97
		Grade 6: A22-A23, C76-C77, C92, C106-C107, C115, D4-D5, D20-D21, D24, D38-D39, E12-E13, E20-E21, E33, E47, E120
	Use observations to make a prediction	**Grade 4:** A39, A55, A71, A77, A83, A108-A109, B5, B10-B11, B34-B35, B59, C34-C35, C40-C41, D5, D11, D19, D38-D39, E55, E63, E93, F17, F23, F33, F63, F69
		Grade 5: A5, B5, B41, B57, B91, B117, C101, C115, F33, F39, F97, F109
		Grade 6: B27, B50-B51, C29, C53, C99, C107, C115, D47, D62-D63, D114-D115, D123, E5, E58-E59, E106-E107, E123, F4-F5
	Associate objects with their source	**Grade 4:** A18-A19, A83, B18-B19
		Grade 5: A96, B8-B9, B14-B15, C34, C40-C45, E38-E43
		Grade 6: D40-D45, D48-D53, D56-D61
	Read a label	**Grade 4:** A71, B30, B77, D19, E42, E55
		Grade 5: B96-B97, C20
		Grade 6: E82-E83, E90-E91, F4-F5